International Series in Operations Research & Management Science

Volume 291

More information about this series at http://www.springer.com/series/6161

Tadeusz Sawik

Supply Chain Disruption Management

Using Stochastic Mixed Integer Programming

Second Edition

Springer

Tadeusz Sawik
Department of Operations Research
AGH University of Science and Technology
Kraków, Poland

Department of Engineering
Reykjavik University
Reykjavik, Iceland

ISSN 0884-8289 ISSN 2214-7934 (electronic)
International Series in Operations Research & Management Science
ISBN 978-3-030-44813-4 ISBN 978-3-030-44814-1 (eBook)
https://doi.org/10.1007/978-3-030-44814-1

This Springer imprint is published by the registered company Springer Nature Switzerland AG.
The registered company address is: Gewerbestrasse 11, 6330 Cham, Switzerland

To Bartek, Siasia, Kalinka, Aleksander,
Diana, Kajetan-Ragnar, and Rokimi
with love

and to the Memory of my Parents

Preface

Scope

This book deals with stochastic combinatorial optimization problems in supply chain disruption management, with a particular focus on the management of disrupted flows in customer-driven supply chains. The problems are modeled using a scenario-based stochastic mixed integer programming to address risk-neutral, risk-averse, and mean-risk decision-making in the presence of supply chain disruption risks. One of the main objectives of this book is to present a computationally efficient portfolio approach to integrated decision-making in global supply chains under disruption risks. In the context of supply chain disruptions, the portfolio is defined as the allocation of demand for different part types among suppliers and the allocation of demand for different product types among production facilities (e.g., assembly plants) of the final manufacturer. The allocation of demand for parts among suppliers is defined as a supply portfolio, whereas the allocation of demand for products among production facilities of the final manufacturer is defined as a demand or capacity portfolio. Unlike most of the reported research on the supply chain risk management which mainly focuses on the risk mitigation decisions taken prior to a disruption, the proposed portfolio approach combines decisions made before, during, and after the disruption. When a disruption occurs, the primary portfolios determined prior to a disruption are replaced by recovery portfolios. The selection of portfolios will be combined with the management of disrupted material flows, i.e., supply, production, and distribution scheduling under disruption risks. The book demonstrates that the developed portfolio approach leads to simple decision-making models and, owing to embedded network flow structures, to computationally efficient stochastic mixed integer programs with a strong LP relaxation.

Moreover,

- integrated vs. hierarchical decision-making is compared depending on the available information on disruption events;

- a multi-objective decision-making is analyzed to trade off between: cost vs. service level objective functions, fairness vs. non-equitability of objective functions, average vs. worst-case performance of a supply chain, etc.;
- a multi-period decision-making is modeled to capture dynamics of disruption and recovery processes, i.e., static vs. dynamic portfolios, scheduling of supply, production, and distribution operations to control disrupted flows under time-varying conditions such as demands, capacities, etc.;
- a multi-level disruption scenarios are modeled to capture partially disrupted flows, partially fulfilled orders, partially recovered facilities, and partially available capacity;
- multi-tier vs. two-tier resilient supply portfolios are modeled to compare resilience of multi-tier supply chain networks vs. two-tier supply chains.

The book also addresses the issue of fundamental understanding of average and worst-case performance of a global supply chain in the presence of flow disruption risks, understanding of the recovery mechanisms as well as understanding of the supply chain resilience.

A straightforward computational approach used in this book is to solve the deterministic equivalent mixed integer program of a two-stage stochastic mixed integer program with recourse, which allows for a direct application of commercially available software for mixed integer programming. In the computational experiments reported throughout the book, an advanced algebraic modeling language AMPL (see, Fourer et al. 2003) and the CPLEX, Gurobi, and XPRESS solvers have been applied.

Content

The book is divided into an introductory Chap. 1, where an overview of supply chain disruption modeling and management is provided with a particular emphasis on an innovative multi-portfolio approach, and the six main parts. Part I addresses selection of a supply portfolio, Part II, integrated selection of supply portfolio and scheduling, Part III, integrated, equitably efficient selection of supply portfolio and scheduling, Part IV integrated selection of primary and recovery supply (and demand) portfolios and scheduling, Part V, selection of supply portfolio in multi-tier supply chain networks, and finally Part VI addresses disruption management of information flows in supply chains.

Part I (Chaps. 2–4) introduces the portfolio approach for supplier selection and order quantity allocation in the presence of supply chain disruption risks, i.e., for determining a supply portfolio. The proposed portfolio approach allows the two popular in financial engineering percentile measures of risk, Value-at-Risk (VaR) and Conditional Value-at-Risk (CVaR) to be applied for managing the risk of supply disruptions. For a finite number of scenarios, CVaR allows the evaluation of worst-case costs (or worst-case service level) and shaping of the resulting cost (service level) distribution through optimal supplier selection and order quantity allocation decisions, i.e., the selection of optimal supply portfolio.

Part I is comprised of these chapters:

- Chapter 2, Selection of Static Supply Portfolio. This chapter deals with the selection of a static supply portfolio under disruption risks, i.e., for determining a single-period allocation of demand for parts among selected suppliers to minimize expected or expected worst-case cost or maximize expected or expected worst-case service level.
- Chapter 3, Selection of Dynamic Supply Portfolio. In this chapter, the static portfolio approach and the stochastic mixed integer programming formulations presented in Chap. 2 are enhanced for a multi-period supplier selection and order quantity allocation in the presence of both the low probability and high impact supply chain disruption risks and the high probability and low impact supply chain delay risks. The suppliers are subject to local delivery delay risks, and both local and regional delivery disruption risks. In the delivery scenario analysis, both types of the supply chain risks are simultaneously considered.
- Chapter 4, Selection of Resilient Supply Portfolio. In this chapter, the portfolio approach and SMIP models presented in Chap. 2 are enhanced for the combined selection and protection of part suppliers and order quantity allocation in a supply chain with disruption risks. The protection decisions include the selection of suppliers to be protected against disruptions and the allocation of risk mitigation inventory of parts to be pre-positioned at the protected suppliers so as to maintain uninterrupted supplies in case of natural or man-made disruptive events.

Part II of the book concerns with integrated selection of supply portfolio and scheduling. The medium- to short-term decisions of the supplier selection and order quantity allocation, driven by the time-varying customer demand, are made along with scheduling of customer orders execution and distribution. The advantage of such a joint decision-making is especially evident in the presence of supply chain disruption risks.

Part II has two chapters:

- Chapter 5, Integrated Selection of Supply Portfolio and Scheduling of Production. This chapter proposes a SMIP approach to integrated supplier selection and customer order scheduling in the presence of supply chain disruption risks for a single, dual, or multiple sourcing strategy. The suppliers are assumed to be located in two or more disjoint geographic regions: in the producer's region (domestic suppliers) and outside the producer's region (foreign suppliers). The supplies are subject to independent random local disruptions that are uniquely associated with a particular supplier and to random regional disruptions that may result in disruption of all suppliers in the same geographic region simultaneously.
- Chapter 6, Integrated Selection of Supply Portfolio and Scheduling of Production and Distribution. The purpose of this chapter is to study the integrated decision-making to simultaneously select suppliers of parts, allocate order quantity, and schedule production and delivery of finished products to customers in a supply

chain under disruption risks. In addition to supplier selection, order quantity allocation and scheduling of customer orders, distribution of finished products to customers is simultaneously considered with different shipping methods to optimize the trade-off between cost and service level. The three different shipping methods will be modeled and compared for the distribution of products: batch shipping with a single shipment of different customer orders, batch shipping with multiple shipments of different customer orders, and individual shipping of each customer order immediately after its completion.

Part III addresses equitably efficient selection of supply portfolio and scheduling. A fair optimization of an average performance of a supply chain with respect to equally important conflicting objective functions and a fair mean-risk optimization of average and worst-case performance are considered in the presence of supply chain disruption risks. The conflicting and equally important objective functions are expected values of cost and customer service level and the corresponding expected and expected worst-case values, respectively. The fairness and the mean-risk fairness reflect the decision-makers common requirement to maintain an equally good performance of a supply chain with respect to equally important objectives and under varying operating conditions.
Part III has two chapters:

- Chapter 7, A Fair Decision-Making under Disruption Risks. In this chapter, the two risk-neutral conflicting criteria: expected cost and expected service level are fairly optimized to achieve an equitably efficient supply portfolio and production schedule in the presence of supply chain disruption risks. In order to obtain an equitably efficient solution, the ordered weighted averaging (OWA) aggregation of the two conflicting objective functions is applied. The equitably efficient solutions obtained for the ordered weighted averaging aggregation of the two conflicting objective functions will be compared with nondominated solutions obtained using the weighted-sum aggregation approach.
- Chapter 8, A Robust Decision-Making under Disruption Risks. In this chapter, we look for an equitably efficient solution with respect to both average-case and worst-case performance measures of a supply chain. Such an equitably efficient average and worst-case solution, or equivalently equitably efficient risk-neutral and risk-averse solution will be called a fair mean-risk solution. The solution will equitably focus on the two objective functions: the expected value (average-case performance measure) and the expected worst-case value (worst-case performance measure), i.e., Conditional Value-at-Risk of the selected optimality criterion, cost or service level. The fair mean-risk decision-making aims at equalizing the distance to optimality both under business-as-usual and under worst-case conditions, which reflects a common requirement to maintain an equally good performance of a supply chain under varying operating conditions. Therefore, the mean-risk fairness, i.e., the equitably efficient performance of a supply chain in the average-case as well as in the worst-case, in this chapter, will be called robustness.

Part IV focuses on selection of primary and recovery portfolios and scheduling. The selection of primary suppliers and order quantity allocation to mitigate the impact of disruption risks is combined with the selection of recovery suppliers and assembly plants to optimize the recovery processes. The two decision-making approaches will be considered and compared: an integrated approach with some information about the future potential disruption scenarios available ahead of time, and a hierarchical approach with no such information available. In the integrated approach, which may account for all potential disruption scenarios, the primary portfolios that will hedge against all potential disruptive events will be determined along with the recovery portfolios for each scenario. In the hierarchical approach, first, the primary portfolios are determined, and then, when the primary portfolios are impacted by a disruptive event, the recovery portfolios are selected to optimize the process of recovery from the disruption. Both the integrated and the hierarchical decision-making account for time and cost of mitigation and recovery processes and aim at optimizing cost and service level as the two equally important, conflicting objective functions. Owing to the embedded network flow structures, computationally efficient stochastic mixed integer programs with a very strong LP relaxation are developed.

Part IV has four chapters:

- Chapter 9, Selection of Primary and Recovery Supply Portfolios and Scheduling. In this chapter, the portfolio approach presented in the previous chapters for the selection of primary suppliers and order quantity allocation to mitigate the impact of disruption risks is enhanced also for the recovery process, i.e., for the selection of both primary and recovery suppliers and order quantity allocation to mitigate the impact of disruption risks and optimize the recovery process. Unlike most of the reported research on the supply chain risk management which focuses on the risk mitigation decisions taken prior to a disruption, this chapter combines decisions made before, during, and after the disruption.
- Chapter 10, Selection of Primary and Recovery Supply and Demand Portfolios and Scheduling. In this chapter, the portfolio approach proposed in Chapter 9 for the selection of primary and recovery suppliers and order quantity allocation to mitigate the impact of disruption risks is enhanced also for the recovery process of the final manufacturer's assembly plants for finished products. Unlike most of the reported research on supply chain disruption management, a disruptive event is assumed to impact both a primary supplier of parts and the final manufacturer's primary assembly plant. Then the manufacturer may choose alternate (recovery) suppliers and move production to alternate (recovery) plants along with transshipment of parts from the impacted primary plant to the recovery plants. The resulting allocation of unfulfilled demand for parts among recovery suppliers and unfulfilled demand for products among recovery assembly plants determines recovery supply portfolio and recovery demand portfolio, respectively. In addition, the allocation of inventory of parts at impacted primary plants among recovery plants defines a recovery transshipment portfolio.
- Chapter 11, Selection of Primary and Recovery Supply and Demand Portfolios and Scheduling: A Two-Period Approach. This chapter presents a novel two-

period modeling approach for supply chain disruption mitigation and recovery, where the planning horizon is divided into two aggregate periods: pre-disruption Period 1 and post-disruption Period 2. The corresponding mitigation and recovery decisions are: (1) primary supply and demand portfolios and production prior to disruption, and (2) recovery supply, transshipment, and demand portfolios and production after disruption. The two variants of the two-period modeling approach are proposed and compared: with best-case and with worst-case capacity constraints, where the available recovery production capacity may be overestimated or underestimated, respectively.
- Chapter 12, Selection of Supply and Demand Portfolios and Production and Inventory Scheduling. In this chapter, the risk-averse integrated selection of supply and demand portfolios and production and inventory scheduling are considered for the period-dependent demand. In order to allow for a delayed satisfaction of demand using product inventory, production is scheduled along with the inventory. To this end, the additional cost of the inventory holding and penalty for delayed demand are introduced. The impact of different demand patterns on supply chain performance is investigated.

Part V addresses selection of resilient supply portfolio in a multi-tier supply chain network, geographically dispersed over multiple regions, where in each region suppliers of the same and/or of different tiers are located. In addition to multi-sourcing and recovery and transshipment supplies, to improve the network resilience, prepositioning of risk mitigation inventory at primary suppliers of parts of different tiers and at primary assembly plant is considered. Static (single-period) and dynamic (two-period) decision-making models are developed and some proven properties of optimal solutions derived from the proposed models provide additional managerial insights. Owing to the embedded network flow structures, computationally efficient stochastic mixed integer programs with a very strong LP relaxation are developed. Part V has two chapters:

- Chapter 13, Selection of Resilient Multi-Tier Supply Portfolio. This chapter develops a multi-portfolio approach to disruption mitigation and recovery in a geographically dispersed multi-tier supply chain network. A static (single-period) model is developed for resilient, risk-neutral, and risk-averse decision-making, where the supply chain resilience is improved by pre-positioning of risk mitigation inventory at different tiers of the network. The proposed approach is a non-trivial enhancement of a two-tier supply portfolio approach introduced in Part I and extended later in Part IV for a combined risk mitigation and recovery in the supply chains.
- Chapter 14, Selection of Resilient Multi-Tier Supply Portfolio: A Two-Period Approach. The two-period approach developed in Chap. 11 for the selection of primary and recovery portfolios in a two-tier supply chain, in this chapter, is enhanced for a multi-tier supply chain network. The static (single-period) decision-making model proposed in Chap. 13 for the selection of resilient multi-tier supply portfolio is extended for a dynamic (two-period) setting, where the planning

horizon is divided into two aggregate periods: pre-disruption Period 1 and post-disruption Period 2.

Part VI deals with disruption management of information flows in supply chains caused by cybersecurity incidents. The supply portfolio approach applied to mitigate the impact of supply disruptions has been modified to select countermeasure portfolio to mitigate the impact of information flow disruptions.
Part VI has one chapter:

- Chapter 15, Selection of Cybersecurity Safeguards Portfolio. This chapter deals with the selection of countermeasures portfolio in cybersecurity planning to prevent or mitigate the impact of information flow disruptions in a supply chain. A scenario-based bi-objective SMIP approach with CVaR as a risk measure is proposed for the decision-making. Given a set of potential threats and a set of available countermeasures, the decision-maker needs to decide which countermeasure to implement under limited budget to minimize potential losses from successful cyber-attacks. The selection of countermeasures is based on their effectiveness of blocking different threats, implementation costs, and probability of potential attack scenarios. The bi-objective trade-off model provides the decision-maker with a simple tool for balancing expected and worst-case losses and for shaping of the resulting cost distribution through the selection of optimal subset of countermeasures for implementation, i.e., the selection of optimal countermeasure portfolio.

The Parts I–VI and the chapters within each part are arranged in the order recommended for reading, while Part VI with Chap. 15 can be read independently of the other chapters. Each chapter ends with a table summarizing major managerial insights and the end-of-chapter problems to help the reader a self-check of material comprehension and to encourage for a further self-study.

The book can be considered a companion as well as a follower of my previous book on scheduling in supply chains using mixed integer programming (Sawik 2011a), where deterministic MIP approaches were developed for integrated scheduling in customer-driven supply chains, in particular, in the electronics supply chains. The reader interested in knowing more about stochastic programming is referred to the monographs by Birge and Louveaux (2011) or Kall and Mayer (2011). For a general introduction to mixed integer programming models and techniques, the reader is referred to the application-oriented book by Chen et al. (2010) or to the seminal work in the field by Nemhauser and Wolsey (1999). The fundamentals of supply chain theory are well presented by Snyder and Shen (2011), and for an engineering-oriented general reference work on supply chains, the reader is referred to the book by Dolgui and Proth (2010). Finally, some books cover supply chain risk management in general, e.g., Kouvelis et al. (2011), Sodhi and Tang (2012), Khojasteh (2018), Ivanov (2018) and Ivanov et al. (2019), and some of these emphasize supply chain disruption management, e.g., Gurnani et al. (2012).

About the Second Edition

Supply chain risk management has been rapidly evolving in recent years, however, its core remains essentially unchanged. All stochastic mixed integer programming models from the first edition and the results of computational study are still relevant. The proposed multi-portfolio approach can be particularly useful for material flow coordination in contemporary supply chain networks, which are cyber-physical systems, where integrated decision-making and control over the entire network is of utmost importance.

Despite the substantial revisions made throughout the text, much of the brand new material is concentrated in Part IV and Part V, where the four new Chaps. 11–14 are added, using the results of my recent research. In particular, Part V deals with supplier selection under disruption risks in a multi-tier supply chain network, which is nearly unreported in the literature. Moreover, the original Chaps. 1, 3, and 10 are expanded and significantly updated. In the second edition, new propositions are formulated to provide better insights into the properties of developed stochastic mixed integer programs and their optimal solutions. Finally, each chapter ends with a table providing the reader with major managerial insights.

Acknowledgment

I wish to express my thanks to Series Editor, Camille Price, who invited me to prepare this extended second edition of my monograph for International Series in Operations Research and Management Science.

Audience

The book is addressed to practitioners and researchers on supply chain risk management and disruption management, and to students in management, industrial engineering, operations research, applied mathematics, computer science, and the like at masters and PhD levels. It is not necessary to have a detailed knowledge of stochastic programming and integer programming in order to go through this book. The knowledge required corresponds to the level of an introductory course in operations research and supply chain management for engineering, management, and economics students.

Kraków, Poland Tadeusz Sawik
January 2020

Acknowledgements to the First Edition

The book is based on the results of my research on supply chain scheduling and disruption management by stochastic mixed integer programming over the last decade. I wish to acknowledge many anonymous reviewers for their comments and suggestions on my submissions to different international journals, in particular, *Omega: The International Journal of Management Science*, *International Journal of Production Research*, *Computers and Operations Research* and *Decision Support Systems*.

The material presented in this book is in part based on the results of different research projects on supply chain optimization and risk management, funded by Polish Ministry of Science and Higher Education and by NCN. This project has been partially supported by NCN research grant #DEC-2013/11/B/ST8/04458. Thanks are also due to the AGH University of Science and Technology for its support of research (grants #11.11.200.273 and #11.11.200.324) on supply chain risk management over the last 5 years.

Finally, I wish to express my thanks to Springer Editor, Camille Price, who invited me to prepare this monograph for International Series in Operations Research and Management Science.

Contents

List of Figures

List of Tables

Acronyms

ABC_E	Model for risk-neutral decision-making
ABC_CV	Model for risk-averse decision-making
ABC_ECV	Model for mean-risk decision-making
CVaR	Conditional Value-at-Risk
CVaR^c	Conditional Cost-at-Risk
CVaR^{sl}	Conditional Service-at-Risk
E^c	Expected cost
E^{sl}	Expected service level
IP	Integer Programming
MIP	Mixed Integer Programming
SMIP	Stochastic Mixed Integer Programming
VaR	Value-at-Risk
VaR^c	Cost-at-Risk
VaR^{sl}	Service-at-Risk

Chapter 1
Introduction

1.1 Overview of Supply Chain Disruption Management

A typical global supply chain in modern industry is a cyber-physical network of multiple part suppliers at different geographical locations and multiple production plants and distribution centers, where supplied parts are assembled into finished products and next distributed to customers. Figure 1.1 shows a schematic diagram of a multi-tier supply chain network, where each vertical level (suppliers, producers, distribution centers, customers) is called a tier or echelon, and the arcs represent material flows. For example, a large and complex multi-tier supply chain network of Ford Motor Company (e.g., Simchi-Levi et al. 2015) consists of over 50 manufacturing plants, ten tiers of suppliers, including 1400 tier 1 supplier companies with 4400 manufacturing sites in over 60 countries. Six million vehicles produced annually, require 55,000 different part types with a complex bill of materials.

In order to achieve a high customer service level at a low cost, a variety of complex, interconnected decision-making problems should be solved. The decision-making problems are strictly associated with the control and optimization of material flows (as well as financial and information flows) in the network, in particular, optimization of disrupted flows. Different types of material flows (e.g., flows of parts from suppliers to producers, flows of semi-finished products at producers, flows of finished products from producers to distribution centers and from the distribution centers to customers) should be coordinated in an efficient manner. In global supply chains, the control and optimization of material flows are accomplished by scheduling of manufacturing and supplies of parts, scheduling of production and customer orders for finished products, and scheduling of deliveries to customers. All those scheduling decisions should be efficiently coordinated to fulfill customer demand, especially in the presence of supply chain disruption risks. The schedule of customer orders immediately depends on the schedule of parts supplies, which in turn depends on supplier selection and order quantity allocation, that is, on supply

T. Sawik, *Supply Chain Disruption Management*, International Series in Operations Research & Management Science 291, https://doi.org/10.1007/978-3-030-44814-1_1

portfolio. On the other hand, the schedule of customer orders implicitly defines the schedule of deliveries of finished products to customers, which in turn depends on customer order allocation among assembly plants, that is, on demand portfolio. In view of the recent trend of outsourcing, offshoring and globalization, coordinated decision-making, e.g., selection of primary and recovery part suppliers and allocation of order quantities, selection of primary and recovery assembly plants and allocation of customer demand, and scheduling of customer orders in the assembly plants may significantly improve performance of a multi-tier supply chain network under disruption risks. However, most work on coordinated supply chain scheduling focuses on coordinating non-disrupted flows of supply and demand over a supply chain network to minimize the inventory, transportation, and shortage costs. The research on quantitative approaches to coordinated scheduling of disrupted flows in supply chains has not been sufficiently reported in the literature and is mostly limited to separate investigation of supply, production, or distribution stage.

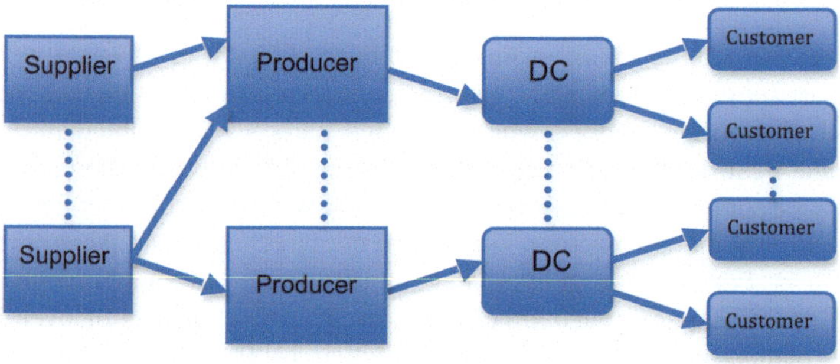

Fig. 1.1 A multi-tier supply chain network

In contemporary global and lean supply chains, disruption risk management has become a vital part of supply chain management strategy. Material flows in supply chains can be disrupted by unexpected natural or man-made disasters such as earthquakes, fires, floods, hurricanes or equipment breakdowns, labor strikes, economic crisis, bankruptcy or by a deliberate sabotage or terrorist attack. The low-probability and high-impact flow disruptions and the resulting losses may threaten financial state of firms. For example, the Taiwan earthquake of September 1999 created huge losses for many electronics companies supplied with components by Taiwanese manufacturers, e.g., Apple lost many customer orders due to supply shortage of DRAM chips (Sheffi 2005). The Philips microchip plant fire of March 2000 in New Mexico resulted in 400 million Euros in lost sales by a major cell phone producer, Ericsson (Norrman and Jansson 2004). The disruptions in the automotive and electronics supply chains that occurred in 2011 after the Great East Japan Tohoku earthquake on March 11 and then the Thailand flooding in October, resulted in huge losses of major automakers and electronics manufacturers, e.g., Haraguchi and Lall (2015), Park et al. (2013). Similar effects were observed after the Kyushu earthquake in April 2016, e.g., Marszewska (2016). Toyota supply chain has been again severely disrupted when two plants of Aisin Seiki, a key supplier of car body and engine components were destroyed, and Renesas, a key supplier of semiconductors for Toyota engines, had to halt production in Kumamoto plant.

The external complexity of outsourcing and offshoring of multinational corporations may additionally contribute to disruption risks of contemporary global multi-tier supply chain networks. For example, Boeing's global supply chain of 787 Dreamliner consists of over 50 tier 1 suppliers, including 28 suppliers outside of the USA. The external complexity disrupted the multinational supply chain and the final assembly of an aircraft, planned for 2–3 days was delayed by 4 years (e.g., Celo et al. 2018).

A special case of global supply chain disruption risk is an outbreak of a pandemic disease that may fast disperse over many geographic regions. The pandemic disease outbreak is characterized by long-term disruption existence and its unpredictable scaling. Simultaneously with epidemic outbreak propagation in the population, propagation of disruptions in global supply chains can be observed. Recent examples include SARS, MERS, Ebola, Swine flu, and most recently, pandemic of coronavirus disease COVID-19, e.g., Ivanov (2020).

The transboundary dynamics of the COVID-19 pandemic, which has caused severe global socioeconomic disruption, may have an impact on the scale of using outsourcing and offshoring in the future supply chains. An overreliance on lean manufacturing and just-in-time methodology may also be significantly limited in the future and pre-positioning of risk mitigation inventory more frequently used in practice.

In order to minimize losses caused by the shortage of material supplies, customer companies (firms) apply different disruption management strategies. For example, the firms may participate in supplier's recovery process after disruption to reduce recovery time. When the Tohoku earthquake and tsunami disrupted Toyota supply chain and, in particular, supply chain of automotive semiconductors, Toyota supported recovery of its suppliers, Fujimoto and Park (2013). The automotive semiconductors were manufactured by Renesas Electronics, who shares over 44% of world-wide automobile microcontroller units, and its main plant in Naka was severely damaged by the earthquake. The shipping of automotive semiconductors was expected to be stopped for 8 months. In order to shorten the expected recovery time, Toyota and other Japanese automotive, electronics and semiconductor equipment manufacturers sent to Naka over 2500 engineers to support plant recovery. As a result, the recovery time to start shipping of automobile microcontroller units was shortened from 8 to 5 months, Matsuo (2015). Another example of helping by customer companies in supplier's recovery process was the case of Riken Corporation, the largest supplier of piston rings to all Japanese automobile manufacturers. In July 2007, Riken plant in city of Kashiwazaki was hit by a strong earthquake and severely damaged. Immediately after the shutdown, the Japanese automakers coordinated by Toyota sent over 650 people including many equipment engineers to help its recovery and as a result the stoppage of piston rings production was shortened to 2 weeks only, Whitney et al. (2014). A similar action has been applied to help recovery of Aisin Seiki plants after the Kyushu earthquake in April 2016, (Marszewska 2016).

The above real-world examples illustrate a well known disruption management strategy of helping a primary supplier recover more quickly. However, when a primary supplier is hit by disruption, the customer company may choose either to support recovery of disrupted primary supplier, rely on a preselected backup supplier or select an alternate (recovery) supplier, non-disrupted or disrupted less severely than the primary supplier. In a similar way, a firm whose primary plant is hit by disruption may either stop production until recovery process is finished or move production to alternate (recovery) plants along with transshipment of parts to the recovery plants. The complex decision-making that involves various characteristics of a supply chain should be supported by optimization models to minimize cost (or maximize profit) and maximize service level as typical objective functions. The objective of this book is to present such optimization models and to stimulate further research on fundamental understanding of various mitigation and recovery policies in the presence of flow disruption risks in global supply chains.

1.2 Value-at-Risk Versus Conditional Value-at-Risk

A common tool for supply chain optimization under disruption risks is stochastic programming, in particular stochastic mixed integer programming (stochastic MIP or SMIP), e.g., Heckmann et al. (2015). Stochastic MIP is an exact mathematical modeling approach that allows for the inclusion of uncertainty by probabilistic scenarios of disruptive events and for finding the optimal solutions with respect to multiple objective functions. In this book the stochastic combinatorial decision-making problems will be formulated as multi-period, multi-objective stochastic mixed integer programs with expected or expected worst-case performance measures and trade-offs between various objective functions. In order to mitigate the impact of supply chain disruptions, the two popular in financial engineering percentile measures of risk, Value-at-Risk (VaR) and Conditional Value-at-Risk (CVaR), will be applied for the decision-making, e.g., Sawik (2008, 2009, 2010, 2011b, 2012a,b,c, 2013a,b, 2016). This section briefly defines and compares VaR and CVaR, e.g., Uryasev (2000), Sarykalin et al. (2008).

Let $F_W(u) = Prob\{W \leq u\}$ be the cumulative distribution function of a random variable W representing cost. The VaR (Value-at-Risk) of W with the confidence level $\alpha \in [0,1)$ is defined as a lower α-percentile of the random variable W

$$VaR_\alpha(W) = \min\{u : F_W(u) \geq \alpha\}.$$

VaR represents the maximum cost associated with a specified confidence level of outcomes (i.e., the likelihood that a given portfolio's costs will not exceed the amount defined as VaR). However, VaR does not account for properties of the cost distribution beyond the confidence level and hence does not explain the magnitude of the cost when the VaR limit is exceeded. Moreover, VaR is not a coherent measure of risk since it fails to hold the sub-additivity property ($f(x+y) \leq f(x) + f(y)$ where $f(.)$ is the risk measure). VaR of a portfolio can be higher than the sum of VaRs of the individual assets in the portfolio.

On the other hand, CVaR focuses on the tail of the cost distribution, that is, on outcomes with the highest cost. Assuming that $F_W(u)$ is a continuous distribution function, the CVaR of W with the confidence level $\alpha \in [0,1)$, $CVaR_\alpha(W)$, equals the expectation of W subject to $W \geq VaR_\alpha(W)$. However, in the general case $CVaR_\alpha(W)$ is not equal to an average of outcomes greater than $VaR_\alpha(W)$ and is defined as the mean of the generalized α-tail distribution

$$CVaR_\alpha(W) = \int_{-\infty}^{\infty} u dF_W^\alpha(u),$$

where

$$F_W^\alpha(u) = \begin{cases} 0 & \text{if } u < VaR_\alpha(W) \\ \frac{F_W(u)-\alpha}{1-\alpha} & \text{if } u \geq VaR_\alpha(W). \end{cases}$$

$CVaR_\alpha(W)$ can be considered as a generalization of the expected value, when $\alpha = 0$ they are equivalent. On the other hand, the higher is the confidence level α, the closer are values of $VaR_\alpha(W)$ and $CVaR_\alpha(W)$.

Alternatively, $CVaR_\alpha(W)$ can be defined as the weighted average of $VaR_\alpha(W)$ and the conditional expectation of W subject to $W > VaR_\alpha(W)$.

When the distribution function has a vertical jump at $VaR_\alpha(W)$ (the probability interval of confidence level α with the same VaR), a probability "atom" is said to be present at $VaR_\alpha(W)$. For example, when the distribution is modeled by scenarios, the probability measure is concentrated in finitely many points and the corresponding distribution function is a step function (constant between jumps) with jumps at those points. Since $CVaR_\alpha(W)$ is obtained by averaging a fractional number of scenarios, the $VaR_\alpha(W)$ atom can be split. When $F_W(VaR_\alpha(W)) > \alpha$, then probability $1 - F_W(VaR_\alpha(W))$ of the cost interval $[VaR_\alpha(W), \infty)$ is smaller than $1-\alpha$.

Note that if $F_W(VaR_\alpha(W)) = 1$, so that $VaR_\alpha(W)$ is the highest cost that may occur, then $CVaR_\alpha(W) = VaR_\alpha(W)$.

Summarizing the above definitions (from now on, VaR and CVaR will be denoted without subscript α, and with superscripts c and sl to denote cost and service level, respectively):

- Cost-at-risk (VaR^c) at a $100\alpha\%$ confidence level is the targeted cost such that for $100\alpha\%$ of the scenarios, the outcome will not exceed VaR^c. In other words, VaR^c is a decision variable based on the α-percentile of costs, i.e., in $100(1-\alpha)\%$ of the scenarios, the outcome may exceed VaR^c.
- Conditional cost-at-risk (CVaR^c) at a $100\alpha\%$ confidence level is the expected cost in the worst $100(1-\alpha)\%$ of the cases. In other words, we allow $100(1-\alpha)\%$ of the outcomes to exceed VaR^c, and the mean value of these outcomes is represented by CVaR^c.

In other words, VaR^c is the acceptable cost level above which we want to minimize the number of outcomes and CVaR^c considers those portfolio outcomes, where costs exceed VaR^c (see, Fig. 1.2).

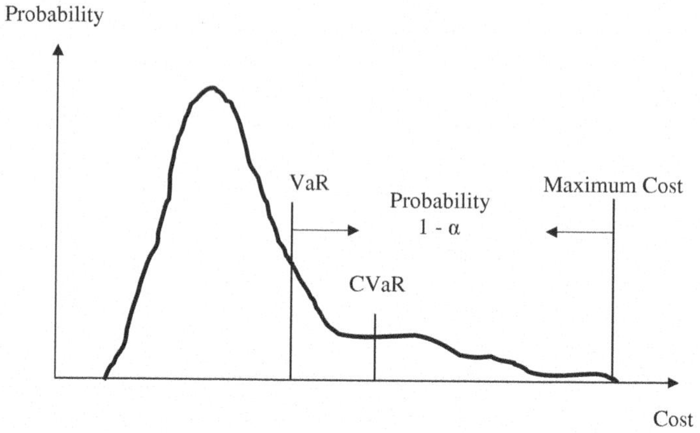

Fig. 1.2 Distribution of cost: VaR^c vs. CVaR^c

Generally, confidence level α indicates the level of conservatism that a decision maker is willing to adopt. As α approaches one, the range of acceptable worst-cases becomes narrower in the corresponding optimization problem. Figure 1.2 clarifies the concept of CVaR and demonstrates that CVaR is the conditional expected value exceeding the VaR.

Similar definitions of VaR and CVaR for service level are given below.

- Service-at-risk (VaR^{sl}) at a $100\alpha\%$ confidence level is the targeted service level such that for $100\alpha\%$ of the scenarios, the outcome will not be below VaR^{sl}. In other words, VaR^{sl} is a decision variable based on the α-percentile of service level, i.e., in $100(1-\alpha)\%$ of the scenarios, the outcome may be below VaR^{sl}.
- Conditional service-at-risk (CVaR^{sl}) at a $100\alpha\%$ confidence level is the expected service level in the worst $100(1-\alpha)\%$ of the cases. In other words, we allow $100(1-\alpha)\%$ of the outcomes to be below VaR^{sl}, and the mean value of these outcomes is represented by CVaR^{sl}.

In other words, VaR^{sl} is the acceptable service level below which we want to maximize the number of outcomes and CVaR^{sl} considers those portfolio outcomes, where service levels do not exceed VaR^{sl} (see, Fig. 1.3).

Since VaR and CVaR measure different parts of the cost (service level) distribution, VaR may be better for optimizing portfolios when good models for tails are not available, otherwise CVaR may be preferred. There are many practical reasons explaining the success of CVaR as a risk measure, e.g., Filippi et al. (2020). CVaR, which is a so-called downside risk measure, penalizes only the negative deviations with respect to a target, set by a decision-maker. CVaR is sensitive to the worst outcomes but it is not as conservative as a minimax approach. However, it simultaneously considers only a fraction of the worst outcomes, while neglecting the others. It is the decision-maker who decides on this fraction by choosing her confidence level. Computationally, CVaR can often be easily embedded in an optimization model by

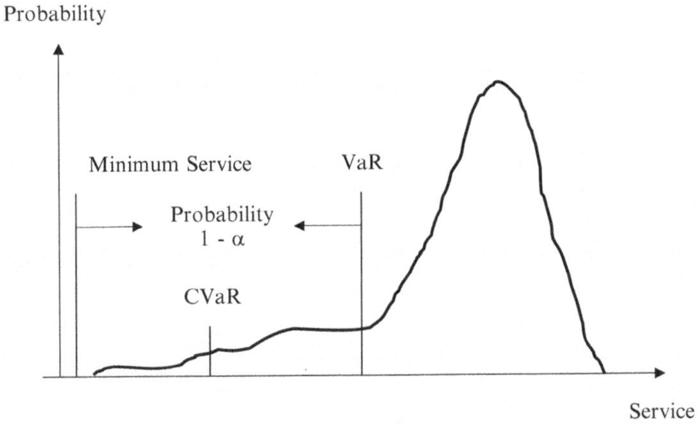

Fig. 1.3 Distribution of service level: VaR^{sl} vs. CVaR^{sl}

adding linear constraints and continuous variables, so that a linear structure of the model can be preserved.

1.3 Local, Regional, and Global Disruptions

Assume that the supply chain consists of $\overline{I}$ interconnected facilities (nodes in this network) that are located in $\overline{R}$ disjoint geographic regions. A facility in a supply chain network may be a supplier of raw material or components, a manufacturer or an assembly plant, a distribution center, or a retailer. Denote by $I^r \subseteq I = \{1,\ldots,\overline{I}\}$ the subset of supply chain facilities in region $r \in R = \{1,\ldots,\overline{R}\}$, where $\bigcup_{r\in R} I^r = I$. The supply chain facilities are subject to random independent local disruptions that are uniquely associated with a particular facility, which may arise from equipment breakdowns, local labor strikes, fires, etc. Denote by p_i the local disruption probability for object $i \in I$, i.e., the state of facility i is "on" (non-disrupted) with probability $(1-p_i)$, or is "off" (disrupted) with probability p_i.

In addition to independent local disruptions of each facility individually, there are also potential regional disruptive events that may result in correlated regional disruption of all facilities in the same geographic region, and global disruptive super events that may simultaneously impact all the facilities, i.e., the entire supply chain. For example, such regional disruptive events may include, flood, earthquake, regional strike in a transportation sector, whereas global disaster super events may include an economic crisis, widespread labor strike in a transportation sector, etc.

Let p^r and p^* be the probability of correlated regional disruption, simultaneously of all facilities $i \in I^r$ in region $r \in R$, and correlated global disruption, simultaneously of all facilities $i \in I$, respectively. The global disruptive super event, the regional disruptive events in each region and the local disruptive events are assumed to be independent events. Thus, the disruption probability, π_i, of every facility $i \in I^r, r \in R$ is

$$\pi_i = p^* + (1-p^*)p^r + (1-p^*)(1-p^r)p_i;\ i \in I^r, r \in R. \quad (1.1)$$

Denote by P_s the probability that disruption scenario s is realized, where each scenario $s \in S$ is comprised of a unique subset $I_s \subset I$ of facilities that are in state "on" (non-disrupted), and $S = \{1, \ldots, \overline{S}\}$ is the index set of all disruption scenarios. Each scenario $s \in S$ can be represented by a 0-1 vector $\lambda_s = \{\lambda_{1s}, \ldots, \lambda_{\overline{I}s}\}$, where $\lambda_{is} = 0$ denotes disruption of facility $i \in I$ under scenario $s \in S$, and $\lambda_{is} = 1$ denotes a normal, non-disruptive state of facility $i \in I$ under scenario $s \in S$. Given the number of facilities $\overline{I}$, the total number of scenarios in which at least one facility is non-disrupted is given by $\sum_{i=1}^{\overline{I}} \binom{\overline{I}}{i} = 2^{\overline{I}} - 1$. Including the scenario in which all facilities are disrupted, there are a total of $\overline{S} = 2^{\overline{I}}$ potential disruption scenarios. For each scenario $s \in S$, the facilities $i \in I \setminus I_s$ can be disrupted either by a local, regional, or global disaster event.

The probability P_s of each disruption scenario $s \in S$ is derived as follows. First, the probability P_s^r of realizing disruption scenario s for suppliers in I^r is determined. If there are non-disrupted suppliers in region r, i.e., $I^r \bigcap I_s \neq \emptyset$, then P_s^r is the product of regional non-disruption probability, $(1-p^r)$, local probabilities of non-disrupted suppliers, $(1-p_i)$, $i \in I^r \bigcap I_s$, and local probabilities of disrupted suppliers, p_i, $i \in I^r \setminus I_s$. Otherwise, i.e., if all suppliers in region r are disrupted, $I^r \bigcap I_s = \emptyset$, then either the entire region is disrupted with probability, p^r, or the region is non-disrupted with probability, $(1-p^r)$, and every supplier $i \in I^r$ is locally disrupted with probability, p_i. Thus, the probability P_s^r is

$$P_s^r = \begin{cases} (1-p^r)\prod_{i \in I^r \bigcap I_s}(1-p_i)\prod_{i \in I^r \setminus I_s} p_i & \text{if } I^r \bigcap I_s \neq \emptyset \\ p^r + (1-p^r)\prod_{i \in I^r} p_i & \text{if } I^r \bigcap I_s = \emptyset. \end{cases} \quad (1.2)$$

The probability P_s for each disruption scenario $s \in S$ with the subset I_s of non-disrupted facilities, and with all possible combinations of different disruptive events considered, is

$$P_s = \begin{cases} (1-p^*)\prod_{r \in R} P_s^r & \text{if } I_s \neq \emptyset \\ p^* + (1-p^*)\prod_{r \in R} P_s^r & \text{if } I_s = \emptyset, \end{cases} \quad (1.3)$$

If global and regional disruption probabilities are negligible, i.e., $p^* = 0$ and $p^r = 0, r \in R$, the probability P_s of disruption scenario s in the presence of independent local disruptive events only, reduces to

$$P_s = \prod_{i \in I_s}(1-p_i) \cdot \prod_{i \notin I_s} p_i. \quad (1.4)$$

1.4 Two-Level Versus Multi-Level Disruptions

In this section the scenarios with the two-level, all-or-nothing disruptive events considered in Sect. 1.3 are enhanced for the multi-level (partial) disruptive events. In contrast to yield uncertainty (e.g., defective products) that occurs, for instance, when the quantity of supply delivered is a random variable, typically modeled as either a random additive or multiplicative quantity, the multi-level disruptions are modeled as events of different level (e.g., partial capacity available) which occur randomly and may have a random length, e.g., Schmitt and Snyder (2012).

Assume that each facility $i \in I$ is subject to random independent local disruptions of different levels, $l \in L_i = \{0, \ldots, \overline{L_i}\}$, where the disruption level refers to the available fraction of full capacity of a facility (e.g., a partial fulfillment of an order by a supplier, a partial fulfillment of a customer order by a producer, etc.).

Level $l = 0$ represents complete shutdown of a facility, i.e., no capacity available, (e.g., no order delivery) while level $l = \overline{L_i}$ represents normal conditions with full capacity available (e.g., full order delivery). The fraction of full capacity of facility i available under disruption level l is described by γ_{il}

$$\gamma_{il} = \begin{cases} 0 & \text{if } l = 0 \\ \in (0,1) & \text{if } l = 1, \ldots, \overline{L_i} - 1 \\ 1 & \text{if } l = \overline{L_i}. \end{cases} \tag{1.5}$$

Denote by $S = \{1, \ldots, \overline{S}\}$ the index set of all disruption scenarios, and by P_s the probability of disruption scenario $s \in S$. Each scenario $s \in S$ can be represented by an integer-valued vector $\lambda_s = \{\lambda_{1s}, \ldots, \lambda_{\overline{I}s}\}$, where $\lambda_{is} \in L_i$ is the disruption level of facility $i \in I$ under scenario $s \in S$. When all potential disruption scenarios are considered, then $\overline{S} = \prod_{i \in I}(\overline{L_i} + 1)$.

Assume that for each scenario $s \in S$, each facility can be disrupted either by a multi-level local disruptive event or by a two-level regional disruptive event. The probability P_s for disruption scenario $s \in S$ with the subset I_s of non-shutdown facilities is given by (1.3). Now, the probability, P_s^r, of realizing of disruption scenario s for facilities in I^r is (e.g., Sawik 2015b)

$$P_s^r = \begin{cases} (1 - p^r)\prod_{i \in I^r}(p_{i,\lambda_{is}}) & \text{if } I^r \bigcap I_s \neq \emptyset \\ p^r + (1 - p^r)\prod_{i \in I^r} p_{i0} & \text{if } I^r \bigcap I_s = \emptyset, \end{cases} \tag{1.6}$$

where $p_{i,\lambda_{is}}$ is the probability of occurrence of the disruption at level $l = \lambda_{is}$ at facility i.

1.5 Risk-Neutral, Risk-Averse, and Mean-Risk Decision-Making

In this book we consider three types of decision-making policies.

1. Risk-neutral that is based on expected value optimization approach, e.g., expected cost minimization or expected service level maximization approach. The risk-neutral policy focuses on an average performance of a supply chain.
2. Risk-averse wherein, rather that optimizing the expected value of an objective function, the decision maker uses a Conditional Value-at-Risk (CVaR) approach to measure and quantify risk and to define what comprises a worst-case scenario. The CVaR methodology allows the decision maker to evaluate worst-case values of an objective function, to specify to what extent worst-case scenarios should be avoided and to shape distribution of the resulting objective function values, associated with such a policy. The risk-averse policy focuses on worst-case performance of a supply chain.
3. Mean-risk wherein the decision maker seeks for a best trade-off between expected value and CVaR of an objective function. The mean-risk policy focuses on both the average and the worst-case performance of a supply chain, simultaneously.

In this book we utilize stochastic mixed integer programming (SMIP) approach, that leads to a two-stage optimization problem. The decisions that are made ahead of time are considered the first stage decisions and are represented by the first stage decision variables.

Typical *first stage decision variables* are

- binary selection variables, such as supplier selection variables, supplier protection variables,
- fractional allocation variables, such as order quantity allocation variables, risk mitigation inventory pre-positioning variables.

In the risk-averse decision-making, VaR (cost-at-risk, service-at-risk, etc.) can also be interpreted as first stage variables.

The second stage decision variables represent decisions that are made after the realization of the random events (e.g., supply disruptions) is known. The second stage variables are dependent on the realized random event.

Typical *second stage decision variables* are

- binary selection variables, such as recovery supplier selection variables, recovery plant selection variables,
- time-indexed binary assignment variables, such as production scheduling variables, distribution scheduling variables,
- fractional allocation variables, such as recovery order quantity allocation variables, recovery demand allocation variables, risk mitigation inventory usage variables, transshipment variables,
- continuous variables, such as tail cost, tail service level.

1.5.1 Risk-Neutral Decision-Making

The stochastic formulation of the risk-neutral decision-making problem aimed at loss (cost) minimization can be written as

$$\min_{\mathbf{x}\in X} \mathbf{c}^T\mathbf{x} + E[Q(\mathbf{x},\xi^s)], \tag{1.7}$$

where $\mathbf{c}^T\mathbf{x} + E[Q(\mathbf{x},\xi^s)]$ is the total cost function of the first stage problem and

$$Q(\mathbf{x},\xi^s) = \min_{\mathbf{y}^s\in Y^s} \{(\mathbf{q}^s)^T\mathbf{y}^s\} \tag{1.8}$$

is the optimal solution of the second-stage problem corresponding to the first stage decision variables $\mathbf{x}$ and the realization of the random data ξ^s for scenario $s \in S$, denoted by $\xi^s = (\mathbf{q}^s, Y^s)$.

$E[Q(\mathbf{x},\xi^s)]$ is the expected "cost" taken with respect to random scenario $s \in S$.

The objective function $Q(\mathbf{x},\xi^s)$ of the second-stage problem (1.8), also known as the recourse (cost) function, is a random variable.

Here $\mathbf{x}$ and $\mathbf{y}^s$ are the vectors of first stage and second stage decision variables, where the first stage decisions are deterministic and the second-stage decisions are dependent on random scenario s. X denotes the feasible set of first stage decisions and Y^s is the feasible set of second stage decisions for random scenario $s \in S$. The second-stage problem (1.8) may be infeasible for some first stage decisions $\mathbf{x} \in X$.

To deal with the uncertainty in the second stage, a scenario-based modeling approach is proposed that has been widely used in stochastic programming. In the second stage, let us consider random scenario $s \in S$ to have a discrete distribution, where P_s is the probability of occurrence for scenario $s \in S$, and S is a finite set of scenarios. Given a finite set of scenarios, S, with associated probabilities, $P_s, s \in S$, the expected value $E[Q(\mathbf{x},\xi^s)]$ can be evaluated as $E[Q(\mathbf{x},\xi^s)] = \sum_{s\in S} P_s Q(\mathbf{x},\xi^s)$. Hence, we can present the deterministic equivalent of the stochastic formulation (1.7).

$$\begin{aligned} \min \quad & \mathbf{c}^T\mathbf{x} + \sum_{s\in S} P_s(\mathbf{q}^s)^T\mathbf{y}^s \\ \text{s.t.} \quad & \mathbf{x} \in X,\ \mathbf{y}^s \in Y^s;\ s \in S. \end{aligned} \tag{1.9}$$

Model (1.9) is also known as the wait-and-see model (e.g., Birge and Louveaux 2011; Kall and Mayer 2011). In contrast to the two-stage approach with the recourse model (1.7), in the wait-and-see approach both, the decisions on the first stage variables $\mathbf{x}$ and the second stage variables $\mathbf{y}^s$, are taken simultaneously only when the outcome of $\xi^s = (\mathbf{q}^s, Y^s)$ is known.

1.5.2 Risk-Averse Decision-Making

In the model proposed below, CVaR is represented by an auxiliary function (1.10) introduced by Rockafellar and Uryasev (2000), for a set of pre-defined scenarios $s \in S$ with corresponding probabilities, P_s. Using the wait-and-see approach, the stochastic formulation of the risk-averse decision-making problem aimed at CVaR of loss (cost) minimization, given confidence level α, can be written as

Minimize

$$CVaR = VaR + (1-\alpha)^{-1} \sum_{s \in S} P_s \tau_s \tag{1.10}$$

subject to

$$\tau_s \geq \mathbf{c}^T \mathbf{x} + (\mathbf{q}^s)^T \mathbf{y}^s - VaR;\ s \in S \tag{1.11}$$

$$\mathbf{x} \in X \tag{1.12}$$

$$\mathbf{y}^s \in Y^s;\ s \in S \tag{1.13}$$

$$\tau_s \geq 0;\ s \in S. \tag{1.14}$$

In the above formulation, constraints (1.11) compute the tail cost, τ_s, for scenario s and condition (1.14) indicates that the scenarios in which the loss exceeds VaR are considered only.

1.5.3 Mean-Risk Decision-Making

In the mean-risk formulation for the wait-and-see approach , λ is a nonnegative trade-off coefficient representing the decision maker risk preference. For a given confidence level α one can construct the mean-risk efficient frontier by the parameterization on λ the weighted-sum program presented below.

Minimize

$$\lambda(\mathbf{c}^T\mathbf{x}+\sum_{s\in S}P_s(\mathbf{q}^s)^T\mathbf{y}^s)+(1-\lambda)(VaR+(1-\alpha)^{-1}\sum_{s\in S}P_s\tau_s), \quad (1.15)$$

where $0 \le \lambda \le 1$
subject to

$$\tau_s \ge \mathbf{c}^T\mathbf{x}+(\mathbf{q}^s)^T\mathbf{y}^s - VaR;\ s\in S \quad (1.16)$$

$$\mathbf{x}\in X \quad (1.17)$$

$$\mathbf{y}^s \in Y^s;\ s\in S \quad (1.18)$$

$$\tau_s \ge 0;\ s\in S. \quad (1.19)$$

The resulting decision vector $\mathbf{x}$ is efficient in the mean-risk sense, i.e., it has the lowest possible CVaR for a given expected cost, and for a given CVaR it has the lowest possible expected cost.

1.6 A Multi-Portfolio Approach

Disruptions in supply chain rarely occur as isolated events. Supply chain disruptions must be contained before they propagate through the supply chain and create greater losses. Supply chain disruption propagation is defined as the spread of the disruption effects beyond the initial disruption location. In the literature, supply chain disruption propagation is also known as domino effect, snowball effect, and more recently as the ripple effect, e.g., Blackhurst et al. (2011), Basole and Bellamy (2014), Scheibe and Blackhurst (2018), Ivanov et al. (2015, 2019). In order to contain disruptions and prevent them from spreading through the supply chain, the decision-making on disruption management should be coordinated and both proactive and reactive decisions, spatially and temporally integrated.

In this book an innovative multi-portfolio approach and SMIP formulations with an embedded network flow problem are developed for spatially and temporally integrated decision-making in a supply chain under disruption risks. The portfolio approach to supply chain disruption management was introduced by Sawik (2011b,c,d, 2013c), first to mitigate the impact of disruption risks, and later for a combined risk mitigation and recovery in supply chains (e.g., Sawik 2017b, 2019a). In the context of supply chain disruptions, the portfolio is defined as the allocation of demand for different part types among suppliers and the allocation of demand for different product types among assembly plants of final manufacturer.

The concept of a multi-portfolio approach to supply chain disruption management is illustrated in Fig. 1.4. Depending on the impact of disruptive events, different types of primary and recovery portfolios should be determined. When a disruptive event hits part suppliers only, then only supply portfolios need to be selected. In a general case, however, a disruptive event may impact both primary suppliers of parts and primary assembly plants of the manufacturer. Then the manufacturer selects recovery suppliers, recovery assembly plants along with the allocation of demand for parts and products among selected facilities and transshipment of the inventory of parts from disabled primary plants to recovery plants.

The multi-portfolio approach, integrated over time and space, allows for a coordinated decision-making in supply chain networks under disruption risks. When the multi-portfolio approach is applied, the proactive and reactive decisions are spatially and temporally integrated: the primary supply and demand portfolios to be implemented before a disruptive event are optimized simultaneously with recovery supply, transshipment, and demand portfolios for each potential disruption scenario for the aftermath period. The portfolios for both part suppliers and product manufacturers in different geographic regions are determined simultaneously. The multi-portfolio approach can be particularly useful for material flow coordination in contemporary multi-tier supply chain networks, which are cyber-physical systems, where integrated decision-making and control over the entire network is of utmost importance. In the SMIP decision-making models, each portfolio type is determined using binary selection variables and fractional demand allocation variables. Without loss of generality, in the definitions provided below we assume single part type and product type environment with a unit requirement of part per product and a single primary assembly plant, $j = 1$. In addition, denote by, D, total demand for parts/products, and by I, J and m, n, respectively, the index set of suppliers, assembly plants and the total number of suppliers, assembly plants.

The primary supply portfolio is determined by the primary supplier selection variables and primary demand for parts allocation variables:

$u_i \in \{0,1\}$, such that the value 1 means that supplier $i \in I$ is selected as a primary supplier; otherwise $u_i = 0$;

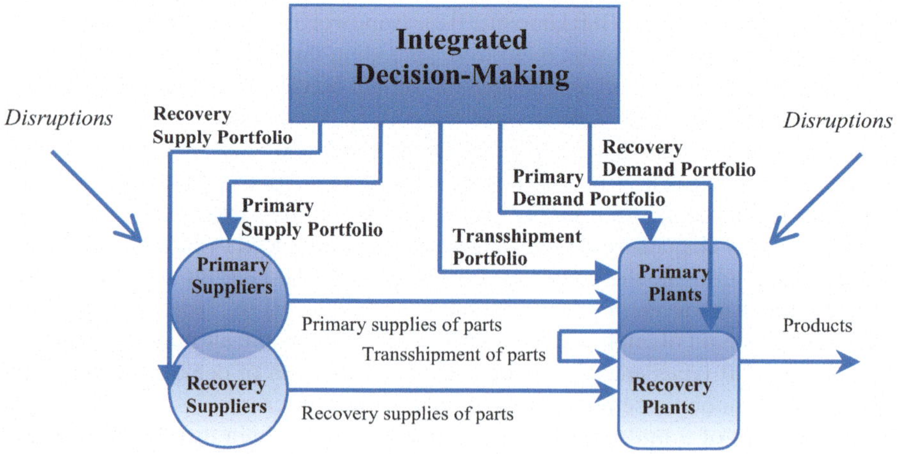

Fig. 1.4 A multi-portfolio approach to supply chain disruption management

$v_i \in [0,1]$, the fraction of total demand for parts ordered from primary supplier $i \in I$ (such that $u_i = 1$) to primary plant $j = 1$.

The recovery demand portfolio is determined by the recovery plant selection variables and recovery demand for products allocation variables:

$q_j^s \in \{0,1\}$, such that value 1 means that assembly plant $j \in J$ is selected as a recovery plant under disruption scenario s; otherwise $q_j^s = 0$;

$Q_j^s \in [0,1]$, the fraction of total demand for products to be completed by recovery plant $j \in J$ (such that $q_j^s = 1$) under disruption scenario s.

The recovery supply portfolio is determined by the recovery supplier selection variables and recovery demand for parts allocation variables:

$U_i^s \in \{0,1\}$, such that value 1 means that supplier $i \in I$ is selected as a recovery supplier under disruption scenario s; otherwise $U_i^s = 0$;

$V_{ij}^s \in [0,1]$, the fraction of total demand for parts ordered from recovery supplier $i \in I$ (such that $U_i^s = 1$) to recovery plant $j \in J$ (such that $q_j^s = 1$), under disruption scenario s.

The recovery transshipment portfolio is determined by the inventory of parts transshipment variables:

$w_j^s \in [0,1]$, the fraction of total demand for parts, transshipped from the disrupted primary plant $j = 1$ to recovery plant $j \in J$ (such that $q_j^s = 1$), under disruption scenario s, where w_1^s represents the parts that remain in primary plant $j = 1$.

In a general case of the supply chain-wide impact of disruption risks, the different types of portfolios that should be determined, may include:

The primary supply portfolio

$$(v_1, \ldots, v_m),$$

which specifies supplies of parts from selected primary suppliers to primary assembly plant $j = 1$, where

$$v_i \in [0,1]; i \in I, \ \sum_{i \in I} v_i = 1.$$

The recovery supply portfolio for each scenario s

$$(V_{1j}^s, \ldots, V_{mj}^s); \ \ j \in J,$$

which specifies supplies of parts from recovery suppliers to recovery assembly plants, where

$$V_{ij}^s \in [0,1]; i \in I, j \in J, \ \sum_{i \in I} (\gamma_i^s v_i + \sum_{j \in J} V_{ij}^s) = 1,$$

$\sum_{i \in I} \gamma_i^s v_i$ denotes actual delivery of parts from primary suppliers to primary plant under disruption scenarios s and γ_i^s is the fraction of an order delivered by supplier i under scenario s.

The recovery transshipment portfolio for each scenario s

$$(w_1^s, \ldots, w_n^s),$$

which specifies transshipment of parts from primary assembly plant, $j = 1$ to recovery plants, where

$$w_j^s \in [0,1]; j \in J, \ \sum_{j \in J} w_j^s = \sum_{i \in I} \gamma_i^s v_i - \sum_{t \in T: t < t_s} x_{1t}^s / D,$$

and $\sum_{t \in T: t < t_s} x_{1t}^s$ denotes production at primary assembly plant $j = 1$, before disruption at time t_s.

The recovery demand portfolio for each scenario s

$$(Q_1^s, \ldots, Q_n^s),$$

which specifies an allocation of unfulfilled demand for products among recovery plants, where

$$Q_j^s \in [0,1]; j \in J, \quad \sum_{t \in T: t < t_s} x_{1t}^s / D + \sum_{j \in J} Q_j^s = 1.$$

In addition, the two decision-making strategies will be considered: an integrated strategy with the perfect information about all potential future disruption scenarios, and a hierarchical approach with no such information available ahead of time. In the integrated approach, which accounts for all potential disruption scenarios, the primary supply portfolio that will hedge against all scenarios is determined along with the recovery supply and demand portfolios and production and inventory scheduling of finished products for each scenario. In the hierarchical approach, first the primary supply portfolio is selected to optimize supplies and production in a deterministic environment (without a disruption). Then, when a primary supplier and/or primary assembly plant is hit by a disruption, the recovery supply and demand portfolios are determined along with transshipment of parts and production and inventory scheduling at recovery plants to optimize the supply chain recovery, given the unfulfilled demand and the inventory of parts available for transshipment.

Part I
Selection of Supply Portfolio

Chapter 2
Selection of Static Supply Portfolio

2.1 Introduction

In a customer-driven supply chain manufacturers should be prepared to produce varieties of products to meet the different customer needs. Each product is typically composed of many common and non-common (custom) parts that can be sourced from different suppliers with different supply capacity. An important issue is how to best allocate the orders for parts among various part suppliers to fulfill all customer orders for products and to achieve a high customer service level at a low cost and, in addition, to mitigate the impact of supply chain disruption risks. Supply chain management, in particular, deals with selection of supply portfolio, i.e., selection of suppliers and order quantity allocation under uncertain quality of supplied materials and reliability of on-time delivery. The decision maker needs to decide from which supplier to purchase parts required to meet customer demand. The decision is based on price, quality, and reliability criteria that may conflict each other. For example, the supplier offering the lowest price may not have the best quality or the supplier with the best quality may not deliver on time. In stochastic supply settings, supplier selection allows the producer to decide whether it should cooperate with a low cost, yet risky suppliers over more expensive but possibly more reliable suppliers. A common risk-neutral objective of minimizing expected cost or maximizing expected service level is therefore influenced by uncertainty and risk. As a result, new non-risk-neutral objectives of minimizing and maximizing the number of outcomes that could occur above an acceptable cost level and below an acceptable service level, respectively, are observed in practice. Furthermore, to reduce the fixed ordering costs of creating contracts and maintaining relationships with suppliers, the number of suppliers and the total number of orders should be minimized. On the other hand, however, the selection of more suppliers may divert the risk of unreliable supplies. In global supply chains a multi-regional suppliers base is a frequent solution, where suppliers from different geographic regions are selected.

T. Sawik, *Supply Chain Disruption Management*,
International Series in Operations Research & Management Science 291,
https://doi.org/10.1007/978-3-030-44814-1_2

Then, in addition to independent local disruptions (i.e., equipment breakdown, fires, etc.) that are uniquely associated with a particular supplier, the supplies of parts are also subject to regional disruptions (e.g., floods, hurricanes, earthquakes, economic crisis, etc.) simultaneously of all suppliers in the same region.

This chapter deals with selection of a static supply portfolio under disruption risks, i.e., for determining a single-period supply portfolio. The proposed portfolio approach allows the two popular in financial engineering percentile measures of risk, Value-at-Risk (VaR) and Conditional Value-at-Risk (CVaR) to be applied for managing the risk of supply disruptions. For a finite number of scenarios, CVaR allows the evaluation of worst-case costs (or worst-case service level) and shaping of the resulting cost (service level) distribution through optimal supplier selection and order allocation decisions, i.e., the selection of optimal supply portfolio. Since common parts can be efficiently managed by material requirement planning methods, the focus in this chapter is on supplies of custom parts that can be critical in make-to-order manufacturing.

The following SMIP models are presented in this chapter:

SP_E(c) for risk-neutral selection of supply portfolio to minimize expected cost;
SP_E(sl) for risk-neutral selection of supply portfolio to maximize expected service level;
SP_CV(c) for risk-averse selection of supply portfolio to minimize CVaR of cost;
SP_CV(sl) for risk-averse selection of supply portfolio to maximize CVaR of service level;
SP_ECV(c) for mean-risk selection of supply portfolio to optimize trade-off between expected cost and CVaR of cost;
SP_ECV(sl) for mean-risk selection of supply portfolio to optimize trade-off between expected service level and CVaR of service level.

In the computational experiments described in Sect. 2.6, single-region and multi-region sourcing subject to local and regional disruption risks are illustrated with numerical examples. The major managerial insights are provided at the end of this chapter.

In the next chapter the static portfolio approach will be enhanced for a multi-period supplier selection and order allocation in the presence of supply chain disruption and delay risks, where in the scenario analysis the low-probability and high-impact supply disruptions are combined with the high-probability and low-impact supply delays. Unlike for a single-period problem considered in this chapter, in a multi-period setting the decision maker needs to decide from which supplier and when to purchase custom parts required for each customer order to meet customer requested due dates at a low cost and a high customer service level and to mitigate the impact of disruption and delay risks.

2.2 Problem Description

In the supply chain under consideration various types of products are assembled by a single producer to satisfy customer orders, using custom parts purchased from multiple suppliers (for notation used, see Table 2.1). Each supplier can provide the producer with custom parts for all customer orders. However, the suppliers have different limited capacity and, in addition, differ in price and quality of offered parts and in reliability of delivery of parts. Let $I=\{1,\ldots,\overline{I}\}$ be the set of $\overline{I}$ suppliers and $J=\{1,\ldots,\overline{J}\}$ the set of $\overline{J}$ customer orders for the products, known ahead of time.

Table 2.1 Notation: static supply portfolio

Indices	
i	= supplier, $i \in I$
j	= customer order, $j \in J$
r	= geographic region, $r \in R$
s	= disruption scenario, $s \in S$
Input Parameters	
c_i	= capacity of supplier i
d_j	= demand for parts required for customer order j
D	= $\sum_{j\in J} d_j$—total demand for parts
e_i	= cost of ordering parts from supplier i
g_j	= per unit penalty cost of unfulfilled customer order j
o_{ij}	= unit price of parts for customer order j purchased from supplier i
p_i	= local disruption probability for supplier i
p^r	= regional disruption probability for all suppliers in region r
α	= confidence level
ρ_i	= expected defect rate of supplier i

Each order $j \in J$ is described by the quantity d_j of required custom parts. Denote by c_i the capacity of supplier $i \in I$, by e_i the cost of ordering parts from supplier $i \in I$, and by o_{ij} the unit purchasing price of parts for customer order $j \in J$ from supplier $i \in I$.

The ordered parts are dispatched to the producer after the completion time of their manufacturing. For each supplier, however, the quality of delivered part may vary randomly. When the suppliers are selected, the risk of defective parts can be considered using past observations. Since quality of parts varies among different suppliers, a different average defect rate can be associated with each supply portfolio. Let ρ_i be the expected defect rate of supplier i.

The suppliers are assumed to be located in $\overline{R}$ disjoint geographic regions. Denote by $I^r \subseteq I$ the subset of suppliers in region $r \in R=\{1,\ldots,\overline{R}\}$, where $\bigcup_{r\in R} I^r = I$. The supplies of parts are subject to random local disruptions that are uniquely associated with a particular supplier, which may arise from equipment breakdowns, local labor strike, fires, etc. Denote by p_i the local disruption probability for supplier i, i.e., the parts ordered from supplier i are delivered without disruptions with probability $(1-p_i)$, or not at all with probability p_i. In addition to independent local disruptions of each supplier individually, the supplies of parts are also subject to regional correlated disruption of all suppliers in the same region simultaneously, with probability p^r for region $r \in R$.

Denote by π_i the disruption probability of every supplier $i \in I^r, r \in R$

$$\pi_i = p^r + (1 - p^r)p_i; \; i \in I^r, r \in R. \tag{2.1}$$

Let $S = \{1, \ldots, \overline{S}\}$ be the index set of $\overline{S} = 2^{\overline{I}}$ disruption scenarios, where each scenario $s \in S$ defines a subset $I_s \subset I$ of non-disrupted suppliers. The supplies from every supplier, $i \in I \setminus I_s$, can be independently disrupted either by a local or by a regional disaster event. The probability P_s of each disruption scenario $s \in S$ is a product over all regions $r \in R$ of probabilities P_s^r of realizing disruption scenario s for suppliers in I^r,

$$P_s = \prod_{r \in R} P_s^r, \tag{2.2}$$

where P_s^r is (cf. Sect. 1.3)

$$P_s^r = \begin{cases} (1-p^r)\prod_{i \in I^r \bigcap I_s}(1-p_i)\prod_{i \in I^r \setminus I_s} p_i & \text{if } I^r \bigcap I_s \neq \emptyset \\ p^r + (1-p^r)\prod_{i \in I^r} p_i & \text{if } I^r \bigcap I_s = \emptyset. \end{cases} \tag{2.3}$$

The producer does not need to pay for ordered and defective or undelivered parts. However, the producer may be charged with a much higher cost of unfulfilled customer orders for products, caused by the shortage of parts, undelivered due to supply disruptions. Let g_j be the per unit penalty cost of unfulfilled customer order j.

The decision maker needs to decide from which suppliers to purchase custom parts required for each customer order to achieve a minimum cost of ordering, purchasing, and shortages and to mitigate the impact of disruption risks by minimizing the potential worst-case cost or maximizing the potential worst-case service level.

2.3 Models for Risk-Neutral Decision-Making

In this section two SMIP models **SP_E(c)** and **SP_E(sl)** are proposed for risk-neutral selection of a static supply portfolio (see, Fig. 2.1), i.e., for determining a single-period supply portfolio to minimize expected cost and maximize expected service level, respectively. The static supply portfolio is defined below (for definition of problem variables, see Table 2.2).

$$(V_1, \ldots, V_{\overline{I}}),$$

where

$$\sum_{i \in I} V_i = 1$$

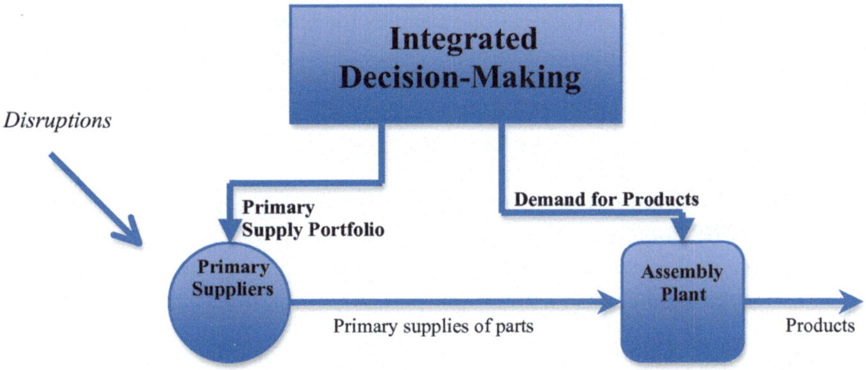

Fig. 2.1 Supply portfolio selection under disruption risks

Table 2.2 Problem variables: static supply portfolio

		First stage variables
u_i	=	1, if an order for parts is placed on supplier i; otherwise $u_i = 0$ (supplier selection)
v_{ij}	=	the fraction of demand for parts required for customer order j ordered from supplier i (allocation of demand for custom parts)
		Auxiliary variables
V_i	=	the fraction of total demand for parts allocated to supplier i (supply portfolio: allocation of total demand for parts)
VaR^c		cost-at-risk, the targeted cost such that for a given confidence level α, for $100\alpha\%$ of the scenarios, the outcome is below VaR^c
VaR^{sl}		service-at-risk, the targeted service level such that for a given confidence level α, for $100\alpha\%$ of the scenarios, the outcome is above VaR^{sl}
$\mathscr{C}_s$		≥ 0, the tail cost for scenario s, i.e., the amount by which costs in scenario s exceed VaR^c
$\mathscr{S}_s$		≥ 0, the tail service level for scenario s, i.e., the amount by which VaR^{sl} exceeds service level in scenario s

and $0 \leq V_i \leq 1$ is the fraction of the total demand for parts ordered from supplier i, and V_i is determined by the custom parts allocation variables v_{ij}

$$V_i = \sum_{j \in J} d_j v_{ij}/D; \; i \in I. \tag{2.4}$$

When deciding on a static supply portfolio it is assumed that the orders for all parts are simultaneously placed on selected suppliers (e.g., at time 0), and each supplier delivers all the ordered parts at the earliest possible delivery date. Therefore, the allocation of orders for parts among the suppliers is not combined with the allocation of orders among the planning periods. Nevertheless, the static portfolio should be checked against the risk of supply disruptions across all potential disruption scenarios.

Notice that Table 2.2 does not explicitly define the second stage variables for the stochastic mixed integer programming problem considered. The second stage variables are simply demand allocation variables for realized disruption scenarios s, $\tilde{v}_{ij}^s$; $i \in I, j \in J, s \in S$, defined as follows:

$$\tilde{v}_{ij}^s = \begin{cases} v_{ij} & \text{if } i \in I_s, j \in J, s \in S \\ 0 & \text{if } i \notin I_s, j \in J, s \in S. \end{cases}$$

In view of the above definition, an explicit introduction of the second stage variables $\tilde{v}_{ij}^s$ into the SMIP model formulations is not required.

In a risk-neutral operating conditions the overall quality of the supply portfolio can be measured by the expected cost per part, E^c, (2.5), or expected service level E^{sl}, (2.6).

$$E^c = \sum_{i \in I} e_i u_i / D + \sum_{i \in I} \sum_{j \in J} o_{ij} d_j v_{ij} / D + \sum_{s \in S} \sum_{i \notin I_s} \sum_{j \in J} P_s (g_j - o_{ij}) d_j v_{ij} / D \tag{2.5}$$

$$E^{sl} = \sum_{s \in S} \sum_{i \in I_s} \sum_{j \in J} P_s d_j v_{ij} / D \tag{2.6}$$

The expected cost E^c includes, cost of ordering,
$\sum_{i \in I} e_i u_i / D$,
cost of purchasing non-defective parts,
$\sum_{i \in I} \sum_{j \in J} o_{ij} d_j v_{ij} / D$,
and cost of shortage of parts due to supply disruptions (cost of unfulfilled customer orders less cost of non-delivered parts),
$\sum_{s \in S} \sum_{i \notin I_s} \sum_{j \in J} P_s (g_j - o_{ij}) d_j v_{ij} / D$.
The purchase orders for parts are assumed to be inflated by the reject rates ρ_i of defective parts, i.e., are equal to $(1 + \rho_i) d_j v_{ij}$ for all $i \in I, j \in J$. However, since the producer does not need to pay for ordered and defective parts in the amount of $\rho_i d_j v_{ij}$, the corresponding purchasing cost per part for delivered parts is simply given by $\sum_{i \in I} \sum_{j \in J} o_{ij} d_j v_{ij} / D - \sum_{s \in S} \sum_{i \notin I_s} \sum_{j \in J} P_s o_{ij} d_j v_{ij} / D$.

The expected cost per part, E^c, (2.5), can also be written as follows:

$$E^c = \sum_{i \in I} e_i u_i / D + \sum_{s \in S} \sum_{i \in I_s} \sum_{j \in J} P_s o_{ij} d_j v_{ij} / D + \sum_{s \in S} \sum_{i \notin I_s} \sum_{j \in J} P_s g_j d_j v_{ij} / D, \tag{2.7}$$

where $\sum_{s \in S} \sum_{i \in I_s} \sum_{j \in J} P_s o_{ij} d_j v_{ij} / D$ is the expected purchasing cost per part for delivered parts.

The expected service level, E^{sl}, (2.6), is a surrogate measure of expected customer demand fulfillment rate and represents the expected fraction of fulfilled demand for required parts.

The SMIP models **SP_E(c)** and **SP_E(sl)** are formulated below. The supply portfolio will be optimized by minimizing expected cost per part, E^c, (2.5) or by maximizing expected service level, E^{sl}, (2.6).

SP_E(c): *Selection of risk-neutral* **S**upply **P**o*rtfolio to minimize* **E***xpected* **c***ost*

Minimize (2.5)
subject to

1. *Supply portfolio selection constraints*:
   - the total demand for parts required for each customer order must be fully allocated among suppliers (2.8),
   - for each selected supplier the total quantity of ordered parts cannot exceed the supplier capacity (2.9),
   - parts cannot be ordered from non-selected suppliers (2.10),
   - at least one customer order should be assigned to each selected supplier (2.11),

$$\sum_{i \in I} v_{ij} = 1; \; j \in J \tag{2.8}$$

$$\sum_{j \in J} (1+\rho_i) d_j v_{ij} \le c_i u_i; \; i \in I \tag{2.9}$$

$$v_{ij} \le u_i; \; i \in I, j \in J \tag{2.10}$$

$$\sum_{j \in J} v_{ij} \ge u_i; \; i \in I \tag{2.11}$$

2. *Non-negativity and integrality conditions*

$$u_i \in \{0,1\}; \; i \in I \tag{2.12}$$

$$v_{ij} \in [0,1]; \; i \in I, j \in J. \tag{2.13}$$

Notice that if $g_j = g \forall j \in J$, i.e., per unit penalty cost of unfulfilled customer order is identical for all orders j, then E^c, (2.7), can be expressed by the following simplified formula:

$$E^c = \sum_{i \in I} e_i u_i / D + \sum_{s \in S} \sum_{i \in I_s} \sum_{j \in J} P_s o_{ij} d_j v_{ij} / D + g(1 - E^{sl}), \tag{2.14}$$

where E^{sl} is the expected service level (2.6).

SP_E(sl): *Selection of risk-neutral* **S**upply **P**o*rtfolio to maximize* **E***xpected* **s***ervice* **l***evel*

Maximize (2.6)
subject to (2.8)–(2.13).

If total available capacity of all suppliers is not greater than total demand for required parts, i.e., $\sum_{i\in I} c_i/(1+\rho_i) \le \sum_{j\in J} d_j$, then the demand allocation equality constraints (2.8) should be replaced by inequalities

$$\sum_{i\in I} v_{ij} \le 1;\ j \in J; \tag{2.15}$$

otherwise no feasible solution exists.

A simple upper bound on the expected service level, E^{sl}, (2.6), is derived below.

Proposition 2.1.

$$E^{sl} \le \min\{1, \sum_{r\in R}\sum_{i\in I^r}(1-p^r)(1-p_i)c_i/(1+\rho_i)D\}. \tag{2.16}$$

Proof. The supply portfolio selection constraints (2.9) imply that

$$\begin{aligned}\sum_{s\in S}\sum_{i\in I_s}\sum_{j\in J} P_s d_j v_{ij}/D \le \\ \sum_{r\in R}\sum_{i\in I^r}\sum_{j\in J}(1-p^r)(1-p_i)d_j v_{ij}/D \le \\ \sum_{r\in R}\sum_{i\in I^r}(1-p^r)(1-p_i)c_i u_i/(1+\rho_i)D \le \\ \sum_{r\in R}\sum_{i\in I^r}(1-p^r)(1-p_i)c_i/(1+\rho_i)D,\end{aligned}$$

where $(1-p^r)(1-p_i) = 1-\pi_i$, (2.1), is a non-disruption probability of supplier $i \subset I^r$.

Since E^{sl} cannot be greater than 1, its upper bound is 1, if $\sum_{r\in R}\sum_{i\in I^r}(1-p^r)(1-p_i)c_i/(1+\rho_i)D > 1$.

In the proposed models the parts required for each customer order are assumed to be partially provided by one or more suppliers and the customer order allocation variable v_{ij} represents the fraction of all parts required for order j provided by supplier i. In some practical cases, all custom parts of the same type that are required for a customer order are purchased from a single supplier. Then, the corresponding continuous allocation variable v_{ij} should be redefined as a binary assignment variable denoting whether or not all parts required for order j are provided by supplier i. If all v_{ij} are defined to be binary variables, then **SP_E(c)** and **SP_E(sl)** become pure stochastic binary programming models.

2.4 Models for Risk-Averse Decision-Making

In the risk-averse selection of supply portfolio under disruption risks, the confidence level α is fixed by the decision maker to control the risk of losses due to supply disruptions. We assume that the decision maker is willing to accept only portfolios for which the total probability of scenarios with costs greater than VaR^c or with service level lower than VaR^{sl} is not greater than $1-\alpha$. Furthermore, a risk aversive decision maker wants to minimize the expected worst-case costs exceeding VaR^c or to maximize the expected worst-case service level below VaR^{sl}.

Define by $\mathscr{C}_s$ the tail cost for scenario s, where tail cost is defined as the nonnegative amount by which costs in scenario s exceed VaR^c. In a similar way, define by $\mathscr{S}_s$ the tail service level for scenario s, where tail service level is defined as the nonnegative amount by which VaR^{sl} exceeds service level in scenario s.

The portfolio will be optimized by calculating VaR^c and minimizing CVaR^c simultaneously or by calculating VaR^{sl} and maximizing CVaR^{sl}, respectively. By measuring CVaR^c or CVaR^{sl}, the magnitude of the tail costs or the tail service level is considered to achieve a more accurate estimate of the risks of minimizing cost or maximizing service level, respectively. When using CVaR^c to minimize worst-case costs and CVaR^{sl} to maximize worst-case service level, CVaR^c is always not less than VaR^c and CVaR^{sl} is always not greater than VaR^{sl}, respectively.

In the proposed model CVaR is represented by an auxiliary function (2.17) and (2.20) introduced by Rockafellar and Uryasev (2000). The SMIP models **SP_CV(c)** and **SP_CV(sl)** for selection of risk-averse supply portfolio to reduce the risk of high costs and the risk of low service level, respectively, are formulated below.

SP_CV(c): *Selection of risk-averse* **S**upply **P**ortfolio *to minimize* **CV**aR *of* **c**ost

Minimize

$$CVaR^c = VaR^c + (1-\alpha)^{-1}\sum_{s\in S} P_s\mathscr{C}_s \tag{2.17}$$

subject to

1. *Supply portfolio selection constraints*: (2.8)–(2.11)
2. *Risk constraints*:
   - the tail cost for scenario s is defined as the nonnegative amount by which cost per part in scenario s exceeds VaR^c,

$$\mathscr{C}_s \geq \sum_{i\in I} e_i u_i/D + \sum_{i\in I}\sum_{j\in J} o_{ij} d_j v_{ij}/D + \sum_{i\notin I_s}\sum_{j\in J}(g_j - o_{ij}) d_j v_{ij}/D - VaR^c;\ s\in S. \tag{2.18}$$

(continued)

3. Non-negativity and integrality conditions: (2.12), (2.13) and

$$\mathscr{C}_s \geq 0;\ s \in S. \tag{2.19}$$

SP_CV(sl): *Selection of risk-averse* **S**upply **P**o*rtfolio to maximize* **CV***aR of* **s**ervice **l**evel

Maximize

$$CVaR^{sl} = VaR^{sl} - (1-\alpha)^{-1} \sum_{s \in S} P_s \mathscr{S}_s \tag{2.20}$$

subject to

1. Supply portfolio selection constraints: (2.8)–(2.11)
2. Risk constraints:
   – the tail service level for scenario s is defined as the nonnegative amount by which VaR^{sl} exceeds service level in scenario s,

$$\mathscr{S}_s \geq VaR^{sl} - \sum_{i \in I_s} \sum_{j \in J} d_j v_{ij}/D;\ s \in S. \tag{2.21}$$

3. *Non-negativity and integrality conditions*: (2.12), (2.13) and

$$\mathscr{S}_s \geq 0;\ s \in S. \tag{2.22}$$

Note that as $\mathscr{C}_s$ and $\mathscr{S}_s$ are constrained of being positive, the model **SP_CV(c)** tries to decrease VaR^c and the model **SP_CV(sl)** tries to increase VaR^{sl}, respectively. Hence they positively impact the objective functions. However, large reduction in VaR^c and large increase in VaR^{sl} may result in more scenarios with positive tail costs and with positive tail service levels, respectively.

If for some customer order j all required parts must be supplied by a single supplier, then the corresponding nonnegative allocation variable v_{ij} should be redefined as a binary assignment variable denoting whether or not all parts required for order j are provided by supplier i, similarly as for the risk-neutral models **SP_E(c)** and **SP_E(sl)**.

2.5 Models for Mean-Risk Decision-Making

In the single objective approach the supply portfolio is selected by minimizing either the expected cost per part, E^c, (2.5), the expected service level, E^{sl}, (2.6), the expected worst-case cost per part, $CVaR^c$, (2.17) or the expected worst-case service level, $CVaR^{sl}$, (2.20). In this section the two cost functions and the two service level functions are considered simultaneously, and a bi-objective selection of supply portfolio is presented aimed at minimizing both objective functions to balance expected costs or expected service level with the risk tolerance. This trade-off model is known as the mean-risk model (e.g., Ogryczak and Ruszczynski 2002), formulated as the optimization of a composite objective consisting of the expected cost (service level) and the CVaR as a risk measure.

The nondominated solution set of the bi-objective supply portfolio can be found by the parameterization on λ the weighted-sum programs **SP_ECV(c)** and **SP_ECV(sl)** presented below. The mean-risk program **SP_ECV(c)** is based on model **SP_CV(c)** with the addition of objective (2.5) of model **SP_E(c)**. Similarly, the mean-risk program **SP_ECV(sl)** is based on model **SP_CV(sl)** with the addition of objective (2.6) of model **SP_E(sl)**.

SP_ECV(c): *Selection of mean-risk* **S**upply **P**ortfolio *to minimize weighted sum of* **E**xpected cost *and* **CV**aR *of* **c**ost

Minimize

$$\lambda E^c + (1-\lambda)CVaR^c \tag{2.23}$$

where $0 \leq \lambda \leq 1$,
subject to (2.5), (2.8)–(2.13), (2.17)–(2.19).

SP_ECV(sl): *Selection of mean-risk* **S**upply **P**ortfolio *to maximize weighted sum of* **E**xpected service level *and* **CV**aR *of* **s**ervice **l**evel

Maximize

$$\lambda E^{sl} + (1-\lambda)CVaR^{sl} \tag{2.24}$$

where $0 \leq \lambda \leq 1$,
subject to (2.6), (2.8)–(2.13), (2.20)–(2.22).

Steuer (1996) proved that for mixed integer programs, there may be portions of the nondominated set (nearby weakly nondominated solution) that the above approach is unable to compute, even if the complete parameterization on λ is attempted.

2.6 Computational Examples

In this section some computational examples are presented to illustrate possible applications of the proposed SMIP approach for selection of static supply portfolio under disruption risks. First, a single-region sourcing case will be illustrated, where all suppliers are located in a single geographic region, and then examples of multi-region sourcing with subsets of suppliers in different geographic regions, each subject to different regional disruption risks. For the single-region sourcing, minimization of cost is considered only, whereas for the multi-region sourcing, both minimization of cost and maximization of service level are considered.

2.6.1 Single-Region Sourcing

In this subsection all suppliers are assumed to be located in the same geographic region, and hence the regional disruption can be called a global disruption. Denote by, p^*, the global disruption probability for the entire region. Now, the probability, P_s (2.3), of each disruption scenario s, can be calculated using the following formula:

$$P_s = \begin{cases} (1-p^*)\hat{P}_s & \text{if } I_s \neq \emptyset \\ p^* + (1-p^*)\prod_{i\in I} p_i & \text{if } I_s = \emptyset, \end{cases} \tag{2.25}$$

where $\hat{P}_s$ is the probability of disruption scenario s in the presence of independent local disruptive events only

$$\hat{P}_s = \prod_{i\in I_s}(1-p_i)\cdot\prod_{i\notin I_s} p_i. \tag{2.26}$$

If the probability of regional disruption $p^* = 0$, then the probability P_s reduces to $\hat{P}_s$ for independent local disruptive events.

The following parameters have been used for the example problems:

- $\bar{I}$, the number of suppliers, was equal to 7, 10, or 14 and the corresponding number $\bar{S} = 2^{\bar{I}}$ of disruption scenarios, was equal to 128, 1024, or 16384, respectively;
- $\bar{J}$, the number of customer orders, was equal to 50;
- d_j, the numbers of required parts for each customer order, were integers uniformly distributed over $[100,500]$, i.e., generated from a U[100;500] distribution;
- c_i, the capacity of each supplier i, was equal to $\lceil 2\sum_{j\in J} d_j/\bar{I}\rceil$ ($\lceil\cdot\rceil$ denotes the smallest integer not less than $\cdot$), i.e., the total capacity of all suppliers was equal to the double total demand for parts;
- e_i, the cost of ordering parts from supplier i, was equal to 500 for each supplier i;
- g_j, the per unit shortage cost for customer order j, was equal to 100 for all customer orders j;
- ρ_i, the expected defect rate of each supplier i, was exponentially distributed, ranging from 0.0003 to 0.03;
- o_{ij}, the unit price of parts required for each customer order j purchased from each supplier i, was uniformly distributed over [10,15] (i.e., drawn from U[10;15]) and reduced by the factor $(1-\rho_i)$ to get a lower price for parts from the suppliers with a higher defect rate;
- p_i, the local disruption probability was uniformly distributed over [0,0.06], i.e., the disruption probabilities were drawn independently from U[0;0.06];
- p^*, the global disruption probability was equal to 0.01;
- α, the confidence level, was equal to 0.50, 0.75, 0.90, 0.95, or 0.99.

For the example problems, the total demand for parts is $D = \sum_{j \in J} d_j = 14750$ parts. Solution results for the risk-neutral model **SP_E(c)** are shown in Table 2.3, and for the risk-averse model **SP_CV(c)** with different confidence levels, in Table 2.4. The size of the mixed integer programs for different number $\bar{I}$ of suppliers is represented by the total number of variables, Var., number of binary variables, Bin., number of constraints, Cons, and number of nonzero coefficients in the constraint matrix, Nonz. Table 2.4 also presents the probability $1 - F(VaR^c)$ of outcomes with worst-case cost above VaR^c. Note that the number of variables and constraints in the mixed integer program **SP_CV** grows exponentially in the number $\bar{I}$ of suppliers. The table demonstrates that the number of selected suppliers increases with the confidence level α, which indicates that the impact of disruption risks is mitigated by diversification of the supply portfolio. Note that VaR^c becomes smaller than expected cost when $\alpha = 0.50$ and $\alpha = 0.75$.

Table 2.3 Risk-neutral solutions for model **SP_E(c)**: single-region sourcing

No.of Suppliers	Expected Cost	No. of Selected Suppliers
7	12.38	4
Var.=364, Bin.=7, Cons.=72, Nonz.=1428[a]		
10	13.18	5
Var.=520, Bin.=10, Cons.=81, Nonz.=2040[a]		
14	12.03	7
Var.=728, Bin.=14, Cons.=93, Nonz.=2856[a]		

[a] Var.=no. of variables, Bin.=no. of binary variables, Cons.=no. of constraints, Nonz.= no. of nonzero coefficients

The optimal risk-neutral supply portfolio for model **SP_E(c)** and 10 suppliers is shown in Fig. 2.2. In addition, the figure presents for each supplier i the expected defect rate ρ_i, the average unit price $\sum_{j \in J} o_{ij}/\bar{J}$, and disruption probabilities, π_i, (2.1). For the optimal supply portfolio the total demand was equally allocated among five suppliers with the lowest disruption probabilities.

In the computational experiments the confidence level α is set at five levels of 0.5, 0.75, 0.90, 0.95, and 0.99, which means that focus is on minimizing the highest 50%, 25%, 10%, 5%, and 1% of all scenario outcomes, i.e., costs per part.

Figure 2.3 shows the probability mass functions and the cumulative distribution functions for the optimal risk-averse portfolios with different confidence levels for 10 suppliers. Figure 2.3 indicates that the mass function of cost per part is concentrated in a few points and the resulting cumulative distribution is a discontinuous step function with jumps at those points. Such results are typical for the scenario-based optimization under uncertainty, where the probability measure is concentrated in finitely many points. The resulting discontinuity (vertical jumps) of the distribution function leads to probability intervals of confidence level α with the same VaR. The discrete distributions of cost per part for the optimal supply portfolios with four different confidence levels and the corresponding probabilities concentrated at each level of cost are presented also in Table 2.5. The table shows that the probabilities are concentrated at 6, 10, 10, 10 points, respectively, for the confidence level $\alpha =$

Table 2.4 Risk-averse solutions: single-region sourcing

Confidence level α	0.50	0.75	0.90	0.95	0.99
Model **SP_CV(c)**: 7 suppliers Var.=493, Bin.=7, Cons.=200, Nonz.=47380[a]					
CVaR^c	14.15	17.69	28.21	39.01	100.14
VaR^c	10.61	10.61	13.90	21.19	100.14
E^c	12.38	12.38	12.59	12.94	14.24
$1-F(VaR^c)$	0.053	0.053	0.039	0.018	0
No. of suppliers selected	4	4	7	7	4
Model **SP_CV(c)**: 10 suppliers Var.=1545, Bin.=10, Cons.=1115, Nonz.=526338[a]					
CVaR^c	16.02	21.69	31.67	42.05	100.17
VaR^c	10.35	10.35	20.29	23.19	100.17
E^c	13.18	13.18	14.06	13.96	15.89
$1-F(VaR^c)$	0.113	0.113	0.041	0.032	0
No. of suppliers selected	5	5	9	10	5
Model **SP_CV(c)**: 14 suppliers Var.=17113, Bin.=14, Cons.=16477, Nonz.=11733800[a]					
CVaR^c	13.60	16.70	25.59	35.94	100.24
VaR^c	10.51	10.51	13.75	16.89	100.24
E^c	12.05	12.05	12.50	12.43	15.50
$1-F(VaR^c)$	0.115	0.097	0.058	0.029	0
No. of suppliers selected	7	7	13	11	7

[a] Var.=no. of variables, Bin.=no. of binary variables, Cons.=no. of constraints, Nonz.= no. of nonzero coefficients

0.5, 0.9, 0.989, 0.99. In the examples, a large probability atom is concentrated at the highest cost. As a consequence, a slight increase of the confidence level from $\alpha = 0.989$ to $\alpha = 0.99$ results in a significant change in VaR from 37.33 to 100.17, while only a slight increase of CVaR^c from 94.81 to 100.17 is observed. Moreover, the optimal portfolio has been changed significantly; for $\alpha = 0.989$ the total demand has been equally allocated among all ten suppliers ($V_i = 0.1$ forall i), whereas for $\alpha = 0.99$ among five suppliers only ($V_i = 0.2$ for $i = 1,2,6,7,8$). This degree of instability of the optimal supply portfolio due to the discontinuity in the distribution function may be distressing in practice, when a slightly higher confidence level is required. Despite the limited change in CVaR^c, the above results demonstrate that the well known misbehavior in the dependence of VaR^c and optimal supply portfolio on the confidence level can as well be encountered when CVaR^c is applied as a risk measure.

The computational results indicate that the smaller is the number of concentration points and the greater are probability atoms concentrated at those points, the greater can be the positive difference $F(VaR^c) - \alpha$. As a result the smaller than $1-\alpha$ can be the probability of outcomes with cost higher than VaR^c. For example (see, Table 2.4), for $\overline{I} = 10$ and $\alpha = 0.5$, $VaR^c = 10.35$ and $F(VaR^c) = 0.88666 > 0.5$, which indicates a high concentration of probability measure at point 10.35 for the optimal supply portfolio. Actually, the probability that cost per part is 10.35 is 0.88666 (see, Table 2.5), which indicates that $VaR^c = 10.35$ is the lowest cost that may occur and that for the confidence level $\alpha = 0.5$, less than 11.33% of the cost outcomes are above VaR^c.

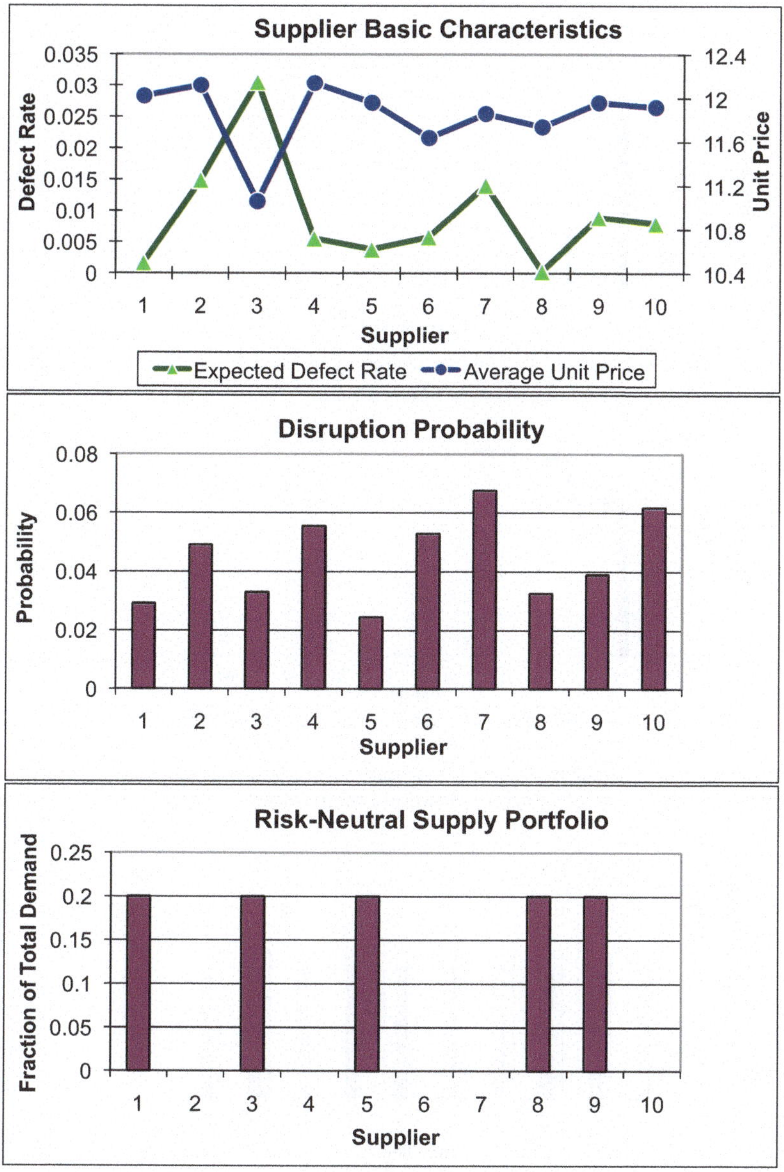

Fig. 2.2 Risk-neutral supply portfolio for model **SP_E(c)**: 10 suppliers

Moreover, if the highest cost probability is greater than $1-\alpha$, then CVaR^c and VaR^c are identical and both equal to the highest cost. In the example for ten suppliers and $\alpha = 0.99$, the highest cost per part is 100.17 and the probability concentrated at 100.17 is $0.01000004 > 1-\alpha$, then $VaR^c = 100.17$ is the highest cost per part that may occur and hence $CVaR^c = VaR^c = 100.17$ (see, Table 2.5 and Fig. 2.3). Similar results $CVaR^c = VaR^c = 100.14$ and $CVaR^c = VaR^c = 100.24$ have been

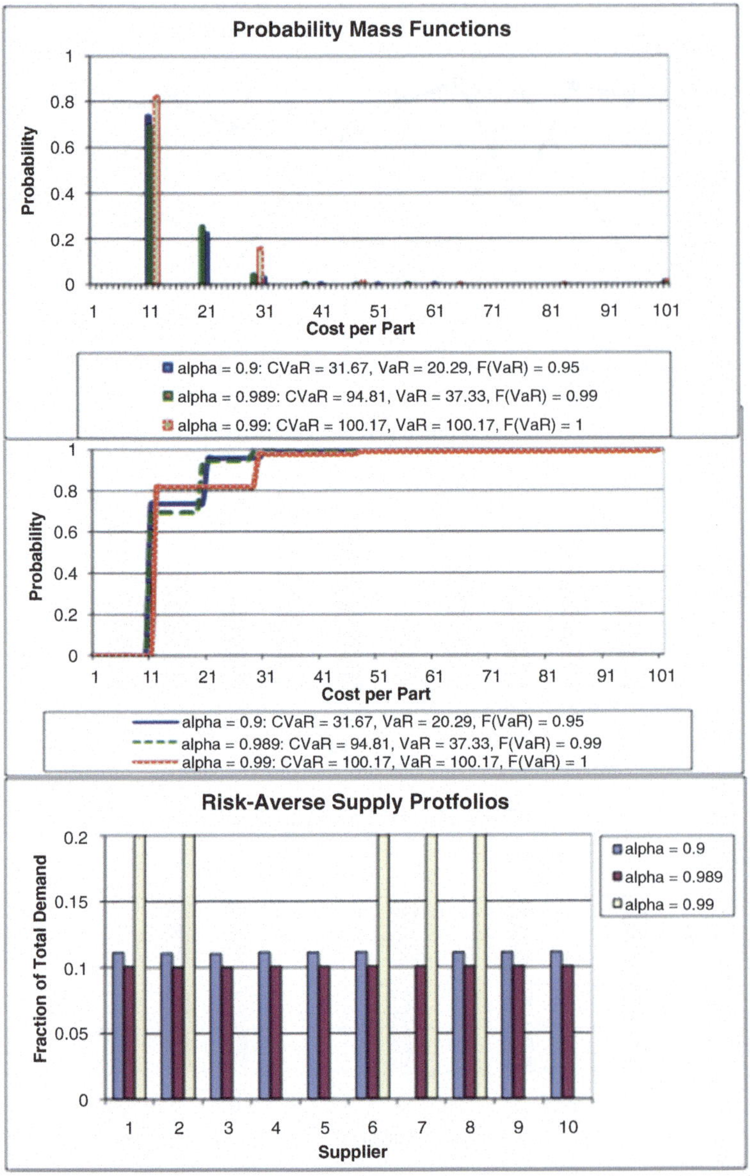

Fig. 2.3 Risk-averse supply portfolios and cost distributions for model **SP_CV(c)**: 10 suppliers

obtained for $\alpha = 0.99$, respectively, for seven and 14 suppliers (see, Table 2.4), which indicates that the corresponding probabilities of the highest cost per part are greater than $1 - \alpha = 0.01$.

If the probability measure is concentrated at the highest cost and is greater than $1 - \alpha$, so that CVaR^c and VaR^c are identical with the highest cost, then for a higher

Table 2.5 Probability of cost per part for optimal risk-averse supply portfolios: 10 suppliers

Cost interval	$\alpha = 0.5$	$\alpha = 0.9$	$\alpha = 0.989$	$\alpha = 0.99$
[10, 11)	0.886661618	0.736768317	0.693881575	0
[11, 12)	0	0	0	0.821176973
[19, 20)	0	0	0.251829646	0
[20, 21)	0	0.221857039	0	0
[28, 29)	0.098887038	0	0.040311352	0
[29, 30)	0	0	0	0.156904596
[30, 31)	0	0.029090556	0	0
[37, 38)	0	0	0.003744839	0
[40, 41)	0	0.002178297	0	0
[46, 47)	0.004355651	0	0.000223405	0
[47, 48)	0	0	0	0.011507829
[50, 51)	0	0.000102579	0	0
[55, 56)	0	0	8.94e−06	0
[60, 61)	0	3.15e−06	0	0
[64, 65)	9.47e−05	0	2.43e−07	0.0004038
[70, 71)	0	6.30e−08	0	0
[73, 74)	0	0	4.41e−09	0
[80, 81)	0	7.91e−10	0	0
[82, 83)	1.01e−06	0	5.14e−11	6.76e−06
[90, 91)	0	5.66e−12	0	0
[91, 92)	0	0	3.46e−13	0
[100, 101)	0.010000004	0.01	0.01	0.010000043

confidence level α, a smaller number of suppliers are selected, which indicates that diversification of the supply portfolio is not necessary any more. For instance, the optimal risk-averse supply portfolio selected for $\alpha = 0.99$ consists of five suppliers only, the same number as that for a much lower α (cf. Table 2.4, Fig. 2.3).

In the computational experiments the local disruption probabilities p_i were assumed to be very low and were drawn from U[0;0.06]. To study the effect of the increasing range of disruption probabilities, the probabilities have been drawn also from U[0;0.25], U[0;0.5], or U[0;1]. The effect of varying distribution of local disruption probabilities is illustrated in Fig. 2.4, where the optimal risk-averse supply portfolios are presented. Figure 2.4 indicates that for a greater range of disruption probabilities, the suppliers with the highest disruption probabilities are not selected. For example, for U[0;1] and $\alpha = 0.75$, suppliers $i = 4, 6, 7, 10$ with the four highest disruption rates were not selected, whereas for $\alpha = 0.985$, suppliers $i = 7, 10$ with the two highest disruption rates were not selected. Similar results are observed for the other distributions of disruption probability.

Finally, Table 2.6 presents solution results for model **SP_CV(c)** applied to optimization of a single sourcing, with binary assignment variables v_{ij}, i.e., when for each customer order, the required parts must be provided by a single supplier only. Comparison of the results shown in Table 2.6 with the corresponding results presented in Table 2.4 for continuous allocation variables v_{ij}, indicates that in the

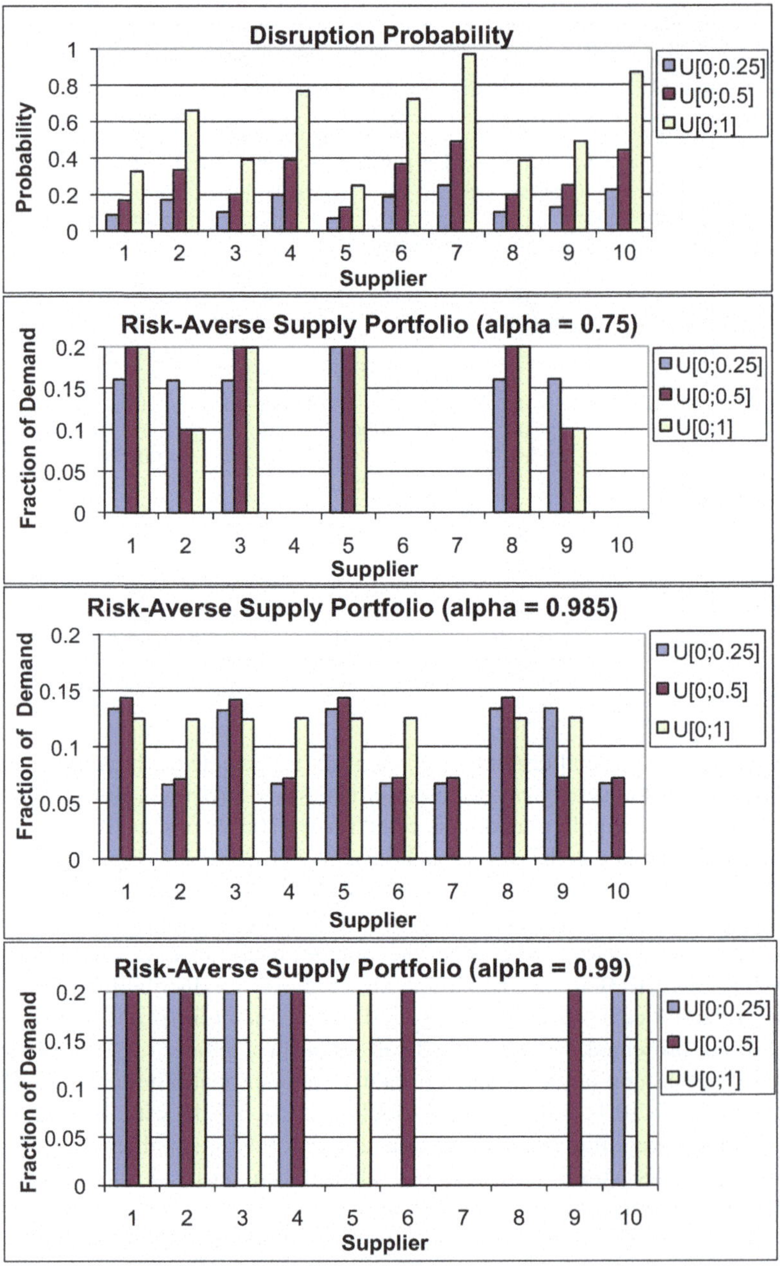

Fig. 2.4 Risk-averse supply portfolios for different local disruption probabilities for model **SP_CV(c)**: 10 suppliers

Table 2.6 Solutions results for model **SP_CV(c)** with binary assignment variables $v_{ij} \in \{0,1\}$: 10 suppliers

Confidence level α	0.50	0.75	0.90	0.95	0.99
Var.=1545, Bin.=510, Cons.=1105, Nonz.=526328[a]					
CVaR^c	16.08	21.80	31.75	42.08	100.20
VaR^c	10.36	10.40	20.16	23.21	100.20
E^c	13.22	13.25	14.07	13.94	15.74
$1-F(VaR^c)$	0.155	0.155	0.094	0.053	0
No. of suppliers selected	6	6	9	10	6

[a] Var.: no. of variables; Bin.: no. of binary variables; Cons.: no. of constraints; Nonz.: no. of nonzero coefficients

former case both the expected cost per part and CVaR^c were slightly higher and, in addition, for a low α the number of selected suppliers was greater. Such results were expected, since model **SP_CV(c)** with the continuous allocation variables is a partial LP relaxation of that model with the binary assignment variables.

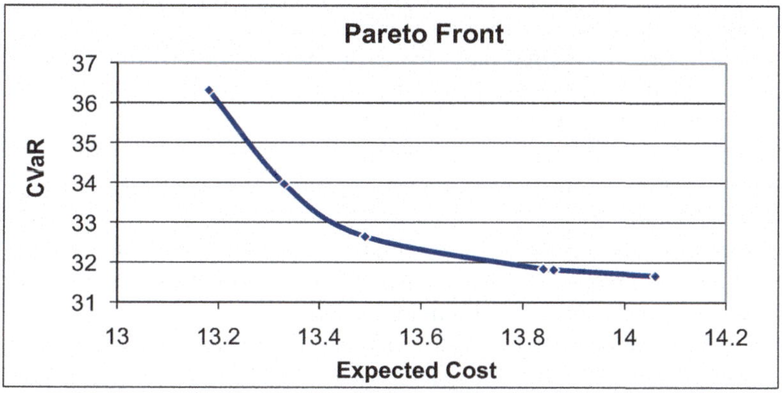

Fig. 2.5 Pareto front for mean-risk model **SP_ECV(c)**: 10 suppliers, $\alpha = 0.9$

For the mean-risk approach, the subsets of nondominated solutions were computed by parameterization on $\lambda \in \{0.01, 0.10, 0.25, 0.50, 0.75, 0.90, 0.99\}$ the weighted-sum program **SP_ECV(c)**. The subset of nondominated solutions found for the selected seven levels of trade-off parameter λ is: $(E^c, CVaR^c)$ = (13.18, 36.32), (13.33, 33.97), (13.49, 32.65), (13.84, 31.84), (13.86, 31.82), (14.06, 31.67). The trade-off between the expected cost and the expected worst-case cost is clearly shown in Fig. 2.5, where the convex efficient front for the mean-risk model **SP_ECV(c)** with $\alpha = 0.9$ is presented. The results emphasize the effect of varying cost/risk preference of the decision maker.

Note that solutions to single objective models **SP_E(c)** and **SP_CV(c)** are equivalent to the nondominated solutions of the weighted-sum program **SP_ECV(c)** for $\lambda = 1$ and $\lambda = 0$, respectively.

The computational experiments were performed using the AMPL programming language and the CPLEX. The solver was capable of finding proven optimal solutions within CPU seconds for all examples.

2.6.2 Multi-Region Sourcing

In this subsection the suppliers are assumed to be located in multiple geographic regions subject to different regional disruption risks. The following parameters used for the example problems are different from those in Sect. 2.6.1:

- $\overline{I}$, the number of suppliers, was equal to 10 and the corresponding number $\overline{S} = 2^{\overline{I}}$ of disruption scenarios, was equal to 1024;
- $\overline{R}$, the number of geographic regions, was equal to 3, and the subsets of suppliers were $I^1 = \{1,2,3\}$, $I^2 = \{4,5,6\}$, and $I^3 = \{7,8,9,10\}$, respectively;
- $\overline{J}$, the number of customer orders, was equal to 25;
- d_j, the numbers of required parts for each customer order, were integers uniformly distributed over $[1000, 15000]$ for all customer orders j and the resulting total demand for parts was $D = 132500$;
- e_i, the cost of ordering parts, were integers in $\{5000, 6000, \ldots, 10000\}$, $\{10000, 11000, \ldots, 15000\}$, and $\{15000, 16000, \ldots, 30000\}$, respectively, for suppliers $i \in I^1$, $i \in I^2$, and $i \in I^3$;
- g_j, the per unit shortage cost for each customer order j, was integer uniformly distributed over $[\max_{i \in I}(o_{ij}), 4\max_{i \in I}(o_{ij})]$, i.e., ranging from one to four times of maximum purchasing cost of required parts;
- o_{ij}, the unit price of parts for customer order j purchased from supplier i, was uniformly distributed over [13,15], [11,13], and [9,11], respectively, for suppliers $i \in I^1$, $i \in I^2$, and $i \in I^3$;
- p_i, the local disruption probability was uniformly distributed over [0.005,0.01], [0.01,0.05], and [0.05;0.10], respectively for suppliers $i \in I^1$, $i \in I^2$, and $i \in I^3$, i.e., the disruption probabilities were drawn independently from U[0.005;0.01], U[0.01,0.05], and U[0.05;0.10], respectively;
- p^r, the regional disruption probability was 0.001, 0.005, and 0.01, respectively, for region $r = 1$, $r = 2$, and $r = 3$;

Basic characteristics of suppliers: average unit price $\sum_{j \in J} o_{ij}/\overline{J}$, and disruption probability, π_i, (2.1), are presented in Fig. 2.6. Figure 2.6 indicates that the most reliable and most expensive are suppliers $i = 1,2,3$ in region $r = 1$, while suppliers $i = 7,8,9,10$ in region $r = 3$ are most competitive and most unreliable. In particular, supplier $i = 8$ is the cheapest and most unreliable among all suppliers.

The solution results for the risk-averse models **SP_CV(c)** and **SP_CV(sl)** with different confidence levels are shown in Table 2.7. For both models the number of

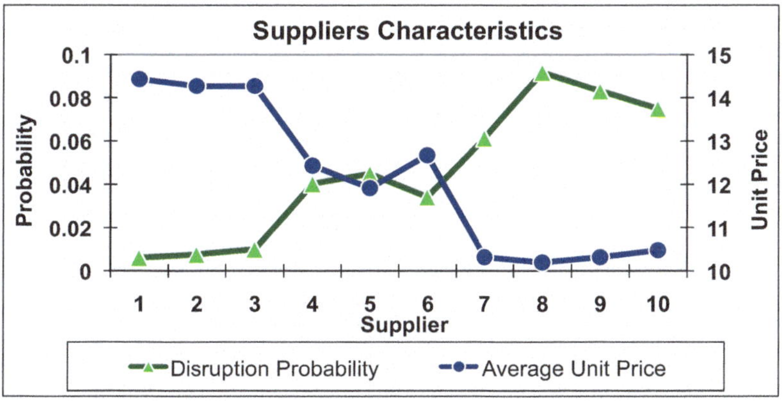

Fig. 2.6 Suppliers

selected suppliers increases with the confidence level. While for model **SP_CV(c)** the cheapest, yet most unreliable supplier $i = 8$ was never selected, for model **SP_CV(sl)** all 10 suppliers are selected for $\alpha = 0.9$, 0.95, and $\alpha = 0.99$.

Figure 2.7 shows the optimal risk-averse supply portfolios for models **SP_CV(c)** and **SP_CV(sl)** and the three confidence levels, $\alpha = 0.75,\ 0.9,\ 0.99$. For both models and $\alpha = 0.75$ the most unreliable suppliers $i = 7, 8, 9, 10$ in region $r = 3$ are not selected and for all confidence levels most demand for parts is allocated among the three most reliable, yet most expensive suppliers, $i = 1, 2, 3$, in region $r = 1$, in particular for $\alpha = 0.75, 0.9$. Similar properties of the risk-averse supply portfolios were observed in case of single-region sourcing (see, Fig. 2.4).

The probability measure for the scenario-based optimization under uncertainty is concentrated in finitely many points and the resulting discrete distribution of cost may have different effect on the optimal portfolio. In particular, the well known misbehavior in the dependence of VaR on the confidence level can as well be encountered when CVaR is applied as a risk measure. For instance, if a large probability atom is concentrated at some cost, a slight increase of the confidence level may result in a significant change in VaR as well as in the optimal portfolio, while only a slight change of CVaR may occur. Such an instability of the optimal portfolio due to the discontinuity in the distribution function may be distressing in practice, when a slightly higher confidence level is required.

On the other hand the computational results indicate that the smaller is the number of concentration points and the greater are probability atoms concentrated at those points, the greater can be the positive difference $F(VaR) - \alpha$. As a result the smaller than $1 - \alpha$ can be the probability of outcomes with cost higher (service level lower) than VaR.

The major managerial insights are summarized in Table 2.8. Overall, the main results of computational study for the risk-neutral and risk-averse decision-making models are in line with other research. The results of computational experiments prove that *the proposed portfolio approach and developed SMIP models provide*

Table 2.7 Risk-averse solutions: multi-region sourcing

Confidence level α	0.50	0.75	0.90	0.95	0.99
Model **SP_CV(c)**					
Var.=1285, Bin.=10, Cons.=1319, Nonz.=269558[a]					
CVaR^c	15.35	17.15	20.49	23.90	30.43
VaR^c	13.29	13.68	16.44	17.19	26.21
$1-F(VaR^c)$	0.149	0.125	0.055	0.036	0.005
E^c	14.32	14.55	14.59	14.51	14.61
$E^{sl\ (b)}$	97.51	97.98	97.30	97.56	96.15
Suppliers selected(% of total demand)	1(20)	1(20)	1(20)	1(20)	1(19)
	2(19)	2(19)	2(20)	2(19)	2(13)
	3(20)	3(20)	3(20)	3(20)	3(12)
		4(14)	4(8)	4(8)	4(6)
	5(8)	5(7)	5(6)	5(9)	5(6)
	6(13)	6(20)	6(6)	6(8)	6(8)
	7(20)		7(12)	7(16)	7(9)
			9(8)		9(12)
					10(15)
Model **SP_CV(sl)**					
Var.=1285, Bin.=10, Cons.=1319, Nonz.=131318[a]					
$\text{CVaR}^{sl}\%$	96.01	92.02	85.29	80.12	69.82
$\text{VaR}^{sl}\%$	100	100	91.87	87.50	77.77
$1-F(VaR^{sl})$	0.125	0.125	0.064	0.033	0.007
$E^{sl\,b}$	98.00	98.00	97.29	97.43	97.04
E^c	15.56	15.48	15.86	15.85	15.61
Suppliers selected(% of total demand)	1(20)	1(20)	1(20)	1(20)	1(20)
	2(19)	2(19)	2(20)	2(20)	2(18)
	3(20)	3(20)	3(20)	3(20)	3(18)
	4(20)	4(20)	4(8)	4(8)	4(10)
	5(1)	5(1)	5(8)	5(9)	5(4)
	6(20)	6(20)	6(8)	6(9)	6(9)
			7(4)	7(3)	7(9)
			8(4)	8(3)	8(4)
			9(4)	9(3)	9(4)
			10(4)	10(3)	10(4)

[a] Var.: no. of variables; Bin.: no. of binary variables;
Cons.: no. of constraints; Nonz.: no. of nonzero coefficients
[b] $(\sum_{s\in S}\sum_{i\in I_s}\sum_{j\in J} P_s d_j v_{ij}/D)100\%$

the decision maker with a simple tool for evaluating the relationship between expected and worst-case costs. For a finite number of scenarios, the proposed models allow the evaluation of worst-case costs and shaping of the resulting cost distribution through the selection of optimal supply portfolio. The approach has proven to be computationally very efficient. The optimal risk-neutral or risk-averse supply portfolio can be found within CPU seconds, using commercially available solvers for MIP.

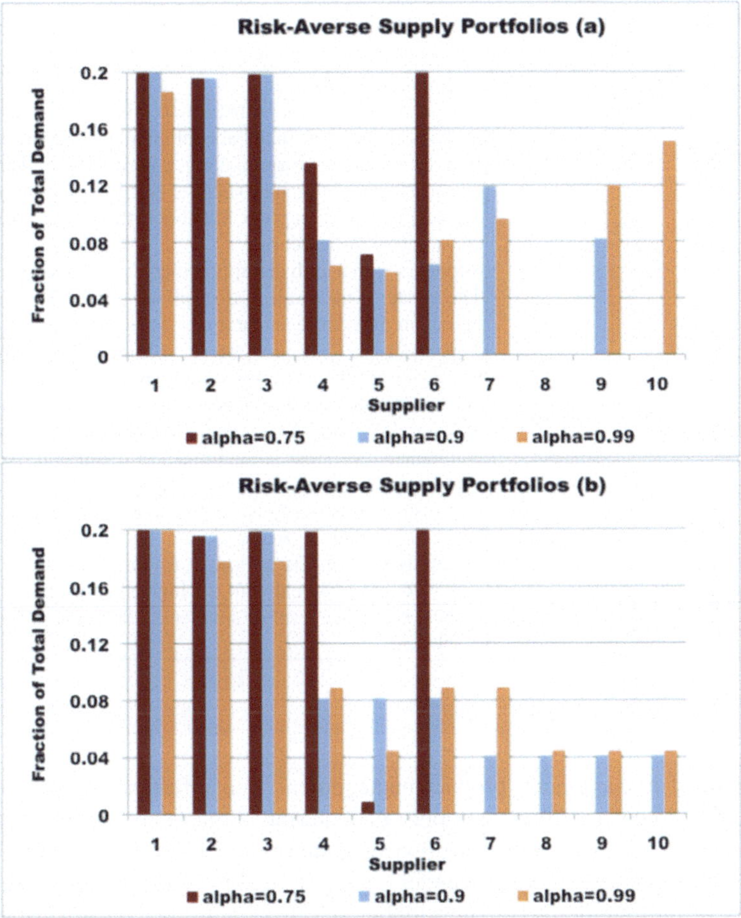

Fig. 2.7 Risk-averse supply portfolios: (**a**) model **SP_CV(c)**, (**b**) model **SP_CV(sl)**

2.7 Notes

The supply chain risk management has been extensively studied over the past decade. Research addresses the two risk levels (e.g., Tang 2006): operational risks or disruption risks. Operational risks are referred to the inherent uncertainties arising from the problems of coordinating supply and demand such as uncertain customer demand, uncertain supply, and uncertain cost. Disruption risks are referred to the major disruptions to normal activities caused by natural and man-made disasters such as earthquakes, floods, hurricanes, etc., or equipment breakdowns, economic crises such as currency evaluation, labor strikes, terrorist attacks. In most cases, the business impact associated with disruption risks is much greater than that of the operational risks. In practice four basic approaches can be applied to mitigate the impact of supply chain risks (Tang 2006; Stecke and Kumar 2009): supply manage-

Table 2.8 Managerial insights: static supply portfolio

Decision-making issue	Decision-making impact
Disruption probability	The supply disruption probability is a key determinant in the decision on demand allocation among suppliers. In a risk-averse decision-making, an order for delivery of parts from a particular supplier is selected more on the supply non-disruption likelihood than on its purchasing cost or defect rate Suppliers with a high disruption rate are rarely selected The greater is the range of disruption probabilities, the higher are both expected and worst-case costs The closer are disruption rates for different suppliers, the closer are the corresponding quantities of ordered parts in the optimal supply portfolio
Confidence level	The number of selected suppliers increases with the confidence level, that is the impact of disruption risks is mitigated by diversification of supply portfolio However, if probability measure is concentrated at the highest cost and is greater than $1 - \alpha$, diversification of supply portfolio is no longer useful. Then, CVaR^c and VaR^c are identical with the highest cost and for a higher confidence level α, smaller number of suppliers is selected

ment, demand management, product management, and information management. In particular, to ensure efficient supply of materials along a supply chain, supply chain management deals with selection of a supply portfolio, i.e., supplier selection and order quantity allocation under uncertain quality of supplied materials and reliability of on-time delivery. The supplier selection and order quantity allocation problem is a complex stochastic combinatorial optimization problem; however, the research on supplier selection under disruption risks is limited. For example, chance-constrained programming models were developed by Kasilingam and Lee (1996) to account for stochastic demand and by Wu and Olson (2008) to consider expected losses from quality acceptance inspection or late delivery. Parlar and Perry (1996) present a continuous time model in which the availability of each of the m suppliers is uncertain because of disruptions such as equipment breakdown. By considering the case that each supplier is either "on" or "off," there are 2^m possible number of states for the whole system. For each of these 2^m states, they analyze a state-specific (s, Q) ordering policy so that the buyer would order Q units when the on-hand inventory reaches s. The risks associated with a supplier network was studied by Berger et al. (2004), who considered catastrophic super events that affect all suppliers, as well as unique events that impact only one single supplier, and then a decision-tree based model was presented to help determine the optimal number of suppliers needed for the buying firm. Ruiz-Torres and Mahmoodi (2007) considered unequal failure probabilities for all the suppliers. Berger and Zeng (2006) studied the optimal supply size in a single or multiple sourcing strategy context, under a number of scenarios that are determined by various financial loss functions, the operating cost functions, and the probabilities of all the suppliers being down. Yu et al. (2009) considered the impacts of supply disruption risks on the choice between the single and dual sourcing methods in a two-tier supply chain with a non-stationary and price-sensitive demand. Dual sourcing along with a backup production was studied by He and Hongyan (2015) and in contract manufacturing by Su and Liu (2015). Yue et al. (2010) introduced frontier sourcing portfolios to support manufacturers sourcing decisions, which consider the cost and probability of finishing the order on time. Ravindran et al. (2012) developed multi-criteria supplier selection models incorporating supplier risks. In the multi-objective formulation, price, lead time, disruption risk due to natural event, and quality risk are explicitly considered as four conflicting objectives that have to be minimized simultaneously. Four different variants of goal programming were used to solve the multi-objective optimization problem. Xanthopoulos et al. (2012) developed newsvendor-type inventory models for capturing the trade-off between inventory policies and disruption risks in a dual sourcing supply chain network, where both supply channels are subject to disruption risks. The models were developed for both risk-neutral and risk-averse decision-making. Li and Zabinsky (2011) developed a two-stage stochastic programming model and a chance-constrained programming model to determine a minimal set of suppliers and optimal order quantities. Both models include several objectives and strive to balance a small number of suppliers with the risk of not being able to meet demand. The stochastic programming model is scenario-based and uses penalty coefficients whereas the chance-constrained programming model assumes a probability distribution and constrains the probability of not meeting demand. Hammami et al. (2014) proposed a scenario-based stochastic model for supplier selection in the presence of uncertain fluctuations of currency exchange rates and price discounts. Merzifonluoglu (2015) considered a two-stage setting for selection supply portfolio under risks of supply shortages and demand volatility. Customer demand and spot prices were random and possibly correlated and firms primary suppliers were subject to complete shutdown and yield uncertainties. In the first stage, the firm determines size of forward (long/fixed price) contracts with primary suppliers and option (reservation of a certain capacity at a certain price) contracts with backup suppliers. Then, after monitoring the risk of primary suppliers, customer demand, and the spot price, in the second stage the firm decides whether to utilize the capacity reserved at the backup suppliers via option contracts or to make purchases from the spot market. Using CVaR as a risk measure, a mean risk decision-making was applied to balance expected profit of the firm and the risk.

The vast majority of the decision models are mathematical programming models either single objective, e.g., Kasilingam and Lee (1996), Basnet and Leung (2005), Sawik (2005) or multiple objectives, e.g., Weber and Current (1993), Xia and Wu (2007), Demirtas and Ustun (2008), Ustun and Demirtas (2008), PrasannaVenkatesan and Goh (2016). The models developed for supplier selection and order allocation can be either single-period models (e.g., Weber and Current 1993; Demirtas and Ustun 2008) that do not consider inventory management or multi-period models (e.g., Ghodsypour and O'Brien 2001; Basnet and Leung 2005; Ustun and Demirtas 2008; Che and Wang 2008) which consider the inventory management by lot-sizing and scheduling of orders.

The material presented in this chapter is based on research reported by Sawik (2011b) and Sawik (2011c), who proposed a portfolio approach for the supplier selection and order quantity allocation under disruption risks and under operational risks, respectively. The author applied the two popular in financial engineering percentile measures of risk, value-at-risk (VaR) and conditional value-at-risk (CVaR) (e.g., Sarykalin et al. 2008) for managing the risk of supply disruptions or supply delays. The proposed models were further enhanced in this chapter for maximization of expected or expected worst-case service level and for a multi-region sourcing subject to regional disruption risks.

Various simplifying assumptions that have been used in the models presented in this chapter can be relaxed. For example, it has been assumed that each supplier is capable of manufacturing all required part types. In a more general setting, each supplier may only be prepared to manufacture a subset of part types and provide with the parts the corresponding subset of customer orders. The proposed models can be enhanced also for a discount environment, where the suppliers offer discounts based on quantity or business volume of ordered parts, e.g., Sawik (2010a). A critical issue that need to be considered before any practical application of the proposed models is attempted, however, is the estimation of probabilities and the resulting costs associated with each type of disaster event, for which different approaches are suggested in the literature, such as expert systems, game theory, utilization of large simulation models, etc. (e.g., Knemeyer et al. 2009).

Problems

2.1. Modify the probability for disruption scenarios (2.2) to account for correlated regional disruptions that may affect simultaneously suppliers in different regions.

2.2. Modify the SMIP models presented in this chapter for multiple part types with subsets of part types required for each customer order and subsets of suppliers available for each part type.

2.3. Enhance the SMIP models presented in this chapter for selection of static supply portfolio

(a) with total quantity discount for all ordered parts
(b) with total business volume discount (see, Sawik 2010a).

2.4. Mixed mean-risk static supply portfolio

(a) Modify model **SP_ECV(c)** to optimize expected cost and CVaR of service level and model **SP_ECV(sl)** to optimize expected service level and CVaR of cost.

(b) How should the values of the optimized objective functions be scaled into the interval [0,1] to avoid dimensional inconsistency among the two objectives and how should the trade-off parameter be selected?
(c) How would you interpret the mixed mean-risk supply portfolio?

2.5. Explain why for a greater range of disruption probabilities, both expected and expected worst-case costs are higher.

Chapter 3
Selection of Dynamic Supply Portfolio

3.1 Introduction

In this chapter the single-period (static) portfolio approach and the SMIP formulations presented in Chapter 2 are enhanced for a multi-period (dynamic) supplier selection and order quantity allocation in the presence of both the low-probability and high-impact supply chain disruption risks and the high-probability and low-impact supply chain delay risks.

The supplier selection problem is highly dynamic in practice. The supplier selected for one period may not necessarily be selected for the next period to supply the same set of parts. While the delay risks may frequently arise from problems in coordinating supply and demand, the disruption risks are referred to the major disruptions to normal activities caused by natural and man-made disasters. In most cases, the business impact associated with disruption risks is much greater than that of the delay risks. It is common in the make-to-order manufacturing to use penalties that are proportional to the amount of days the delivery exceeded the delivery date. Moreover, a long delay in supply activities can be considered a supply disruption, and vice versa a long recovery time from a disruption leads to long supply delays.

The following stochastic IP (integer programming) and MIP models are presented in this chapter:

DSP_E(c) for risk-neutral selection of dynamic supply portfolio to minimize expected cost;
DSP_E(sl) for risk-neutral selection of dynamic supply portfolio to maximize expected service level;
DSP_CV(c) for risk-averse selection of dynamic supply portfolio to minimize CVaR of cost;

(continued)

T. Sawik, *Supply Chain Disruption Management*,
International Series in Operations Research & Management Science 291,
https://doi.org/10.1007/978-3-030-44814-1_3

DSP_CV(sl) for risk-averse selection of dynamic supply portfolio to maximize CVaR of service level;
DSP_ECV(c) for mean-risk selection of dynamic supply portfolio to optimize trade-off between expected cost and CVaR of cost;
DSP_ECV(sl) for mean-risk selection of dynamic supply portfolio to optimize trade-off between expected service level and CVaR of service level.

The suppliers are subject to local delivery delay risks, and both local and regional delivery disruption risks. In the delivery scenario analysis, both types of the supply chain risks are simultaneously considered.

In the computational experiments described in Sect. 3.6, the general "on-time, delay, or never" delivery scenarios are illustrated with numerical examples and compared with "longest delay-or-never" worst-case scenarios. The major managerial insights are provided at the end of this chapter.

3.2 Problem Description

Table 3.1 Notation: dynamic supply portfolio

	Indices
i	= supplier, $i \in I$
j	= customer order, $j \in J$
r	= geographic region, $r \in R$
s	= delivery scenario, $s \in S$
t	= planning period, $t \in T$
	Input Parameters
c_{it}	= capacity of supplier i in period t
d_j	= demand for parts required for customer order j
D	= $\sum_{j \in J} d_j$—total demand for parts
$\underline{\delta}_j$	= the earliest delivery date of parts for customer order j
δ_j	= the latest delivery date of parts for customer order j
e_i	= cost of ordering parts from supplier i
g_j	= per unit penalty cost of unfulfilled customer order j caused by shortage of required parts
h_j	= per unit and per period penalty cost of delayed customer order j caused by delayed delivery of required parts
o_{ij}	= unit price of part for customer order j purchased from supplier i
ε_{ij}	= per unit price reduction for each delayed part for customer order j from supplier i
$p_{i,\tau}$	= probability of τ periods delay for delivery of parts from supplier i, $\tau \in \{0,\dots,\overline{\tau}\}$
$p_{i,\overline{\tau}+1}$	= local disruption probability of supplier i
p^r	= regional disruption probability of all suppliers $i \in I^r$ in region r
α	= confidence level
τ	= $0,\dots,\overline{\tau}$, delivery delay
ρ_i	= expected defect rate of supplier i

The approach proposed in this chapter can be applied for a very common type of supply chain, with a producer of single product, who obtains raw materials from several different suppliers with limited supply capacity to meet customer orders by customer requested due dates. However, the approach is also applicable to the case of a single producer, who assembles different types of products using product-specific parts purchased from multiple suppliers (for notation used, see Table 3.1). In order to simplify further considerations it is assumed that for each product type, one product-specific part type (e.g., a critical custom part type) needs to be supplied in required amount by custom parts manufacturers. For example, in the electronics industry producer of different electronic devices needs to be supplied by electronics manufacturers with printed wiring boards of different device-specific design. However, the last assumption can be easily relaxed to consider supplies of different product-specific part types required for each product type.

Let $I = \{1,\ldots,\overline{I}\}$ be the set of $\overline{I}$ suppliers and $J = \{1,\ldots,\overline{J}\}$ the set of $\overline{J}$ customer orders for the finished products, known ahead of time. Each customer order $j \in J$ is described by the quantity d_j of required custom parts and their latest delivery date, δ_j, to ensure meeting the customer requested due date for products. The planning horizon consists of $\overline{T}$ periods and denote by $T = \{1,\ldots\overline{T}\}$ the set of planning periods.

Each supplier can provide the producer with custom parts for all customer orders. However, the suppliers have different capacity and, in addition, differ in price and quality of purchased parts and in reliability of delivery. Let c_{it} be the capacity of supplier i in period t, e_i, cost of ordering parts from supplier i, o_{ij}, per unit price of custom parts for customer order j purchased from supplier i, and ρ_i, the expected defect rate for supplier i.

The suppliers are assumed to be located in $\overline{R}$ disjoint geographic regions. Denote by $I^r \subseteq I$ the subset of suppliers in region $r \in R = \{1,\ldots,\overline{R}\}$, where $\bigcup_{r \in R} I^r = I$. The supplies of parts are subject to random local delays or disruptions that are uniquely associated with a particular supplier, which may arise from equipment breakdowns, local labor strike, fires, etc. In addition to independent local delays and independent local disruptions of each supplier individually there are also potential regional disasters that may result in correlated regional disruption of all suppliers in the same region simultaneously. For example, such regional disaster events may include floods, hurricanes, earthquakes, widespread labor strikes in a transportation sector, etc.

The delivery of parts from each supplier $i \in I$ is subject to random delay of different length, $\tau \in \mathscr{T} = \{0,\ldots,\overline{\tau},\overline{\tau}+1\}$, i.e., parts ordered for period t are delivered in period $t+\tau$, where $\tau = 0$ represents on-time delivery and $\tau = \overline{\tau}$ represents maximum delay that may occur. The delivery of parts is also subject to random disruptions, i.e., parts are not delivered at all. By convention, the dummy delay $\tau = \overline{\tau}+1$ represents disruption of supplies, i.e., no delivery of parts. If the delivery of parts for customer order j occurs in period $t+\tau > \delta_j$, $\tau \leq \overline{\tau}$, then the delivery is delayed and the producer is charged by the customer with contractual penalty, $h_j d_j(t+\tau-\delta_j)$, where h_j is the per unit and per period penalty cost of delayed customer order j caused by the delayed delivery of required parts. If supplies of parts for customer order j

are disrupted (i.e., dummy delay of $\tau = \overline{\tau} + 1$ periods), then the producer is charged with the contractual penalty, $g_j d_j$, for unfulfilled order j, where g_j is the per unit penalty cost of unfulfilled customer order j caused by the shortage of required parts.

Denote by $S = \{1, \ldots, \overline{S}\}$ the index set of all delivery scenarios, and by P_s the probability of delivery scenario $s \in S$. Each scenario $s \in S$ can be represented by an integer-valued vector $\tau_s = \{\tau_{1s}, \ldots, \tau_{\overline{I}s}\}$, where $\tau_{is} \in \mathscr{T}$ is the length of delay from supplier $i \in I$ under scenario $s \in S$. There are a total of, $\overline{\tau} + 2$, potential delivery delays, $\tau = 0, \ldots, \overline{\tau}, \overline{\tau} + 1$, from each of the $\overline{I}$ suppliers, including delivery on time, $\tau = 0$, and delivery disruption, $\tau = \overline{\tau} + 1$. When all potential delivery scenarios are considered, then $\overline{S} = (\overline{\tau} + 2)^{\overline{I}}$. For each scenario $s \in S$, the supplies from every supplier can be delayed or disrupted either by a local or a regional event. Denote by $I_s \subset I$ the subset of non-shutdown (non-disrupted) suppliers, who can deliver orders on time or delayed under scenario s. The probability P_s for delivery scenario $s \in S$ with the subset I_s of non-shutdown suppliers is

$$P_s = \prod_{r \in R} P_s^r. \tag{3.1}$$

P_s^r is the probability of realizing delivery scenario s for suppliers in I^r

$$P_s^r = \begin{cases} (1 - p^r) \prod_{i \in I^r} (p_{i,\tau_{is}}) & \text{if } I^r \bigcap I_s \neq \emptyset \\ p^r + (1 - p^r) \prod_{i \in I^r} p_{i,\overline{\tau}+1} & \text{if } I^r \bigcap I_s = \emptyset, \end{cases} \tag{3.2}$$

where $p_{i,\tau_{is}}$ is the probability of occurrence of the delay of length τ_{is} periods from supplier i under scenario s. By convention, $p_{i,\overline{\tau}+1}$ is the probability of local disruption of supplier i. The parts ordered from supplier i can be delivered on time with probability, p_{i0}, delayed with probability $\sum_{\tau=1}^{\overline{\tau}} p_{i\tau}$, or not delivered at all with probability $p_{i,\overline{\tau}+1} = 1 - \sum_{\tau=0}^{\overline{\tau}} p_{i\tau}$.

Denote by π_i the total disruption (shutdown) probability of every supplier $i \in I^r, r \in R$

$$\pi_i = p^r + (1 - p^r) p_{i,\overline{\tau}+1}; \; i \in I^r, r \in R. \tag{3.3}$$

The decision maker needs to decide on the selection of part suppliers, on quantities of various custom parts to be ordered from each selected supplier and on delivery dates to minimize expected cost or maximize expected service level or to mitigate the impact of disruption risks by minimizing the potential worst-case cost or maximizing the potential worst-case service level. Hence, the decision maker needs to select a risk-neutral or risk-averse dynamic supply portfolio, i.e., the allocation of demand for parts among the suppliers and among the planning periods.

Table 3.2 Problem variables: dynamic supply portfolio

		First stage variables
u_i	=	1, if an order for parts is placed on supplier i; otherwise $u_i = 0$ (supplier selection)
v_{ijt}	=	1, if parts required for customer order j are to be delivered by supplier i in period t; otherwise $v_{ijt} = 0$ (order assignment)
		Auxiliary variables
V_{it}	=	the fraction of total demand for parts ordered from supplier i to be delivered in period t (dynamic supply portfolio: allocation of total demand for parts among suppliers and over time)
VaR^c		cost-at-risk, the targeted cost such that for a given confidence level α, for $100\alpha\%$ of the scenarios, the outcome is below VaR^c
VaR^{sl}		service-at-risk, the targeted service level such that for a given confidence level α, for $100\alpha\%$ of the scenarios, the outcome is above VaR^{sl}
$\mathscr{C}_s$	$\geq$	0, the tail cost for scenario s, i.e., the amount by which costs in scenario s exceed VaR^c
$\mathscr{S}_s$	$\geq$	0, the tail service level for scenario s, i.e., the amount by which VaR^{sl} exceeds service level in scenario s

3.3 Models for Risk-Neutral Decision-Making

In this section two stochastic IP models **DSP_E(c)** and **DSP_E(sl)** are presented for selection of a risk-neutral multi-period supply portfolio in the presence of supply delay and disruption risks to minimize expected cost or to maximize expected service level, respectively.

The decision maker needs to select a dynamic supply portfolio, i.e., the allocation of orders for parts among the suppliers and among the planning periods. The dynamic supply portfolio is defined below, (for definition of problem variables, see Table 3.2).

$$\{V_{it} : i \in I, t \in T\},$$

where

$$\sum_{i \in I} \sum_{t \in T} V_{it} = 1$$

and $0 \leq V_{it} \leq 1$ is the fraction of the total demand for parts ordered from supplier i for period t and V_{it} is determined by the order for custom parts assignment variables v_{ijt}

$$V_{it} = \sum_{j \in J} d_j v_{ijt} / D; \; i \in I. \tag{3.4}$$

The supply delays and disruptions may result in the shortage of required parts and the corresponding delay and shortage penalty costs of delayed or unfulfilled customer orders should be incorporated into the model. Clearly, the producer does not need to pay for ordered and defective or undelivered parts, whereas parts delivered late may be paid for at a reduced price. However, the producer can be charged with a much higher penalty cost for delayed or unfulfilled customer orders for products, caused by the shortage of required parts due to defective, delayed, or undelivered parts. In a make-to-order manufacturing no inventory of custom parts can be kept on hand and the parts are requisitioned with each customer order. In addition, custom parts required for each customer order j are to be delivered within a time window $[\underline{\delta}_j, \delta_j]$ derived from the customer requested due date to further reduce any inventory holding cost. The delivery cannot be earlier than the earliest date $\underline{\delta}_j$ and the required parts are assumed to be processed as soon as they are delivered. As a result no overage costs can be considered. On the other hand, the required parts can be delivered late, beyond the latest date δ_j, or not delivered at all. Then, delay or disruption penalty costs represent the underage costs.

For each delivery scenario $s \in S$ and each customer order $j \in J$ to be supplied with parts by non-disrupted supplier $i \in I_s$ in period t (i.e., with $v_{ijt} > 0$), the penalty cost of delivery delayed by $1 \leq \tau_{is} \leq \overline{\tau}$ periods is given by

$$h_j \max\{0, t + \tau_{is} - \delta_j\} d_j,$$

i.e., deliveries later than δ_j are penalized only.

The resulting total expected delay penalty cost over all delivery scenarios can be expressed by

$$\sum_{s \in S} \sum_{i \in I_s} \sum_{t \in T} \sum_{j \in J: \delta_j < t + \tau_{is}} P_s(h_j(t + \tau_{is} - \delta_j) - \varepsilon_{ij}) d_j v_{ijt},$$

where the producer does not need to pay full price for late delivery of parts and hence the purchasing costs of those parts are reduced by ε_{ij} per unit for each delayed part for customer order j from supplier i.

The total expected penalty cost of unfulfilled customer orders due to shortage of required parts caused by supply disruptions is

$$\sum_{s \in S} \sum_{i \notin I_s} \sum_{j \in J} \sum_{t \in T} P_s(g_j - o_{ij}) d_j v_{ijt},$$

where the producer does not need to pay for parts ordered and non-delivered and hence the purchasing costs of those parts are deducted from the cost of unfulfilled customer orders.

In a risk-neutral operating conditions the overall quality of the supply portfolio can be measured by the expected cost per part, E^c, (3.5), or expected service level E^{sl}, (3.6).

$$\begin{aligned} E^c = & \sum_{i \in I} e_i u_i / D + \sum_{i \in I} \sum_{j \in J} \sum_{t \in T} o_{ij} d_j v_{ijt} / D \\ & + \sum_{s \in S} \sum_{i \in I_s} \sum_{t \in T} \sum_{j \in J: \delta_j < t + \tau_{is}} P_s(h_j(t + \tau_{is} - \delta_j) - \varepsilon_{ij}) d_j v_{ijt} / D \\ & + \sum_{s \in S} \sum_{i \notin I_s} \sum_{j \in J} \sum_{t \in T} P_s(g_j - o_{ij}) d_j v_{ijt} / D \end{aligned} \tag{3.5}$$

$$E^{sl} = \sum_{s \in S} \sum_{i \in I_s} \sum_{j \in J} \sum_{t \in T: \underline{\delta}_j \leq t + \tau_{is} \leq \delta_j} P_s d_j v_{ijt} / D. \tag{3.6}$$

The expected cost E^c includes, cost of ordering,
$\sum_{i\in I} e_i u_i/D$,
cost of purchasing non-defective parts,
$\sum_{i\in I}\sum_{j\in J}\sum_{t\in T} o_{ij} d_j v_{ijt}/D$,
cost of delayed deliveries of parts,
$\sum_{s\in S}\sum_{i\in I_s}\sum_{t\in T}\sum_{j\in J:\delta_j<t+\tau_{is}} P_s(h_j(t+\tau_{is}-\delta_j)-\varepsilon_{ij})d_j v_{ijt}/D$,
and cost of shortage of parts due to supply disruptions (cost of unfulfilled customer orders less cost of non-delivered parts),
$\sum_{s\in S}\sum_{i\notin I_s}\sum_{j\in J}\sum_{t\in T} P_s(g_j-o_{ij})d_j v_{ijt}/D$.

The purchase orders for parts are assumed to be inflated by the reject rates $\rho_i, i\in I$ of defective parts, i.e., are equal to $\sum_{i\in I}\sum_{j\in J}\sum_{t\in T}(1+\rho_i)d_j v_{ijt}$. However, since the producer does not need to pay for ordered and defective parts in the amount of $\sum_{i\in I}\sum_{j\in J}\sum_{t\in T}\rho_i d_j v_{ijt}$, and pays a reduced price for delayed parts in the amount of $\sum_{i\in I}\sum_{t\in T}\sum_{j\in J:\delta_j<t+\tau} d_j v_{ijt}$, the corresponding purchasing cost per part for delivered parts is simply given by

$$\sum_{i\in I}\sum_{j\in J}\sum_{t\in T} o_{ij} d_j v_{ijt}/D$$
$$-\sum_{s\in S}\sum_{i\in I_s}\sum_{t\in T}\sum_{j\in J:\delta_j<t+\tau_{is}} P_s\varepsilon_{ij} d_j v_{ijt}/D - \sum_{s\in S}\sum_{i\notin I_s}\sum_{j\in J}\sum_{t\in T} P_s o_{ij} d_j v_{ijt}/D.$$

The expected service level, E^{sl}, (3.6), is a surrogate measure of expected customer demand fulfillment rate and represents the expected fraction of fulfilled demand for required parts.

The stochastic IP models **DSP_E(c)** and **DSP_E(sl)** for selection of risk-neutral dynamic supply portfolio to minimize expected cost or maximize expected service level, respectively, are presented below.

DSP_E(c): *Selection of risk-neutral* **D***ynamic* **S***upply* **P***ortfolio to minimize* **E***xpected* **c***ost*

Minimize (3.5)
subject to

1. *Dynamic supply portfolio selection constraints*:
   - for each customer order the required parts must be delivered exactly once not earlier than the earliest and not later than the latest delivery date, (3.7),
   - parts cannot be ordered from non-selected suppliers, (3.8),
   - for each selected supplier the total quantity of ordered parts cannot exceed the supplier capacity, (3.9),
   - at least one order should be assigned to each selected supplier, (3.10),

(continued)

$$\sum_{i\in I}\sum_{t\in T:\underline{\delta}_j\le t\le\delta_j} v_{ijt}=1;\ j\in J \tag{3.7}$$

$$v_{ijt}\le u_i;\ i\in I, j\in J, t\in T \tag{3.8}$$

$$\sum_{j\in J}(1+\rho_i)d_jv_{ijt}\le c_{it}u_i;\ i\in I, t\in T \tag{3.9}$$

$$\sum_{j\in J}\sum_{t\in T}v_{ijt}\ge u_i;\ i\in I \tag{3.10}$$

2. *Integrality conditions*:

$$u_i\in\{0,1\};\ i\in I, t\in T \tag{3.11}$$

$$v_{ijt}\in\{0,1\};\ i\in I, j\in J, t\in T. \tag{3.12}$$

DSP_E(sl): *Selection of risk-neutral* **D**ynamic **S**upply **P**ortfolio *to maximize* **E**xpected **s**ervice **l**evel

Maximize (3.6)

subject to (3.7)–(3.12).

A simple upper bound on the expected service level, E^{sl}, (3.6), is derived below.

Proposition 3.1.

$$E^{sl}\le\min\{1,\sum_{r\in R}\sum_{i\in I^r}\sum_{t\in T}(1-p^r)(1-p_{i,\overline{\tau}+1})c_{it}/(1+\rho_i)D\}. \tag{3.13}$$

Proof. The dynamic supply portfolio selection constraints (3.9) imply that

$$\sum_{s\in S}\sum_{i\in I_s}\sum_{j\in J}\sum_{t\in T:\underline{\delta}_j\le t+\tau_{is}\le\delta_j}P_sd_jv_{ijt}/D\le$$

$$\sum_{r\in R}\sum_{i\in I^r}\sum_{j\in J}\sum_{t\in T:\underline{\delta}_j\le t+\tau_{is}\le\delta_j}(1-p^r)(1-p_{i,\overline{\tau}+1})d_jv_{ijt}/D\le$$

$$\sum_{r\in R}\sum_{i\in I^r}\sum_{t\in T}(1-p^r)(1-p_{i,\overline{\tau}+1})c_{it}u_i/(1+\rho_i)D\le$$

$$\sum_{r\in R}\sum_{i\in I^r}\sum_{t\in T}(1-p^r)(1-p_{i,\overline{\tau}+1})c_{it}/(1+\rho_i)D,$$

where $(1-p^r)(1-p_{i,\overline{\tau}+1})=1-\pi_i$, (3.3) is a non-disruption probability of supplier $i\in I^r$.

Since E^{sl} cannot be greater than 1, its upper bound is 1, if $\sum_{r\in R}\sum_{i\in I^r}\sum_{t\in T}(1-p^r)(1-p_{i,\overline{\tau}+1})c_{it}/(1+\rho_i)D>1$.

Models **DSP_E(c)** and **DSP_E(sl)** are deterministic equivalent integer programs of two-stage stochastic integer programs with recourse (Birge and Louveaux 2011). The primary portfolio selection variables, u_i, v_{ijt}, are referred to as first stage decisions. The second stage variables are simply demand allocation variables for realized disruption scenarios s, $\tilde{v}^s_{ijt}$; $i \in I, j \in J, t \in T, s \in S$, defined as follows:

$$\tilde{v}^s_{ijt} = \begin{cases} v_{ij,t-\tau_{is}} & \text{if } i \in I_s, j \in J, t \in T, s \in S \\ 0 & \text{if } i \notin I_s, j \in J, t \in T, s \in S. \end{cases}$$

In view of the above definition, an explicit introduction of the second stage variables $\tilde{v}^s_{ijt}$ into the stochastic IP formulations is not required.

Notice that if for each customer order, all custom parts can be provided by multiple delivery from different suppliers, then the binary order assignment variables v_{ijt} should be redefined as fractional allocation variables denoting the fraction of total demand for parts required for customer order j to be delivered by supplier i in period t. Then, if total available capacity of all suppliers is not greater than total demand for required parts, i.e., $\sum_{i\in I}\sum_{t\in T} c_{it}/(1+\rho_i) \le \sum_{j\in J} d_j$, the demand allocation equality constraints (3.7) should be replaced by inequalities

$$\sum_{i\in I}\sum_{t\in T:\underline{\delta}_j\le t\le \delta_j} v_{ijt} \le 1;\ j \in J; \tag{3.14}$$

otherwise no feasible solution exists.

3.3.1 Dynamic Supply Portfolio for All-or-Nothing Delivery Scenarios

The dynamic supply portfolio approach is particularly useful when supply delay risks need to be considered. Otherwise, the static supply portfolio presented in Chapter 2 may be a more appropriate approach. For the pure all-or-nothing delivery scenarios (i.e., for disruption scenarios), with supply disruptions only and no delays (see (2.2), (2.3) in Chapter 2), the proposed multi-period models for $\overline{I}$ suppliers and $\overline{T}$ planning periods can be transformed into equivalent single-period models for $m = (\overline{I})(\overline{T})$ suppliers. To this end, each pair of indices $(i,t), i \in I, t \in T$ should be replaced with a single index $k \in K$, where $K = \{1,\dots,m\}$ represents the set of equivalent m single-period suppliers. In the proposed stochastic IP models, variables u_i, v_{ijt} should be replaced by u_k, v_{jk}, respectively. Then, the dynamic supply portfolio selection constraints (3.7) are replaced by

$$\sum_{k\in K_j} v_{jk} = 1;\ j \in J,$$

where K_j is the subset of equivalent single-period suppliers that are capable of delivering parts required for customer order j and it represents the subset $\{(i,t) : i \in I, \underline{\delta}_j \le t \le \delta_j\}$ of pairs (i,t) feasible for customer order j.

3.4 Models for Risk-Averse Decision-Making

In the risk-averse selection of supply portfolio under disruption risks, the confidence level α is fixed by the decision maker to control the risk of losses due to supply disruptions. We assume that the decision maker is willing to accept only portfolios for which the total probability of scenarios with costs greater than VaR^c or with service level lower than VaR^{sl} is not greater than $1-\alpha$. Furthermore, a risk aversive decision maker wants to minimize the expected worst-case costs exceeding VaR^c or to maximize the expected worst-case service level below VaR^{sl}.

Define by $\mathscr{C}_s$ the tail cost for scenario s, where tail cost is defined as the nonnegative amount by which cost in scenario s exceeds VaR^c. In a similar way, define by $\mathscr{S}_s$ the tail service level for scenario s, where tail service level is defined as the nonnegative amount by which VaR^{sl} exceeds service level in scenario s.

The portfolio will be optimized by calculating VaR^c and minimizing CVaR^c simultaneously or by calculating VaR^{sl} and maximizing CVaR^{sl}, respectively.

In the proposed models CVaR is represented by an auxiliary function (3.15) and (3.18) introduced by Rockafellar and Uryasev (2000). The SMIP models **DSP_CV(c)** and **DSP_CV(sl)** for the risk-averse selection of supply portfolio to reduce the risk of high costs and the risk of low service level, respectively, is formulated below.

DSP_CV(c): *Selection of risk-averse* **D***ynamic* **S***upply* **Po***rtfolio to minimize* **CV***aR of* **cost**

Minimize

$$CVaR^c = VaR^c + (1-\alpha)^{-1}\sum_{s\in S} P_s\mathscr{C}_s \tag{3.15}$$

subject to

1. *Dynamic supply portfolio selection constraints*: (3.7)–(3.10)
2. *Risk constraints*:
   - the tail cost for scenario s is defined as the nonnegative amount by which cost per part in scenario s exceeds VaR^c,

$$\mathscr{C}_s \geq \sum_{i\in I} e_i u_i/D + \sum_{i\in I}\sum_{j\in J}\sum_{t\in T} o_{ij}d_j v_{ijt}/D + \sum_{i\in I_s}\sum_{t\in T}\sum_{j\in J:\delta_j<t+\tau_{is}} (h_j(t+\tau_{is}-\delta_j)-\varepsilon_{ij})d_j v_{ijt}/D + \sum_{i\notin I_s}\sum_{j\in J}\sum_{t\in T}(g_j-o_{ij})d_j v_{ijt}/D - VaR^c. \tag{3.16}$$

3. *Non-negativity and integrality conditions*: (3.11), (3.12) and

$$\mathscr{C}_s \geq 0;\ s\in S. \tag{3.17}$$

DSP_CV(sl): *Selection of risk-averse* **D***ynamic* **S***upply* **P***ortfolio to maximize* **CV***aR of* **s***ervice* **l***evel*

Maximize

$$CVaR^{sl} = VaR^{sl} - (1-\alpha)^{-1} \sum_{s \in S} P_s \mathscr{S}_s \tag{3.18}$$

subject to

1. *Dynamic supply portfolio selection constraints*: (3.7)–(3.10)
2. *Risk constraints*:
   - the tail service level for scenario s is defined as the nonnegative amount by which VaR^{sl} exceeds service level in scenario s,

$$\mathscr{S}_s \geq VaR^{sl} - \sum_{i \in I_s} \sum_{j \in J} \sum_{t \in T: \underline{\delta}_j \leq t + \tau_{is} \leq \delta_j} d_j v_{ijt}/D; \; s \in S. \tag{3.19}$$

3. *Non-negativity and integrality conditions*: (3.11), (3.12) and

$$\mathscr{S}_s \geq 0; \; s \in S. \tag{3.20}$$

Note that variables VaR^c and VaR^{sl} do not need to be constrained of being non-negative. As $\mathscr{C}_s$ and $\mathscr{S}_s$ are constrained of being positive, the model tries to decrease VaR^c and increase VaR^{sl}, respectively, and hence positively impact the objective function. However, large reduction in VaR^c and increase in VaR^{sl} may result in more scenarios with positive tail costs and service levels, respectively.

3.5 Models for Mean-Risk Decision-Making

In the single objective approach the supply portfolio is selected by minimizing either the expected cost per part, E^c, (3.5), or the expected worst-case cost per part, $CVaR^c$, (3.15) or by maximizing either the expected service level, E^{sl}, (3.6), or the expected worst-case service level, $CVaR^{sl}$, (3.18). In this section the two cost functions and the two service level functions are considered simultaneously, and a bi-objective selection of supply portfolio is presented aimed at minimizing both objective functions to balance expected costs or expected service level with the risk tolerance. This trade-off model is known as the mean-risk model, formulated as the optimization of a composite objective consisting of the expected cost (service level) and the CVaR as a risk measure.

A subset of nondominated solutions for the bi-objective dynamic supply portfolio can be found by the parameterization on λ the weighted-sum programs **DSP_ECV(c)** and **DSP_ECV(sl)** presented below. The mean-risk program **DSP_ECV(c)** is based on model **DSP_CV(c)** with the addition of objective (3.5) of model **DSP_E(c)**. Similarly, the mean-risk program **DSP_ECV(sl)** is based on model **DSP_CV(sl)** with the addition of objective (3.6) of model **DSP_E(sl)**.

DSP_ECV(c): *Selection of mean-risk* **D**ynamic **S**upply **P**ortfolio *to minimize weighted sum of* **E**xpected cost *and* **CV***aR of* **c**ost

Minimize

$$\lambda E^c + (1-\lambda)CVaR^c, \tag{3.21}$$

where $0 \le \lambda \le 1$

subject to (3.5), (3.7)–(3.12), (3.15)–(3.17).

DSP_ECV(sl): *Selection of mean-risk* **D**ynamic **S**upply **P**ortfolio *to maximize weighted sum of* **E**xpected service level *and* **CV***aR of* **s**ervice **l**evel

Maximize

$$\lambda E^{sl} + (1-\lambda)CVaR^{sl} \tag{3.22}$$

where $0 \le \lambda \le 1$,

subject to (3.6)–(3.12), (3.18)–(3.20).

3.6 Computational Examples

In this section some computational examples are presented to illustrate possible applications of the proposed stochastic IP and MIP models for selection of a dynamic supply portfolio in the presence of both delay and disruption risks. The problem is rarely considered in the literature and to the best of our knowledge there is no solution algorithm or benchmark problem that can be used for comparison of the proposed portfolio approach and stochastic models. Although the input data for the examples are hypothetical, their relations to each other are real and in part they have been taken from real case studies (Sawik 2011a). In the cost-oriented supplier selection, a combination of disruption and delay risks directly impacts both shortage and delay penalties, while in the service-oriented decisions both demand fulfillment rate (the fraction of demand for parts fulfilled during the planning horizon) and service level (the fraction of demand for parts fulfilled within due windows) are simultaneously impacted. Therefore, the solution values of service level are reported along with the associated demand fulfillment rates, in particular for the service-oriented models. Furthermore, to emphasize the impact of supply delays on selection of a dynamic supply portfolio under combined disruption and delay risks, selected solutions with neglected delay penalties are reported for a comparison.

The following parameters have been used for the example problems:

- $\bar{I}$, the number of suppliers, was equal to 5.
- $\bar{R}$, the number of geographic regions, was equal to 2, and the subsets of suppliers were $I^1 = \{1,2\}$ and $I^2 = \{3,4,5\}$, respectively.

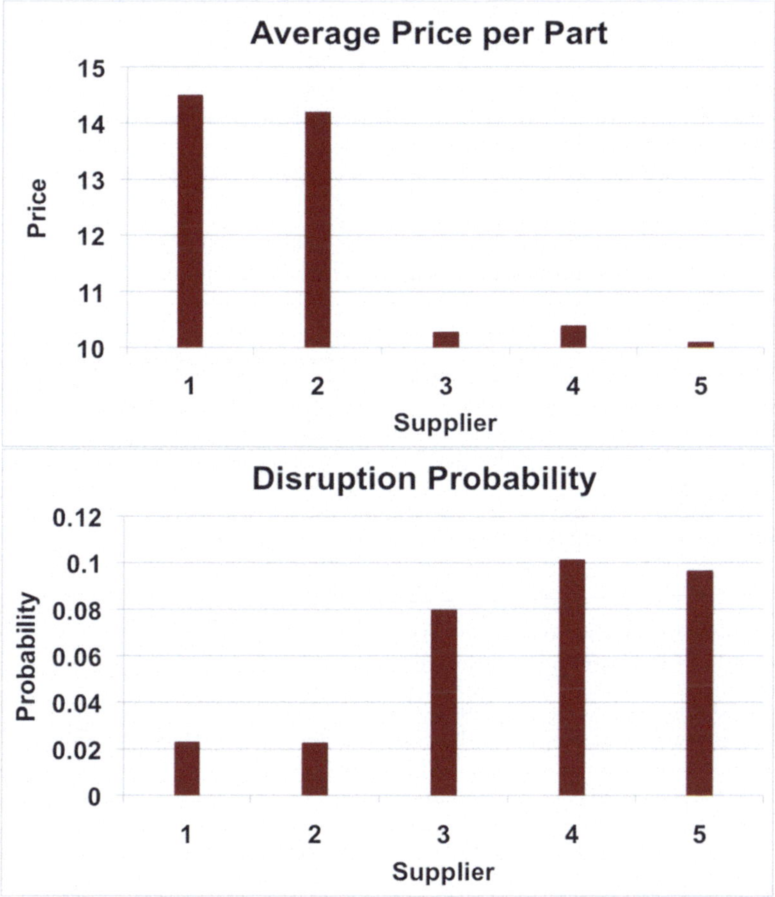

Fig. 3.1 Suppliers

- $\overline{J}$, the number of customer orders, was equal to 50.
- $\overline{T}$, the planning horizon, was equal to 14 periods.
- $\overline{\tau}$, maximum delivery delay was equal to 4 periods for each supplier, so that the set of all possible delays is $\mathscr{T} = \{0,1,2,3,4,5\}$, where a dummy delay, $\overline{\tau}+1 = 5$, represents local disruption of a supplier. The corresponding number of delivery scenarios was equal to $\overline{S} = 6^5 = 7776$.
- d_j, the numbers of required parts for each customer order, were integers uniformly distributed over $[500,5000]$ for all customer orders j and the resulting total demand for parts was $D = 138000$.
- e_i, the cost of ordering parts, were integers in $\{8000,9000\}$ and $\{12000,13000,14000\}$, respectively, for suppliers $i \in I^1$ and $i \in I^2$.

- o_{ij}, the unit price of parts for customer order j purchased from supplier i, was uniformly distributed over [13,15] and [9,11], respectively, for suppliers $i \in I^1$ and $i \in I^2$.
- ε_{ij}=0.05o_{ij}, per unit price reduction for each delayed part of customer order j from supplier i was 5% of the unit price.
- $h_j = 1$, per unit and per period delay penalty cost was equal to 1 for all customer orders j.
- $g_j = 5\max_{i \in I}(o_{ij})$, per unit shortage cost for each customer order j, was equal to five times of the maximum purchasing cost of required parts.
- $c_{it} = 10000; \forall i \in I, t \in T$.
- p_{i5}, the local disruption probability (dummy delay, $\overline{\tau}+1 = 5$, represents local disruption) was uniformly distributed over [0.01,0.05] and [0.05;0.10], respectively, for suppliers $i \in I^1$ and $i \in I^2$, i.e., the disruption probabilities were drawn independently from U[0.01,0.05] and U[0.05;0.10], respectively.
  Given local disruption probabilities, p_{i5}, $i \in I$, the probabilities for different delay length $\tau = 0,1,2,3,4$ were calculated as follows:

  probability of on-time delivery ($\tau = 0$), $p_{i0} = 0.3(1 - p_{i5})$;
  probability of 1-period delay ($\tau = 1$), $p_{i1} = 0.25(1 - p_{i5})$;
  probability of 2-period delay ($\tau = 2$), $p_{i2} = 0.2(1 - p_{i5})$;
  probability of 3-period delay ($\tau = 3$), $p_{i3} = 0.15(1 - p_{i5})$;
  probability of maximum, 4-period delay ($\tau = \overline{\tau} = 4$), $p_{i4} = 0.1(1 - p_{i5})$;

  for all suppliers $i \in I$, i.e., the longer delay the lower probability of its occurrence. Notice that for each supplier i, the expected delay is equal to 2 periods, i.e.,

$$\lceil (0.25(1-p_{i5}) + 2(0.2(1-p_{i5})) + 3(0.15(1-p_{i5})) + 4(0.1(1-p_{i5})))/(1-p_{i5}) \rceil = 2.$$

- p^r, the regional disruption probability was 0.005 and 0.01, respectively, for region $r = 1$ and $r = 2$.
- δ_j, the latest delivery date of parts required for customer order j, was integer uniformly distributed over $[3, \overline{T}]$, i.e., generated from a U[3;14] distribution.
- $\underline{\delta}_j = \delta_j - 2$, the earliest delivery date of parts required for each customer order j, was two periods earlier (i.e., the expected delay) than the latest delivery date.
- ρ_i, the expected defect rate of each supplier i was exponentially distributed.
- α, the confidence level, was equal to 0.50, 0.75, 0.90, 0.95, or 0.99.

Note that cost of lost customer orders, g_j, is set to be much higher than the corresponding cost, h_j, of delayed orders, which is typical for industrial practice.

Figure 3.1 shows basic characteristics of each supplier i, average price per part, $\sum_{j \in J} o_{ij}/\overline{J}$, and disruption probability, π_i, (3.3). Suppliers $i = 1,2$ in region $r = 1$ are most reliable and most expensive, supplier $i = 4$ is most unreliable, while supplier $i = 5$ is the cheapest one.

The risk-neutral solutions for models **DSP_E(c)** and **DSP_E(sl)** are shown in Table 3.3, and the risk-averse solutions for models **DSP_CV(c)** and **DSP_CV(sl)** with different confidence levels in Table 3.4. For each model Table 3.3 shows the associated values of both risk-neutral objective functions, expected cost, E^c, and expected service levels, E^{sl}. Moreover, for model **DSP_E(c)**, the shortage and delay penalties are separately indicated. In Table 3.4, conditional cost-at-risk, CVaR^c, and

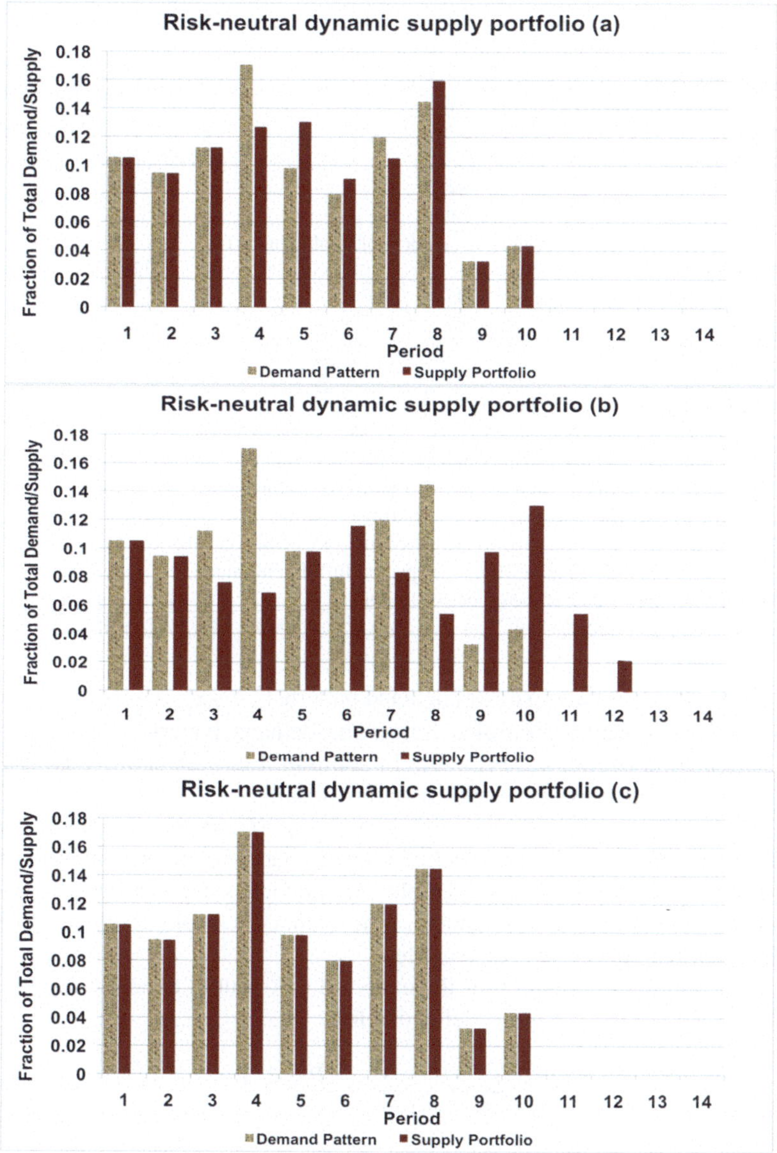

Fig. 3.2 Risk-neutral dynamic supply portfolio: (**a**) model **DSP_E(c)**, (**b**) model **DSP_E(c)** with delay penalty neglected, (**c**) model **DSP_E(sl)**

conditional service-at-risk, CVaR^{sl}, are shown along with the associated expected cost, E^c, and expected service level, E^c.

Table 3.3 indicates that the risk-neutral supply portfolios do not include the most unreliable suppliers $i = 4,5$. The corresponding risk-neutral dynamic supply portfolios, $(\sum_{i\in I}\sum_{j\in J} d_j v_{ijt}/D, t \in T)$, are shown in Fig. 3.2 for cost-oriented and service-

Table 3.3 Risk-neutral solutions

Model **DSP_E(c)**	Model **DSP_E(c)**[a]	Model **DSP_E(sl)**
	Var.=3505, Bin.=3505, Cons.=3675, Nonz.=18325[b]	
$E^c = 15.47$, $(3.73^c, 0.36^d)$	$E^c = 15.85$, $(4.46^c, 0.88^d)$	$E^{sl} = 73.08\%$
$E^{sl} = 69.91\%$	$E^{sl} = 48.73\%$	$E^c = 16.04$, $(1.89^c\ 0.34^d)$
94.93[e]	93.96[e]	97.44[e]
15.47[f]	15.23[f]	73.11[f]
	Suppliers selected(% of total demand)[g]	
1(14)		1(32)
2(37)	2(34)	2(63)
3(49)	3(66)	3(5)

[a] model **DSP_E(c)** without delay penalty,
$\sum_{s\in S}\sum_{i\in I_s}\sum_{t\in T}\sum_{j\in J:\delta_j<t+\tau_{is}} P_s(h_j(t+\tau_{is}-\delta_j)-\varepsilon_{ij})d_j v_{ijt}/D$, in (3.5)
[b] Var.: total number of variables; Bin.: number of binary variables;
Cons.: number of constraints; Nonz.: number of nonzero coefficients
[c] $\sum_{s\in S}\sum_{i\notin I_s}\sum_{j\in J}\sum_{t\in T} P_s(g_j - o_{ij})d_j v_{ijt}/D$—shortage penalty
[d] $\sum_{s\in S}\sum_{i\in I_s}\sum_{t\in T}\sum_{j\in J:\delta_j<t+\tau_{is}} P_s(h_j(t+\tau_{is}-\delta_j))d_j v_{ijt}/D$—delay penalty
[e] $\sum_{s\in S}\sum_{i\in I_s}\sum_{j\in J}\sum_{t\in T} P_s d_j v_{ijt}/D \times 100\%$—expected fulfillment rate
[f] LP relaxation solution value of the objective function
[g] $1(\sum_{j\in J}\sum_{t\in T} d_j v_{1jt}/D\times 100\%)$, $2(\sum_{j\in J}\sum_{t\in T} d_j v_{2jt}/D\times 100\%)$, $3(\sum_{j\in J}\sum_{t\in T} d_j v_{3jt}/D\times 100\%)$

oriented objectives. For comparison, demand pattern, $(\sum_{j\in J:\underline{\delta}_j=t} d_j/D, t\in T)$, is also presented. Both dynamic portfolios follow the demand pattern, in particular the service-oriented portfolio, see Fig. 3.2c. In addition, Fig. 3.2b shows cost-oriented dynamic supply portfolio with neglected delay penalty,
$\sum_{s\in S}\sum_{i\in I_s}\sum_{t\in T}\sum_{j\in J:\delta_j<t+\tau_{is}} P_s(h_j(t+\tau_{is}-\delta_j))d_j v_{ijt}/D$, in the objective function (3.5). Comparison of Fig. 3.2a, b clearly indicates that the dynamic supply portfolio with no delay risk considered is more unleveled and leads to greater supply fluctuations and manufacturing delays. Table 3.3 shows that the most reliable supplier 1 is not selected when delay risk is disregarded.

Figure 3.3 shows distribution of the expected risk-neutral dynamic supply portfolio, $\tilde{V}_{it}, i\in I, t\in\{1,\dots,\overline{T}+\overline{\tau}\}$, defined below.

$$\tilde{V}_{it} = \sum_{s\in S}\sum_{j\in J}\sum_{t'\in T:t=t'+\tau_{is}} P_s d_j \tilde{v}^s_{ijt'},$$

where

$$\tilde{v}^s_{ijt'} = \begin{cases} v_{ijt'} & \text{if } \tau_{is}\le\overline{\tau} \\ 0 & \text{if } \tau_{is}=\overline{\tau}+1. \end{cases}$$

The expected fraction of total demand for parts is unevenly distributed over the planning horizon. For both objective functions, the highest delivery levels appear in mid-horizon and the lowest levels at the beginning and at the end of the horizon.

In the computational experiments for the risk-averse portfolios, the confidence level α is set at five levels of 0.5, 0.75, 0.90, 0.95, and 0.99, which means that focus

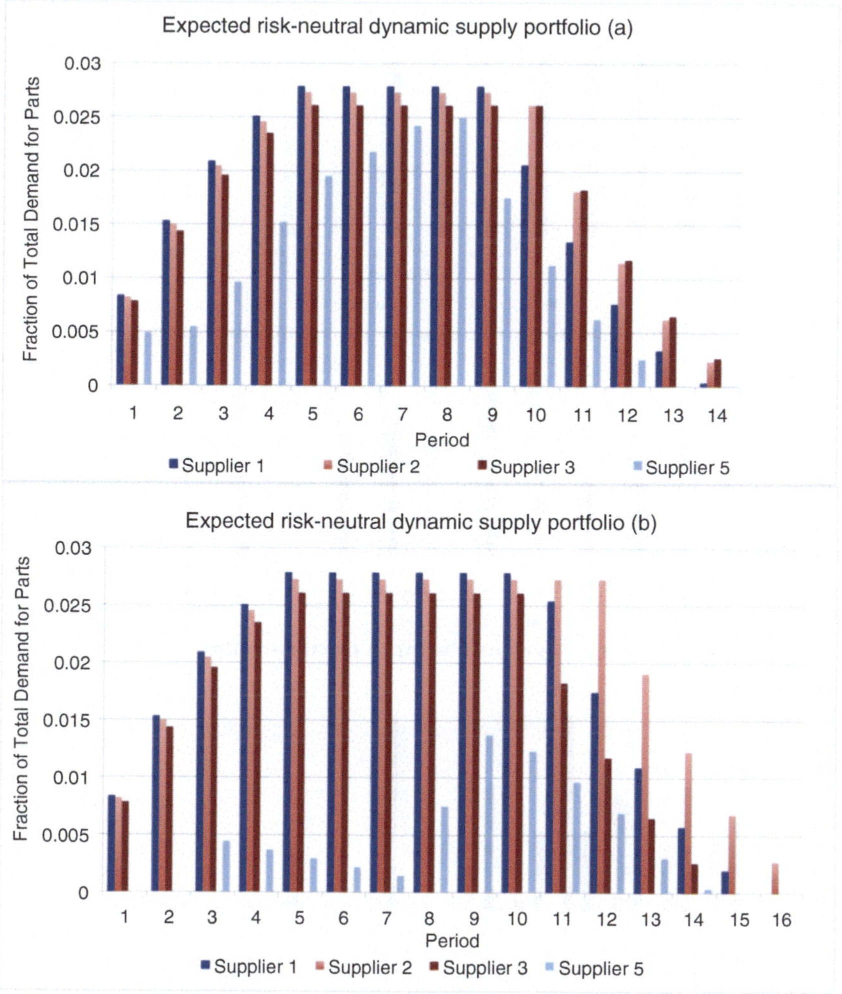

Fig. 3.3 Expected risk-neutral dynamic supply portfolio: (**a**) model **DSP_E(c)**, (**b**) model **DSP_E(sl)**

is on minimizing the highest 50%, 25%, 10%, 5%, and 1% of all scenario outcomes, i.e., costs per part or service level. When α increases, a more risk aversive decision-making focuses on a smaller set of outcomes and the number of selected suppliers also is increasing to mitigate the impact of disruption risks by diversification of the supply portfolio. Table 3.4 demonstrates that for model **DSP_CV(c)** and low confidence levels, $\alpha = 0.5, 0.75, 0.9$, only the most reliable suppliers 1 and 2 are selected to minimize the dominating worst-case shortage cost. However, for a larger

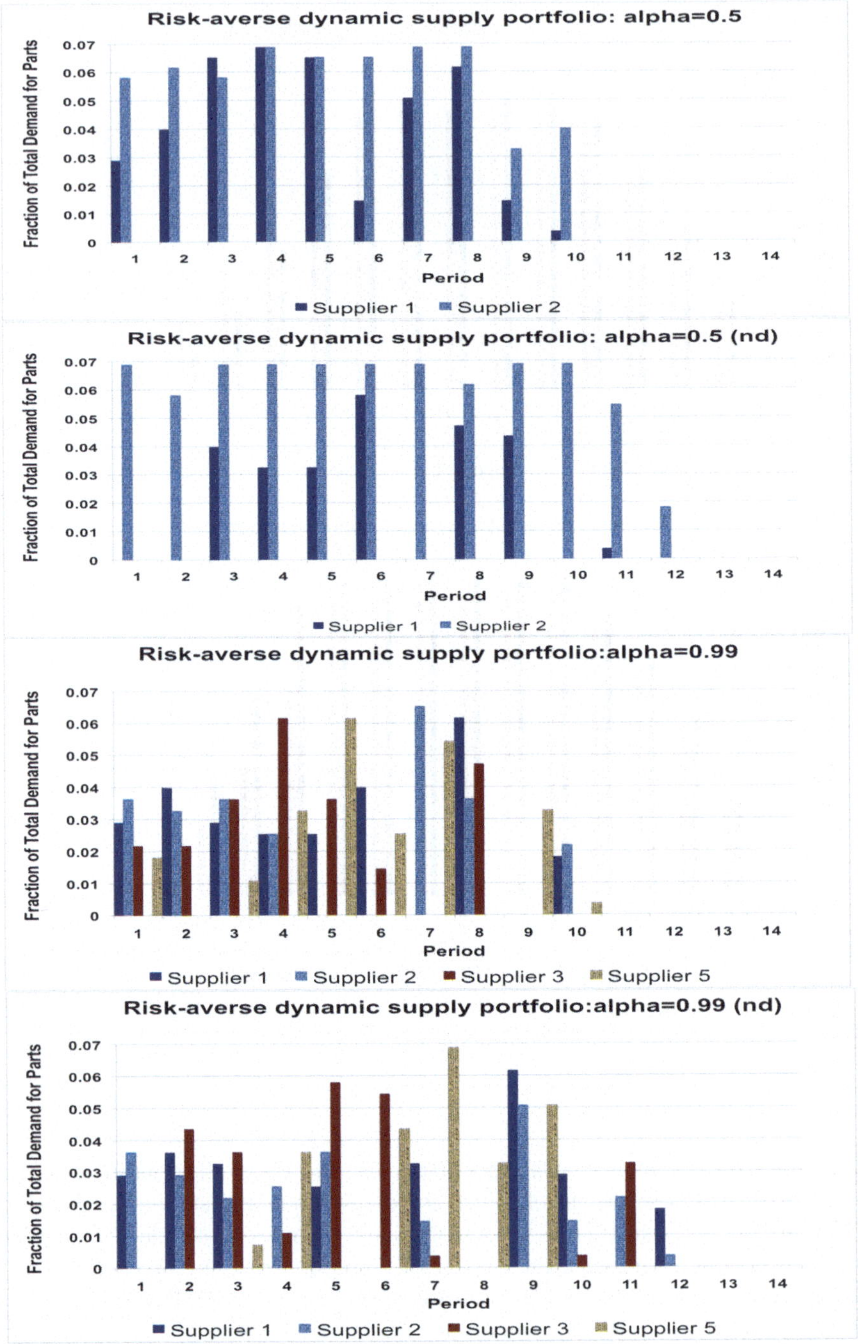

Fig. 3.4 Risk-averse dynamic supply portfolio for model **DSP_CV(c)**: $\alpha = 0.5$ and $\alpha = 0.99$ (nd—neglected delay penalty)

Table 3.4 Risk-averse solutions

Confidence level α	0.50	0.75	0.90	0.95	0.99
Model **DSP_CV(c)**					
Var.=11282, Bin.=3505, Cons.=11451, Nonz.=27288757[a]					
CVaR^c	17.29	20.24	28.41	35.82	45.59
VaR^c	14.19	14.38	14.87	30.04	43.07
E^c	15.74	15.75	15.75	15.80	15.68
E^{sl}%	71.80	72.36	72.36	71.10	70.47
Expected fulfillment rate[b]	97.71	97.71	97.71	94.88	94.58
LP[c]	17.27	20.19	28.29	35.74	45.47
Suppliers selected(% of total demand)[d]	1(41)	1(43)	1(49)	1(30)	1(27)
	2(59)	2(57)	2(51)	2(29)	2(25)
				3(14)	3(24)
				4(14)	
				5(13)	5(24)
Model **DSP_CV(sl)**					
Var.=11282, Bin.=3505, Cons.=11451, Nonz.=22713877[a]					
CVaR^{sl}%	53.97	43.87	32.66	25.46	13.99
VaR^{sl}%	71.74	59.78	39.85	39.85	19.93
E^{sl}%	71.02	70.16	70.16	70.16	70.14
Expected fulfillment rate[b]	94.70	93.54	93.54	93.54	93.52
E^c	16.42	16.68	16.68	16.47	16.78
LP[c]	53.97	43.93	32.75	25.49	14.04
Suppliers selected(% of total demand)[d]	1(28)				1(20)
	2(28)				2(20)
	3(15)				3(20)
	4(14)				4(20)
	5(15)				5(20)

[a] Var.: total number of variables; Bin.: number of binary variables;
Cons.: number of constraints, Nonz.: number of nonzero coefficients

[b] $\sum_{s\in S}\sum_{i\in I_s}\sum_{j\in J}\sum_{t\in T} P_s d_j v_{ijt}/D \times 100\%$

[c] LP relaxation solution value of the objective function

[d] $1(\sum_{j\in J}\sum_{t\in T} d_j v_{1jt}/D \times 100\%)$, $2(\sum_{j\in J}\sum_{t\in T} d_j v_{2jt}/D \times 100\%)$, $3(\sum_{j\in J}\sum_{t\in T} d_j v_{3jt}/D \times 100\%)$, $4(\sum_{j\in J}\sum_{t\in T} d_j v_{4jt}/D \times 100\%)$,$5(\sum_{j\in J}\sum_{t\in T} d_j v_{5jt}/D \times 100\%)$

α, these suppliers are supported by the less reliable suppliers 3, 4, and 5 to better mitigate the impact of part shortages by a more diversified portfolio.

For model **DSP_CV(sl)** and $\alpha \geq 0.75$, the total demand is equally allotted among all suppliers to better mitigate the impact of disruption risks on worst-case service level. Notice that for the lowest confidence level α, VaR^c is smaller than expected cost, E^c, and VaR^{sl} is greater than expected service level, E^{sl}.

The cost-oriented risk-averse dynamic supply portfolios for the lowest and highest confidence levels, respectively $\alpha = 0.5$ and $\alpha = 0.99$, are shown in Fig. 3.4. For comparison, Fig. 3.4 also shows the risk-averse dynamic supply portfolio with no delay penalty, $\sum_{i\in I_s}\sum_{t\in T}\sum_{j\in J:\delta_j<t+\tau_{is}}(h_j(t+\tau_{is}-\delta_j))d_j v_{ijt}/D$, in the risk constraints (3.16). While for the risk-averse dynamic portfolio the same suppliers are selected, the allocation of orders among suppliers and over time is different when

delay risk is disregarded. For example, for $\alpha = 0.5$, the percentage of demand allocation is changing from (Supplier 1: 41%, Supplier 2: 59%, see Table 3.4) to (Supplier 1: 25%, Supplier 2: 75%), that is more demand is moved to the cheaper supplier 2. The results clearly indicate that when delay risk is neglected, the dynamic portfolios are spread over more periods and may lead to greater manufacturing delays. Thus, neglecting potential delay risks might negatively impact material supplies.

In addition to expected service level, E^{sl}, Tables 3.3 and 3.4 show the associated expected fraction of fulfilled demand (on time or delayed), $\sum_{s\in S}\sum_{i\in I_s}\sum_{j\in J}\sum_{t\in T} P_s d_j v_{ijt}/D$. Comparison of the latter solution values with E^{sl} clearly indicates that a large fraction of fulfilled demand is delayed. For all models the expected fractions of fulfilled demand are very similar and close to 95%. A high expected fulfillment rate indicates that the dynamic supply portfolios well mitigate the impact of disruption risks. The expected service levels are similar (close to 70%) for all confidence levels, α, only for model **DSP_CV(sl)**. For the cost-oriented models, however, E^{sl} depends on the confidence level, α, for model **DSP_CV(c)** and attains its lower value (48.73%) for model **DSP_E(c)** with neglected delay penalty in the objective function.

In order to provide more insights, additional computational experiments were conducted with different shortage-to-delay unit penalty ratios: for unit delay penalty, $h_j = h = 1, \forall j \in J$, and unit shortage penalty, $g_j = g \in \{1, 10, 25, 50, 75, 100\}, \forall j \in J$. For different unit penalty ratio, g/h, Table 3.5 presents risk-neutral solutions for model **DSP_E(c)** and risk-averse solutions for model **DSP_CV(c)** with the highest confidence level $\alpha = 0.99$.

For the risk-neutral decision-making, the solution results are in line with an intuitive point of view. As shortage-to-delay unit penalty ratio increases, more demand is moved from unreliable, low-cost suppliers to reliable, high-cost suppliers to minimize shortage cost. For a low ratio, the risk-neutral supply portfolio does not include the most reliable and most expensive suppliers 1 and 2 to minimize the dominating purchasing costs, whereas for a high ratio, the most reliable suppliers 1 and 2 are selected only to minimize the dominating shortage costs. For the risk-averse decision-making, the supply portfolio for low unit penalty ratios also does not include the most reliable and most expensive suppliers 1 and 2; however for higher ratios, these suppliers are selected along with the less reliable supporting suppliers to better mitigate the impact of shortage risk by portfolio diversification. Except for the most unreliable supplier 4, which is not selected at all for all higher ratios, $g/h = 50, 75, 100$.

Table 3.5 Solution results for different shortage-to-delay unit penalty ratio

Unit penalty ratio, g/h	1	10	25	50	75	100
Model **DSP_E(c)**						
E^c	9.35	10.20	11.55	13.80	15.54	16.33
$E^{sl}\%$	66.57	66.74	67.04	66.40	70.56	71.23
Expected fulfillment rate[a]	90.51	90.57	90.97	91.11	95.08	97.71
LP[b]	9.33	10.19	11.55	13.77	15.53	16.32
Suppliers selected(% of total demand)[c]					1(14)	1(40)
					2(40)	2(60)
	3(23)	3(25)	3(44)	3(51)	3(46)	
	4(42)	4(39)	4(21)	4(18)		
	5(35)	5(36)	5(35)	5(31)		
Model **DSP_CV(c)** with $\alpha = 0.99$						
CVaR^c	11.10	11.13	19.61	32.85	46.17	59.72
VaR^c	11.00	11.00	18.85	31.27	43.76	56.26
E^c	9.36	10.20	13.22	14.38	15.73	17.10
$E^{sl}\%$	66.86	66.86	70.60	70.78	70.72	70.69
Expected fulfillment rate[a]	90.73	90.73	94.85	94.64	94.56	94.52
LP[b]	11.08	11.10	19.40	32.75	46.27	59.60
Suppliers selected(% of total demand)[c]			1(27)	1(27)	1(26)	1(26)
			2(30)	2(26)	2(26)	2(25)
	3(34)	3(33)	3(17)	3(24)	3(24)	3(25)
	4(34)	4(34)	4(9)			
	5(32)	5(33)	5(17)	5(23)	5(24)	5(24)

[a] $\sum_{s\in S}\sum_{i\in I_s}\sum_{j\in J}\sum_{t\in T} P_s d_j v_{ijt}/D \times 100\%$

[b] LP relaxation solution value of the objective function

[c] $1(\sum_{j\in J}\sum_{t\in T} d_j v_{1jt}/D \times 100\%)$, $2(\sum_{j\in J}\sum_{t\in T} d_j v_{2jt}/D \times 100\%)$, $3(\sum_{j\in J}\sum_{t\in T} d_j v_{3jt}/D \times 100\%)$, $4(\sum_{j\in J}\sum_{t\in T} d_j v_{4jt}/D \times 100\%)$,$5(\sum_{j\in J}\sum_{t\in T} d_j v_{5jt}/D \times 100\%)$

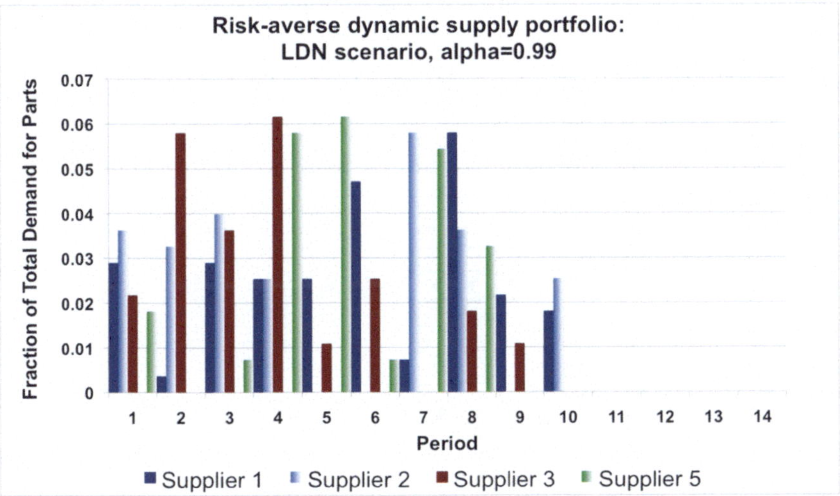

Fig. 3.5 Risk-averse dynamic supply portfolios for model **DSP_CV(c)** and LDN scenario

Table 3.6 Risk-neutral solutions for LDN scenarios

Model **DSP_E(c)**	Model **DSP_E(sl)**
Var.=3505, Bin.=3505, Cons.=3675, Nonz.=18325[a]	
$E^c = 16.64$	$E^{sl} = 97.71\%$[b]
$E^{sl} = 94.99\%$[b]	$E^c = 18.02$
16.62[c]	97.71[c]
Suppliers selected(% of total demand)[d]	
1(11)	1(26)
2(41)	2(74)
3(48)	

[a] Var.: total number of variables; Bin.: number of binary variables;
Cons.: number of constraints; Nonz.: number of nonzero coefficients

[b] $\sum_{s\in S}\sum_{i\in I_s}\sum_{j\in J}\sum_{t\in T} P_s d_j v_{ijt}/D \times 100\%$—expected fulfillment rate

[c] LP relaxation solution value of the objective function

[d] $1(\sum_{j\in J}\sum_{t\in T} d_j v_{1jt}/D \times 100\%)$, $2(\sum_{j\in J}\sum_{t\in T} d_j v_{2jt}/D \times 100\%)$, $3(\sum_{j\in J}\sum_{t\in T} d_j v_{3jt}/D \times 100\%)$

3.6.1 Worst-Case Scenarios with Longest Delays and Disruptions

In this subsection a special subset of delivery scenarios is considered such that delivery of parts from each supplier is either delayed by maximum delay length, $\overline{\tau}$, or parts are not delivered at all, because of supply disruptions. Let us call such binomial delivery scenarios, LDN (Longest Delay-or-Never) scenarios. The total number of LDN scenarios is $2^{\overline{I}} = 32$. Given local disruption probabilities, $p_{i5},\ i \in I$, the probability for the longest delay of delivery is simply $p_{i4} = (1 - p_{i5})$, while the remaining probabilities, $p_{i0}, p_{i1}, p_{i2}, p_{i3}$ are zero for all $i \in I$. Notice that binomial LDN delivery scenarios can be considered as pure all-or-nothing disruption scenarios, where ordered parts are either delivered with a fixed delay, $\overline{\tau}$, or not delivered at all.

For LDN scenarios the fraction of demand fulfilled within delivery due windows is always zero because the longest delay, $\overline{\tau} = 4$, is always greater than each due window length, $\delta_j - \underline{\delta_j} = 2,\ j \in J$. Therefore, in this subsection the service level is measured by demand fulfillment rate, i.e., the fraction of demand fulfilled during the planning horizon.
Equation (3.6) is replaced by

$$E^{sl} = \sum_{s\in S}\sum_{i\in I_s}\sum_{j\in J}\sum_{t\in T} P_s d_j v_{ijt}/D,$$

and risk constraints (3.19) by

$$\mathscr{S}_s \geq VaR^{sl} - \sum_{i\in I_s}\sum_{j\in J}\sum_{t\in T} d_j v_{ijt}/D;\ s \in S.$$

The risk-neutral solutions for LDN scenarios and models **DSP_E(c)** and **DSP_E(sl)** are summarized in Table 3.6. The solution results for model **DSP_E(c)**

Table 3.7 Risk-averse solutions for LDN scenarios

Confidence level α	0.50	0.75	0.90	0.95	0.99
Model **DSP_CV(c)**					
Var.=3538, Bin.=3505, Cons.=3739, Nonz.=186613 [a]					
CVaR^c	18.21	20.88	28.89	36.53	46.01
VaR^c		15.54		30.99	43.73
E^c		16.87		16.95	16.83
Expected fulfillment rate[b]		97.71		94.85	94.56
LP[c]	18.19	20.86	28.87	36.44	45.96
Suppliers selected(% of total demand)[d]		1(43)		1(30)	1(26)
		2(57)		2(29)	2(26)
				3(13)	3(24)
				4(14)	
				5(14)	5(24)
Model **DSP_CV(sl)**					
Var.=3538, Bin.=3505, Cons.=3707, Nonz.=74389[a]					
CVaR%	95.43	90.86	77.14	62.26	46.23
VaR%	100	100	100	71.38	50
Expected fulfillment rate[b]	97.71	97.71	97.71	94.73	94.44
E^c	17.97	18.11	18.00	19.08	18.27
LP[c]	95.43	90.86	77.14	62.27	46.23
Suppliers selected(% of total demand)[d]	1(25)	1(28)	1(26)	1(29)	1(25)
	2(75)	2(72)	2(74)	2(29)	2(25)
				3(14)	3(25)
				4(14)	
				5(14)	5(25)

[a]Var.: total number of variables; Bin.: number of binary variables; Cons.: number of constraints; Nonz.: number of nonzero coefficients

[b] $\sum_{s\in S}\sum_{i\in I_s}\sum_{j\in J}\sum_{t\in T}P_s d_j v_{ijt}/D\times 100\%$

[c] LP relaxation solution value of the objective function

[d] $1(\sum_{j\in J}\sum_{t\in T}d_j v_{1jt}/D\times 100\%)$, $2(\sum_{j\in J}\sum_{t\in T}d_j v_{2jt}/D\times 100\%)$, $3(\sum_{j\in J}\sum_{t\in T}d_j v_{3jt}/D\times 100\%)$, $4(\sum_{j\in J}\sum_{t\in T}d_j v_{4jt}/D\times 100\%)$,$5(\sum_{j\in J}\sum_{t\in T}d_j v_{5jt}/D\times 100\%)$

are similar to the corresponding results for the general delivery scenarios, since demand fulfillment rate is independent of delivery delays. Table 3.6 shows that the expected costs are greater for LDN scenarios (cf. Table 3.3).

The risk-averse solutions for LDN scenarios are summarized in Table 3.7. Table 3.7 demonstrates that for both models **DSP_CV(c)** and **DSP_CV(sl)** and low confidence levels the two most reliable and expensive suppliers 1 and 2 are selected only, while for a greater α the portfolio is more diversified. Suppliers 1 and 2 are supported by the less reliable suppliers 3, 4, and 5 to better mitigate the impact of disruption risk and minimize the worst-case cost or maximize the worst-case fulfillment rate. The results indicate that in the example problems the shortage costs are predominating worst-case costs.

Table 3.8 Probability of cost per part for model **DSP_CV(c)**: LDN scenarios

Cost interval	$\alpha = 0.5, 0.75, 0.9$	$\alpha = 0.95$	$\alpha = 0.99$
[13, 14]	0	0.731265	0.805646
[15, 16]	0.959466	0	0
[22, 23]	0	0.199973	0
[28, 29]	0	0	0.167932
[30, 31)	0	0.037792	0
[31, 32]	0	0.007137	0
[39, 40)	0	0.017474	0
[40, 41]	0.017756	0	0
[43, 44]	0	0	0.025001
[48, 49)	0	0.004721	0
[49, 50]	0.017455	0	0
[56, 57]	0	0.001481	0
[58, 59]	0	0	0.001335
[65, 66]	0	0.000100	0
[73, 74]	0.005323	0	0

Table 3.9 Probability of fulfillment rate for model **DSP_CV(sl)**: LDN scenarios

Fulfillment rate interval %	$\alpha = 0.5, 0.75, 0.9$	$\alpha = 0.95$	$\alpha = 0.99$
[14, 15]	0	0.000101	0
[24, 25)	0	0	0.001335
[25, 26)	0.017455	0	0
[26, 27]	0	0.017455	0
[28, 29]	0	0.001481	0
[42, 43)	0	0.004459	0
[43, 44]	0	0.000262	0
[49, 50]	0	0	0.025001
[56, 57)	0	0.002034	0
[57, 58]	0	0.015440	0
[71, 72]	0	0.044929	0
[73, 74)	0	0.017756	0
[74, 75)	017756	0	0
[75, 76]	0	0	0.167932
[85, 86]	0	0.199973	0
[99, 100]	0.959466	0.731265	0.805646

Table 3.7 also shows that solution values for model **DSP_CV(c)** and binomial LDN delivery scenarios are greater than the corresponding values for the general multinomial delivery scenario, while the corresponding supply portfolios are similar (cf. Table 3.4). An example of optimal risk-averse dynamic supply portfolio for model **DSP_CV(c)** and the highest confidence level $\alpha = 0.99$ is presented in Fig. 3.5. Despite the nearly equal allocation of total demand for parts among four suppliers (cf. Table 3.7), the order allocation over the planning horizon is unbalanced to better meet the varying customer demand pattern (cf. Fig. 3.4).

For optimal risk-averse supply portfolios and LDN scenarios, Tables 3.8 and 3.9 present discrete distributions of cost per part and demand fulfillment rate, respec-

Table 3.10 Managerial insights: dynamic supply portfolio

Decision-making issue	Decision-making impact
Cost- vs. service-oriented decision-making	Both cost- and service-oriented dynamic supply portfolios that simultaneously mitigate the impact of low-probability disruption risk and high-probability delay risk lead to a high expected demand fulfillment rate
Decision-making with disregarded potential delay risks	The decision-making may lead to greater supply fluctuations and manufacturing delays, which is particularly harmful in a make-to-order and just-in-time environment. When the delay risk is disregarded the most reliable suppliers are not selected or are allotted less demand and the dynamic supply portfolio is spread over more periods
Risk-neutral decision-making: shortage-to-delay penalty ratio	In order to minimize expected shortage cost, as shortage-to-delay unit penalty ratio increases, more demand is moved from unreliable, low-cost suppliers to reliable, high-cost suppliers
Risk-averse decision-making: shortage-to-delay penalty ratio	The dynamic supply portfolio for a low shortage-to-delay unit penalty ratio does not include most reliable and expensive suppliers, whereas for the higher ratios such suppliers are selected along with the less reliable supporting suppliers to better mitigate the impact of shortage risk by portfolio diversification
Risk-averse decision-making: binomial vs. multinomial delivery scenarios	For binomial (LDN) scenarios the probability measure for optimal risk-averse dynamic supply portfolio is concentrated in a fewer number of points and contains larger probability atoms at higher costs

tively. The tables show corresponding probabilities concentrated at each level of cost and fulfillment rate. The probabilities are concentrated at a few points only, which is typical for the scenario-based optimization under uncertainty, where the probability measure is concentrated in finitely many points.

The cost distributions for optimal risk-averse supply portfolio and the highest confidence level $\alpha = 0.99$ for worst-case LDN binomial disruption scenarios and general multinomial delivery scenarios are compared in Fig. 3.6. For the worst-case LDN binomial scenarios the probability measure is concentrated in a smaller number of points than for the general multinomial scenarios and contains larger probability atoms at higher costs. This indicates that the impact of LDN binomial scenarios is more difficult to mitigate than that of general multinomial delay and disruption scenarios.

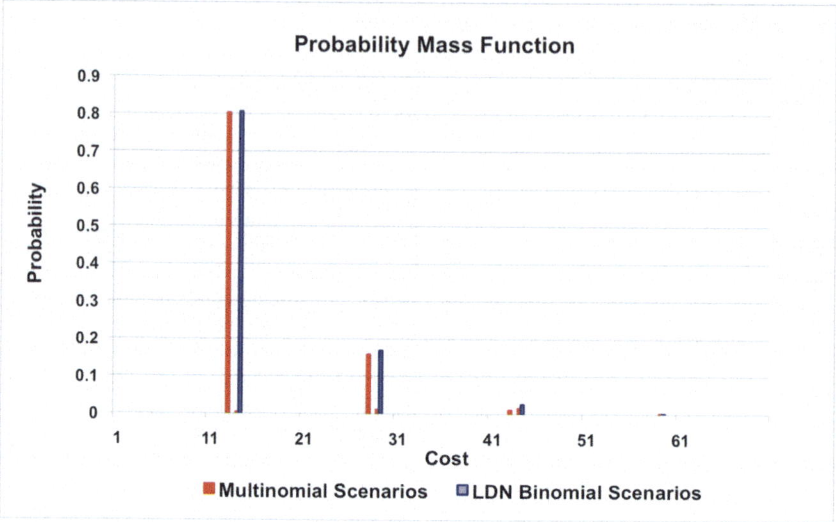

Fig. 3.6 Probability mass function for model **DSP_CV(c)** and $\alpha = 0.99$: multinomial vs. binomial scenarios

The major managerial insights are summarized in Table 3.10. Overall, the computational results demonstrate that: *the portfolio approach leads to a time-indexed SMIP formulation with a strong LP relaxation, which has proven to be computationally very efficient.*

The computational experiments were performed using the AMPL programming language and the Gurobi 7.5.0 solver on a MacBookPro laptop with Intel Core i7 processor running at 2.8 GHz and with 16 GB RAM. The solver was capable of finding proven optimal solution for all examples with CPU time ranging from fraction of a second for risk-neutral solutions to several seconds for risk-averse solutions.

3.7 Notes

The supplier selection problem is a dynamic decision-making problem. For example, the customer demand, supplier capacity, lead time, price, discounts, and quality of supplying parts may vary over time. The delay risks frequently arise from problems in coordinating supply and demand (e.g., Oke and Gopalakrishnan 2009; Sawik 1977, 2011c). When all suppliers are similarly exposed to high-impact disruption risks, then the low-impact delay risk may predominate the supplier selection decision. The integrated scheduling of orders over time with the supplier selection may significantly reduce cost for a multi-period planning horizon. However, the problem of a multi-period supplier selection and order quantity allocation in the presence of supply chain delay and disruption risks is very rarely reported in the literature. The reason for that can be due to a more complex decision-making that requires to simultaneously consider the high-probability and low-impact supply delays and the low-probability and high-impact supply disruptions.

The majority of models developed for supplier selection and order quantity allocation are static (single-period) models (e.g., Weber and Current 1993; Demirtas and Ustun 2008) that do not consider inventory management. The dynamic (multi-period) models are capable of considering the inventory management by lot-sizing and scheduling of orders (e.g., Ghodsypour and O'Brien 2001; Basnet and Leung 2005; Ustun and Demirtas 2008; Che and Wang 2008). For example, Sen et al. (2014) developed a scenario-based multi-stage stochastic optimization model for multi-period supplier selection under price uncertainty that allows to consider random events such as a drop in price, a price change in the spot market, or a new discount offer. The model was used to purchase a large volume of multiple items over multiple periods from multiple suppliers that provide base prices and discounts. For custom-engineered products, however, no inventory of custom parts can be kept on hand. Instead, the custom parts often need to be requisitioned with each customer order and hence the custom parts inventory need not to be considered. In a parallel research area of energy supply chains, a scenario-based CVaR optimization of electricity supply portfolio under price uncertainty in the spot market was recently investigated by Canelas et al. (2020).

Unlike disruptions, delays are referred to relatively periodic events, originating directly from internal activities of companies and/or relationships within partners in the supply chain. In particular, the high-probability supply delays may have a significant impact in a make-to-order manufacturing and just-in-time environment. A delay in the supply of raw materials, parts, or components can lead to longer lead times, rescheduling decisions, congestion at the shop floor, decrease of service level, and consequently to lower profits. Moreover, if any delays occur, waste may increase due to shutdown of production lines. A successful just-in-time implementation requires a multi-period supplier selection and supplier performance evaluation tools, e.g., Aksoy and Ozturk (2011).

Quang and Hara (2018) define delays in supply chain activities as time risk that may detrimentally affect supply chain performance. The authors emphasize that the delay risk should be considered in terms of time series, which is useful in forecasting how many delays and of what length are incurred and likely to occur, and quantifying them within a specific period of time (e.g., transportation delays due to seasonal weather conditions). Delays in materials and equipment procurement are often a contributory cause of cost overruns in construction projects, e.g., Manavazhi and Adhikari (2002). In contrast to project management, where a joint impact of disruption and delay is often studied (e.g., Eden et al. 2000), in the literature on supplier selection the two types of risks are usually considered separately.

The idea of a dynamic supply portfolio approach in the presence of both the low-probability and high-impact supply disruptions and the high-probability and low-impact supply delays was presented by Sawik (2011d) for a multi-period supplier selection and order quantity allocation in a make-to-order environment. For the selection of a dynamic supply portfolio, a SMIP was proposed to incorporate risks via scenario analysis. In the scenario analysis, both types of the supply chain risks are simultaneously considered. The disruption and delay risks were incorporated utilizing the concepts of percentile measures of risk, VaR and CVaR. In Sawik (2011d), the cost-based objective function was considered only and general delivery scenarios with on-time, delayed, or no supplies. The delays were modeled as statistically independent, discrete multivariate random variables. In addition to the general scenarios, the two special subsets of the scenarios were considered, with at most one disruption of each supplier over the horizon and with multiple consecutive disruptions of each supplier over the horizon since its first disruption. The best-case and worst-case scenarios, respectively, with no delays or with the longest delays were analyzed. Rabbani et al. (2014) proposed a combination of two models by Sawik (2010a, 2011d) to consider a multi-period supplier selection and order allocation problem. The combined model finds the optimal dynamic supply portfolio, and also accounts for the quantity and business volume discounts offered by suppliers. A dynamic supplier selection problem was also considered by Ware et al. (2014), who developed a mixed-integer nonlinear programming model that incorporates risk associated with fluctuating demand, supply disruption, quality failure probability, and delivery delay. In a related study, Tsai (2016) presented a formulation that helps to mitigate supply failure by considering a dynamic model-based optimization approach, in which the risk of supply interruption is time-dependent. The proposed model supports decision-making on how many standby suppliers are ideal in the presence of supply chain disruption risk.

The material presented in this chapter has been updated using results presented in Sawik (2018a), where both cost- and service-oriented dynamic supply portfolios were considered to simultaneously mitigate the impact of low-probability disruption risk and high-probability delay risk.

Problems

3.1. Modify the probability for delivery scenarios (3.2) to account for regional disruptions that may affect suppliers during a fixed time interval.

3.2. Modify the stochastic IP models presented in this chapter for multiple part types with subsets of part types required for each customer order and subsets of suppliers available for each part type.

3.3. Modify the stochastic IP models presented in this chapter to replace a single delivery by multiple deliveries at a fixed delivery date or delivery time window.

3.4. Mixed mean-risk dynamic supply portfolio

(a) Modify model **DSP_ECV(c)** to optimize expected cost and CVaR of service level and model **DSP_ECV(sl)** to optimize expected service level and CVaR of cost.

(b) How should the values of the optimized objective functions be scaled into the interval [0,1] to avoid dimensional inconsistency among the two objectives and how should the trade-off parameter be selected?
(c) How would you interpret the mixed mean-risk dynamic supply portfolio?

3.5. In the computational examples in Sect. 3.6, Table 3.4 indicates that the service-optimal risk-averse dynamic supply portfolio for a high confidence level is perfectly balanced. Why do the optimal risk-averse service-based dynamic supply portfolios have such property for the example problem?

Chapter 4
Selection of Resilient Supply Portfolio

4.1 Introduction

In this chapter, the portfolio approach and SMIP models presented in Chap. 2 are enhanced for the combined selection and protection of part suppliers and order quantity allocation in a supply chain with disruption risks. The protection decisions include the selection of suppliers to be protected against disruptions and the allocation of risk mitigation inventory of parts to be pre-positioned at the protected suppliers so as to maintain uninterrupted supplies in case of natural or man-made disruptive events. The decision maker needs to decide which supplier to select for parts delivery and how to allocate orders quantity among the selected suppliers, and which of the selected suppliers to protect against disruptions and how to allocate risk mitigation inventory among the protected suppliers. The problem objective is to achieve a minimum cost of suppliers protection, risk mitigation inventory pre-positioning, parts ordering, purchasing, transportation and shortage and to mitigate the impact of disruption risks by minimizing the potential worst-case cost. The resulting supply portfolio can be called resilient, with the supplier's resiliency defined as its capability of supplying parts in the face of disruptive events. The portfolio includes protected suppliers that are capable of fully or partially supplying parts in the face of disruptive events as well as the risk mitigation inventory pre-positioned at the protected suppliers. Depending on the level of supplier protection, the capacity of each protected supplier remains fully or partially available under a disruptive event. The risk mitigation inventory is used to compensate for the loss of capacity of suppliers hit by disruptions, unprotected or insufficiently protected, and to partially or fully replace non-delivered parts ordered from the disrupted suppliers. For the selection of risk-neutral, risk-averse or mean-risk supply portfolio, SMIP formulations are developed with single- or multi-level protection of fortified suppliers. In the former case, full remaining capacity of a disrupted supplier is maintained by its fortification (protection against disruptions), whereas in the latter case the remaining fraction of

T. Sawik, *Supply Chain Disruption Management*,
International Series in Operations Research & Management Science 291,
https://doi.org/10.1007/978-3-030-44814-1_4

full capacity depends on the protection level applied. A simple protection index is introduced to evaluate the trade-off between the cost of suppliers protection and the estimated losses caused by supply disruptions, if no protective countermeasure is applied.

The following SMIP models are presented in this chapter:

ReSP_E for risk-neutral selection of resilient supply portfolio to minimize expected cost;
ReSP_CV for risk-averse selection of resilient supply portfolio to minimize CVaR of cost;
ReSP_ECV for mean-risk selection of resilient supply portfolio to optimize trade-off between expected cost and CVaR of cost.
ReSP(mlp)_E model **ReSP_E** for multi-level protection;
ReSP(mlp)_CV model **ReSP_CV** for multi-level protection;
ReSP(mlp)_ECV model **ReSP_ECV** for multi-level protection.

The resilient supply portfolio is selected ahead of time to optimize performance of a supply chain under disruption risks: average performance (risk-neutral models, **ReSP_E** and **ReSP(mlp)_E**), worst-case performance (risk-averse models, **ReSP_CV** and **ReSP(mlp)_CV**) or combined, average—worst-case performance (mean-risk models **ReSP_ECV** and **ReSP(mlp)_ECV**).

In order to illustrate the proposed SMIP approach, computational examples are presented in Sect. 4.6, and major managerial insights are provided at the end of this chapter.

4.2 Problem Description: Single-Level Protection

In the supply chain under consideration various types of products are assembled by a single producer to satisfy customer orders, using different part types purchased from multiple suppliers (for notation used, see Table 4.1). The suppliers have different limited capacity and, in addition, differ in price and quality of offered parts. Let $I = \{1,\dots\bar{I}\}$ be the set of $\bar{I}$ suppliers and $J = \{1,\dots\bar{J}\}$ the set of $\bar{J}$ part types required for the products. Denote by d_j the demand for each part type $j \in J$ and assume that d_j is known ahead of time.

The usage per part of supplier's capacity is assumed to be different for different part types and suppliers. Let c_i be the total capacity of supplier $i \in I$ (e.g., the total number of available machine-hours) and a_{ij} the unit capacity consumption of supplier i for part type j, i.e., the amount of capacity of supplier i used to manufacture one part type j (e.g., the unit processing time).

Denote by o_{ij} the unit purchasing price, including shipping cost of part type $j \in J$ from supplier $i \in I$ (assume that for suppliers incapable of providing some part types, the corresponding unit price and unit capacity consumption are very large numbers).

Table 4.1 Notation: resilient supply portfolio, single-level protection

	Indices
i	$=$ supplier, $i \in I$
j	$=$ part type, $j \in J$
s	$=$ disruption scenario, $s \in S$
	Input Parameters
a_{ij}	$=$ per unit capacity consumption of supplier i for part type j
c_i	$=$ capacity of supplier i
d_j	$=$ demand for part type j
D	$= \sum_{j \in J} d_j$—total demand for parts
e_i	$=$ cost of ordering parts from supplier i
f_i	$=$ protection cost for supplier i
g_j	$=$ per unit shortage cost of part type j
h_{ij}	$=$ per unit cost of pre-positioning risk mitigation inventory of parts type j at supplier i
o_{ij}	$=$ unit price of part type j purchased and shipped from supplier i
p_i	$=$ local disruption probability for supplier i
p^*	$=$ global disruption probability for all suppliers
α	$=$ confidence level
v	$=$ minimum order size
ρ_{ij}	$=$ expected defect rate of supplier i for part type j

Let ρ_{ij} be the expected defect rate (reject rate) of supplier i for part type j. The rate is based on past observations. The fixed cost of ordering parts from supplier $i \in I$ is denoted by e_i.

The supplies of parts are subject to independent random local disruptions that are uniquely associated with a particular supplier, which may arise from equipment breakdowns, local labor strike, bankruptcy, terrorist attack, from local natural disasters such as earthquakes, fires, floods, hurricanes, etc. Denote by p_i the local disruption probability for supplier i, i.e., the parts ordered from supplier i are delivered without disruptions with probability $(1-p_i)$, or not at all with probability p_i.

In addition to independent local disruptions of each supplier, there are potential global disasters that may result in all suppliers disruption simultaneously. For example, such global super events may include economic crisis, widespread labor strike in a transportation sector, etc. Although the probability of such disaster events usually is very low, its consequences may be very high. Denote by p^* the probability of simultaneous correlated global disruption of all suppliers due to some disaster super event.

Let P_s be the probability that disruption scenario s is realized, where each scenario $s \in S$ is comprised of a unique subset $I_s \subset I$ of suppliers who deliver parts without disruptions, and $S = \{1, \dots, \overline{S}\}$ is the index set of all scenarios. There are a total of $\overline{S} = 2^{\overline{I}}$ potential disruption scenarios.

The global disaster and the local disruptive events at each supplier are assumed to be independent events, therefore the probability P_s of each disruption scenario $s \in S$ under the risks of both type of events is

$$P_s = \begin{cases} (1-p^*)\hat{P}_s & \text{if } I_s \neq \emptyset \\ p^* + (1-p^*)\prod_{i \in I} p_i & \text{if } I_s = \emptyset, \end{cases}$$

where $\hat{P}_s$ is the probability of disruption scenario s in the presence of independent local disruptive events only

$$\hat{P}_s = \prod_{i \in I_s}(1 - p_i) \cdot \prod_{i \notin I_s} p_i.$$

If the probability of global disruption $p^* = 0$, then the probability P_s reduces to $\hat{P}_s$ for independent local disruptive events.

The producer does not need to pay for ordered and undelivered parts. However, the producer can be charged with a much higher penalty cost of unfulfilled customer orders for products, caused by the shortage of parts, undelivered due to supply disruptions. Let g_j be the per unit cost of shortage of part type j.

In order to mitigate the impact of disruption risks, managers may consider protective countermeasures to fully fortify suppliers against disruptions and to maintain the suppliers normal capability. The capacity of a protected supplier is assumed to remain unchanged under a disruptive event. Therefore, for a protected supplier i its disruption probability p_i is in practice changed to zero.

The protective countermeasure of each supplier may be combined with pre-positioning of risk mitigation inventory of parts manufactured by the protected supplier. The risk mitigation inventory is used to compensate for the loss of capacity of the other suppliers, unprotected and hit by disruptions, and to fulfill non-delivered orders placed on the disrupted suppliers. The risk mitigation inventory pre-positioned at a protected supplier is limited, e.g., by available budget, by available storage space, or by its base capacity, if the inventory is an overtime production. The inventory is replenished once it is used. Assume that the inventory quantity is linked to capacity and cannot be greater than the protected supplier normal capacity. As a result each protected supplier is capable of supplying twice as many parts as its base capacity.

Let f_i be the cost required for full protection of supplier i against disruptions, and denote by h_{ij} the per unit cost of pre-positioning the risk mitigation inventory of part type j at the protected supplier i.

The decision maker needs to decide which supplier to select for purchasing the required parts, which of the selected suppliers to protect against disruptive events, and how to allocate order quantity among the selected suppliers and risk mitigation inventory among the protected suppliers to achieve a minimum total cost of suppliers protection, risk mitigation inventory pre-positioning, parts ordering, purchasing, transportation, defects and shortage and to mitigate the impact of disruption risks by minimizing the potential worst-case cost.

Table 4.2 Problem variables: resilient supply portfolio, single-level protection

	First stage variables
q_i	$=$ 1, if selected supplier i is protected against disruptions; otherwise $q_i = 0$ (supplier protection)
u_i	$=$ 1, if an order for parts is placed on supplier i; otherwise $u_i = 0$ (supplier selection)
v_{ij}	$=$ the fraction of demand for parts type j ordered from unprotected supplier i (unprotected allocation of demand for part types)
w_{ij}	$=$ the fraction of demand for parts type j ordered from protected supplier i (protected allocation of demand for part types)
x_{ij}	$=$ risk mitigation inventory of parts type j pre-positioned at protected supplier i, in fraction of supplier's capacity c_i (pre-positioning of risk mitigation inventory)
	Second stage variables
y_{ijs}	$=$ risk mitigation inventory of parts type j pre-positioned at protected supplier i, used under disruption scenario s (usage of risk mitigation inventory)
z_{js}	$=$ inventory shortage of parts type j under disruption scenario s
	Auxiliary variables
VaR^c	cost-at-risk, the targeted cost such that for a given confidence level α, for $100\alpha\%$ of the scenarios, the outcome is below VaR^c
$\mathscr{C}_s$	≥ 0, the tail cost for scenario s, i.e., the amount by which costs in scenario s exceed VaR^c

4.3 Resilient Supply Portfolio with Single-Level Protection

In this section three SMIP models are formulated for supplier selection and protection, and order quantity allocation problem, i.e., for determining a resilient supply portfolio. For definition of problem variables, see Table 4.2.

The resilient supply portfolio is selected ahead of time in such a way as to minimize the potential average or worst-case cost (or a combination of both) under different disruption scenarios. The capacity of protected suppliers and the risk mitigation inventory pre-positioned at the protected suppliers are used to reduce the potential highest costs.

When deciding on a supply portfolio it is assumed that the orders for all parts are simultaneously placed on selected suppliers, and each protected and each non-disrupted unprotected supplier delivers all the ordered parts. However, unprotected suppliers hit by disruptions fail to deliver the ordered parts, and then the non-delivered orders are partially or fully replaced by the risk mitigation inventory of parts pre-positioned at the protected suppliers. The decision on the protection of selected suppliers and the pro-positioning of risk mitigation inventory at the protected suppliers is a part of the resilient supply portfolio decision-making.

4.3.1 Risk-Neutral Model

In this subsection a SMIP model **ReSP_E** is presented for selection of risk-neutral resilient supply portfolio. In the risk-neutral decision-making the overall quality of the supply portfolio can be measured by the total expected cost per part, E^c (4.1). The total expected cost includes: cost of suppliers protection, $\sum_{i\in I} f_i q_i/D$, cost of pre-positioning risk mitigation inventory, $\sum_{i\in I}\sum_{j\in J} h_{ij}c_i x_{ij}/a_{ij}D$, cost of parts ordering, $\sum_{i\in I} e_i u_i/D$, cost of parts purchasing and shipping, including defects, plus cost of replacement delivery of parts pre-positioned, in which the cost of a defective part is assumed to be identical with its price, $\sum_{i\in I}\sum_{j\in J} o_{ij}(d_j v_{ij}/(1-\rho_{ij}) + d_j w_{ij}/(1-\rho_{ij}) + \sum_{s\in S} P_s c_i y_{ijs}/a_{ij})/D$, and finally cost of shortage of parts due to supply disruptions less cost of non delivered parts, $\sum_{s\in S} P_s(\sum_{j\in J} g_j z_{js} - \sum_{i\notin I_s}\sum_{j\in J} o_{ij} d_j v_{ij}/(1-\rho_{ij}))/D$.

The purchase orders for parts are assumed to be inflated by the reject rates ρ_{ij} of defective parts, i.e., are equal to $d_j v_{ij}/(1-\rho_{ij})$ and $d_j w_{ij}/(1-\rho_{ij})$, for all $i \in I, j \in J$, respectively, for unprotected and protected suppliers.

Notice that if an unprotected supplier is subject to a disruptive event and fails to deliver the ordered parts, the non-delivered parts can be replaced by the risk mitigation inventory pre-positioned at protected suppliers. Then, the amount $d_j v_{ij}/(1-\rho_{ij})$ of part type j ordered from unprotected and disabled supplier $i \notin I_s$ can be fully or partially met with the risk mitigation inventory $\sum_{i\in I} c_i x_{ij}/a_{ij}$. The resilient supply portfolio aims at reducing potential losses from non-delivered parts through the optimal selection of suppliers for protection, the allocation among them the risk mitigation inventory and the allocation of orders for parts among both unprotected and protected suppliers.

The SMIP model **ReSP_E** for selection of risk-neutral resilient supply portfolio is formulated below. In the proposed model, the portfolio will be optimized by minimizing expected cost per part E^c, (4.1).

ReSP_E: *Selection of risk-neutral* **Res**ilient **S**upply **P**ortfolio *to minimize* **E**xpected *cost*

Minimize Expected Cost per Part

$$E^c = \sum_{i\in I} e_i u_i/D + \sum_{i\in I} f_i q_i/D + \sum_{i\in I}\sum_{j\in J} h_{ij}c_i x_{ij}/a_{ij}D$$
$$+\sum_{i\in I}\sum_{j\in J} o_{ij}d_j(v_{ij}+w_{ij})/(1-\rho_{ij})D$$
$$+\sum_{s\in S} P_s(\sum_{j\in J} g_j z_{js} + \sum_{i\in I}\sum_{j\in J} o_{ij}c_i y_{ijs}/a_{ij} - \sum_{i\notin I_s}\sum_{j\in J} o_{ij}d_j v_{ij}/(1-\rho_{ij}))/D, \tag{4.1}$$

subject to

1. *Supplier selection and protection constraints*:
   - only selected suppliers can be protected, and risk mitigation inventory can be pre-positioned only at protected suppliers (4.2),
   - the orders for parts can be allocated among selected suppliers only (4.3),
   - each selected supplier is either protected or unprotected and so are the corresponding types of orders for parts allocated among the suppliers (4.4), (4.5),

(continued)

$$\sum_{j\in J} x_{ij} \le q_i \le u_i;\ i \in I, \tag{4.2}$$

$$v_{ij} \le u_i;\ i \in I, j \in J, \tag{4.3}$$

$$v_{ij} \le 1 - q_i;\ i \in I, j \in J, \tag{4.4}$$

$$w_{ij} \le q_i;\ i \in I, j \in J. \tag{4.5}$$

2. *Order quantity allocation constraints*:
   - orders for all required parts of each type must be allocated among selected suppliers (4.6),
   - for each selected supplier (unprotected and protected), the total capacity required to manufacture the ordered quantities of parts cannot exceed available capacity (4.7), (4.8),
   - the ordered quantity assigned to each selected supplier (unprotected or protected) cannot be less than the minimum order size (4.9), (4.10),

$$\sum_{i\in I} (v_{ij} + w_{ij}) = 1;\ j \in J, \tag{4.6}$$

$$\sum_{j\in J} a_{ij} d_j v_{ij}/(1-\rho_{ij}) \le c_i(u_i - q_i);\ i \in I, \tag{4.7}$$

$$\sum_{j\in J} a_{ij} d_j w_{ij}/(1-\rho_{ij}) \le c_i q_i;\ i \in I, \tag{4.8}$$

$$\sum_{j\in J} d_j v_{ij}/(1-\rho_{ij}) \ge v(u_i - q_i);\ i \in I, \tag{4.9}$$

$$\sum_{j\in J} (d_j w_{ij}/(1-\rho_{ij}) + c_i y_{ijs}/a_{ij}) \ge v q_i;\ i \in I, s \in S. \tag{4.10}$$

3. *Risk mitigation inventory pre-positioning and usage constraints*:
   - the risk mitigation inventory used to replace non-delivered units cannot exceed the pre-positioned risk mitigation inventory (4.11),
   - for each part type and disruption scenario, non-delivered orders can (no inventory shortage, $z_{js} = 0$) or cannot (inventory shortage, $z_{js} > 0$) be fully replaced by using the risk mitigation inventory (4.12),

$$y_{ijs} \le x_{ij};\ i \in I, j \in J, s \in S, \tag{4.11}$$

$$z_{js} = \sum_{i\notin I_s} d_j v_{ij}/(1-\rho_{ij}) - \sum_{i\in I} c_i y_{ijs}/a_{ij};\ j \in J, s \in S. \tag{4.12}$$

4. *Non-negativity and integrality conditions*

$$q_i \in \{0,1\};\ i \in I, \tag{4.13}$$

$$u_i \in \{0,1\};\ i \in I, \tag{4.14}$$

(continued)

$$v_{ij} \in [0,1];\ i \in I, j \in J, \tag{4.15}$$
$$w_{ij} \in [0,1];\ i \in I, j \in J, \tag{4.16}$$
$$x_{ij} \in [0,1];\ i \in I, j \in J, \tag{4.17}$$
$$y_{ijs} \in [0,1];\ i \in I, j \in J, s \in S, \tag{4.18}$$
$$z_{js} \geq 0;\ j \in J, s \in S. \tag{4.19}$$

Denote by

$$(V_1, \ldots, V_{\bar{I}}),$$

the supply portfolio, where $\sum_{i \in I} V_i = 1$ and $0 \leq V_i \leq 1$ is the fraction of the total demand for parts ordered from supplier i. The resilient supply portfolio is based on demand allocation, does not account for reject rates, and is determined by the order quantity allocation variables v_{ij}, w_{ij}

$$V_i = \sum_{j \in J} d_j(v_{ij} + w_{ij})/D;\ i \in I. \tag{4.20}$$

Alternatively, to account for the actual supplies of parts and the reject rates, the supply portfolio $(V_1', \ldots, V_{\bar{I}}')$ can be calculated as below.

$$V_i' = \frac{\sum_{j \in J} d_j(v_{ij} + w_{ij})/(1 - \rho_{ij})}{\sum_{i \in I} \sum_{j \in J} d_j(v_{ij} + w_{ij})/(1 - \rho_{ij})};\ i \in I. \tag{4.21}$$

Note that the nonnegative inventory shortage variables z_{js} can be eliminated from the model using Eqs. (4.12). Then, variables z_{js} in the objective function (4.1) should be replaced by

$$\sum_{i \notin I_s} d_j v_{ij}/(1 - \rho_{ij}) - \sum_{i \in I} c_i y_{ijs}/a_{ij},$$

and equality constraints (4.12), replaced by inequality constraints

$$\sum_{i \in I} c_i y_{ijs}/a_{ij} \leq \sum_{i \notin I_s} d_j v_{ij}/(1 - \rho_{ij});\ j \in J, s \in S.$$

The inventory shortage variables z_{js} are slack variables for the last constraints.

Similarly, slack variables for inequality constraints (4.11) represent surplus of the pre-positioned inventory $(x_{ij} - y_{ijs})$ of each part type j at each supplier i, under each disruption scenario $s \in S$. Given optimal supply portfolio, the actual usage of risk mitigation inventory depends on the realized disruption scenario and the corresponding needs for a replacement of non-delivered orders from disrupted suppliers. Depending on the realized disruption scenario, the pre-positioned inventory can be used fully, partially, or not at all.

In the proposed model the required parts of each type are assumed to be partially provided by one or more suppliers and the order allocation variables v_{ij} or w_{ij}

represent the fraction of all required parts of type j provided, respectively, by unprotected or protected supplier i. In some practical cases all parts of the same type are purchased from a single supplier. Then, the corresponding continuous allocation variables v_{ij} and w_{ij} should be redefined as binary assignment variables denoting whether or not all parts of type j are provided, respectively, by unprotected or protected supplier i.

4.3.2 Risk-Averse Model

In the selection of a resilient supply portfolio under disruption risks, the decision maker controls the risk of high losses due to supply disruptions by choosing the confidence level α. We assume that the decision maker is willing to accept only portfolios for which the total probability of scenarios with costs greater than VaR^c is not greater than $1-\alpha$. The greater the confidence level α, the more risk aversive is the decision maker and the smaller percent of the highest cost outcomes is focused on. Moreover, a risk aversive decision maker wants to minimize the expected worst-case costs exceeding VaR^c, by minimizing CVaR^c. When using CVaR^c to minimize worst-case costs, CVaR^c is always not less than VaR^c.

Define $\mathscr{C}_s$ as the tail cost for scenario s, where tail cost is defined as the nonnegative amount by which costs in scenario s exceed VaR^c. The portfolio will be optimized by calculating VaR^c and minimizing CVaR^c simultaneously. By measuring CVaR^c, the magnitude of the tail costs is considered to achieve a more accurate estimate of the risks of minimizing cost. In the proposed model, CVaR^c is represented by an auxiliary function (4.22) introduced by Rockafellar and Uryasev (2000). The SMIP model **ReSP_CV** for selection of risk-averse resilient supply portfolio to reduce the risk of high costs is formulated below.

ReSP_CV: *Selection of risk-averse* **Re***silient* **S***upply* **P***ortfolio to minimize* **CV***aR of cost*

Minimize

$$CVaR^c = VaR^c + (1-\alpha)^{-1}\sum_{s\in S} P_s\mathscr{C}_s, \qquad (4.22)$$

subject to

1. *Supplier selection and protection constraints*: (4.2)–(4.5)
2. *Order quantity allocation constraints*: (4.6)–(4.10)
3. *Risk mitigation inventory pre-positioning and usage constraints*: (4.11)–(4.12)
4. *Risk constraints*:
   - the tail cost for scenario s is defined as the nonnegative amount by which cost per part in scenario s exceeds VaR^c,

(continued)

$$\begin{aligned}\mathscr{C}_s \geq \sum_{i\in I} e_i u_i/D + \sum_{i\in I} f_i q_i/D + \sum_{i\in I}\sum_{j\in J} h_{ij} c_i x_{ij}/a_{ij} D \\ + \sum_{i\in I}\sum_{j\in J} o_{ij} d_j (v_{ij} + w_{ij})/(1-\rho_{ij}) D \\ +(\sum_{j\in J} g_j z_{js} + \sum_{i\in I}\sum_{j\in J} o_{ij} c_i y_{ijs}/a_{ij} - \sum_{i\notin I_s}\sum_{j\in J} o_{ij} d_j v_{ij}/(1-\rho_{ij}))/D \\ -VaR^c;\ s\in S. \qquad (4.23)\end{aligned}$$

4. *Non-negativity and integrality conditions*: (4.13)–(4.19) and

$$\mathscr{C}_s \geq 0;\ s\in S. \qquad (4.24)$$

Note that, if for some part type j all required parts must be supplied by a single supplier, then the corresponding nonnegative allocation variables v_{ij}, w_{ij}, $i\in I$ should be redefined to be binary assignment variables, similarly as for **ReSP_E** model.

4.3.3 Mean-Risk Model

In the single objective approach the resilient supply portfolio is selected by minimizing either the expected cost per part, E^c, (4.1) or the expected worst-case cost per part, $CVaR^c$, (4.22). In this subsection the two cost functions are considered simultaneously, and a bi-objective selection of resilient supply portfolio is presented aimed at minimizing both objective functions to balance expected costs with the risk tolerance. This type of trade-off model is known as the mean-risk model (e.g., Ogryczak and Ruszczynski 2002), formulated as the optimization of a composite objective consisting of the expected cost and the conditional-cost-at-risk as a risk measure.

The nondominated solution set of the bi-objective resilient supply portfolio can be found by the parameterization on λ the weighted-sum program **ReSP_ECV** presented below. The scalarizing SMIP model is based on **ReSP_CV** model with the addition of objective (4.1) of model **ReSP_E**.

ReSP_ECV: *Selection of mean-risk* **Re***silient* **S***upply* **P***ortfolio to minimize weighted sum of* **E***xpected cost and* **CV***aR of cost*

Minimize

$$\lambda E^c + (1-\lambda) CVaR^c, \qquad (4.25)$$

where $0 \leq \lambda \leq 1$,

subject to (4.1)–(4.19), (4.22)–(4.24).

4.4 Protection Index

There are many alternative countermeasures that decision maker must consider to manage the risk of supply disruptions. For each protective countermeasure being considered, it is important to determine the cost of the countermeasure and the estimated loss to be incurred if a supply disruption occurs. For example, if a protective wall is built to protect a supplier against flooding, the decision maker must consider the construction cost against the estimated loss caused by potential flooding that may disrupt supplies of materials, if no protective wall is build. The trade-off between the cost of suppliers protection against potential disruptions and the losses caused by supply disruptions can be evaluated, for example, by the following unit protection cost:

$$\varphi = \frac{\sum_{i\in I} f_i}{\sum_{j\in J} g_j d_j}. \qquad (4.26)$$

The unit protection cost is the ratio of total protection cost to total loss resulted from the shortage of parts caused by the potential supply disruptions. Given demand for parts, the unit protection cost, φ, increases with suppliers protection costs, $f_i, i \in I$, and decreases with parts shortage costs, $g_j, j \in J$.

To ensure delivery of all required parts under disruption risks it is not necessary to protect all suppliers against disruptions and only a subset of suppliers needs to be protected. As a result the actual value of unit protection cost can be less than (4.26), and its optimal value φ^o can be found as a solution to the following mixed integer program:

$$\varphi^o = \min\{\frac{\sum_{i\in I} f_i q_i}{\sum_{j\in J} g_j d_j} : \sum_{j\in J} a_{ij} d_j w_{ij}/(1-\rho_{ij}) \le 2c_i q_i; i \in I,$$
$$\sum_{i\in I} w_{ij} = 1; j \in J,\ w_{ij} \le q_i; i \in I, j \in J,$$
$$q_i \in \{0,1\}; i \in I,\ w_{ij} \in [0,1]; i \in I, j \in J\}. \qquad (4.27)$$

The optimal value φ^o is calculated under global disruption scenario, assuming no supplies from unprotected suppliers and a full level of risk mitigation inventory prepositioned at each protected supplier. As a result a double capacity is available at each protected supplier and no capacity at each unprotected supplier.

The unit protection cost, however, does not involve any information on the estimated probability of potential disruptive events. One may assume that the higher the probability of a disruptive event and the higher the potential losses, the higher the protection cost the decision maker is willing to cover. In addition to the above unit protection cost, φ, the following protection index Φ is introduced in which the protection cost of each supplier i, f_i, is additionally divided by its disruption probability, $(p^* + (1-p^*)p_i)$

$$\Phi = \frac{\sum_{i \in I} f_i/(p^* + (1-p^*)p_i)}{\sum_{j \in J} g_j d_j}. \tag{4.28}$$

The protection index is the ratio of total weighted (multiplied by $1/(p^* + (1-p^*)p_i)$) protection cost to total loss resulted from the shortage of parts caused by supply disruptions. Given demand for parts, the protection index, Φ, increases with supplier protection cost per disruption probability, $f_i/(p^* + (1-p^*)p_i), i \in I$, and decreases with parts shortage costs, $g_j, j \in J$.

The optimal value Φ^o of protection index can be found as a solution to the following mixed integer program, in which no supplies from unprotected suppliers and a full level of risk mitigation inventory pre-positioned at each protected supplier are assumed.

$$\begin{aligned}
\Phi^o = \min\{ & \frac{\sum_{i \in I}(f_i/(p^* + (1-p^*)p_i))q_i}{\sum_{j \in J} g_j d_j} : \\
& \sum_{j \in J} a_{ij} d_j w_{ij}/(1-\rho_{ij}) \leq 2c_i q_i; i \in I, \\
& \sum_{i \in I} w_{ij} = 1; j \in J,\ w_{ij} \leq q_i; i \in I, j \in J, \\
& q_i \in \{0,1\}; i \in I,\ w_{ij} \in [0,1]; i \in I, j \in J\}.
\end{aligned} \tag{4.29}$$

Both the unit protection cost and the protection index can be used for a rough evaluation of different protection strategies.

4.5 Resilient Supply Portfolio with Multi-Level Protection

One of the main assumptions in the models presented in Sect. 4.3 is that the capacity of a fortified supplier remains unchanged after any disruptive event and hence the impacted supplier is capable of full delivering all ordered parts. However, this assumption may not be always realistic in practice. For example, fortifying a supplier against an earthquake may not protect it sufficiently well against a much stronger earthquake or a combined earthquake and flooding, etc., and hence the capacity of an impacted supplier can be actually reduced.

In this section, an enhancement of proposed models is considered assuming that the impact of disruptive events on supplier's capacity may not be fully mitigated by the fortification of a supplier, based on the level of protection investments. In the models presented below, the suppliers can be fortified at different discrete levels of protection, each of which has its own fortification cost and the associated capacity available after a disruptive event. The remaining capacity available depends on the protection level applied. The higher the protection level, the higher the remaining capacity in the aftermath of a disruptive event. However, the maximum amount of risk mitigation inventory pre-positioned at a fortified supplier is bounded by its full capacity, independent of the protection level.

The new parameters and variables are introduced in Table 4.3. Denote by $l \in L_i = \{1, \ldots, \overline{L_i}\}$, the protection level of supplier i, which refers to the fortification cost and the fraction of supplier's full capacity available after occurrence of a disruptive event. Level $l = \overline{L_i}$ represents the highest available protection for supplier i against disruptions and refers to the highest fortification cost and the highest fraction of remaining capacity. Let $\hat{f}_{il}$ and γ_{il} be the fortification cost and the remaining fraction of full capacity for supplier i protected at level l, where

$$0 < \hat{f}_{i1} < \hat{f}_{i2} < \cdots < \hat{f}_{i\overline{L_i}}$$

and

$$0 < \gamma_{i1} < \gamma_{i2} < \cdots < \gamma_{i\overline{L_i}} \leq 1.$$

Table 4.3 Notation: resilient supply portfolio, multi-level protection

	Indices
l	= protection level, $l \in L_i, i \in I$
	Parameters
$\hat{f}_{il}$	= fortification cost for supplier i protected at level l
γ_{il}	= the fraction of full capacity of supplier i protected at level l, available under disruptive event
	First stage variables
Q_{il}	= 1, if selected supplier i is protected at level l; otherwise $Q_{il} = 0$ (supplier protection level variable)
W_{ijl}	= the fraction of demand for parts type j ordered from supplier i protected at level l (part type demand allocation variable)

A supplier i protected at level l delivers fraction γ_{il} of all ordered parts, under a disruptive event. The non-delivered parts can be replaced by the risk mitigation inventory pre-positioned at protected suppliers. The non-delivered amount, $\sum_{i \in I} d_j(1-\gamma_{il})W_{ijl}/(1-\rho_{ij})$, of each part type j can be fully or partially met with the risk mitigation inventory $\sum_{i \in I} c_i x_{ij}/a_{ij}$.

The enhancement **ReSP(mlp)_E** of model **ReSP_E** is shown below.

ReSP(mlp)_E: *Selection of risk-neutral* **Res***ilient* **S***upply* **P***ortfolio with* **m***ulti-***l***evel* **p***rotection*

Minimize

$$\hat{E}^c = \sum_{i \in I} e_i u_i/D + \sum_{i \in I}\sum_{l \in L_i} \hat{f}_{il} Q_{il}/D + \sum_{i \in I}\sum_{j \in J} h_{ij} c_i x_{ij}/a_{ij} D$$
$$+ \sum_{i \in I}\sum_{j \in J} o_{ij} d_j (v_{ij} + \sum_{l \in L_i} W_{ijl})/(1-\rho_{ij}) D$$
$$+ \sum_{s \in S} P_s \Big(\sum_{j \in J} g_j z_{js} + \sum_{i \in I}\sum_{j \in J} o_{ij} c_i y_{ijs}/a_{ij} - \sum_{i \notin I_s}\sum_{j \in J} o_{ij} d_j v_{ij}/(1-\rho_{ij})$$
$$- \sum_{i \notin I_s}\sum_{j \in J}\sum_{l \in L_i} o_{ij} d_j (1-\gamma_{il}) W_{ijl}/(1-\rho_{ij})\Big)/D, \quad (4.30)$$

(continued)

subject to (4.3), (4.11), (4.14), (4.15), (4.17)–(4.19) and

$$\sum_{j\in J} x_{ij} \le \sum_{l\in L_i} Q_{il} \le u_i;\ i\in I, \quad (4.31)$$

$$v_{ij} \le 1-\sum_{l\in L_i} Q_{il};\ i\in I, j\in J, \quad (4.32)$$

$$W_{ijl} \le Q_{il};\ i\in I, j\in J, l\in L_i, \quad (4.33)$$

$$\sum_{i\in I}(v_{ij}+\sum_{l\in L_i} W_{ijl}) = 1;\ j\in J, \quad (4.34)$$

$$\sum_{j\in J} a_{ij}d_jv_{ij}/(1-\rho_{ij}) \le c_i(u_i-\sum_{l\in L_i} Q_{il});\ i\in I, \quad (4.35)$$

$$\sum_{j\in J} a_{ij}d_jW_{ijl}/(1-\rho_{ij}) \le c_iQ_{il};\ i\in I, l\in L_i, \quad (4.36)$$

$$\sum_{j\in J} d_jv_{ij}/(1-\rho_{ij}) \ge v(u_i-\sum_{l\in L_i} Q_{il});\ i\in I, \quad (4.37)$$

$$\sum_{j\in J}\sum_{l\in L_i}(d_jW_{ijl}/(1-\rho_{ij})+c_iy_{ijs}/a_{ij}) \ge v\sum_{l\in L_i} Q_{il};\ i\in I, s\in S, \quad (4.38)$$

$$z_{js} = \sum_{i\notin I_s} d_jv_{ij}/(1-\rho_{ij}) + \sum_{i\notin I_s}\sum_{l\in L_i} d_j(1-\gamma_{il})W_{ijl}/(1-\rho_{ij})$$
$$-\sum_{i\in I} c_iy_{ijs}/a_{ij};\ j\in J, s\in S, \quad (4.39)$$

$$Q_{il}\in\{0,1\};\ i\in I, l\in L_i, \quad (4.40)$$

$$W_{ijl}\in[0,1];\ i\in I, j\in J, l\in L_i. \quad (4.41)$$

In the objective function (4.30), $\sum_{i\in I}\sum_{l\in L_i}\hat{f}_{il}Q_{il}$ is the total fortification cost of protected suppliers, and the last subtracted term, $\sum_{s\in S}P_s\sum_{i\notin I_s}\sum_{j\in J}\sum_{l\in L_i}o_{ij}d_j(1-\gamma_{il})W_{ijl}/(1-\rho_{ij})$ is the expected cost of non-delivered parts by protected suppliers.

In model **ReSP(mlp)_E**, constraints (4.31) and (4.33) ensure that each supplier can be protected at most at one level, and the allocation of demand for parts among the fortified suppliers accounts for their protection levels. Since each supplier i can be protected at most at one level l, (4.31), at most one variable W_{ijl}, (4.33), may take on a positive value for each part type j. Equation (4.39) defines shortage of each part type inventory under each disruptive event, due to undelivered parts by both unprotected and protected suppliers. Equation (4.39) is equivalent to (4.12) in model **ReSP_E**, where a fortified supplier was capable of full delivering of ordered parts under disruptive event.

Notice that model **ReSP(mlp)_E** can be derived from **ReSP_E** by replacing variables $q_i, i\in I$ and $w_{ij}, i\in I, j\in J$ by sums over all protection levels $l\in L_i$ of the new variables Q_{il} and W_{ijl}: $q_i=\sum_{l\in L_i}Q_{il}$; $i\in I$ and $w_{ij}=\sum_{l\in L_i}W_{ijl}$; $i\in I, j\in J$. In addition, the objective function (4.1) and constraint (4.12) have been modified to account for undelivered parts by protected suppliers, and Eqs. (4.5), (4.8) redefined for each protection level.

The enhancements, **ReSP(mlp)_CV** and **ReSP(mlp)_ECV**, respectively, of risk-averse model **ReSP_CV** and mean-risk model **ReSP_ECV** are presented below.

ReSP(mlp)_CV: *Selection of risk-averse* **Res***ilient* **S***upply* **P***ortfolio with* **m***ulti-***l***evel* **p***rotection*

Minimize (4.22)

subject to (4.3), (4.11), (4.14), (4.15), (4.17)–(4.19), (4.24), (4.31)–(4.41) and

$$\begin{aligned}\mathscr{C}_s \geq \sum_{i\in I} e_i u_i/D + \sum_{i\in I}\sum_{l\in L_i} \hat{f}_{il} Q_{il}/D + \sum_{i\in I}\sum_{j\in J} h_{ij} c_i x_{ij}/a_{ij} D \\ + \sum_{i\in I}\sum_{j\in J} o_{ij} d_j (v_{ij} + \sum_{l\in L_i} W_{ijl})/(1-\rho_{ij}) D \\ +(\sum_{j\in J} g_j z_{js} + \sum_{i\in I}\sum_{j\in J} o_{ij} c_i y_{ijs}/a_{ij} - \sum_{i\notin I_s}\sum_{j\in J} o_{ij} d_j v_{ij}/(1-\rho_{ij}) \\ - \sum_{i\notin I_s}\sum_{j\in J}\sum_{l\in L_i} o_{ij} d_j (1-\gamma_{il}) W_{ijl}/(1-\rho_{ij}))/D - VaR^c;\ s\in S. \quad (4.42)\end{aligned}$$

ReSP(mlp)_ECV: *Selection of mean-risk* **Res***ilient* **S***upply* **P***ortfolio with* **m***ulti-***l***evel* **p***rotection*

Minimize

$$\lambda \hat{E}^c + (1-\lambda) CVaR^c, \quad (4.43)$$

where $0 \leq \lambda \leq 1$,

subject to (4.3), (4.11), (4.14), (4.15), (4.17)–(4.19), (4.22), (4.24), (4.30)–(4.41).

4.6 Computational Examples

In this section some computational examples are presented to illustrate possible applications of the proposed SMIP approach for the selection and protection of suppliers and order quantity allocation in a supply chain under disruption risks. First, the examples with single protection levels are presented and then, for comparison, applications of models for multi-level protection are illustrated. The following parameters have been used for the example problems:

- $\overline{I}$, the number of suppliers, was equal to 10 and the corresponding number of disruption scenarios, was equal to, $\overline{S} = 2^{\overline{I}} = 1024$;

- $\bar{J}$, the number of part types, was equal to 25;
- a_{ij}, the unit capacity consumptions were integers in $\{1,2,3\}$ drawn from int(U[1;3]) distribution, for all suppliers i and part types j;
- c_i, the capacity of each supplier i, was integer drawn from $1000\lceil(\sum_{j\in J} a_{ij}d_j/\bar{I})U[0.75;1.25]/1000\rceil$ distribution ($\lceil\cdot\rceil$ denotes the smallest integer not less than $\cdot$);
- d_j, the required numbers of parts, were integers in $\{5000, 10{,}000, 15{,}000, 20{,}000, 25{,}000, 30{,}000, 35{,}000, 40{,}000, 45{,}000, 50{,}000\}$ drawn from 5000int(U[1;10]) distribution, for all part types j;
- e_i, the cost parts, were integers in $\{500, 600, 700, 800, 900, 1000\}$ drawn from 100int(U[5;10]) distribution, for all suppliers i;
- f_i, the protection costs, were integers in $\{500000, 550000, 600000, 650000, 700000, 750000, 800000, 850000, 900000\}$ drawn from 50000int(U[10;18]) distribution, for all suppliers i;
- g_j, the unit shortage costs, were integers in $\{70, 80, 90, 100\}$ drawn from 10int(U[7;10]) distribution, for all part types j;
- h_{ij}, the unit cost of risk mitigation inventory of part type j pre-positioned at supplier i, was equal to $0.4o_{ij}$ for all suppliers i and part types j;
- o_{ij}, the unit price (including shipping cost) of part type j purchased and transported from each supplier i, was uniformly distributed over [10,15] (i.e., drawn from U[10;15]) and reduced by the factor $(1-\rho_{ij})$ to get a lower price for parts from the suppliers with a higher defect rate;
- ρ_{ij}, the expected defect rate of each supplier i for each part type j, was exponentially distributed, ranging from 0.0003 to 0.03;
- v, the minimum order size, was equal to 500;
- p_i, the local disruption probability, was uniformly distributed over [0,0.06] or over [0.06,0.15], i.e., the disruption probabilities were drawn independently from U[0;0.06] or from U[0.06;0.15]. The global disruption probability p^* was either 0.001 or 0.01, respectively;
- α, the confidence level, was equal to 0.50, 0.75, 0.90, 0.95, or 0.99.

The computational experiments were performed for the same replication of the above input data set. The resulting total demand for parts was $D = \sum_{j\in J} d_j = 740{,}000$, for all test examples. The two different combinations of global and local disruption probabilities were considered: $p^* = 0.001$ and $p_i \in U[0;0.06], i \in I$ for reliable suppliers and $p^* = 0.01$ and $p_i \in U[0.06;0.15], i \in I$ for unreliable suppliers.

4.6.1 Single-Level Protection

The solution results are presented in Tables 4.4 and 4.5, respectively, for the risk-neutral model **ReSP_E** and the risk-averse model **ReSP_CV** with different confidence levels. The confidence level α is set at five levels of 0.5, 0.75, 0.90, 0.95, and 0.99, which means that focus is on minimizing the highest 50%, 25%, 10%, 5%, and 1% of all scenario outcomes, i.e., costs per part. The size of the corresponding mixed integer programs is represented by the total number of variables, Var., number of binary variables, Bin., number of constraints, Cons, and number of nonzero coefficients in the constraint matrix, Nonz. Table 4.5 also presents the probability $1 - F(VaR^c)$ of outcomes with worst-case cost above VaR^c as well as the expected cost for the optimal risk-averse supply portfolio.

Table 4.4 Solution results for model **ReSP_E**

Disruption probability	$p^* = 0.001,\ p_i \in [0, 0.06], i \in I$	$p^* = 0.01,\ p_i \in [0.06, 0.15], i \in I$
Var. = 282,105, Bin. = 20, Cons. = 282,151, Nonz. = 924,155[a]		
Expected cost	12.63	14.97
No. of suppliers selected	9	10
(incl. suppliers protected)	(0)	(2)
Pre-positioned risk mitigation inventory	0	141,888
CPU[b]	5	3243

[a]Var.: number of variables; Bin.: number of binary variables; Cons.: number of constraints; Nonz.: number of nonzero coefficients

[b]CPU seconds for proving optimality on a MacBookPro6.2, Intel Core i7, 2.66 GHz, RAM 8 GB/CPLEX 12.4

*global disruption probability for all suppliers

The optimal risk-neutral supply portfolio, $(V_1, \ldots, V_{\bar{I}})$, (4.20), for model **ReSP_E** that aims at minimization of total expected cost per part is shown in Fig. 4.1 for two scenarios:

(a) with global disruption probability $p^* = 0.001$ and reliable suppliers with local disruption probabilities $p_i \in [0, 0.06], i \in I$, and
(b) with global disruption probability $p^* = 0.01$ and unreliable suppliers with local disruption probabilities $p_i \in [0.06, 0.15], i \in I$.

In addition, total disruption probability $p^* + (1 - p^*)p_i$ for each supplier i is also presented. The optimal supply portfolio for scenario with reliable suppliers allocates the total demand for parts among nine unprotected suppliers with the lowest disruption probabilities, except for supplier 4 with the highest disruption probability. The lower disruption probability of a supplier, the higher percentage of the total demand allocated. The above observations indicate that when the shortage cost of parts dominates the purchasing cost, disruption probability becomes a key determinant in the decision of demand allocation among the suppliers to minimize total expected cost. Similar results are observed for scenario with unreliable suppliers. However, the total demand is now allocated among all ten suppliers, with the largest orders placed on protected suppliers 1 and 10, where the risk mitigation inventory was pre-positioned (see, Table 4.4).

The optimal risk-averse supply portfolios for **ReSP_CV** model and the three confidence levels: 0.90, 0.95, and 0.99 are shown in Fig. 4.2. Table 4.5 and Fig. 4.2 indicate that when α increases and a more risk-averse decision-making focuses on a smaller set of outcomes, the number of protected suppliers also increases to mitigate the impact of disruption risks. Simultaneously, the pre-positioned risk mitigation inventory at the protected suppliers increases, while the total number of all selected

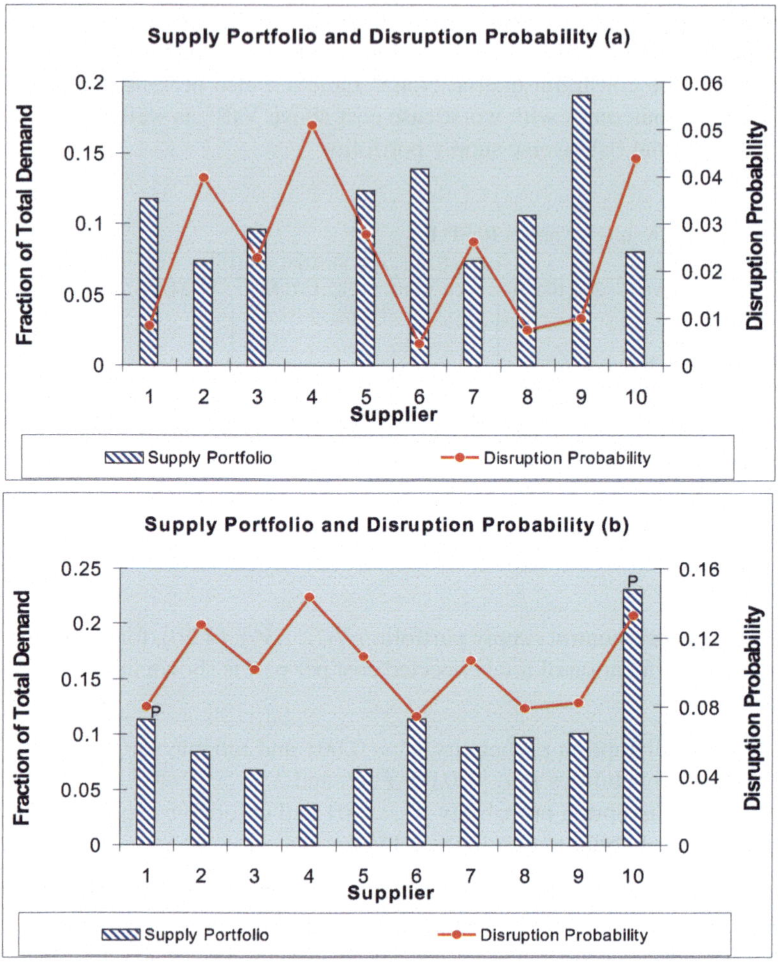

Fig. 4.1 Optimal supply portfolio for model **ReSP_E**: (**a**) $p^* = 0.001$, $p_i \in [0, 0.06], i \in I$, (**b**) $p^* = 0.01$, $p_i \in [0.06, 0.15], i \in I$ (P—protected supplier)

suppliers decreases and more orders are placed on the protected suppliers. This indicates that instead of further diversification of supplies by selecting of more suppliers, the impact of disruption risks is rather mitigated by selecting of less suppliers and by protecting most of the selected suppliers and, in addition, by the pre-positioning of risk mitigation inventory at the protected suppliers.

Unless protected, the suppliers with the highest disruption probability are rarely selected. For example (see, Fig. 4.2), in the optimal supply portfolio for scenario with reliable suppliers and $\alpha = 0.99$ or for scenario with unreliable suppliers with $\alpha = 0.9, 0.95, 0.99$, supplier 4 (cf. Fig. 4.1) is not selected, while supplier 10 is selected and protected.

Table 4.5 Solutions results for model **ReSP_CV**

Confidence level α	0.50	0.75	0.90	0.95	0.99
Var. = 283,130, Bin. = 20, Cons. = 283,175, Nonz. = 1,868,008[a]					
$p^* = 0.001$, $p_i \in U[0;0.06]$					
CVaR^c	13.43	14.02	14.68	15.21	**16.00**[b]
VaR^c	12.78	12.93	14.01	14.36	**16.00**
E^c	13.11	13.20	14.08	14.35	**16.00**
$1-F(VaR^c)$	0.151	0.151	0.033	0.025	0.000
No. of suppliers selected (incl. protected)	10(1)	10(1)	10(2)	10(2)	8(4)
Pre-positioned risk mitigation inventory	70 139	65 652	128 192	133 349	193 950
CPU[c]	1037	3079	4770	16,752	1852
$p^* = 0.01$, $p_i \in U[0.06;0.15]$					
CVaR^c	15.55	15.89	**16.00**	**16.00**	**16.00**
VaR^c	15.02	15.44	**16.00**	**16.00**	**16.00**
E^c	15.28	15.54	**16.00**	**16.00**	**16.00**
$1-F(VaR^c)$	0.302	0.067	0.000	0.000	0.000
No. of suppliers selected (incl. protected)	10(3)	10(3)	8(4)	8(4)	8(4)
Pre-positioned risk mitigation inventory	167 357	213 402	193 950	193 950	193 950
CPU[c]	23,931	28,856	2354	4030	4357

[a]Var.: number of variables; Bin.: number of binary variables; Cons.: number of constraints; Nonz.: number of nonzero coefficients

[a]$VaR^c = CVaR^c = E^c = 16.00$ is the highest cost that may occur for the optimal supply portfolio, $1-F(16)=0$

[c]CPU seconds for proving optimality on a MacBookPro 6.2, Intel Core i7, 2.66 GHz, RAM 8 GB/CPLEX 12.4

*global disruption probability for all suppliers

The discrete distributions of cost per part for the optimal risk-averse supply portfolios (Fig. 4.2a) with three different confidence levels, for global disruption probability $p^* = 0.001$ and reliable suppliers with local disruption probabilities $p_i \in [0, 0.06], i \in I$ are presented in Table 4.6. The table demonstrates that the probability mass function is concentrated in a few points, which is typical for the scenario-based optimization under uncertainty, where the probability measure is concentrated in finitely many points. Large probability atoms are concentrated at eight points, six points, one point, respectively, for the confidence level $\alpha = 0.9, 0.95, 0.99$. In particular, for $\alpha = 0.99$ the whole probability is concentrated at one point only: at cost 16.00. Similar results were obtained for the global disruption probability $p^* = 0.01$ and the unreliable suppliers with local disruption probabilities $p_i \in [0.06, 0.15], i \in I$. For the optimal resilient supply portfolio (Fig. 4.2b), the whole probability is concentrated at a single cost level 16.00 for each confidence level $\alpha = 0.9, 0.95, 0.99$, i.e., the cost distributions were identically shaped by the optimal resilient supply portfolio for the three different confidence levels. The optimal resilient supply portfolios for scenario with reliable suppliers with $\alpha = 0.99$ as well as for scenario with unreliable suppliers with $\alpha = 0.9, 0.95, 0.99$ are identical (see, Table 4.5 and Fig. 4.2).

The computational results indicate that the smaller is the number of concentration points and the greater are probability atoms concentrated at those points, the greater

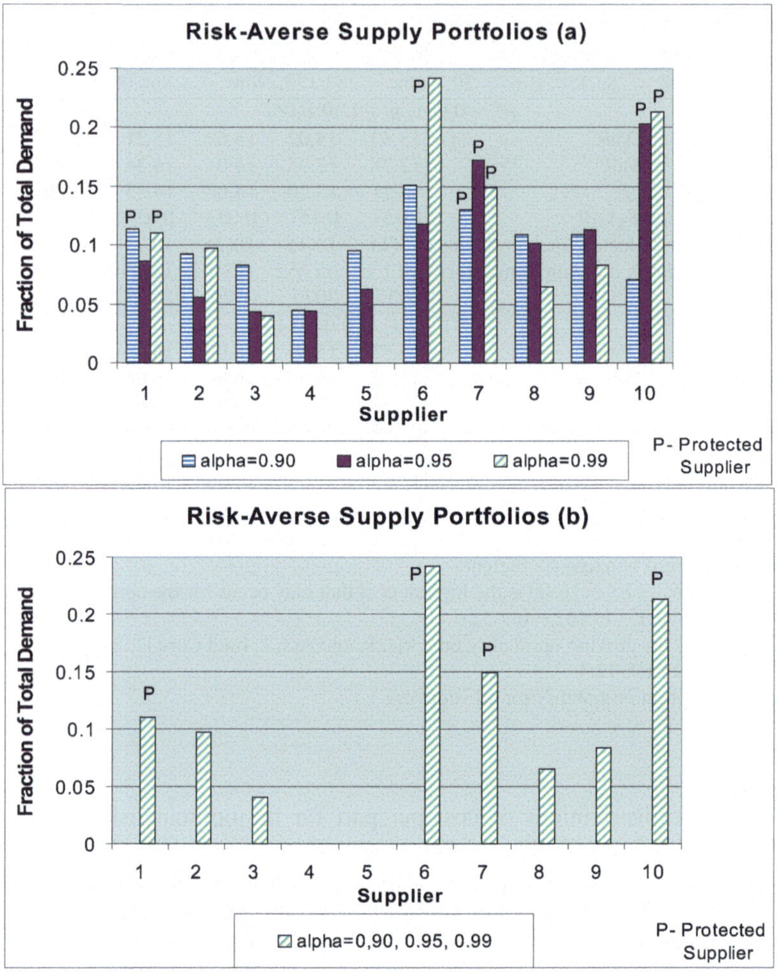

Fig. 4.2 Optimal supply portfolio for model **ReSP_CV**: (**a**) $p^* = 0.001$, $p_i \in [0, 0.06], i \in I$, (**b**) $p^* = 0.01$, $p_i \in [0.06, 0.15], i \in I$

can be the positive difference $F(VaR^c) - \alpha$, i.e., the smaller than $1 - \alpha$ can be the probability $1 - F(VaR^c)$ of outcomes with worst-case cost higher than VaR^c. For example (see, Table 4.5), for $\alpha = 0.9$, VaR^c=14.01, and $1 - F(VaR^c) = 0.033 < 1 - \alpha = 0.1$, which indicates a high concentration of probability measure at point 14.01 for the optimal supply portfolio. Actually, the probability that cost per part is 14.01 is 0.967, which indicates that $VaR^c = 14.01$ is the lowest cost that may occur and that for the confidence level $\alpha = 0.9$, less than 3.3% of the cost outcomes are above VaR^c.

Moreover, if the highest cost probability is greater than $1 - \alpha$, then CVaR^c and VaR^c are identical and both equal to the highest cost. In the example for scenario

Table 4.6 Probability of cost per part for optimal supply portfolios: $p^* = 0.001$, $p_i \in [0, 0.06], i \in I$

Cost interval	$\alpha = 0.90$	$\alpha = 0.95$	$\alpha = 0.99$
$[14, 15)$	0.9899	0.9950	0
$[15, 16)$	0.0032	0.0014	1
$[16, 17)$	0.0040	0.0020	0
$[17, 18)$	0.0005	0.0003	0
$[18, 19)$	0.0011	0.0003	0
$[19, 20)$	0.0002	0	0
$[20, 21)$	0.0001	0	0
$[47, 48)$	0	0.0010	0
$[53, 54)$	0.0010	0	0

*global disruption probability for all suppliers

with reliable suppliers and $\alpha = 0.99$ (see, Table 4.6), the highest cost per part is 16.00 and the probability concentrated at 16.00 is $0.999 > 1 - \alpha = 0.01$, then $VaR^c = 16.00$ is the highest cost per part that may occur ($1 - F(VaR^c) = 0.000$, in Table 4.5) and hence $CVaR^c = VaR^c = 16.00$. Similar results with $CVaR^c = VaR^c = 16.00$ were obtained for the optimal supply portfolios with $\alpha = 0.9, 0.95, 0.99$, for scenarios with unreliable suppliers with global and local disruption probabilities, respectively, $p^* = 0.01$, $p_i \in [0.06, 0.15], i \in I$, (see, Table 4.5). The results presented in Table 4.5 indicate that when for the optimal supply portfolio with a confidence level α, the probability measure is concentrated at the highest cost and hence is greater than $1 - \alpha$, so that CVaR^c and VaR^c are identical with the highest cost, then the number of selected suppliers is smaller than for a lower confidence level, however, more suppliers are protected.

Table 4.5 also demonstrates that the higher the protection index Φ (4.28) (or its optimal value Φ^o (4.29)), the less the number of protected suppliers in the optimal supply portfolio. For example, for scenario with reliable suppliers ($\Phi = 8.353$ and $\Phi^o = 1.033$) only a few of the selected suppliers are protected and only for higher confidence levels α, whereas for scenario with unreliable suppliers ($\Phi = 1.141$ and $\Phi^o = 0.295$) more of the selected suppliers are protected, even for lower confidence levels.

For the bi-objective mean-risk approach, the subsets of nondominated solutions were computed by parameterization on $\lambda \in \{0.01, 0.10, 0.20, 0.30, 0.40, 0.50, 0.60, 0.70, 0.80, 0.90, 0.99\}$ the weighted-sum program **ReSP_ECV**. The results obtained for the confidence level $\alpha = 0.99$ and reliable suppliers with global and local disruption probability, respectively, $p^* = 0.001$ and $p_i \in [0, 0.06], i \in I$, are presented in Table 4.7. The subsets of nondominated solutions that were found for the selected 11 levels of weight λ consists of eight different nondominated solutions. The trade-off between the expected cost and the expected worst-case cost is clearly shown in Fig. 4.3, where the convex efficient front of Mean Cost—CVaR, for the confidence level $\alpha = 0.99$ is presented. The results emphasize the effect of varying average/worst-case cost preference of the decision maker; the lower the trade-off parameter λ, the more risk-oriented the decision-making. Note that solutions to single

Table 4.7 Nondominated solutions for mean-risk model **ReSP_ECV**: $\alpha = 0.99$, $p^* = 0.001$, $p_i \in [0, 0.06], i \in I$

λ	0.01	0.10	0.20	0.30	0.40	0.50	0.60	0.70	0.80	0.90	0.99
$p^* = 0.001$, $p_i \in U[0;0.06], i \in I$ ($\varphi^o = 0.032$, $\Phi^o = 1.033$)											
E^c	16.00				15.98	15.43	15.03	14.39	14.36	13.15	12.63
CVaR^c	16.00				16.00	16.44	16.89	18.29	18.40	25.71	33.70
VaR^c	16.00				15.99	15.42	15.07	14.55	14.52	19.08	23.77
No. of suppliers selected	8				8	10	10	10	10	10	9
(incl. protected)	(4)				(4)	(3)	(3)	(2)	(2)	(1)	(0)
Risk mitigation inventory	193,950				193,997	218,988	177,867	144,381	141,565	70,594	0
CPU[a]	5186	4200	1531	1659	4720	6902	7961	10,345	8250	2997	217

[a]CPU seconds for proving optimality on a MacBookPro 6.2, Intel Core i7, 2.66 GHz, RAM 8 GB/CPLEX 12.4

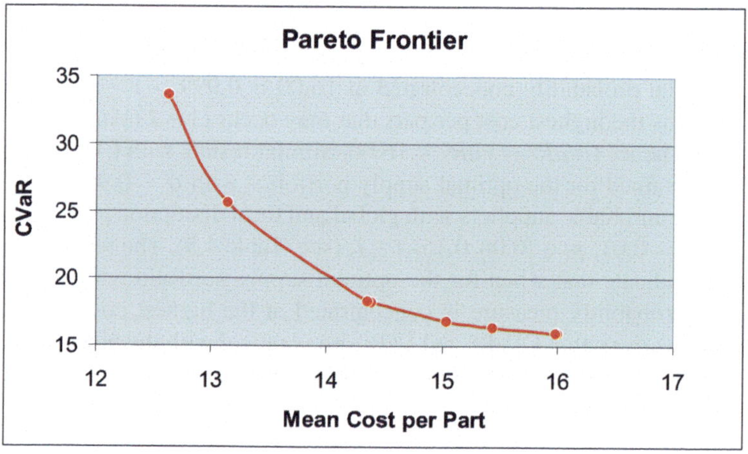

Fig. 4.3 Pareto front for model **ReSP_ECV**: $\alpha = 0.99$, $p^* = 0.001$, $p_i \in [0, 0.06], i \in I$

objective models **ReSP_E** and **ReSP_CV** are equivalent to the nondominated solutions of the weighted-sum program **ReSP_ECV** with $\lambda = 1$ and $\lambda = 0$, respectively.

Similar computations for the confidence level $\alpha = 0.99$ and unreliable suppliers with global and local disruption probability, respectively, $p^* = 0.01$ and $p_i \in [0.06, 0.15], i \in I$, produce the three nondominated solutions only, (E^c, CVaR^c): (16.00,16,00) for $\lambda \in \{0.01, 0.10, 0.20, 0.30, 0.40\}$, (15.98,16.01) for $\lambda \in \{0.50, 0.60, 0.70, 0.80, 0.90\}$, and (14.97,47.25) for $\lambda = 0.99$. For the first two solutions eight suppliers were selected, including two protected, and for the last solution ten suppliers were selected of them two were protected. Figure 4.4 presents the supply portfolio for the last nondominated solution as well as the corresponding optimal cost distribution, which indicates a large probability atom of 0.01 concentrated at the highest cost of 47.25. Note that the risk-averse supply portfolio for $\lambda = 0.99$ is an average performance-oriented portfolio (cf. objective function (4.25)) and virtually neglects the risk of high costs (0.01 weight of CVaR^c in (4.25)). Therefore, it is very close to the risk-neutral portfolio presented in Fig. 4.1b. The cost distribution shaped with a risky supply portfolio that focuses on an average performance of supply chain more often includes probability atoms at high costs.

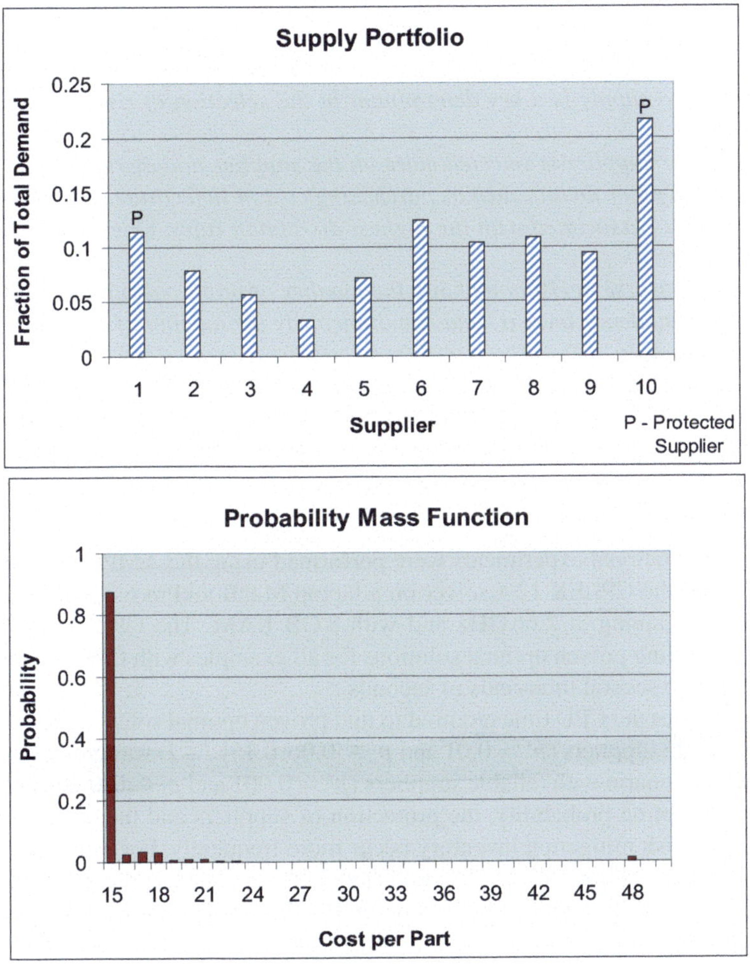

Fig. 4.4 Optimal supply portfolio and probability mass function for model **ReSP_ECV**: $p^* = 0.01$, $p_i \in [0.06, 0.15], i \in I$, $\alpha = 0.99$, $\lambda = 0.99$

Comparison of risk-neutral and risk-averse supply portfolios indicates that most risk-averse portfolios include protected suppliers and hence, can be more resilient. In contrast, the risk-neutral decision-making that focuses on the expected cost only, rarely selects a resilient portfolio with protected suppliers. Moreover, the mean-risk decision-making that aims at minimizing the weighted sum of the expected cost and the expected worst-case cost produces a subset of nondominated supply portfolios that contains both protected and unprotected suppliers.

The computational experiments with the single-level protection models indicate that:

- *when the shortage cost of parts dominates the purchasing cost, probability of disrupting a supply is a key determinant in the selection of risk-neutral supply portfolio.*
- *a particular supplier is selected more on the supplier non-disruption likelihood than on the other factors such as purchasing cost or defect rate,*
- *the suppliers associated with the highest disruption rates, unless protected, are rarely selected.*
- *in the risk-averse decision-making, the number of protected suppliers increases with the confidence level α, and simultaneously the number of all selected suppliers decreases,*
- *in the risk-averse decision-making instead of diversification of supplies by selecting of more suppliers, the impact of disruption risks is rather mitigated by selecting of less suppliers and by protecting most of the selected suppliers and, in addition, by the pre-positioning of risk mitigation inventory at the protected suppliers.*

The computational experiments were performed using the AMPL programming language and the CPLEX 12.4 solver on a laptop MacBookPro 6.2 with Intel Core i7 processor running at 2.66 GHz and with 8 GB RAM. The CPLEX solver was capable of finding proven optimal solutions for all examples with CPU time ranging from several to several thousands of seconds.

In most cases the CPU time required to find proven optimal solutions for scenario with unreliable suppliers ($p^* = 0.01$ and $p_i \in [0.06, 0.15], i \in I$) was greater than that required for scenario with reliable suppliers ($p^* = 0.001$ and $p_i \in [0, 0.06], i \in I$). For a higher disruption probability, the protection of suppliers and the pre-positioning and usage of risk mitigation inventory occur more frequently. For example, the optimal risk-neutral solution with the protected suppliers and the pre-positioned risk mitigation inventory required much longer CPU time than the low-probability case with no protected suppliers and no risk mitigation inventory (see, Table 4.4).

4.6.2 Multi-Level Protection

In this subsection simple computational examples are provided to illustrate the proposed SMIP approach for selection of resilient supply portfolio with multiple protection levels available for suppliers. For comparison, the basic input data used are the same as for the examples with single protection levels. The local disruption probabilities of suppliers, p_i, were distributed exactly the same way as for the examples with single protection levels, however, slightly different values were generated for

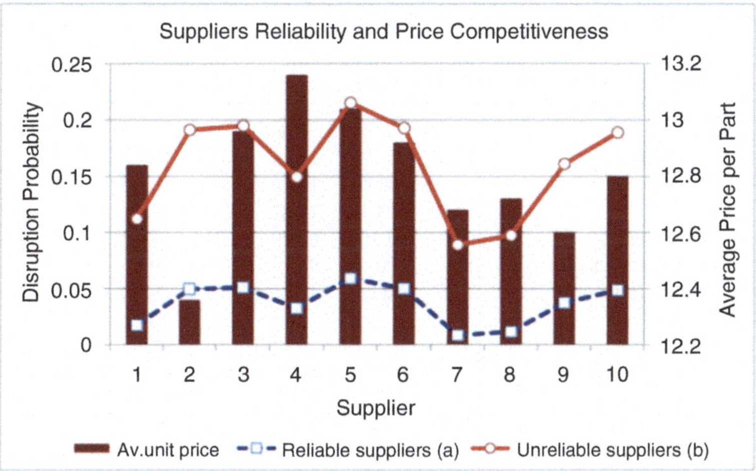

Fig. 4.5 Suppliers

Table 4.8 Solution results for model **ReSP(mlp)_E**

Disruption probability	$p^* = 0.001$, $p_i \in [0, 0.06], i \in I$	$p^* = 0.01$, $p_i \in [0.06, 0.15], i \in I$
Var. = 282,875, Bin. = 50, Cons. = 282,920, Nonz. = 1,439,505[a]		
Expected cost	12.79	14.53
No.of suppliers selected	10	10
Protected suppliers (protection level)	9(1)	1(1), 6(1), 9(1)
Pre-positioned risk mitigation inventory	85,030	261,909
CPU[b]	17	1044

[a]Var.: number of variables; Bin.: number of binary variables; Cons.: number of constraints; Nonz.: number of nonzero coefficients

[b]CPU seconds for proving optimality on a MacBookPro, Intel Core i7, 2.8 GHz, RAM 16 GB/Gurobi 7

*global disruption probability for all suppliers

the examples in this subsection. Figure 4.5 shows basic characteristics of each supplier i: average price per part, $\sum_{j \in J} o_{ij}/\overline{J}$, and disruption probability, $p^* + (1 - p^*)p_i$, for two scenarios:

(a) with global disruption probability $p^* = 0.001$ and reliable suppliers with local disruption probabilities $p_i \in [0, 0.06], i \in I$, and
(b) with global disruption probability $p^* = 0.01$ and unreliable suppliers with local disruption probabilities $p_i \in [0.06, 0.15], i \in I$.

The only new parameters, the number of protection levels, $\overline{L}_i$, the fortification cost, $\hat{f}_{il}$, and the remaining fraction of capacity, γ_{il}, for supplier i protected at level l, are shown below.

$\overline{L}_i = 4$ for all suppliers $i \in I$.

$\hat{f}_{i1} = f_i/4$, $\hat{f}_{i2} = f_i/2$, $\hat{f}_{i3} = 3f_i/4$, $\hat{f}_{i4} = f_i$ for all suppliers $i \in I$, where f_i is the fortification cost for the case with single protection levels.

Table 4.9 Solutions results for model **ReSP(mlp)_CV**

Confidence level α	0.50	0.75	0.90	0.95	0.99
Var. = 283,900, Bin. = 50, Cons. = 283,944, Nonz. = 3,182,078[a]					
$p^* = 0.001$, $p_i \in U[0;0.06]$					
CVaR^c	13.32	13.77	14.45	14.93	16.17
VaR^c	12.34	13.00	13.80	13.90	16.15
E^c	12.83	13.19	13.87	13.94	16.12
$1 - F(VaR^c)$	0.281	0.162	0.029	0.019	0.006
No. of suppliers selected	10	10	9	10	10
Protected suppliers (protection level)		1(1)	1(1)	1(1)	1(2)
					2(1)
					4(1)
			6(1)	6(1)	
					7(1)
	9(1)	9(1)	9(1)	9(1)	9(2)
		10(1)			
Pre-positioned risk mitigation inventory	91,574	173,542	259,787	264,526	551,749
CPU[b]	565	1271	6304	4230	2190
$p^* = 0.01$, $p_i \in U[0.06;0.15]$					
CVaR^c	15.21	15.99	16.17	16.17	**16.18**[♯]
VaR^c	14.34	14.89	16.16	16.18	**16.18**[♯]
E^c	14.77	15.16	16.14	16.16	**16.18**[♯]
$1 - F(VaR^c)$	0.222	0.134	0.047	0.005	0.000
No. of suppliers selected	10	9	9	10	8
Protected suppliers (protection level)	1(1)	1(1)	1(2)	1(2)	1(2)
			2(1)	2(1)	2(1)
			4(1)	4(1)	4(1)
	6(1)	6(1)			
			7(1)	7(1)	7(1)
	9(1)	9(2)	9(2)	9(2)	9(2)
	10(1)	10(1)			
Pre-positioned risk mitigation inventory	336 809	377 519	543 054	541 012	538 451
CPU[b]	6865	11,133	1598	4084	6035

[a]Var.: number of variables; Bin.: number of binary variables, Cons.: number of constraints; Nonz.: number of nonzero coefficients

[b]CPU seconds for proving optimality on a MacBookPro, Intel Core i7, 2.8 GHz, RAM 16 GB/Gurobi 7

[♯]$VaR^c = CVaR^c = E^c = 16.18$ is the highest cost that may occur for the optimal supply portfolio, $1 - F(16.18) = 0$

[*]global disruption probability for all suppliers

$\gamma_{i1} = 0.2$, $\gamma_{i1} = 0.5$, $\gamma_{i3} = 0.7$, $\gamma_{i4} = 0.9$ for all suppliers $i \in I$.

Thus, the fortification cost for the highest protection level, $l = 4$, is identical with the fortification cost, f_i, for the examples with single protection levels. However, unlike for single protection levels, where 100% of protected supplier's capacity is available after disruption, now only 90% is the highest fraction of remaining capacity.

The solution results are presented in Tables 4.8 and 4.9, respectively, for the risk-neutral model **ReSP(mlp)_E** and the risk-averse model **ReSP(mlp)_CV** with different confidence levels. The tables show results for two scenarios: (a) with reliable suppliers and (b) with unreliable suppliers.

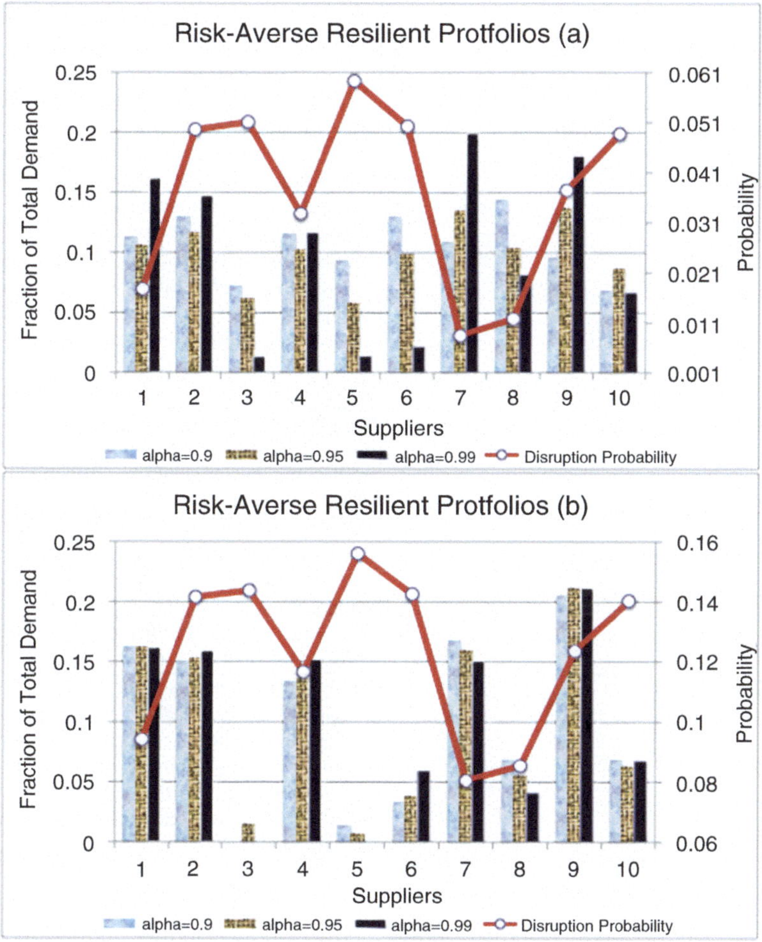

Fig. 4.6 Optimal supply portfolio for model **ReSP(mlp)_CV**: (**a**) $p^* = 0.001$, $p_i \in [0, 0.06], i \in I$, (**b**) $p^* = 0.01$, $p_i \in [0.06, 0.15], i \in I$

In general, risk-neutral supply portfolios for single- and multi-level protection are similar, (cf. Tables 4.4 and 4.8), as well as the corresponding risk-averse portfolios, (cf. Tables 4.5 and 4.9). For both the risk-neutral and the risk-averse portfolio, the number of protected suppliers is greater for scenario (b) with unreliable suppliers, while it additionally increases with the confidence level for the risk-averse portfolio. For example, for scenario (b) with unreliable suppliers, the risk-averse subset of protected suppliers for confidence levels, $\alpha = 0.9, 0.95, 0.99$, is identical with the

Table 4.10 Managerial insights: resilient supply portfolio

Decision-making issue	Decision-making impact
Risk-neutral decision-making	In the risk-neutral decision-making, a supplier is selected more on its non-disruption likelihood than purchasing cost or defect rate Unlike resilient risk-averse portfolio, the resilient risk-neutral supply portfolio rarely includes protected suppliers
Risk-averse decision-making	In the risk-averse decision-making, the number of protected suppliers increases with the confidence level and simultaneously the total number of selected suppliers decreases Instead of diversification of supply portfolio by selecting more suppliers, the impact of disruption risks is rather mitigated by selecting less suppliers, protecting most of them and pre-positioning risk mitigation inventory at protected suppliers
Disruption probability	In supply chains with more unreliable suppliers, protection of suppliers and pre-positioning and usage of risk mitigation inventory occur more frequently In both risk-neutral and risk-averse decision-making, the number of protected suppliers is greater for supply chains with more unreliable suppliers Suppliers with a high disruption probability, unless protected, are rarely selected
Multi-level protection	Lower protection levels are mostly selected. Instead of costly, higher protection levels, the supply portfolio resilience is improved rather by pre-positioning of more risk mitigation inventory at fortified suppliers, for both **ReSP(mlp)_E** and **ReSP(mlp)_CV** decision-making models

subset of protected suppliers for scenario (a) with reliable suppliers and the highest confidence level, $\alpha = 0.99$. The corresponding solution values are very close to each other (see, Table 4.9). The optimal risk-averse supply portfolios, $(V_1, \ldots, V_{\bar{I}})$, (4.20), for scenario (a) with reliable suppliers and scenario (b) with unreliable suppliers and confidence levels, $\alpha = 0.9, 0.95, 0.99$, as well as the disruption probability, $p^* + (1 - p^*)p_i$, for each supplier i, are shown in Fig. 4.6. The resilient portfolios for scenario (b) and different confidence levels are very close to each other. The largest portion of demand for parts has been allocated among the five protected suppliers $i = 1, 2, 4, 7, 9$, while remaining small orders for parts are placed among unprotected suppliers.

The major managerial insights are summarized in Table 4.10. Overall, comparison of risk-neutral and risk-averse decision-making indicates that *most risk-averse portfolios include protected suppliers and hence improve supply chain resilience. In contrast, the risk-neutral decision-making that focuses on the expected cost only, rarely selects a resilient portfolio with protected suppliers. Moreover, the mean-risk decision-making that minimizes the weighted sum of the expected cost and the expected worst-case cost produces a subset of nondominated supply portfolios that contains both protected and unprotected suppliers.* The proposed resilient supply portfolio approach has proven to be computationally very efficient.

4.7 Notes

Supply chain resilience is a relatively new concept that can be defined as "the adaptive capability of the supply chain to prepare for unexpected events, respond to disruptions, and recover from them by maintaining continuity of operations at the desired level of connectedness and control over structure and function" (Ponomarov and Holcomb 2009). Falasca et al. (2008) defined resilience as the ability of a supply chain system to reduce the probabilities of disruptions, to reduce the consequences of those disruptions and the time to recover disrupted operations to their normal performance. Also Fiksel (2006) refers resiliency to a firm's capacity to survive, adapt, and grow in the face of change and uncertainty. According to Sheffi (2005), the companies can develop the resilience in three general ways: (1) creating redundancies throughout the supply chain; for example, with holding extra inventory, maintain low capacity utilization, and contracting with multiple suppliers, (2) increasing supply chain flexibility; for example, with adoption of standardized processes, using concurrent instead of sequential processes, plan to postpone, align procurement strategy with supplier relationships, and (3) changing the corporate culture.

Despite the abundant literature on supplier selection and order quantity allocation problems (e.g., Aissaoui et al. 2007, Ho et al. 2010), the research on quantitative approaches for building a resilient portfolio of suppliers has not often been reported in the literature. Torabi et al. (2015) proposed a bi-objective mixed possibilistic, two-stage stochastic programming model for supplier selection and order allocation problem to build the resilient supply base under operational and disruption risks. The model accounts for uncertainty of critical data and applies several proactive strategies such as suppliers business continuity plans, fortification of suppliers, and contracting with backup suppliers to enhance the resilience level of the selected supply base.

The issue of linking risk assessment with risk mitigation for low-probability high-consequence events such as disruptions of supplies is discussed by Kleindorfer and Saad (2005), Cohen and Kunreuther (2007) and Madadi et al. (2014). The need to build resiliency to disruptive events in supply chains is discussed by Knemeyer et al. (2009), who considered a proactive planning, based on methodology used by the insurance industry. They quantify the risk of multiple types of catastrophic events on key supply chain locations. The proposed proactive planning process involves four critical steps: identification of key supply chain locations and threats, estimation of probabilities and loss for each location, evaluation of alternative countermeasures for each location, and selection of countermeasures that may prevent or mitigate disruption risks. Examples of such countermeasures include relocation of facility away from high risk location, e.g., moving a warehouse to a hurricane-free area, redesigning of facility to increase storm preparedness, building storm walls to help protect against flooding, maintaining excess inventory, etc. In practice, the fortification of suppliers to protect them against disruptions is recently observed. For example, to prevent flooding during monsoons, a giant flood wall with sealable aluminum flood barriers across entrance points has been constructed in Thailand, around the perimeter of Nava Nakorn industrial zone, where over 220 factories of electronics and computer components suppliers are located (Fuller 2012).

In a related stream of research, the design of supply chain networks that are resilient to disruptions is considered and the fortification models are developed to improve the reliability of the existing infrastructure systems for which a complete reconfiguration would be cost prohibitive, e.g., Snyder et al. (2014). The objective of such fortification models is to identify optimal strategies for allocating limited resources among possible mitigation investments. For example, Church and Scaparra (2006) introduced the r-interdiction median problem with fortification in a distribution network with p operating facilities and a set of system users who receive service from their nearest facility. The problem objective is to determine the optimal allocation of a limited amount of protective resources in such a way that the accessibility reduction due to a worst-case loss of r unprotected facilities is minimized. They formulate the problem as a mixed integer program. The proposed model presents the limitation of requiring a complete enumeration of all possible ways of interdicting r of the p facilities. In order to alleviate the size restrictions, in a subsequent work Scaparra and Church (2008) presented a bi-level formulation of the r-interdiction median problem with fortification. The top level problem involves the decisions about which facilities to fortify in order to minimize the worst-case efficiency reduction due to the loss of unprotected facilities and worst-case scenario losses are modeled in the lower-level interdiction problem.

In the disaster management literature a limited number of studies focus on the pre-positioning of the different types of risk mitigation supplies in the presence of uncertainty. For example, Rawls and Turnquist (2010) and Noyan (2012) consider the problem of determining the locations of the response facilities and the inventory levels of disaster relief supplies at each facility to effectively manage the response operations when a disaster occurs. There is also a growing body of literature addressing the quantities of pre-positioned risk mitigation inventory, also called, strategic inventory reserves, Schmitt (2011) or just-in-case inventory, Sheffi (2005), Sheffi and Rice (2005), which should be held throughout the supply chain to protect against disruption risks.

The material presented in this chapter is based in part on research reported in Sawik (2013c), where SMIP models were proposed for risk-neutral, risk-averse and mean-risk selection of resilient supply portfolio under disruption risks. In the proposed models a single protection level was considered only with a full remaining capacity of a fortified supplier available after disruption. In this chapter, however, the SMIP approach was enhanced for multi-level protection with the amount of capacity remaining after disruption, dependent on the fortification cost and the protection level applied. In Sawik (2013d) a resilient supply portfolio was considered with fortified suppliers and regular inventory pre-positioned at the fortified suppliers. The regular inventory can be fully used under each disruption scenario to fulfill regular orders placed on the protected suppliers.

It is worth to note that the number of variables and constraints in the proposed models grow exponentially in the number $\overline{I}$ of suppliers, if all $\overline{S} = 2^{\overline{I}}$ potential scenarios are considered. The 10-supplier examples for 25-part types have approximately 280,000 variables and 280,000 constraints, while for a 20-supplier problem with all potential disruption scenarios considered and a single part type only, this increases to over 20,000,000 variables and 20,000,000 constraints. Solving the problem with such huge number of scenarios is cumbersome. Even the construction of problems of this size may become intractable, e.g., Chahara and Taaffe (2009). However, the total number of potential disruption scenarios that need to be considered can be reduced by eliminating scenarios that are unlikely to realize. Moreover, using of scenario management approaches (e.g., Jenkins 2000) can also be considered to identify a subset of potential disruption scenarios for which a detailed analysis of their impact on supply chain performance may provide information on the impact of all potential disruption scenarios. Another approach used to reduce the number of random disruption scenarios is the Fuzzy C-Mean clustering technique (e.g., Izakian and Abraham 2011) and the possibilistic scenario-based model (e.g., Torabi et al. 2015). In order to reduce the number of scenarios, the FCM technique is used to cluster possible disruptive events at suppliers to different clusters by which the centers of clusters are used as representatives of disruptive events.

In this chapter, the local and regional supply disruptions are assumed to occur independently. The future research should consider new resilience strategies for dependent disruptive events (e.g., Li et al. 2013) or multiple concurrent disruptions (Zobel and Khansa 2014).

Problems

4.1. Modify the probability, P_s, for disruption scenarios to account for suppliers located in different regions, subject to regional disruptions that may affect all suppliers in the same region simultaneously. Modify the SMIP models presented in this chapter for a joint fortification of all suppliers in one region.

4.2. Define service level for the SMIP models presented in this chapter and modify the models for the service level objective function.

4.3. Limited storage space for risk mitigation inventory

(a) Modify the SMIP models presented in Sect. 4.3 for the resilient supply portfolio with single protection levels to account for limited storage space for risk mitigation inventory at protected suppliers.

(b) Modify the SMIP models presented in Sect. 4.5 for the resilient supply portfolio with multiple protection levels to account for limited storage space for risk mitigation inventory at protected suppliers, dependent on protection level.

4.4. Modify the protection index introduced in Sect. 4.4 for the resilient supply portfolios with multiple protection levels.

4.5. In the computational examples for unreliable suppliers, Table 4.9 and Fig. 4.6(b) indicate that the risk-averse resilient solutions for confidence levels $\alpha = 0.9, 0.95, 0.99$ are very close to each other. How, would you explain the reason for that, and what would be the explanation for the case with identical solutions?
Hint: compare with solution results presented in Table 4.5 for a single protection level.

Part II
Integrated Selection of Supply Portfolio and Scheduling

Chapter 5
Integrated Selection of Supply Portfolio and Scheduling of Production

5.1 Introduction

The supplier selection and order quantity allocation are a medium- to short-term decision, driven by the time-varying customer demand. The scheduling horizon for supplies of parts coincides with the scheduling horizon for customer orders and to achieve the best results the supplier selection and order quantity allocation decisions should also be made for the same time horizon. The advantage of a joint decision-making can be proven especially in the presence of supply chain disruption risks. This chapter proposes a SMIP approach to integrated supplier selection and customer order scheduling in the presence of supply chain disruption risks, for a single, dual, or multiple sourcing strategy. The suppliers are assumed to be located in two or more disjoint geographic regions: in the producer's region (domestic suppliers) and outside the producer's region (foreign suppliers). The supplies are subject to independent random local disruptions that are uniquely associated with a particular supplier and to random regional disruptions that may result in disruption of all suppliers in the same geographic region simultaneously. The domestic suppliers are relatively reliable but more expensive, while the foreign suppliers offer competitive prices. However, the foreign suppliers are more prone to breakdowns and material flows from these suppliers are more exposed to unexpected disruptions due to natural or man-made disasters and longer shipping time and distance. Given a set of customer orders for products, the decision maker needs to decide which single supplier, which two suppliers (one from each region) or which multiple suppliers to select for purchasing parts required to complete the customer orders and how to schedule the orders over the planning horizon, to mitigate the impact of disruption risks. The problem objective is either to minimize total cost of ordering and purchasing of parts plus penalty cost of delayed and unfulfilled customer orders due to the parts shortages or to maximize customer service level. The resulting allocation of total demand for parts among the selected suppliers and the schedule of customer

T. Sawik, *Supply Chain Disruption Management*, International Series in Operations Research & Management Science 291, https://doi.org/10.1007/978-3-030-44814-1_5

orders for every potential disruption scenario should be determined ahead of time, either to minimize the average or worst-case cost or to maximize the average or worst-case customer service level.

The following time-indexed SMIP models are presented in this chapter:

SPSm_E(c), SPS2_E(c) and **SPS1_E(c)** for risk-neutral selection of supply portfolio and scheduling of customer orders to minimize expected cost: multiple sourcing, dual sourcing and single sourcing, respectively;

SPSm_E(sl), SPS2_E(sl) and **SPS1_E(sl)** for risk-neutral selection of supply portfolio and scheduling of customer orders to maximize expected service level: multiple sourcing, dual sourcing and single sourcing, respectively;

SPSm_CV(c), SPS2_CV(c) and **SPS1_CV(c)** for risk-averse selection of supply portfolio and scheduling of customer orders to minimize CVaR of cost: multiple sourcing, dual sourcing and single sourcing, respectively;

SPSm_CV(sl), SPS2_CV(sl) and **SPS1_CV(sl)** for risk-averse selection of supply portfolio and scheduling of customer orders to maximize CVaR of service level: multiple sourcing, dual sourcing and single sourcing, respectively;

SPSm_E(c)CV(sl) for mixed mean-risk selection of supply portfolio and scheduling of customer orders to trade-off expected cost and CVaR of service level: multiple sourcing.

For a single, dual, or multiple sourcing strategy and for the two different objective functions, the risk-neutral, risk-averse, and mean-risk solutions that optimize, respectively, average, worst-case, and trade-off between average and worst-case performance of a supply chain are illustrated with computational examples and compared in Sect. 5.6. In addition, the two risk-averse service level measures: expected worst-case order fulfillment rate and expected worst-case demand fulfillment rate were computationally compared in Sect. 5.6.3. The major managerial insights are provided at the end of this chapter.

5.2 Problem Description

In this section the problem of integrated supplier selection, order quantity allocation, and customer orders scheduling in the presence of supply chain disruption risks is described. While the supplier selection is considered to be a long-term strategic decision, the order quantity allocation and customer order scheduling are short- to medium-term tactical decisions. In particular, in a make-to-order manufacturing, all

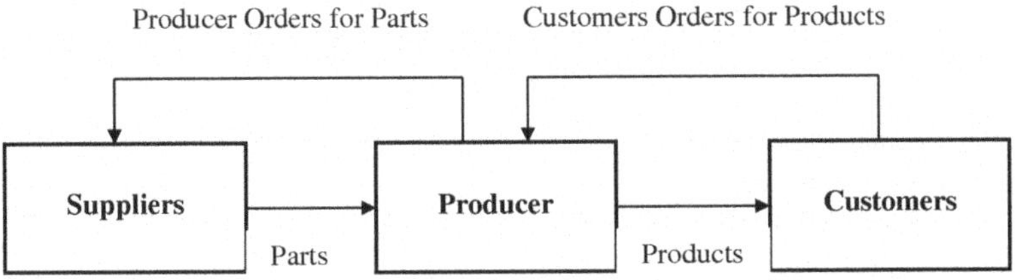

Fig. 5.1 A two-tier supply chain

the above decisions can be made for a short- to medium-term planning horizon. Given a set of part suppliers, the supply portfolio determines an allocation of demand for parts among a subset of selected suppliers and simultaneously, for each disruption scenario an assignment of customer orders to time periods over the planning horizon is found. To justify the integration of supplier selection, order quantity allocation, and customer order aggregate scheduling, assume that all these decisions are made for a medium-term planning horizon. However, the order quantity allocation among the suppliers selected for a longer time horizon in a make-to-order environment can also be made for a short-term horizon with no orders placed on some suppliers, if a shorter planning horizon needs to be considered.

Consider a two-tier customer-driven supply chain (Fig. 5.1) in which various types of products are assembled by a single producer to meet customer demand, using the same critical part type that can be manufactured and provided by many suppliers (for notation used, see Table 5.1).

Let $I = \{1,\dots,\overline{I}\}$ be the set of $\overline{I}$ suppliers, $J = \{1,\dots,\overline{J}\}$ the set of $\overline{J}$ customer orders for products, and $T = \{1,\dots,\overline{T}\}$ the set of $\overline{T}$ planning periods.

Denote by b_j and d_j, respectively, the size and the due date of customer order $j \in J$, i.e., the number units of ordered product type and the latest period of their completion required to deliver the products to the customer by requested date. The customer orders are single-period orders such that each order can be completed in one planning period, e.g., Sawik (2007).

Let a_j be the unit requirement for the critical part of each product in customer order $j \in J$. The total demand for all parts is $A = \sum_{j \in J} a_j b_j$ and the total demand for all products is $B = \sum_{j \in J} b_j$.

The orders for parts are assumed to be placed at the start of the planning horizon, when all customer orders for products are known. Let o_i be the unit purchasing price of parts from supplier $i \in I$ and denote by e_i the fixed ordering cost of creating contracts and maintaining relationships with supplier $i \in I$. Each supplier has sufficient capacity to meet total demand for parts and to complete and prepare orders for shipping. Then, all parts ordered from a supplier are shipped together in a single delivery. The order preparation and transportation time of a shipment from supplier $i \in I$ to the producer is constant and equals to τ_i periods so that the parts ordered from supplier $i \in I$ are delivered in period τ_i and then can be used for the assembly of products in period $\tau_i + 1$, at the earliest.

Table 5.1 Notation: supply portfolio selection and production scheduling

	Indices	
i	=	Supplier, $i \in I$
j	=	Customer order, $j \in J$
r	=	Geographic region, $r \in R$
s	=	Disruption scenario, $s \in S$
t	=	Planning period, $t \in T$
	Input Parameters	
a_j	=	Per unit requirement for parts of each product in customer order j
b_j	=	Size (number of products) of customer order j
A	=	Total demand for parts
B	=	Total demand for products
c_j	=	Per unit capacity consumption of producer for customer order j
C_t	=	Capacity of producer in period t
d_j	=	Due date for customer order j
e_i	=	Fixed cost of ordering parts from supplier i
g_j	=	Per unit penalty cost of unfulfilled customer order j
h_j	=	Per unit and per period penalty cost of delayed customer order j
I^r	=	The subset of suppliers in geographic region r
I_s	=	The subset of suppliers non-disrupted under disruption scenario s
o_i	=	Per unit price of parts purchased from supplier i
p_i	=	Local disruption probability for supplier i
p^r	=	Regional disruption probability for all suppliers in region r
p^*	=	Global disruption probability for all suppliers
α	=	Confidence level
τ_i	=	Delivery lead time from supplier i

The suppliers are assumed to be located in $\overline{R}$ disjoint geographic regions. Denote by $I^r \subseteq I$ the subset of suppliers in region $r \in R = \{1,\ldots,\overline{R}\}$, where $\bigcup_{r\in R} I^r = I$. The supplies of parts are subject to random independent local disruptions that are uniquely associated with a particular supplier, which may arise from equipment breakdowns, local labor strike, fires, etc. Denote by p_i the local disruption probability for supplier i, i.e., the parts ordered from supplier i are delivered without disruptions with probability $(1-p_i)$, or not at all with probability p_i.

In addition to independent local disruptions of each supplier individually, there are also potential regional disasters that may result in correlated regional disruption of all suppliers in the same region and global disruptive super events that may simultaneously impact all suppliers. For example, such global disaster super events may include an economic crisis, widespread labor strike in a transportation sector.

Denote by p^r and p^* the probability of regional disruption, simultaneously of all suppliers $i \in I^r$ in region $r \in R$, and global disruption, simultaneously of all suppliers $i \in I$, respectively. The global disaster, the regional disasters in each region, and the local disruptive events at each supplier are assumed to be independent events. Let π_i be the disruption probability of every supplier $i \in I^r, r \in R$

$$\pi_i = p^* + (1-p^*)p^r + (1-p^*)(1-p^r)p_i;\ i \in I^r, r \in R. \tag{5.1}$$

Denote by P_s the probability that disruption scenario s is realized, where each scenario $s \in S$ is comprised of a unique subset $I_s \subset I$ of suppliers who can deliver parts without disruptions, and $S = \{1,\ldots,\overline{S}\}$ is the index set of all disruption scenarios (there are a total of $\overline{S} = 2^{\overline{I}}$ potential disruption scenarios). For each scenario $s \in S$, the supplies from every supplier, $i \in I \setminus I_s$, can be disrupted either by a local, regional, or global disruptive event.

The probability P_s for each disruption scenario $s \in S$ with the subset I_s of non-disrupted suppliers, and with all possible combinations of different disaster events considered, is (cf. Sect. 1.3)

$$P_s = \begin{cases} (1-p^*)\prod_{r\in R} P_s^r & \text{if } I_s \neq \emptyset \\ p^* + (1-p^*)\prod_{r\in R} P_s^r & \text{if } I_s = \emptyset, \end{cases} \tag{5.2}$$

where P_s^r is the probability of realizing of disruption scenario s for suppliers in I^r

$$P_s^r = \begin{cases} (1-p^r)\prod_{i\in I^r\bigcap I_s}(1-p_i)\prod_{i\in I^r\setminus I_s} p_i & \text{if } I^r\bigcap I_s \neq \emptyset \\ p^r + (1-p^r)\prod_{i\in I^r} p_i & \text{if } I^r\bigcap I_s = \emptyset. \end{cases} \tag{5.3}$$

Assume that the producer has limited time-varying capacity, and denote by C_t the producer capacity available in planning period $t \in T$, and by c_j the unit capacity consumption for each product in customer order $j \in J$.

The producer can be charged with a contractual, order specific penalty cost for delayed or unfulfilled customer orders, caused by the shortage of parts, that are delivered late or not at all due to supply disruptions. Let g_j and h_j be, respectively, the per unit penalty cost of unfulfilled customer order $j \in J$ and the per unit and per period penalty cost of delayed customer order $j \in J$. The penalty cost g_j is assumed to also include the producer lost profit. Therefore, the unit penalty costs h_j and g_j are selected in such a way that each product in an unfulfilled order is penalized much higher than the corresponding product in a delayed order, i.e., $g_j \gg h_j,\ j \in J$. In some cases the contractual penalty cost for unfulfilled customer orders can be shared between the supplier and the producer to compensate the latter for the lost profit due to the undelivered parts. This can be modeled by choosing appropriate lower values of the corresponding unit penalty costs, g_j, to reduce the producer direct losses. Then, the remaining part of the producer penalty cost is assumed to be covered by the contractual compensation from the disrupted supplier.

The producer has three sourcing alternatives to select: single, dual, or multiple sourcing. The objective of the integrated supplier selection and customer order scheduling is to select a single supplier (single sourcing), two suppliers from two different regions (dual sourcing), or multiple suppliers (multiple sourcing), allocate the total demand for parts among the selected suppliers, and schedule the customer orders over the planning horizon to complete the orders and mitigate the impact of disruption risks. The three sourcing strategies will be compared with respect to the two alternative optimality criteria: cost and service level. The objective function is either to minimize expected cost, expected worst-case cost, or maximize expected service level or expected worst-case service level. The supply portfolio (the allocation of total demand for parts among the selected suppliers) and the schedule of customer orders for every potential disruption scenario are determined at the beginning of the planning horizon.

5.3 Models for Risk-Neutral Decision-Making

In this section six time-indexed SMIP models, **SPS1_E(c)**, **SPS2_E(c)**, **SPSm_E(c)**, **SPS1_E(sl)**, **SPS2_E(sl)**, **SPSm_E(sl)**, are proposed for the integrated supplier selection and customer order scheduling to optimize average performance of a supply chain in the presence of disruption risks. The objective of models **SPS1_E(c)**, **SPS2_E(c)**, **SPSm_E(c)** and **SPS1_E(sl)**, **SPS2_E(sl)**, **SPSm_E(sl)** is to minimize the expected cost per product ("E(c)" in the model name) and to maximize the expected service level ("E(sl)" in the model name), respectively.

While the objective of supplier selection is to determine a supply portfolio, i.e., an allocation of demand for parts among the suppliers, the objective of customer orders scheduling is to determine an aggregate production schedule, i.e., an assignment of orders to planning periods over a planning horizon, subject to capacity and parts availability constraints. For the selected supply portfolio and each disruption scenario the optimal schedule of customer orders is determined in such a way as to minimize the penalty cost for delayed and unscheduled (rejected) orders, and by this minimize total cost or maximize service level, respectively.

The following three different sourcing strategies will be considered:

- Single sourcing ("1" in the model name), where the total demand for parts is assigned to exactly one supplier.
- Dual sourcing ("2" in the model name), where the total demand for parts is either assigned to one supplier or allocated between two suppliers, from two different geographic regions. For example, the total demand for parts is split between the two suppliers, one more expensive and more reliable domestic supplier, and one less expensive and less reliable foreign supplier.
- Multiple sourcing ("m" in the model name), where the total demand for parts is allocated among many suppliers.

The problem variables are defined in Table 5.2:

5.3.1 Minimization of Cost

In this subsection the three time-indexed SMIP models are presented for the risk-neutral selection of supply portfolio and scheduling of customer orders to minimize expected cost: model **SPSm_E(c)** for multiple sourcing, model **SPS2_E(c)** for dual sourcing and model **SPS1_E(c)** for single sourcing. The single sourcing strategy can be considered to be a special case of a dual sourcing, while the dual sourcing can be considered a special case of a multiple sourcing. Therefore, first model **SPSm_E(c)** is presented, then model **SPS2_E(c)** and finally model **SPS1_E(c)**.

Table 5.2 Problem variables: supply portfolio selection and production scheduling

	First stage variables
u_i	$=$ 1, if supplier i is selected; otherwise $u_i = 0$ (supplier selection)
v_i	$\in [0,1]$, the fraction of total demand for parts ordered from supplier i (supply portfolio for dual or multiple sourcing)
	Second stage variables
x^s_{jt}	$=$ 1, if under disruption scenario s customer order j is scheduled for period t; otherwise $x^s_{jt} = 0$ (production scheduling)
	Auxiliary variables
VaR^c	cost-at-risk, the targeted cost such that for a given confidence level α, for $100\alpha\%$ of the scenarios, the outcome is below VaR^c
VaR^{sl}	service-at-risk, the targeted service level such that for a given confidence level α, for $100\alpha\%$ of the scenarios, the outcome is above VaR^{sl}
$\mathscr{C}_s$	≥ 0, the tail cost for scenario s, i.e., the amount by which costs in scenario s exceed VaR^c
$\mathscr{S}_s$	≥ 0, the tail service level for scenario s, i.e., the amount by which VaR^{sl} exceeds service level in scenario s

In the risk-neutral decision-making, the overall quality of the supply portfolio and the schedule of customer orders can be measured by the expected cost per product, (5.4), of parts ordering, $\sum_{i\in I} e_i u_i/B$, and purchasing $(\sum_{s\in S} P_s(\sum_{i\in I_s} Ao_i v_i)/B)$, where the producer is not charged with ordered and undelivered parts, plus penalty cost of delayed and unfulfilled (rejected) customer orders due to delays and disruptions of part supplies, $\sum_{s\in S} P_s(\sum_{j\in J}\sum_{t\in T:t>d_j} h_j b_j (t-d_j) x^s_{jt})/B + \sum_{s\in S} P_s(\sum_{j\in J} g_j b_j (1-\sum_{t\in T} x^s_{jt}))/B$.

The SMIP model **SPSm_E(c)** for a multiple sourcing selection of supply portfolio and scheduling of customer orders to minimize expected cost is formulated below.

SPSm_E(c): *Risk-neutral selection of* **S**upply **P**ortfolio *and* **S**cheduling *of customer orders to minimize expected cost:* **m***ultiple sourcing*

Minimize

$$
\begin{aligned}
E^c = \sum_{i\in I} e_i u_i/B + \sum_{s\in S} P_s(\sum_{i\in I_s} Ao_i v_i)/B \\
+\sum_{s\in S} P_s(\sum_{j\in J}\sum_{t\in T:t>d_j} h_j b_j (t-d_j) x^s_{jt})/B \\
+\sum_{s\in S} P_s(\sum_{j\in J} g_j b_j (1-\sum_{t\in T} x^s_{jt}))/B,
\end{aligned}
\tag{5.4}
$$

subject to

Demand allocation constraints:

(continued)

– the total demand for parts must be fully allocated among the suppliers,

$$\sum_{i \in I} v_i = 1. \tag{5.5}$$

Multiple sourcing strategy constraints:

– demand for parts cannot be assigned to non-selected suppliers,

$$v_i \le u_i;\ i \in I. \tag{5.6}$$

Order-to-period assignment constraints:

– for each disruption scenario s, each customer order j is either scheduled during the planning horizon ($\sum_{t \in T} x_{jt}^s = 1$), or unscheduled and rejected ($\sum_{t \in T} x_{jt}^s = 0$) (5.7),
– for each disruption scenario s and each planning period t, the cumulative demand for parts of all customer orders scheduled in periods 1 through t cannot exceed the cumulative deliveries of parts in periods 1 through $t-1$, from the non-disrupted suppliers $i \in I_s$ (5.8),
– for each disruption scenario s, the total requirement for parts of scheduled customer orders is not greater than the total supplies from the non-disrupted suppliers $i \in I_s$ (5.9),

$$\sum_{t \in T} x_{jt}^s \le 1;\ j \in J, s \in S, \tag{5.7}$$

$$\sum_{j \in J} \sum_{t' \in T: t' \le t} a_j b_j x_{jt'}^s \le A \sum_{i \in I_s: \tau_i \le t-1} v_i;\ t \in T, s \in S, \tag{5.8}$$

$$\sum_{j \in J} \sum_{t \in T} a_j b_j x_{jt}^s \le A \sum_{i \in I_s} v_i;\ s \in S. \tag{5.9}$$

Producer capacity constraints:

– for any period t and each disruption scenario s, the total demand on capacity of all customer orders scheduled in period t must not exceed the producer capacity available in this period,

$$\sum_{j \in J} b_j c_j x_{jt}^s \le C_t;\ t \in T, s \in S. \tag{5.10}$$

Non-negativity and integrality conditions:

$$u_i \in \{0,1\};\ i \in I, \tag{5.11}$$

$$v_i \in [0,1];\ i \in I, \tag{5.12}$$

$$x_{jt}^s \in \{0,1\};\ j \in J, t \in T, s \in S. \tag{5.13}$$

In the SMIP model **SPS2_E(c)** formulated below for a dual sourcing, all suppliers are assumed to be located in two different geographic regions and at most one supplier can be selected from each region.

SPS2_E(c): *Risk-neutral selection of supply portfolio and scheduling of customer orders to minimize expected cost: dual sourcing*
Minimize (5.4)
subject to (5.5)–(5.13) and
Dual sourcing strategy constraint:

- at most one supplier can be selected from each region,

$$\sum_{i \in I^r} u_i \le 1; \; r = 1,2. \tag{5.14}$$

Note that dual and multiple sourcing allows, respectively, two and more suppliers to be selected; however, an optimal solution may assign total demand for parts to a single supplier only, if such a solution optimizes the objective function.

The single sourcing model **SPS1_E(c)** is a special case of the dual sourcing model **SPS2_E(c)**. In order to derive **SPS1_E(c)** from **SPS2_E(c)**, it is sufficient to replace inequality $v_i \le u_i;\ i \in I$, (5.6), with an equality constraint $v_i = u_i;\ i \in I$ and by this eliminate variables $v_i,\ i \in I$. Simultaneously, constraint (5.14) becomes weaker than (5.5) and can be removed, while constraints (5.8) and (5.9) should be replaced by

$$\sum_{j \in J} \sum_{t' \in T: t' \le t} x^s_{jt'} \le \bar{J} \sum_{i \in I_s: \tau_i \le t-1} u_i; \; t \in T, s \in S$$

or

$$\sum_{t' \in T: t' \le t} x^s_{jt'} \le \sum_{i \in I_s: \tau_i \le t-1} u_i; \; j \in J, t \in T, s \in S,$$

and

$$\sum_{j \in J} \sum_{t \in T} x^s_{jt} \le \bar{J} \sum_{i \in I_s} u_i; \; s \in S$$

or

$$\sum_{t \in T} x^s_{jt} \le \sum_{i \in I_s} u_i; \; j \in J, s \in S,$$

respectively.

Constraints (5.8) can also be replaced by the following alternative constraints:

$$\sum_{t \in T} t x^s_{jt} \ge \sum_{i \in I_s} (\tau_i + 1) u_i; \; j \in J, s \in S,$$

which sometimes may appear to be more efficient computationally.

The stochastic binary programming model **SPS1_E(c)** for a single sourcing selection of supply portfolio and scheduling of customer orders is formulated below.

SPS1_E(c): *Risk-neutral selection of supply portfolio and scheduling of customer orders to minimize expected cost: single sourcing*

Minimize

$$\sum_{i\in I} e_i u_i/B + \sum_{s\in S} P_s(\sum_{i\in I_s} Ao_i u_i)/B$$
$$+\sum_{s\in S} P_s(\sum_{j\in J}\sum_{t\in T:t>d_j} h_j b_j (t-d_j) x^s_{jt})/B$$
$$+\sum_{s\in S} P_s(\sum_{j\in J} g_j b_j (1-\sum_{t\in T} x^s_{jt}))/B, \tag{5.15}$$

subject to (5.10), (5.11), (5.13) and

Supplier selection constraints:

- exactly one supplier of parts is selected,

$$\sum_{i\in I} u_i = 1. \tag{5.16}$$

Order-to-period assignment constraints: (5.7) and

- for each disruption scenario s, all customer orders can be scheduled only after the delivery of required parts, purchased from a non-disrupted supplier $i \in I_s$,
- each customer order j can be scheduled during the planning horizon under disruption scenario s, only if required parts are ordered from a non-disrupted supplier $i \in I_s$,

$$\sum_{t'\in T:t'\leq t} x^s_{jt'} \leq \sum_{i\in I_s:\tau_i\leq t-1} u_i;\ j\in J, t\in T, s\in S, \tag{5.17}$$

$$\sum_{t\in T} x^s_{jt} \leq \sum_{i\in I_s} u_i;\ j\in J, s\in S. \tag{5.18}$$

The solution to supplier selection and customer order scheduling problem determines for every disruption scenario $s \in S$, the aggregate production schedule $\{\sum_{j\in J} b_j x^s_{jt};\ t\in T\}$ as well as the corresponding service level, $\sum_{j\in J}\sum_{t\in T:t\leq d_j} x^s_{jt}/J$, (the fraction of customer orders fulfilled by their due dates). In addition, for every disruption scenario $s \in S$, the subset of customer orders scheduled by the requested due dates d_j, $\{j\in J: 1+\min_{i\in I}\tau_i \leq \sum_{t\in T} t x^s_{jt} \leq d_j\}$, the subset of delayed customer orders, $\{j\in J: \sum_{t\in T} t x^s_{jt} > d_j\}$, and the subset of rejected (unscheduled) customer orders, $\{j\in J: \sum_{t\in T} x^s_{jt} = 0\}$, are determined.

5.3.2 Maximization of Customer Service Level

The customer service level can be measured either by *order fulfillment rate* or by *demand fulfillment rate*, where the order fulfillment rate and the demand fulfillment rate is the fraction of customer orders (irrespective of their size) and the fraction of customer demand, respectively, that is fulfilled by the customer requested due dates. When the focus is on fulfilling customer orders rather than total demanded quantity, the order fulfillment rate would be the preferred service level measure. Otherwise, the demand fulfillment rate would be selected. Since the order fulfillment rate does not account for the size of customer orders, a high service level can be achieved by fulfilling a large number of small size orders, while leaving the unfulfilled demand relatively high. On the other hand, a high demand fulfillment rate can be achieved by fulfilling a few large size orders, while leaving relatively high, the number of unfulfilled small size orders.

The aim of the next three SMIP models is to achieve the best average customer service level by maximizing the expected fraction of customer orders fulfilled by their due dates, i.e., by maximizing the expected order fulfillment rate.

SPSm_E(sl): *Risk-neutral selection of supply portfolio and scheduling of customer orders to maximize expected service level: multiple sourcing*

Maximize

$$\sum_{s \in S} P_s \sum_{j \in J} \sum_{t \in T: t \leq d_j} x_{jt}^s / \bar{J}, \tag{5.19}$$

subject to (5.5)–(5.13).

SPS2_E(sl): *Risk-neutral selection of supply portfolio and scheduling of customer orders to maximize expected service level: dual sourcing*

Maximize (5.19)

subject to (5.5)–(5.14).

SPS1_E(sl): *Risk-neutral selection of supply portfolio and scheduling of customer orders to maximize expected service level: single sourcing*

Maximize (5.19)

subject to (5.7), (5.10), (5.11), (5.13), (5.16)–(5.18).

Note that maximization of the customer service level simultaneously leads to reduction of the penalty costs for delayed and unfulfilled customer orders that represent an important part of the total cost structure. However, as the objective function (5.19) does not directly include any cost components, the optimal solution to

SPSm_E(sl), **SPS2_E(sl)**, and **SPS1_E(sl)** that maximizes the customer service level depends only on the delivery lead time from suppliers and distribution among the suppliers of local and regional disruption probabilities. Therefore, the solution can be considered to be an ideal solution with respect to the customer service level in a risk-neutral decision-making, and may be used for comparisons with the optimal cost-based solutions.

If instead of the number of customer orders fulfilled on time (i.e., the order, fulfillment rate) customer service level is measured by the fraction of customer demand fulfilled on time (i.e., by the demand fulfillment rate), then (5.19) in models **SPSm_E(sl)**, **SPS2_E(sl)**, and **SPS1_E(sl)** should be replaced by the following objective function to be maximized:

$$\sum_{s \in S} P_s \sum_{j \in J} \sum_{t \in T: t \leq d_j} b_j x_{jt}^s / B. \tag{5.20}$$

5.4 Models for Risk-Averse Decision-Making

In the selection of supply portfolio and scheduling of customer orders under disruption risks, the decision maker controls the risk of high losses due to supply disruptions by choosing the confidence level α. For a given confidence level, let VaR^c be the acceptable cost level above which we want to minimize the number of outcomes and CVaR^c considers those portfolio outcomes, where costs exceed VaR^c. In a similar way, denote by VaR^{sl} the acceptable service level below which we want to minimize the number of outcomes and CVaR^{sl} considers those portfolio outcomes, where service levels are below VaR^{sl}. We assume that the decision maker is willing to accept only portfolios for which the total probability of scenarios with costs greater than VaR^c (or with service levels lower than VaR^{sl}) is not greater than $1-\alpha$. The greater the confidence level α, the more risk aversive is the decision maker and the smaller percent of the highest cost (of the lowest service level, respectively) outcomes is focused on.

In this section the six time-indexed SMIP models: **SPSm_CV(c)**, **SPS2_CV(c)**, **SPS1_CV(c)**, **SPSm_CV(sl)**, **SPS2_CV(sl)**, **SPS1_CV(sl)** are proposed for the integrated supplier selection and customer order scheduling to optimize worst-case performance of a supply chain in the presence of disruption risks. The objective of models **SPSm_CV(c)**, **SPS2_CV(c)**, **SPS1_CV(c)** and **SPSm_CV(sl)**, **SPS2_CV(sl)**, **SPS1_CV(sl)** is to minimize CVaR^c of cost per product ("CV(c)" in the model name) and to maximize CVaR^{sl} of service level ("CV(sl)" in the model name), respectively. The models are risk-averse equivalents, respectively, of **SPSm_E(c)**, **SPS2_E(c)**, **SPS1_E(c)** and **SPSm_E(sl)**, **SPS2_E(sl)**, **SPS1_E(sl)** formulations presented in Sect. 5.3 for the risk-neutral decision-making.

5.4.1 Minimization of Cost

Define $\mathscr{C}_s$ as the tail cost for scenario s, where tail cost is defined as the amount by which costs in scenario s exceed VaR^c. The risk-averse supply portfolio and production schedule will be optimized by calculating VaR^c and minimizing CVaR^c simultaneously. In the proposed model, CVaR^c is represented by an auxiliary function (5.21) introduced by Rockafellar and Uryasev (2000).

The SMIP models **SPSm_CV(c)**, **SPS2_CV(c)**, and **SPS1_CV(c)** for supplier selection and customer order scheduling to optimize worst-case performance of supply chain and reduce the risk of high costs, respectively, for multiple, dual and single sourcing, are formulated below.

SPSm_CV(c): *Risk-averse selection of supply portfolio and scheduling of customer orders to minimize CVaR of cost: multiple sourcing*

Minimize

$$CVaR^c = VaR^c + (1-\alpha)^{-1} \sum_{s \in S} P_s \mathscr{C}_s, \tag{5.21}$$

subject to (5.5)–(5.13) and

Risk constraints:

- the tail cost for scenario s is defined as the nonnegative amount by which cost in scenario s exceeds VaR^c,

$$\mathscr{C}_s \ge \sum_{i \in I} e_i u_i / B + \sum_{i \in I_s} A o_i v_i / B + \sum_{j \in J} \sum_{t \in T: t > d_j} h_j b_j (t - d_j) x_{jt}^s / B + \sum_{j \in J} g_j b_j (1 - \sum_{t \in T} x_{jt}^s) / B - VaR^c; \; s \in S, \tag{5.22}$$

$$\mathscr{C}_s \ge 0. \tag{5.23}$$

SPS2_CV(c): *Risk-averse selection of supply portfolio and scheduling of customer orders to minimize CVaR of cost: dual sourcing*

Minimize (5.21)

subject to (5.5)–(5.14), (5.22), (5.23).

SPS1_CV(c): *Risk-averse selection of supply portfolio and scheduling of customer orders to minimize CVaR of cost: single sourcing*

Minimize (5.21) subject to (5.7), (5.10), (5.11), (5.13), (5.16)–(5.18), (5.23) and

(continued)

Risk constraints:

– the tail cost for scenario s is defined as the nonnegative amount by which cost in scenario s exceeds VaR^c,

$$\mathscr{C}_s \geq \sum_{i \in I} e_i u_i / B + \sum_{i \in I_s} A o_i u_i / B$$
$$+ \sum_{j \in J} \sum_{t \in T: t > d_j} h_j b_j (t - d_j) x^s_{jt} / B + \sum_{j \in J} g_j b_j (1 - \sum_{t \in T} x^s_{jt}) / B$$
$$- VaR^c; \; s \in S. \quad (5.24)$$

5.4.2 Maximization of Customer Service Level

Define $\mathscr{S}_s$ as the tail service level for scenario s, where tail service level is defined as the amount by which VaR^{sl} exceeds service level in scenario s. The risk-averse supply portfolio and production schedule will be optimized by calculating VaR^{sl} and maximizing CVaR^c simultaneously.

The SMIP models **SPSm_CV(sl)**, **SPS2_CV(sl)**, and **SPS1_CV(sl)** for supplier selection and customer order scheduling to optimize worst-case performance of supply chain and reduce the risk of low service levels, respectively, for multiple, dual and single sourcing, are formulated below.

SPSm_CV(sl): *Risk-averse selection of supply portfolio and scheduling of customer orders to maximize CVaR of service level: multiple sourcing*

Maximize

$$CVaR^{sl} = VaR^{sl} - (1 - \alpha)^{-1} \sum_{s \in S} P_s \mathscr{S}_s, \quad (5.25)$$

subject to (5.5)–(5.13) and

Risk constraints:

– the tail service level for scenario s is defined as the nonnegative amount by which VaR^{sl} exceeds service level in scenario s,

$$\mathscr{S}_s \geq VaR^{sl} - \sum_{j \in J} \sum_{t \in T: t \leq d_j} x^s_{jt} / \bar{J}; \; s \in S, \quad (5.26)$$

$$\mathscr{S}_s \geq 0. \quad (5.27)$$

SPS2_CV(sl): *Risk-averse selection of supply portfolio and scheduling of customer orders to maximize CVaR of service level: dual sourcing*
Maximize (5.25)
subject to (5.5)–(5.14), (5.26), (5.27).

SPS1_CV(sl): *Risk-averse selection of supply portfolio and scheduling of customer orders to maximize CVaR of service level: single sourcing*
Maximize (5.25)
subject to (5.7), (5.10), (5.11), (5.13), (5.16)–(5.18), (5.26), (5.27).

If instead of the number of customer orders fulfilled on time (i.e., the order, fulfillment rate) customer service level is measured by the fraction of customer demand fulfilled on time (i.e., by the demand fulfillment rate), then (5.26) in models **SPSm_CV(sl)**, **SPS2_CV(sl)**, and **SPS1_CV(sl)** should be replaced by the following constraints:

$$\mathscr{S}_s \geq VaR^{sl} - \sum_{j \in J} \sum_{t \in T: t \leq d_j} b_j x^s_{jt}/B;\ s \in S. \tag{5.28}$$

In scenario-based optimization under uncertainty, the probability measure is concentrated in finitely many points, called "probability atoms", i.e., singletons, which have positive probability measure. The smaller is the number of concentration points and the greater are probability atoms, the smaller than $(1-\alpha)$ can be the probability, $\sum_{s \in S: \mathscr{S}_s > 0} P_s$, of outcomes with service level below VaR^{sl}, e.g., Sawik (2011b), see also Chap. 2. The following proposition accounts for that property of stochastic optimal solutions.

Proposition 5.1. *The optimized value of* $CVaR^{sl}$ *(5.25), for model* **SPSm_CV(sl)** *and model* **SPSm_CV(sl)** *with (5.26) replaced by (5.28), can be greater than, respectively:*

- *the actual expected worst-case fraction of customer orders fulfilled on time,*

$$\frac{\sum_{s \in S: \mathscr{S}_s > 0} P_s \sum_{j \in J} \sum_{t \in T: t \leq d_j} x^s_{jt}/\overline{J}}{\sum_{s \in S: \mathscr{S}_s > 0} P_s}, \tag{5.29}$$

and

- *the actual expected worst-case fraction of demand fulfilled on time,*

$$\frac{\sum_{s \in S: \mathscr{S}_s > 0} P_s \sum_{j \in J} \sum_{t \in T: t \leq d_j} b_j x^s_{jt}/B}{\sum_{s \in S: \mathscr{S}_s > 0} P_s}. \tag{5.30}$$

(continued)

Proof. In the risk-averse models for maximization of service level, CVaR^{sl}, for a given confidence level, α, is represented by an auxiliary function (5.25), where the probability of tail distribution (i.e., the probability of outcomes with worst-case service level below VaR^{sl}) is fixed to $(1-\alpha)$. However, the actual probability of the tail distribution of service level, $\sum_{s\in S:\mathscr{S}_s>0} P_s$, can be less than $(1-\alpha)$. As a result, the optimized value of CVaR^{sl} (5.25) can be greater than the actual values calculated using (5.29) and (5.30).

Model Enhancements

The SMIP models **SPSm_CV(sl)**, **SPS2_CV(sl)**, and **SPS1_CV(sl)** can be strengthened by the addition of valid inequalities to precisely determine VaR^{sl} for a given confidence level α. As a result, tighter LP relaxations of the corresponding mixed integer programs are achieved.

Proposition 5.2. *The following inequalities that define* VaR^{sl} *are valid inequalities for SMIP models* **SPSm_CV(sl)**, **SPS2_CV(sl)**, *and* **SPS1_CV(sl)**.

$$z_s \geq \sum_{j\in J} \sum_{t\in T:t\leq d_j} x^s_{jt}/\bar{J} - VaR^{sl};\ s\in S, \tag{5.31}$$

$$z_s \leq 1 + \sum_{j\in J} \sum_{t\in T:t\leq d_j} x^s_{jt}/\bar{J} - VaR^{sl};\ s\in S, \tag{5.32}$$

$$\sum_{s\in S} P_s z_s \geq \alpha, \tag{5.33}$$

Proof. Equations (5.31) and (5.32) determine scenarios $s\in S$ with customer service level not less than VaR^{sl}, and (5.33) ensures that the total probability of all such scenarios is not less than the confidence level α. The additional scenario selection binary variable z_s is defined as follows: $z_s = 1$, if for scenario $s\in S$, customer service level, $\sum_{j\in J}\sum_{t\in T:t\leq d_j} x^s_{jt}/\bar{J}$, is not less than VaR^{sl}; otherwise $z_s = 0$.

The introduction of valid inequalities will increase the size of the mixed integer programs **SPSm_CV(sl)**, **SPS2_CV(sl)** and **SPS1_CV(sl)**. The total number of additional binary variables z_s and constraints (5.31)–(5.33) is, respectively, $2^{\bar{I}}$ and $2^{\bar{I}+1}+1$, and hence they grow exponentially with the number $\bar{I}$ of suppliers.

Note that both VaR^{sl} and $\mathscr{S}_s$ can be restricted to being not greater that one and, in addition, $\mathscr{S}_s \leq 1 - z_s, \forall s \in S$, which is equivalent to (5.31).

Denote by **SPSm_CV(sl)+**, **SPS2_CV(sl)+**, and **SPS1_CV(sl)+**, the stochastic mixed integer programs **SPSm_CV(sl)**, **SPS2_CV(sl)**, and **SPS1_CV(sl)** strengthened with added valid inequalities (5.31)–(5.33), respectively.

If in models **SPSm_CV(sl)**, **SPS2_CV(sl)**, and **SPS1_CV(sl)**, customer service level is measured by the fraction of customer demand fulfilled on time (i.e., by demand fulfillment rate), then in the definition of scenario selection variable z_s and in constraints (5.31), (5.32), the term $\sum_{j\in J}\sum_{t\in T:t\leq d_j} x^s_{jt}/\overline{J}$ should be replaced by $\sum_{j\in J}\sum_{t\in T:t\leq d_j} b_j x^s_{jt}/B$.

5.5 Model for Mixed Mean-Risk Decision-Making

In this section a trade-off model **SPSm_E(c)CV(sl)** for a multiple sourcing strategy is developed to evaluate the impact on average cost of optimization the expected worst-case service level.

The expected cost per product, E^c (5.4), and the expected worst-case service level, $CVaR^{sl}$ (5.25), will be simultaneously optimized using the weighted-sum approach.

In order to avoid dimensional inconsistency among the two conflicting objective functions, their values are scaled into the interval [0,1].

Denote by $f^c = \frac{E^c - \underline{E^c}}{\overline{E^c} - \underline{E^c}}$, the normalized expected cost per product ($\underline{E^c}, \overline{E^c}$ are the minimum, the maximum values of E^c, respectively), and by $f^{sl} = \frac{\overline{CVaR^{sl}} - CVaR^{sl}}{\overline{CVaR^{sl}} - \underline{CVaR^{sl}}}$, the normalized expected worst-case service level ($\underline{CVaR^{sl}}, \overline{CVaR^{sl}}$ are the minimum, the maximum values of $CVaR^{sl}$ for a given confidence level α, respectively).

The normalized objective functions f^c and f^{sl} are defined below

$$f^c = (\sum_{i\in I} e_i u_i + \sum_{s\in S} P_s(\sum_{i\in I_s} Ao_i v_i + \sum_{j\in J}\sum_{t\in T:t>d_j} h_j b_j (t-d_j) x^s_{jt}$$

$$+ \sum_{j\in J} g_j b_j (1 - \sum_{t\in T} x^s_{jt})))/B - \underline{E^c})/(\overline{E^c} - \underline{E^c}), \tag{5.34}$$

$$f^{sl} = \frac{\overline{CVaR^{sl}} - (VaR - (1-\alpha)^{-1}\sum_{s\in S} P_s \mathscr{S}_s)}{\overline{CVaR^{sl}} - \underline{CVaR^{sl}}}. \tag{5.35}$$

The maximum value of expected worst-case service level (5.25), $\overline{CVaR^{sl}}$, for a given confidence level α, and the associated maximum value of expected cost (5.4), $\overline{E^c}$, are obtained as the optimal solution of **SPSm_CV(sl)** (with constraint (5.26)/(5.28), respectively, for order/demand fulfillment rate as a service level measure). The minimum value of expected cost (5.4), $\underline{E^c}$, and the associated minimum value of expected worst-case service level, $\underline{CVaR^{sl}}$, for a given confidence level α are obtained as the optimal solution to the following SMIP problem (with constraint (5.26)/(5.28), respectively, for order/demand fulfillment rate as a service level measure).

Minimize $0.99999E^{c}$-$0.00001CVaR^{sl}$
subject to (5.5)–(5.13), (5.27), (5.26)/(5.28).

The much greater weight assigned to the objective function, E^c, reflects the priority given to minimization of expected cost.

Finally, the trade-off model **SPSm_E(c)CV(sl)** for a mixed mean-risk optimization of expected cost per product and expected worst-case service level is shown below (with constraint (5.26)/(5.28), respectively, for order/demand fulfillment rate as a service level measure).

SPSm_E(c)CV(sl): *Mixed mean-risk selection of supply portfolio and scheduling of customer orders to trade-off expected cost and CVaR of service level: multiple sourcing*

Minimize

$$\lambda f^{c} + (1-\lambda) f^{sl}, \tag{5.36}$$

where $0 \leq \lambda \leq 1$,
subject to (5.5)–(5.13), (5.27), (5.26)/(5.28), (5.34), (5.35).

5.6 Computational Examples

In this section some computational examples are presented to illustrate possible applications of the proposed SMIP approach for the selection of suppliers, order quantity allocation, and customer orders scheduling for a single, dual, and multiple sourcing strategy and the two different service level measures: order fulfillment rate and demand fulfillment rate. First, single vs. dual sourcing and single vs. multiple sourcing strategy are compared, then order fulfillment rate vs. demand fulfillment rate are compared as worst-case service level performance measures and finally the impact of optimization the worst-case service on average cost is evaluated.

5.6.1 Single Versus Dual Sourcing

We assume that suppliers are located in two different geographic regions $r = 1,2$. Suppliers $i \in I^1$ are domestic suppliers, relatively reliable but more expensive. Suppliers $i \in I^2$ are located outside the producer's geographic region and offer competitive prices. For the domestic suppliers $i \in I^1$, the unit prices, o_i, are higher than for the foreign suppliers $i \in I^2$, while the fixed ordering costs e_i are lower. However, the foreign suppliers are more prone to breakdowns and material flows from these suppliers are more exposed to unexpected disruptions due to natural or man-made disasters and longer shipping times and distance to the producer. The regional disruption probability, p^1, for the producer domestic region $r = 1$, is much lower than the probability, p^2, for the foreign region $r = 2$.

The probability P_s of each disruption scenario $s \in S$ under the risks of local and regional disaster events is

$$P_s = \begin{cases} p^1p^2 + (1-p^1)(1-p^2)\prod_{i\in I} p_i & \text{if } I_s = \emptyset \\ (1-p^1)p^2\prod_{i\in I_s}(1-p_i)\prod_{i\in I^1\setminus I_s} p_i + (1-p^1)(1-p^2)\hat{P}_s & \text{if } I_s \subseteq I^1 \\ p^1(1-p^2)\prod_{i\in I_s}(1-p_i)\prod_{i\in I^2\setminus I_s} p_i + (1-p^1)(1-p^2)\hat{P}_s & \text{if } I_s \subseteq I^2 \\ (1-p^1)(1-p^2)\hat{P}_s & \text{if } I_s\bigcap I^1 \neq \emptyset, I_s\bigcap I^2 \neq \emptyset, \end{cases} \quad (5.37)$$

where $\hat{P}_s$ is the probability of disruption scenario s in the presence of independent local disruptive events only

$$\hat{P}_s = \prod_{i\in I_s}(1-p_i)\cdot\prod_{i\notin I_s} p_i. \quad (5.38)$$

The following parameters have been used for the example problems:

- $\overline{I}$, the number of suppliers, was equal to 10 and the number of disruption scenarios, was equal to the total number of all potential scenarios $\overline{S} = 2^{\overline{I}} = 1024$;
- $\overline{J}$, the number of customer orders, was equal to 25;
- $\overline{R}$, the number of geographic regions, was equal to 2, and the subsets of domestic and foreign suppliers were $I^1 = \{1,2,3,4,5\}$ and $I^2 = \{6,7,8,9,10\}$, respectively;
- $\overline{T}$, the number of planning periods, was equal to 10;
- a_j, the unit requirements for parts of products in customer orders, were integers in $\{1,2,3\}$ drawn from int(U[1;3]) distribution, for all orders j;
- b_j, the size of customer orders (required numbers of products), were integers in $\{500, 1000, \ldots, 5000\}$ drawn from 500int(U[1;10]) distribution, for all customer orders j;
- c_j, the unit capacity consumptions of producer, were integers in $\{1,2,3\}$ drawn from int(U[1;3]) distribution, for all customer orders j;
- C_t, the capacity of producer in each period t, was integer drawn from $1000\lceil (2\sum_{j\in J} b_jc_j/(\overline{T}-\max_{i\in I}\tau_i))U[0.75;1.25]/1000\rceil$ distribution, i.e., in each period the producer capacity was from 75% to 125% of the double capacity required to complete all customer orders during the planning horizon, after the latest delivery of parts;
- d_j, the due dates for customer orders, were integers in $\{1+\min_{i\in I}(\tau_i), \ldots, \overline{T}\}$ drawn from int(U[2;10]) distribution, for all customer orders j;
- e_i, the cost of ordering parts, were integers in $\{5000, 6000, \ldots, 10000\}$ and integers in $\{15000, 16000, \ldots, 30000\}$, respectively, for domestic suppliers $i \in I^1$ and for foreign suppliers $i \in I^2$;

- g_j, the unit penalty cost of unfulfilled customer orders, was to $2\lceil a_j \max_{i\in I}(o_i)\rceil$ for all orders j, i.e., was approximately twice as large as the maximum unit price of required parts;
- h_j, the unit daily penalty cost of delayed customer orders, was equal to $\lceil a_j \max_{i\in I}(o_i)/350\rceil$ for all orders j, i.e., was approximately 0.28% of the maximum unit price of required parts;
- o_i, the unit price of parts purchased from supplier i, was uniformly distributed over [11,16] and over [1,6], respectively, for domestic suppliers $i \in I^1$ and for foreign suppliers $i \in I^2$.
- α, the confidence level, was equal to 0.50, 0.75, 0.90, 0.95 or 0.99.
- p_i, the local disruption probability , was uniformly distributed over [0.005,0.01] for domestic suppliers $i \in I^1$ and over [0.05,0.10] for foreign suppliers $i \in I^2$, i.e., the disruption probabilities were drawn independently from U[0.005;0.01] and from U[0.05;0.10], respectively, for domestic and foreign suppliers.
  The regional disruption probability was $p^1 = 0.001$ for domestic suppliers $i \in I^1$ and $p^2 = 0.01$ for foreign suppliers $i \in I^2$.
  p^*, the global disruption probability, was 0, i.e., no global disaster super event is considered.
- τ_i, delivery lead time from domestic suppliers $i \in I^1$, were integers in $\{1,2\}$ drawn from int(U[1;2]) distribution, and from foreign suppliers $i \in I^2$ were integers in $\{2,3,4\}$ drawn from int(U[2;4]) distribution.

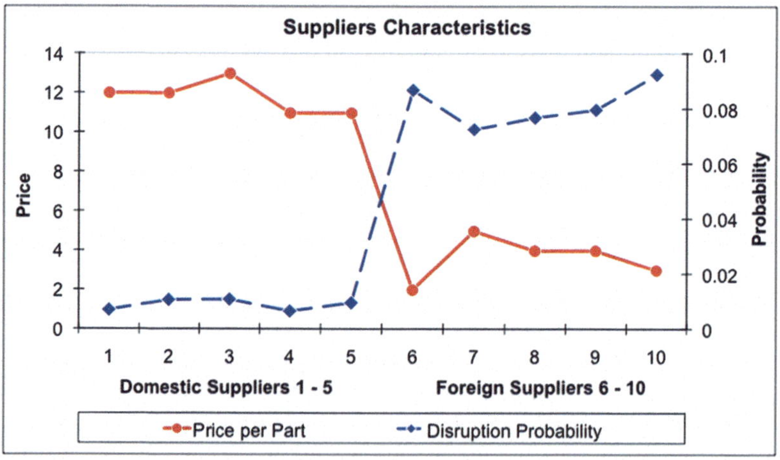

Fig. 5.2 Suppliers

The computational experiments were performed for the same replication of the above input data set. For all test examples, the resulting total demand for parts and products is $A = 132,500$ and $B = 66,000$, respectively. The unit price per part o_i

and disruption probability $p^r + (1 - p^r)p_i$ of each supplier $i \in I^r, r = 1,2$ are shown in Fig. 5.2.

Table 5.3 Risk-neutral solutions: single vs. dual sourcing

Sourcing strategy	Single sourcing	Dual sourcing
Minimization of expected cost (5.4) and (5.15)		
Expected cost	11.92	11.92
Expected service level[a]	61.80	61.80
Selected supplier	6	6
CPU[b]	144	45
Maximization of expected service level (5.19)		
Expected service level	99.35	99.35
Expected cost	23.19	23.19
Selected supplier	4	4
CPU[b]	61	89

[a] $(\sum_{s\in S} P_s \sum_{j\in J} \sum_{t\in T: t\le d_j} x_{jt}^s / J)100\%$
[b] CPU seconds for proving optimality on a MacBookPro 6.2, Intel Core i7, 2.66GHz, RAM 8GB/CPLEX 12.5

Table 5.4 Risk-averse minimization of cost: single vs. dual sourcing

Confidence level α	0.50	0.75	0.90	0.95	0.99
Model **SPS1_CV(c)**: Var.=230,060, Bin.=229,035, Cons.=45,945, Nonz.=2,115,890[a]					
CVaR^c	16.05	23.78	24.96	26.92	42.64
VaR^c	7.79	22.99	22.99	22.99	22.99
Supplier selected	6	4	4	4	4
Expected cost	11.92	23.19	23.19	23.19	23.19
Expected service level[b]	61.80	99.35	99.35	99.35	99.35
CPU[c]	88	103	204	60	880
Model **SPS2_CV(c)**: Var.=230,070, Bin.=229,035, Cons.=45,957, Nonz.=2,121,040[a]					
CVaR^c	16.05	23.78	24.96	26.92	38.29
VaR^c	7.79	22.99	22.99	22.99	37.20
Suppliers selected (% of total demand)	6(100)	4(100)	4(100)	4(100)	4(60) 10(40)
Expected cost	11.92	23.19	23.19	23.19	23.19
Expected service level[b]	61.80	99.35	99.35	99.35	55.32
CPU[c]	34	363	55	44	3473

[a] Var.=number of variables, Bin.=number of binary variables, Cons.=number of constraints, Nonz.= number of nonzero coefficients
[b] $(\sum_{s\in S} P_s \sum_{j\in J} \sum_{t\in T: t\le d_j} x_{jt}^s / J)100\%$
[c] CPU seconds for proving optimality on a MacBookPro 6.2, Intel Core i7, 2.66GHz, RAM 8GB/CPLEX 12.5

The solution results for the two sourcing strategies and the two objective functions are presented in Tables 5.3, 5.4 and 5.5, respectively, for the risk-neutral models and the risk-averse models with different confidence levels, where for the

Table 5.5 Risk-averse maximization of service level: single vs. dual sourcing

Confidence level α	0.50	0.75	0.90	0.95	0.99
Model **SPS1_CV(sl)+**: Var.=231,082, Bin.=230,053, Cons.=49,035, Nonz.=2,242,779[a]					
CVaR^{sl}%	98.70	97.39	93.48	86.95	34.77
VaR^{sl}%	100	100	100	100	100
Expected service level[b]	99.35	99.35	99.35	99.35	99.35
Expected cost	23.19	23.19	23.19	23.19	23.19
Supplier selected	4	4	4	4	4
CPU[c]	10	10	12	14	258
Model **SPS1_CV(sl)**: Var.=231,060, Bin.=229,035, Cons.=46,028, Nonz.=1,934,798[a]					
CPU[c]	19	20	15	18	293
Model **SPS2_CV(sl)+**: Var.=231,088, Bin.=230,053, Cons.=48,017, Nonz.=2,240,749[a]					
CVaR^{sl}%	98.70	97.39	93.48	87.35	66.17
VaR^{sl}%	100	100	100	96	72
Expected service level[b]	99.35	99.35	99.35	99.16	95.01
Expected cost	23.19	23.19	23.19	26.73	25.04
Suppliers selected (% of total demand)	4(100)	4(100)	4(100)	4(89)	4(45)
				7(11)	7(55)
CPU[c]	354	153	192	5900	349
Model **SPS2_CV(sl)**: Var.=230,070, Bin.=229,035, Cons.=45,968, Nonz.=1,996,339[a]					
CPU[c]	465	4380	1409	11,920	3725

[a]Var.=number of variables, Bin.=number of binary variables, Cons.=number of constraints, Nonz.= number of nonzero coefficients

[b]$(\sum_{s\in S} P_s \sum_{j\in J} \sum_{t\in T: t\le d_j} x^s_{jt}/\bar{J})100\%$

[c]CPU seconds for proving optimality on a MacBookPro 6.2, Intel Core i7, 2.66GHz, RAM 8GB/CPLEX 12.5

risk-averse maximization of service level the enhanced models **SPS1_CV(sl)+** and **SPS2_CV(sl)+** were applied. The confidence level α is set at five levels of 0.5, 0.75, 0.90, 0.95, and 0.99, which means that focus is on minimizing the highest (maximizing the lowest) 50%, 25%, 10%, 5%, and 1% of all scenario outcomes, i.e., costs per product (service levels, respectively). The size of the corresponding mixed integer programs is represented by the total number of variables, Var., number of binary variables, Bin., number of constraints, Cons., and number of nonzero coefficients in the constraint matrix, Nonz. In addition to the optimal solution values for the primary objective functions and the allocation of demand among the selected suppliers, Tables 5.3, 5.4 and 5.5 present the associated values of the other objective function.

Table 5.3 indicates that for the risk-neutral models the same optimal solution was found for each sourcing strategy, and the obtained risk-neutral solutions are identical with the corresponding risk-averse solutions for $\alpha = 0.5$ (see, Tables 5.4 and 5.5). Note that a low price, unreliable foreign supplier $i = 6$ was selected to minimize the expected cost per product, while a high price, reliable domestic supplier $i = 4$ was selected to maximize the expected service level.

For the risk-averse cost minimization, Table 5.4 indicates that the solution results for single and dual sourcing are identical for all confidence levels α except for the highest level $\alpha = 0.99$, where for a dual sourcing the total demand is allocated between two suppliers to reduce the worst-case cost outcomes. For the low confidence level $\alpha = 0.5$, a less reliable, low price foreign supplier $i = 6$ is selected, while for a higher α more reliable and more expensive domestic supplier $i = 4$ is chosen. When $\alpha = 0.99$, the expected worst-case cost is lower for a dual sourcing, for which the total demand is allocated between two suppliers.

For the risk-averse service level maximization, for which the supplier selection is independent on any cost parameters and the solution depends only on disruption probability, Table 5.5 demonstrates that for a single sourcing, the same expensive but reliable domestic supplier $i = 4$ is selected for all confidence levels. The same results are obtained for a dual sourcing and $\alpha = 0.5, 0.75, 0.9$, while for a higher confidence level, the total demand is allocated between two suppliers to reduce the worst-case service level outcomes. Then, the expected worst-case service level is higher than the corresponding level for a single sourcing. In particular, for the highest confidence level $\alpha = 0.99$, the expected worst-case service level for a dual sourcing is twice as large as that for a single sourcing. Table 5.5 also shows the results for the original models **SPS1_CV(sl)** and **SPS2_CV(sl)**. The comparison of CPU times clearly indicates the advantage of the enhanced models, in particular, model **SPS2_CV(sl)+**.

To identify the possible impact of the basic cost parameter setting on the achieved solution results, the additional computational experiments were performed with the much smaller differences of the fixed ordering cost, e_i, and the unit price, o_i, for the two groups of suppliers. The following values of parameters e_i and o_i, $i = 1, \ldots, 10$, were selected: $e = (6000, 8000, 7000, 8000, 8500, 8500, 9500, 9000, 10000, 9500)$ and $o = (6, 6, 5, 7, 6, 4, 5, 4, 4, 3)$. Since the service level-based objective function is independent on any cost parameters, the optimal solutions have not changed, while for the cost-based objective function only slight changes were observed (see, optimal supply portfolios in Table 5.6). The cheapest supplier, $i = 3$, from among the most reliable domestic suppliers is selected as the main one (for a dual sourcing) or the only one (for a single sourcing). The results have indicated that the supplier reliability is a key selection parameter, even for a cost-based objective function.

For the optimal risk-averse supply portfolio with $\alpha = 0.99$ and the two objective functions, Figs. 5.3 and 5.4 show the distribution of cost per product and the distribution of customer service level, respectively. Figure 5.3 demonstrates that for both objective functions, the distribution of cost is concentrated at the lowest cost per product for both single and dual sourcing, which indicates that a risk-averse maximization of service level that aims at reducing the expected worst-case fraction

Table 5.6 Risk-averse supply portfolios for suppliers offering similar prices

Confidence level α	0.50	0.75	0.90	0.95	0.99
Model **SPS1_CV(c)**					
Supplier selected	3	3	3	3	3
Model **SPS2_CV(c)**					
Suppliers selected (% of total demand)	3(100)	3(100)	3(100)	3(100)	3(55%) 10(45)

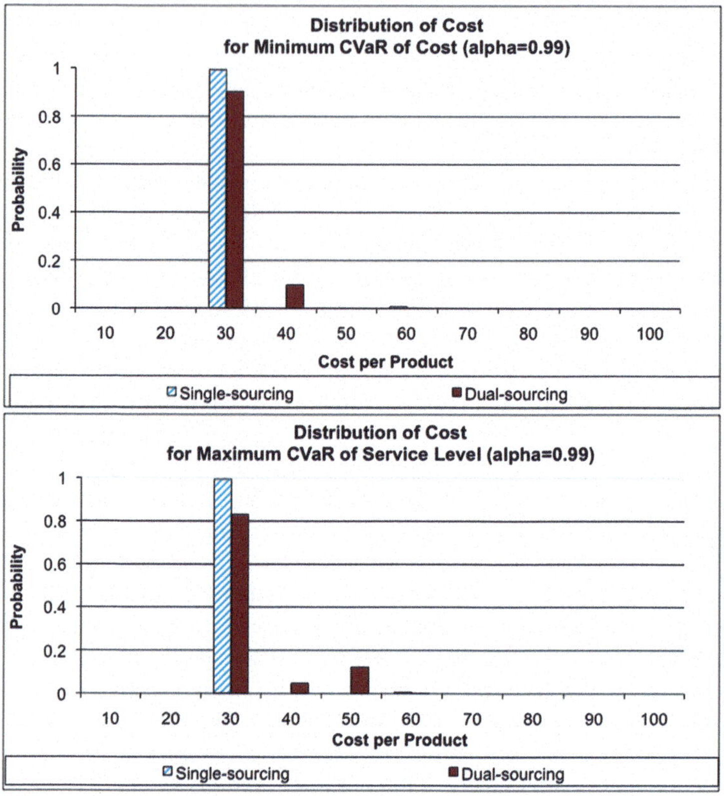

Fig. 5.3 Distribution of cost per product for risk-averse supply portfolio with $\alpha = 0.99$

of delayed and unscheduled customer orders, implicitly reduces the corresponding expected worst-case penalty costs. Figure 5.4 shows that for a single sourcing, the distribution of service level is concentrated at the highest percent of customer orders fulfilled by their due dates, because for both objective functions the same single, reliable supplier (supplier 4) was selected to minimize the worst-case cost of unfulfilled customer orders or maximize the worst-case customer service level, respectively. For a dual sourcing, however, the probability measure is concentrated at the highest service level, only if the CVaR of service level is maximized. When the CVaR of cost is minimized, however, the highest probability measure is concentrated at a lower service level, which indicates that a risk-averse dual sourcing solution that selects a low price, less reliable supporting supplier (supplier 10 vs. supplier 7, see Fig. 5.2 and Tables 5.3 and 5.4) to reduce the worst-case costs may simultaneously decrease the worst-case service levels. For a single sourcing, however, small probability atoms are concentrated at the highest cost and at the lowest service level (0%—all customer orders delayed or unscheduled), respectively, for the risk-averse minimization of cost and maximization of service level. This ad-

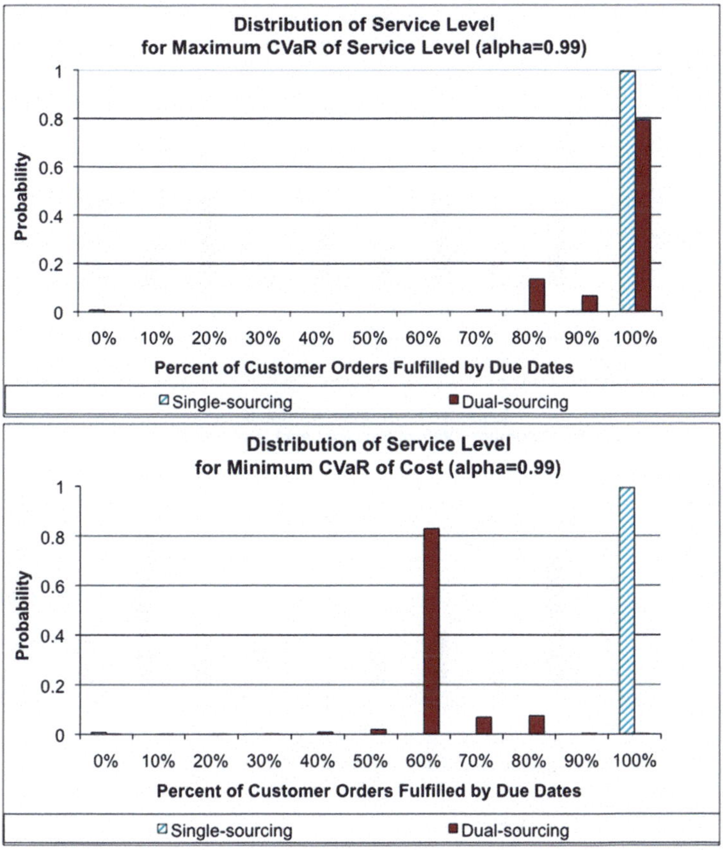

Fig. 5.4 Distribution of customer service level for risk-averse supply portfolio with $\alpha = 0.99$

ditionally indicates that for a single sourcing, all customer orders may be either fulfilled by their due dates or delayed/unscheduled, with a small probability for the latter event, whereas for a dual sourcing, the probability measure is distributed over a range of cost or service level outcomes.

Figure 5.5 presents aggregated demand pattern for products, $\sum_{j \in J: d_j = t} b_j, t \in T$, and expected aggregated production schedule, $\sum_{s \in S} P_s \sum_{j \in J} b_j x^s_{jt}, t \in T$ for the optimal risk-averse supply portfolios, the two objective functions and the two sourcing strategies. For a dual sourcing and maximum CVaR of service level when no cost components are included in the objective function, the expected production schedule approximately follows the aggregated demand and for a single sourcing and both objective functions, the production schedules begin with the highest production level that may help to reduce the penalty cost or the fraction of delayed customer orders.

Finally, the impact of penalty costs imposed on the producer by the customers is evaluated, assuming that a disrupted supplier is contractually obliged to partially

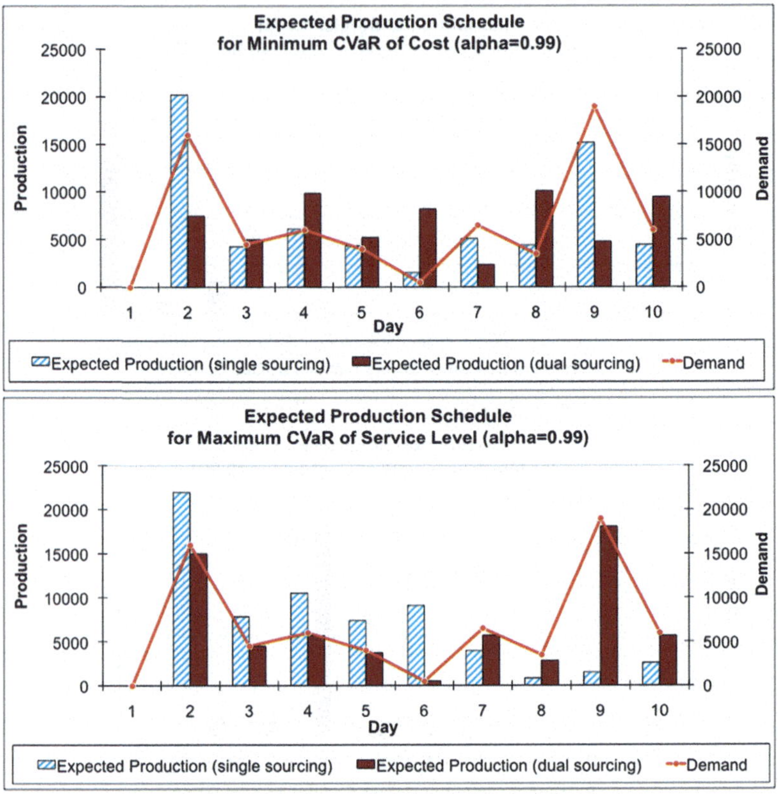

Fig. 5.5 Expected production schedule for risk-averse supply portfolio with $\alpha = 0.99$

cover the producer charges for unfulfilled customer orders due to undelivered parts. Table 5.7 shows the solution results for the unit penalty cost of unfulfilled customer orders, g_j, reduced by half with respect to their original values, i.e., for g_j approximately as large as the maximum unit price of required parts. Thus, the disrupted suppliers are assumed to cover the remaining 50% of the producer penalty charges. Table 5.7 demonstrates that for a lower producer penalty for unfulfilled customer orders, in a dual sourcing no supporting supplier is needed for the highest confidence level. For both single and dual sourcing, the same single supplier was selected. This indicates that the lower the penalty for unfulfilled customer orders, the less reliable and cheaper supplier can be selected by the producer.

Comparison of single and dual sourcing strategies indicates that for both the risk-neutral and the risk-averse solutions with a low confidence level, the same single supplier is selected only; a low price, risky supplier to minimize cost or an expensive, reliable supplier to maximize customer service level. In order to minimize the expected cost per product or CVaR of cost per product for low confidence levels, the cheapest supplier is usually selected. In contrast, to maximize the expected service

Table 5.7 Risk-averse solutions for reduced penalty costs: single vs. dual sourcing

Confidence level α	0.50	0.75	0.90	0.95	0.99
Model **SPS1_CV(c)**					
CVaR^c	11.52	15.25	23.25	23.52	25.61
VaR^c	7.79	7.79	22.99	22.99	22.99
Suppliers selected	6	6	4	4	4
Expected cost	9.65	9.65	23.02	23.02	23.02
Expected service level[a]	61.62	59.46	99.35	99.35	99.35
Model **SPS2_CV(c)**					
CVaR^c	11.52	15.25	23.25	23.52	25.61
VaR^c	7.79	7.79	22.99	22.99	22.99
Suppliers selected	6	6	4	4	4
Expected cost	9.65	9.65	23.02	23.02	23.02
Expected service level[a]	61.62	59.46	99.35	99.35	99.35

[a] $(\sum_{s\in S} P_s \sum_{j\in J} \sum_{t\in T:t\le d_j} x^s_{jt}/\overline{J})100\%$

level or CVaR of service level for low confidence levels, the most reliable supplier (with the lowest disruption probability) is mostly selected. For a higher confidence level, both single and dual sourcing model selects a single, reliable supplier to minimize worst-case cost of unfulfilled customer orders or maximize worst-case customer service level. A difference between single and dual sourcing solutions arises only for the highest confidence levels. Then, for a dual sourcing and the risk-averse solutions with the highest confidence levels, an expensive, reliable supplier selected for lower confidence levels is additionally supported with a low price, risky supplier to allocate the total demand for parts between the two suppliers. However, the selection of the supporting risky supplier depends on the objective function: a cheaper and less reliable supplier is selected to reduce the risk of high costs and a more expensive and more reliable supplier is selected to reduce the risk of low service level.

In general, the computational results indicate that the supplier reliability is a key selection parameter. In order to maximize service level the most reliable supplier is selected as the main one (for a dual sourcing) or the only one (for a single sourcing), whereas to minimize cost, the cheapest one is selected from among most reliable suppliers, respectively.

For the limited number of scenarios considered, the proven optimal solution can be found, using the CPLEX solver for mixed integer programming. However, in the proposed models the number of scheduling variables x^s_{jt} is $O((\overline{J})(\overline{S})(\overline{T}))$ and the number of constraints is $O((\overline{J}+\overline{T})\overline{S})$, i.e., they grow linearly in the number $\overline{S}$ of disruption scenarios and hence exponentially in the number $\overline{I}$ of suppliers, if all, $\overline{S} = 2^{\overline{I}}$, potential scenarios are considered.

The computational experiments were performed using the AMPL programming language and the CPLEX 12.5 solver on a laptop MacBookPro 6.2 with Intel Core i7 processor running at 2.66GHz and with 8GB RAM. The CPLEX solver was capable of finding proven optimal solutions for all examples with CPU time ranging from several seconds for the cost-based objectives to around 6000 s for the service level objectives. However, the enhanced models **SPS1_CV(sl)+** and **SPS2_CV(sl)+** had to be applied for the risk-averse maximization of service level, otherwise the much longer CPU time (up to 12,000 s) was required to prove optimality.

5.6.2 Single Versus Multiple Sourcing

In this subsection the risk-averse strategies are compared for single and multiple sourcing. The following parameters used for the example problems are different from those in Sect. 5.6.1:

- $\overline{R}$, the number of geographic regions, was equal to 3, and the subsets of suppliers were $I^1 = \{1,2,3\}$, $I^2 = \{4,5,6\}$ and $I^3 = \{7,8,9,10\}$, respectively;
- τ_i, the order preparation and shipping times from suppliers, were 2, 3, and 4 time periods, respectively, for suppliers $i \in I^1$, $i \in I^2$, and $i \in I^3$;
- d_j, the due dates for customer orders, were integers in $\{3,\ldots,\overline{T}\}$ drawn from int(U[3;10]) distribution, for all customer orders j;
- e_i, the cost of ordering parts, were integers in $\{5000, 6000, \ldots, 10000\}$, $\{10000, 11000, \ldots, 15000\}$, and $\{15000, 16000, \ldots, 30000\}$, respectively, for suppliers $i \in I^1$, $i \in I^2$, and $i \in I^3$;
- o_i, the unit price of parts purchased from supplier i, was uniformly distributed over $[11,16]$, $[6,11]$, and $[1,6]$, respectively, for suppliers $i \in I^1$, $i \in I^2$, and $i \in I^3$;
- p_i, the local disruption probability was uniformly distributed over $[0.005, 0.01]$, $[0.01, 0.05]$, and $[0.05;0.10]$, respectively, for suppliers $i \in I^1$, $i \in I^2$, and $i \in I^3$, i.e., the disruption probabilities were drawn independently from U$[0.005;0.01]$, U$[0.01, 0.05]$ and U$[0.05;0.10]$, respectively;
- p^r, the regional disruption probability was 0.001, 0.005, and 0.01, respectively, for region $r = 1$, $r = 2$, and $r = 3$;

The computational experiments were performed for the same replication of the above input data set. The input data set generated for all test examples is presented in Table 5.8.

All potential disruption scenarios were considered and to calculate the disruption probability for each scenario, the formulae (5.2) and (5.3) were applied. The unit price per part o_i and the disruption probability π_i, (5.1), of each supplier $i \in I^r, r = 1,2,3$ are shown in Fig. 5.6. The figure indicates that the most reliable (with the lowest disruption probability) is supplier 1, the least reliable (with the highest disruption probability) is supplier 8, the most expensive (with the highest price per part) is supplier 1, and the cheapest (with the lowest price per part) is supplier 7. Note that geographic regions are numbered in such a way that the unit prices are nonincreasing with r, while the fixed ordering costs and the disruption probabilities are nondecreasing with r, i.e.,

$$o_{i_1} \geq o_{i_2} \geq o_{i_3}, \quad e_{i_1} \leq e_{i_2} \leq e_{i_3} \quad \text{and} \quad \pi_{i_1} \leq \pi_{i_2} \leq \pi_{i_3}; \forall i_1 \in I^1, i_2 \in I^2, i_3 \in I^3.$$

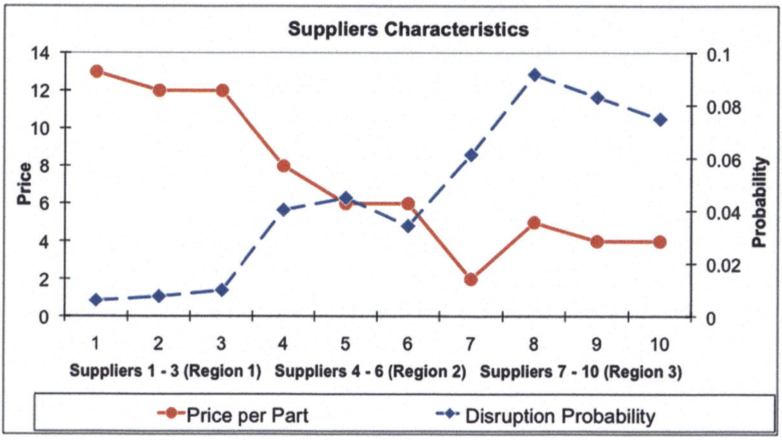

Fig. 5.6 Suppliers

Table 5.8 Input parameters: supply portfolio selection and production scheduling, single vs. multiple sourcing

$a = (2,1,3,3,1,3,2,1,2,2,2,2,3,2,1,3,2,1,3,3,2,1,1,2,1)$	unit requirement for parts
$b = (1,2,9,7,8,5,1,7,5,4,7,4,10,6,8,1,4,2,4,8,6,3,8,7,3) \times 500$	customer order size
$A = 132500$	total demand for parts
$B = 66000$	total demand for products
$c = (2,1,1,2,3,3,1,3,2,1,2,1,3,1,1,3,2,3,1,1,3,2,2,1,2)$	unit capacity consumption
$C_t = C = 38000, \forall t = 1,\dots,10$	producer capacity
Local disruption probabilities of suppliers:	
p = (0.00513571, 0.00666354, 0.00902974, 0.0356206, 0.040175, 0.0294692, 0.0519967, 0.0827215, 0.0739062, 0.0656449)	
Regional disruption probability:	
$p^1 = 0.001$, $p^2 = 0.005$, $p^3 = 0.01$	
Total disruption probabilities of suppliers (5.1):	
π=(0.00613057, 0.00765688, 0.0100207, 0.0404425, 0.0449741, 0.0343219, 0.0519967, 0.0827215, 0.0739062, 0.0656449)	
Cost parameters:	
$e = (8,7,10,13,14,11,19,17,19,27) \times 1000$	fixed ordering cost
$o = (13,12,12,8,6,6,2,5,4,4)$	unit purchasing price
$h_j = 1 \forall j = 1,\dots,25$	unit penalty for delayed demand
$g = (52,26,78,78,26,78,52,26,52,52,52,52,78,52,26,78,52,26,78,78,52,26,26,52,26)$	unit penalty for unfulfilled demand
Time parameters:	
$d = (7,4,10,10,3,8,7,5,5,6,8,7,10,5,3,9,6,4,9,9,7,4,5,7,4)$	customer due dates

The solution results for the two sourcing strategies and the two objective functions are presented in Tables 5.9 and 5.10, respectively. For the risk-averse maximization of service level, models **SPS1_CV(sl)** and **SPSm_CV(sl)** were enhanced with the valid inequalities (5.29)–(5.31). The confidence level α is set at five levels of 0.5, 0.75, 0.90, 0.95, and 0.99, which means that focus is on minimizing the highest (maximizing the lowest) 50%, 25%, 10%, 5%, and 1% of all scenario outcomes, i.e., costs per product (service levels, respectively). In addition to the optimal solution values for the primary objective functions and the allocation of demand among the selected suppliers, Tables 5.9 and 5.10 present the associated values of the other objective function. In particular, the expected cost and the expected service level associated with the optimal risk-averse solutions are presented to evaluate an average performance of the supply chain.

Table 5.9 Risk-averse minimization of cost: single vs. multiple sourcing

Confidence level α	0.50	0.75	0.90	0.95	0.99
Model **SPS1_CV(c)**: Var.=202,095, Bin.=201,060, Cons.=43,717, Nonz.=1,754,761[a]					
$CVaR^c$	10.60	16.47	26.07	28.50	42.22
VaR^c	4.73	4.73	12.33	24.20	26.22
Supplier selected	7	7	6	2	1
Expected cost	7.66	7.66	13.70	24.42	26.38
Expected service level[b]	67.55	67.55	88.84	99.23	99.39
CPU[c]	969	416	1143	613	2456
Model **SPSm_CV(c)**: Var.=202,095, Bin.=201,060, Cons.=43,717, Nonz.=1,754,761[a]					
$CVaR^c$	10.60	16.47	23.53	26.51	30.74
VaR^c	4.73	4.73	19.00	23.33	29.27
Suppliers selected (% of total demand)	7	7		2(26.04)	2(21.89)
					3(20.38)
			4(17.35)	4(20.38)	
			5(17.73)	5(18.87)	5(15.47)
			6(18.12)	6(18.87)	6(15.47)
			7(14.72)	7(15.85)	7(12.83)
			9(16.23)		
			10(15.85)		10(13.96)
Expected cost	7.66	7.66	16.60	22.21	26.65
Expected service level[b]	67.55	67.55	76.03	38.22	31.43
CPU[c]	183	1035	2600	1965	13,640

[a] Var.=number of variables, Bin.=number of binary variables, Cons.=number of constraints, Nonz.= number of nonzero coefficients

[b] $(\sum_{s\in S} P_s \sum_{j\in J} \sum_{t\in T: t\le d_j} x^s_{jt}/\overline{J})100\%$

[c] CPU seconds for proving optimality on a MacBookPro 6.2, Intel Core i7, 2.66GHz, RAM 8GB/CPLEX 12.5

For the minimization of worst-case cost, Table 5.9 indicates that for low confidence levels α, the solution results for single and multiple sourcing are identical, while for the higher levels a subset of suppliers is selected for the multiple sourcing. For a single sourcing and low confidence level $\alpha = 0.5, 0.75$, the cheapest supplier $i = 7$ is selected, then for a higher α, more reliable and expensive suppliers are chosen and finally, the most reliable and expensive supplier $i = 1$ is selected for

Table 5.10 Risk-averse maximization of service level: single vs. multiple sourcing

Confidence level α	0.50	0.75	0.90	0.95	0.99
Model **SPS1_CV(sl)**: Var.=202,095, Bin.=201,060, Cons.=43,727, Nonz.=1,649,427[a]					
$CVaR^{sl}$%	98.76	97.54	93.86	87.73	38.68
VaR^{sl}%	100	100	100	100	100
Supplier selected	1	1	1	1	1
Expected service level[b]	99.39	99.39	99.39	99.39	99.39
Expected cost	26.38	26.38	26.38	26.38	26.38
Model **SPSm_CV(sl)**: Var.=202,095, Bin.=201,060, Cons.=43,727, Nonz.=1,649,427[a]					
$CVaR^{sl}$%	99.21	98.47	96.22	92.45	86.19
VaR^{sl}%	100	100	100	100	92
Suppliers selected (% of total demand)	1(44.15)	1(47.92)	1(47.92)	1(47.92)	1(21.51)
	2(37.74)	2(30.57)	2(30.57)	2(30.57)	2(21.51)
	3(18.11)	3(21.51)	3(21.51)	3(21.51)	3(14.34)
					4(6.79)
					5(7.17)
					6(7.17)
					7(7.17)
					9(7.17)
					10(7.17)
Expected service level[b]	99.61	99.62	99.62	99.62	95.64
Expected cost	25.68	25.64	25.64	25.66	22.76

[a] Var.=number of variables, Bin.=number of binary variables,
Cons.=number of constraints, Nonz.= number of nonzero coefficients
[b] $(\sum_{s\in S} P_s \sum_{j\in J} \sum_{t\in T: t\le d_j} x^s_{jt}/\overline{J})100\%$

$\alpha = 0.99$. For a multiple sourcing and the high confidence levels $\alpha = 0.9, 0.95, 0.99$, the total demand is allocated among five or six suppliers, including the cheapest one.

For the maximization of worst-case service level, where the supplier selection is independent on any cost parameters and the solution depends only on the distribution of disruption probabilities, Table 5.10 demonstrates that for a single sourcing, the same most reliable supplier $i = 1$ is selected for all confidence levels. For a multiple sourcing, however, and the confidence level $\alpha = 0.5, 0.75, 0.9, 0.95$, similar optimal solutions were found with the total demand for parts allocated among the three most reliable and most expensive suppliers $i = 1,2,3$ in region $r = 1$, while to reduce the worst-case service level outcomes for the highest confidence level $\alpha = 0.99$, total demand is allocated among nine suppliers, including the most reliable $i = 1,2,3$ and except for the least reliable, $i = 8$.

Figures 5.7 and 5.8 present the distribution of cost per product and the distribution of customer service level for the two sourcing strategies and the confidence level $\alpha = 0.99$. For a single sourcing and the highest confidence level $\alpha = 0.99$, the highest probability measure of 0.9977 is concentrated at 100% of customer orders fulfilled by their due dates (see, Fig. 5.8), because for both objective functions, the same, most reliable supplier $i = 1$ was selected to minimize the worst-case cost of unfulfilled customer orders or maximize the worst-case customer service level, respectively. Both the distribution of cost and the distribution of service level contain

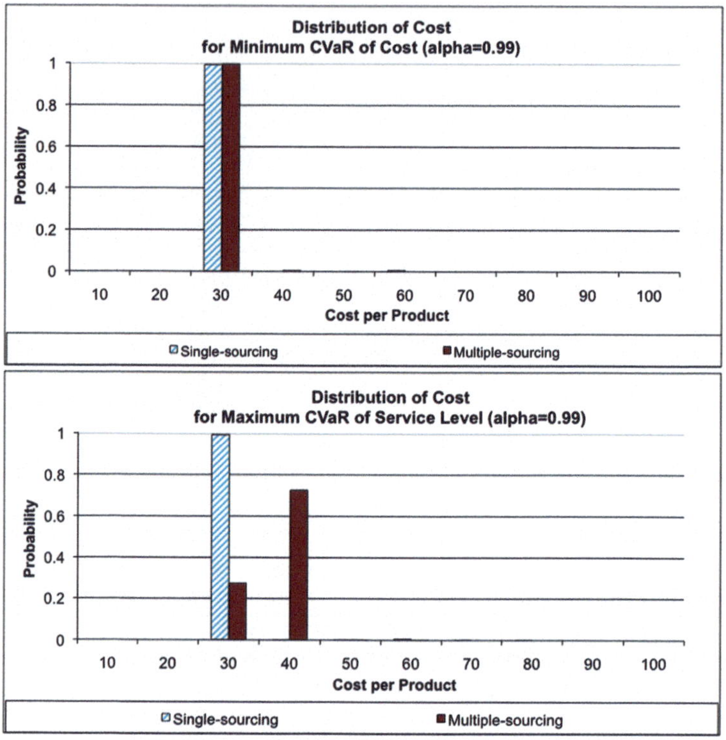

Fig. 5.7 Distribution of cost per product for risk-averse solution with $\alpha = 0.99$

also large probability atoms of 0.0061 at the highest cost, 52.32 (Fig. 5.7) and at the lowest service level, 0% (Fig. 5.8). Similar results are not observed for a multiple sourcing, which eliminates the probability atoms at the highest cost and at the lowest service level. As a consequence, the corresponding optimal solution values are much better than for a single sourcing (CVaR of cost 30.74 vs. 42.22, Table 5.9, and CVaR of service level 86.19% vs. 38.68%, Table 5.10). On the other hand, for a multiple sourcing and the highest confidence level $\alpha = 0.99$, the optimal solution with the minimum, $CVaR^c = 30.74$, simultaneously produces the lowest expected service level, $E^{sl} = 31.43\%$, (see, Table 5.9), while the optimal solution with the maximum, $CVaR^{sl} = 86.19\%$, generates the average expected cost $E^c = 22.76$ (see, Table 5.10). The above conflicting results for a multiple sourcing are not observed for lower confidence levels α, for which additional supporting suppliers are not selected.

It should be pointed out that the cost and the service level objectives are in conflict, which is clearly indicated by the results in Tables 5.9 and 5.10. The expected service level is much lower for the cost-based objective function (Table 5.9), and vice versa the expected cost is much higher for the service level-based objective function (Table 5.10). The low-cost and low-reliability suppliers dominate the cost-based supply portfolios, while the high-reliability and high-cost suppliers dominate

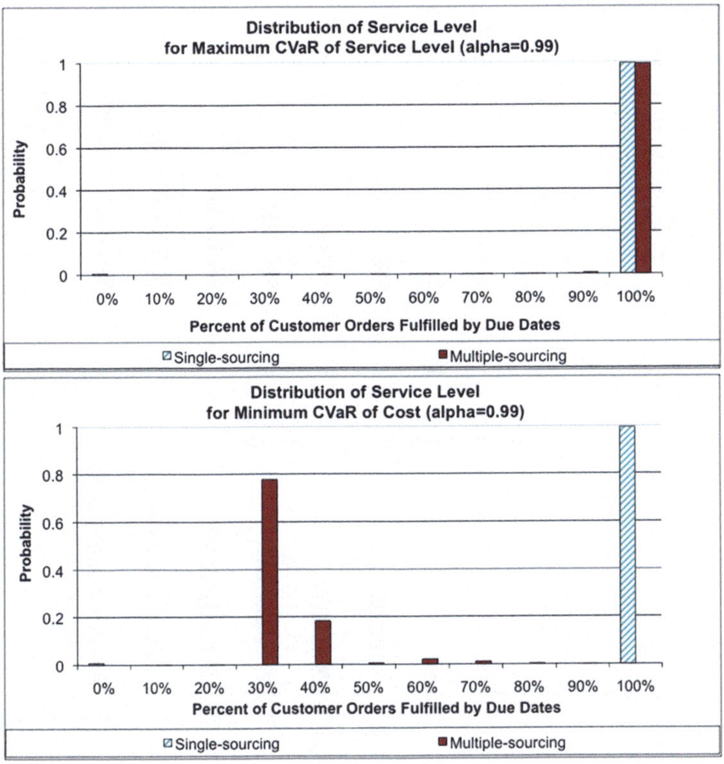

Fig. 5.8 Distribution of customer service level for risk-averse solution with $\alpha = 0.99$

the service level-based portfolios. Furthermore, the average and the worst-case performance measures and the corresponding optimal solutions are also in conflict. The higher the confidence level α, the more risk aversive the decision-making and the smaller percent of the highest cost (or the lowest service level) outcomes is focused on. As a result, the average outcomes are getting neglectful and the average performance associated with the optimal risk-averse solution may become worse. The optimal risk-averse solution with a low worst-case cost may produce a high average cost. Similarly, the optimal risk-averse solution with a high worst-case service level may yield a low average service level. For a high confidence level α, the impact of disruption risks is usually mitigated by diversification of the supply portfolio, i.e., by selecting of more suppliers. In particular, more expensive or less reliable supporting suppliers may be added to the risk-averse supply portfolio, which deteriorates the associated average performance measures, i.e., the expected values of cost or service level.

As an illustrative example, Fig. 5.9 presents the cumulative demand pattern for products, $\sum_{t'\in T:t'\le t}\sum_{j\in J:d_j=t'} b_j, t \in T$, and the expected cumulative production schedules, $\sum_{s\in S} P_s \sum_{t'\in T:t'\le t}\sum_{j\in J} b_j x^s_{jt'}, t \in T$ for the optimal risk-averse solutions with the confidence level $\alpha = 0.9$. Figure 5.9 indicates that for a minimum worst-

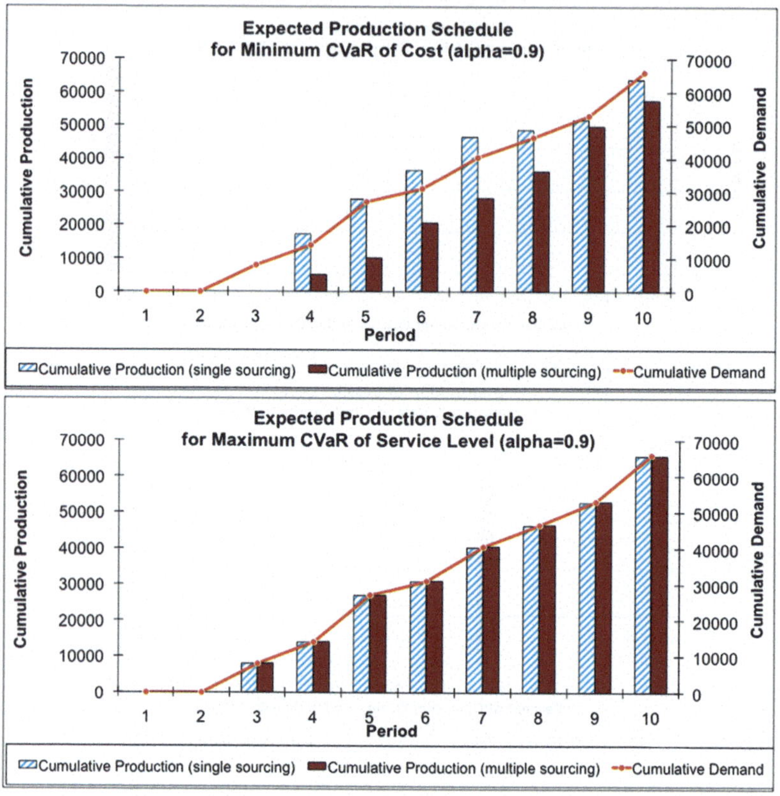

Fig. 5.9 Expected cumulative production for risk-averse solution with $\alpha = 0.9$

case cost objective, the demand of customers is met with a small expected fraction of the rejected demand (0.034) for the single sourcing, while for the multiple sourcing, the expected fraction is much greater (0.122). The expected rejected demand, however, is much smaller for the worst-case service level objective and both sourcing strategies, when no cost components are included in the objective function and more reliable suppliers are selected. Then, the expected production follows the demand pattern with a very small expected fraction of the rejected demand, 0.0061 and 0.0038, respectively, for single and multiple sourcing.

The risk-averse solutions have been compared for the two sourcing strategies and the two objective functions. Some of the basic features of the obtained solutions are discussed below (see, Table 5.11).

Table 5.11 Single vs. multiple sourcing solution

Objective function	Single sourcing	Multiple sourcing
Worst-case cost	The higher is the confidence level, the more reliable supplier is selected	For low confidence levels, the solution for a single sourcing
Worst-case service level	The most reliable supplier is selected for all confidence levels	The higher is the confidence level the more suppliers are selected

For the worst-case cost objective function and a single sourcing, the higher the confidence level, the more reliable supplier is selected to reduce a higher risk of penalty cost for unfulfilled customer orders. In order to minimize the expected worst-case cost per product for low confidence levels, the cheapest suppliers are usually selected. For the worst-case cost objective function and low confidence levels, the optimal solutions for a single and a multiple sourcing are identical. A difference between single and multiple sourcing solutions arises only for high confidence levels, whereas for the worst-case service level objective, the corresponding single and multiple sourcing solutions are different for all confidence levels.

In order to maximize the expected worst-case service level when no cost components are considered in the objective function, the most reliable supplier (with the lowest disruption probability) is selected for a single sourcing, even for low confidence levels. For a multiple sourcing and the worst-case service level, the total demand for parts is allocated among more suppliers for each confidence level to reduce the risk of unfulfilled customer orders. For high confidence levels and both cost and service level objectives, more suppliers are selected to mitigate the impact of disruption risks by diversification of the supply portfolio. The most reliable suppliers are selected to minimize the expected worst-case cost of unfulfilled customer orders or to maximize the expected worst-case customer service level. Furthermore, for the expected worst-case service level objective and both sourcing strategies, the expected production schedule approximately follows the product demand pattern with a very small expected fraction of rejected demand.

Comparison of the probability mass functions for the two sourcing strategies and the two objective functions has indicated that a multiple sourcing strategy better shapes the distribution of cost or customer service level. The multiple sourcing eliminates large probability atoms at the highest cost or at the lowest service level that otherwise may occur, if a single sourcing strategy is applied. Thus, a multiple sourcing strategy better mitigates the risk of high costs or low service levels.

The computational experiments were performed using the AMPL programming language and the CPLEX 12.5 and Gurobi 5.1 solvers on a laptop MacBookPro 6.2 with Intel Core i7 processor running at 2.66GHz and with 8GB RAM. The CPLEX solver outperformed the Gurobi solver when the worst-case cost was minimized, while the Gurobi solver outperformed the CPLEX solver when the worst-case service level was maximized. The solvers were capable of finding proven optimal solutions for all examples with CPU time ranging from several seconds to over four hours. However, if valid inequalities (5.29)–(5.31) were not added to tighten models **SPS1_CV(sl)** and **SPSm_CV(sl)** for optimization of the worst-case service level, then CPU time required to find proven optimal solutions for the example problems was up to 15% and up to 6% longer, respectively.

5.6.3 Order Versus Demand Fulfillment Rate

This subsection focuses on comparison of the two alternative risk-averse service level measures: the expected worst-case order fulfillment rate and the expected worst-case demand fulfillment rate.

The solution results for the two models **SPSm_CV(sl)** and **SPSm_CV(sl)** with (5.26) replaced by (5.28) are compared in Table 5.12. For the service level defined by order fulfillment rate (model **SPSm_CV(sl)**), Table 5.12 demonstrates that for the confidence level $\alpha = 0.5, 0.75, 0.9, 0.95$, similar optimal solutions were found with the total demand for parts allocated among the three most reliable suppliers $i = 1,2,3$ in region $r = 1$, and for the highest confidence level $\alpha = 0.99$, total demand is allocated among nine suppliers, except for the least reliable supplier $i = 8$. For the service level defined by demand fulfillment rate (model **SPSm_CV(sl)** with (5.26) replaced by (5.28)), Table 5.12 demonstrates that for the confidence level $\alpha = 0.5, 0.75, 0.9, 0.95$, similar optimal solutions were found with the total demand for parts allocated among the two most reliable suppliers $i = 1,2$ in region $r = 1$, while to reduce the worst-case service level outcomes for the highest confidence level $\alpha = 0.99$, total demand is allocated among all ten suppliers.

Figure 5.10 presents the distribution of demand fulfillment rate, for the two objective functions of model **SPSm_CV(sl)** and the two confidence levels $\alpha = 0.9$ and $\alpha = 0.99$. The probability mass functions are concentrated in a few points, which is typical for the scenario-based optimization under uncertainty.

For $\alpha = 0.9$, the distribution of service level is similar for both objective functions. The highest probability measure, 0.987, is concentrated at 100% of customer demand fulfilled by due dates. The most reliable suppliers $i = 1,2$ and $i = 1,2,3$ were selected to maximize the expected worst-case service level, for maximization of demand fulfillment rate and order fulfillment rate, respectively. However, the distribution of service level contains also large probability atoms at lower service levels: 0.012 at 70% and 0.001 at the lowest service level, 0%, for maximization of demand fulfillment rate, and 0.009 at 90% 0.006 at 80%, 0.005 at 70% and 0.001 at 0%, for maximization of order fulfillment rate. Thus, for $\alpha = 0.9$, the solution maximizing demand fulfillment rate with a greater probability atom at 70% may be outperformed by the solution maximizing order fulfillment rate.

For $\alpha = 0.99$, the distribution of service level depends on the objective function selected. For maximization of demand fulfillment rate, the highest probability measure of 0.71 is concentrated at 100% and 0.29 at 90% of customer demand fulfilled by their due dates, while for maximization of order fulfillment rate, 0.938 is concentrated at 90%, 0.0562 at 100%, and 0.005 at 80% of customer demand fulfilled by due dates. However, for $\alpha = 0.99$, no probability atoms occur at lower service levels.

Table 5.12 Order vs. demand fulfillment rate: risk-averse multiple sourcing

Confidence level α	0.50	0.75	0.90	0.95	0.99
Model **SPSm_CV(sl)** for order fulfillment rate					
$CVaR^{sl}\%$	99.21	98.47	96.22	92.45	86.19
$VaR^{sl}\%$	100	100	100	100	92
Suppliers selected (% of total demand)	1(44.15)	1(47.92)	1(47.92)	1(47.92)	1(21.51)
	2(37.74)	2(30.57)	2(30.57)	2(30.57)	2(21.51)
	3(18.11)	3(21.51)	3(21.51)	3(21.51)	3(14.34)
					4(6.79)
					5(7.17)
					6(7.17)
					7(7.17)
					9(7.17)
					10(7.17)
Expected service level[a]	99.61	99.62	99.62	99.62	95.64
Expected cost	25.68	25.64	25.64	25.66	22.76
Expected fulfilled demand[c]	99.44	99.45	99.45	99.45	95.86
Expected worst-case fulfilled orders (5.29)	81.93	82.45	82.49	82.58	83.35
Model **SPSm_CV(sl)** for demand fulfillment rate					
$CVaR^{sl}\%$	98.97	97.98	95.00	90.05	78.82
$VaR^{sl}\%$	100	100	100	100	84.85
Suppliers selected (% of total demand)	1(50.19)	1(50.19)	1(51.32)	1(51.32)	1(21.13)
	2(49.81)	2(49.81)	2(48.68)	2(48.68)	2(16.98)
					3(16.60)
					4(5.66)
					5(5.66)
					6(11.32)
					7(5.66)
					8(5.66)
					9(5.66)
					10(5.66)
Expected service level[b]	99.49	99.49	99.50	99.50	87.90
Expected cost	25.52	25.52	25.52	25.54	24.90
Expected fulfilled demand[c]	99.49	99.49	99.50	99.50	87.90
Expected worst-case fulfilled demand (5.30)	60.70	60.84	60.94	61.03	75.56

[a] $(\sum_{s\in S} P_s \sum_{j\in J} \sum_{t\in T:t\le d_j} x^s_{jt}/J)100\%$
[b] $(\sum_{s\in S} P_s \sum_{j\in J} \sum_{t\in T:t\le d_j} b_j x^s_{jt}/B)100\%$
[c] $(\sum_{s\in S} P_s \sum_{j\in J} \sum_{t\in T} b_j x^s_{jt}/B)100\%$

For a high confidence level α, the impact of disruption risks is usually mitigated by diversification of the supply portfolio, i.e., by selecting more suppliers. In particular, less reliable supporting suppliers may be added to the risk-averse supply portfolio, which deteriorates the associated average performance measure, i.e., the expected value of service level.

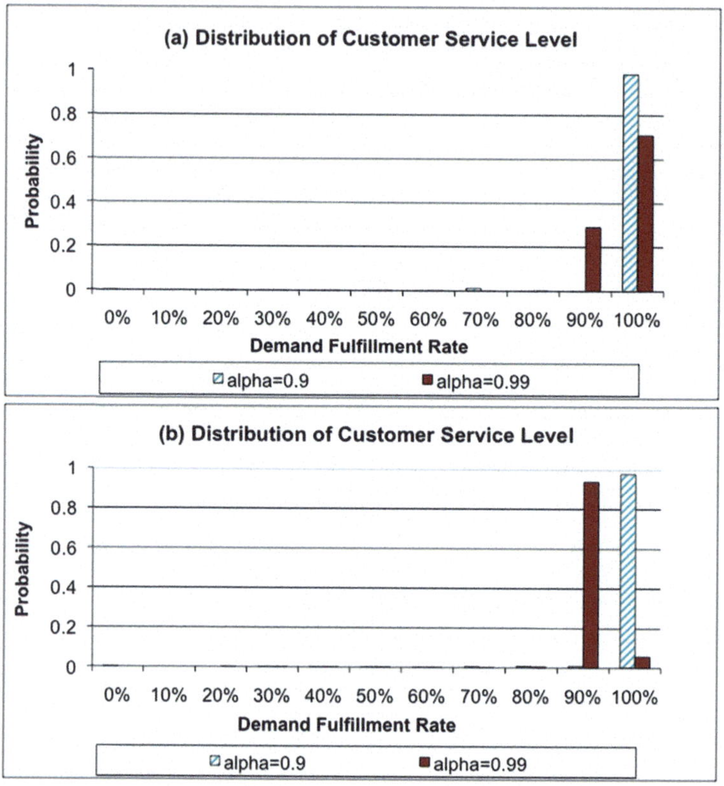

Fig. 5.10 Distribution of demand fulfillment rate: (**a**) model **SPSm_CV(sl)** for demand fulfillment rate, (**b**) model **SPSm_CV(sl)** for order fulfillment rate

5.6.4 Conditional Service-at-Risk Versus Expected Cost

In this subsection the impact on the cost in the process of optimization of worst-case service level is illustrated with additional computational examples. The cost includes the cost of ordering and purchasing of parts plus penalty cost of delayed and unfulfilled customer orders due to parts shortages. For example, the impact on the cost in the process of optimization of worst-case service level is illustrated in Table 5.12 with the expected cost per product, E^c, associated with the risk-averse solution for model **SPSm_CV(sl)**. The expected costs associated with the risk-averse solutions for $\alpha = 0.5$, 0.75, 0.9, 0.95 are similar. To maximize the worst-case service level, the parts are purchased from high-reliability and high-cost suppliers only. As a result the purchasing cost dominates, while the cost of parts shortage is not significant. However, for the highest confidence level $\alpha = 0.99$, when the supply portfolio is most diversified to mitigate the impact of worst 1% of all scenarios outcomes, the orders for parts are also placed on low-cost and low-reliability suppliers. The resulting total purchasing cost decreases as well as the resulting total expected

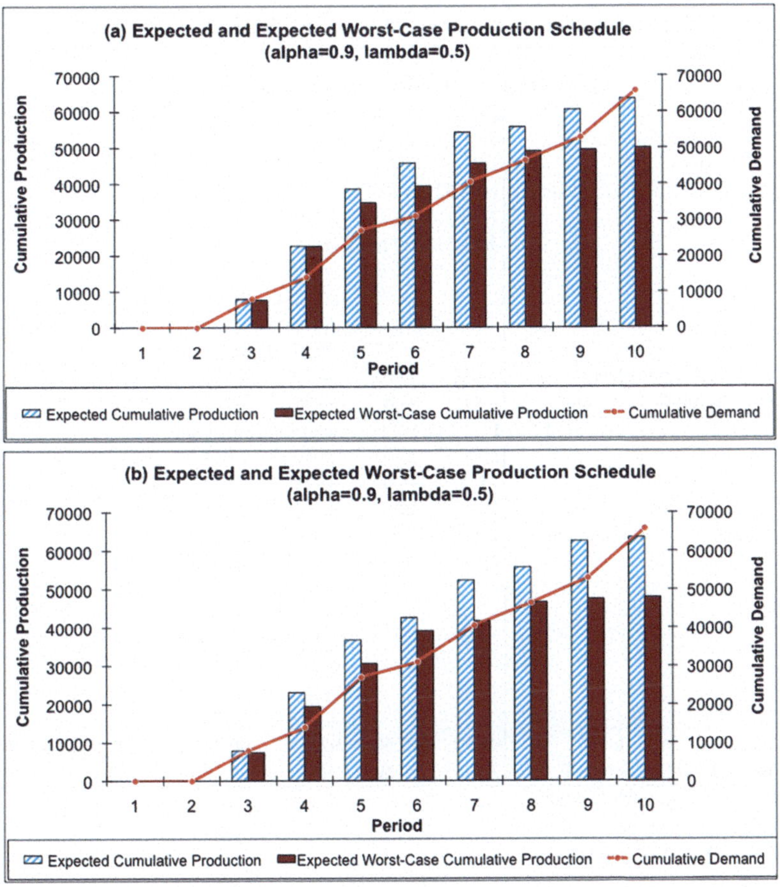

Fig. 5.11 Expected and expected worst-case cumulative production for $\alpha = 0.9$ and $\lambda = 0.5$: (**a**) model **SPSm_E(c)CV(sl)** for demand fulfillment rate, (**b**) model **SPSm_E(c)CV(sl)** for order fulfillment rate

cost per product, $E^c = 22.76$ for model **SPSm_CV(sl)** and $E^c = 24.90$ for model **SPSm_CV(sl)** with (5.26) replaced by (5.28).

Table 5.13 presents a subset of nondominated solutions for the mixed mean-risk model **SPSm_E(c)CV(sl)** and confidence level $\alpha = 0.9$, obtained for a subset of trade-off parameter, $\lambda = 0, 0.01, 0.25, 0.5, 0.75, 0.99, 1$. Note that a single-objective solution with maximum conditional service-at-risk (cf. Table 5.12) and with minimum expected cost is obtained for $\lambda = 0$ and $\lambda = 1$, respectively.

Table 5.13 clearly shows that the two objective functions: expected worst-case service level and expected cost are in conflict. The higher the trade-off parameter λ, the more cost-oriented the decision-making and the lower the expected cost, E^c, the lower the conditional service-at-risk, $CVaR^{sl}$, as well the associated expected service

Table 5.13 Nondominated solutions for mean-risk model **SPSm_E(c)CV(sl)**: $\alpha = 0.9$

Trade-off parameter λ	0	0.01	0.25	0.50	0.75	0.99	1
Model **SPSm_E(c)CV(sl)** for order fulfillment rate							
E^c	25.64	25.50	14.55	12.94	8.49	7.77	7.66
$\text{CVaR}^{sl}\%$	96.22	96.16	88.19	84.88	53.63	28.94	27.66
$\text{VaR}^{sl}\%$	100	100	92	92	92	76	72
Suppliers selected (% of total demand)	1(47.92)	1(40.75)					
	2(30.57)	2(37.74)	2(7.55)	2(6.04)			
	3(21.51)	3(21.51)	3(6.41)				
			5(21.51)	5(20.38)	5(4.90)		
			6(21.51)	6(21.51)	6(5.67)		
			7(21.51)	7(30.56)	7(89.43)	7(100)	7(100)
			10(21.51)	10(21.51)			
Expected service level[a]	99.62	99.62	98.02	97.74	88.16	71.29	67.56
Expected fulfilled demand[c]	99.45	99.44	96.70	96.31	93.94	93.82	93.85
Model **SPSm_E(c)CV(sl)** for demand fulfillment rate							
E^c	25.52	25.52	18.52	13.11	8.23	7.67	7.66
$\text{CVaR}^{sl}\%$	95.00	95.00	85.79	77.44	43.02	29.91	28.08
$\text{VaR}^{sl}\%$	100	100	90.91	84.85	87.88	78.79	73.48
Suppliers selected (% of total demand)	1(51.32)	1(50.19)	1(4.91)				
	2(48.68)	2(49.81)	2(16.98)	2(6.04)			
			3(13.58)				
			4(13.21)				
			5(13.58)	5(22.64)			
			6(13.58)	6(21.51)	6(10.57)		
			7(13.21)	7(27.17)	7(89.43)	7(100)	7(100)
			9(10.95)	9(22.64)			
Expected service level[b]	99.50	99.50	97.27	96.26	83.39	73.90	68.96
Expected fulfilled demand[c]	99.50	99.50	97.27	96.26	94.38	93.84	93.85

[a] $(\sum_{s\in S} P_s \sum_{j\in J} \sum_{t\in T: t\le d_j} x^s_{jt}/J)100\%$
[b] $(\sum_{s\in S} P_s \sum_{j\in J} \sum_{t\in T: t\le d_j} b_j x^s_{jt}/B)100\%$
[c] $(\sum_{s\in S} P_s \sum_{j\in J} \sum_{t\in T} b_j x^s_{jt}/B)100\%$

level for both measures: order fulfillment rate and demand fulfillment rate. As λ increases from 0 to 1, the supply portfolio is changing from the one based on a subset of most reliable and expensive suppliers, $i = 1,2,3$ (model **SPSm_E(c)CV(sl)**) and $i = 1,2$ (model **SPSm_E(c)CV(sl)** with (5.26) replaced by (5.28)) for $\lambda = 0, 0.01$, followed with more diversified subsets of less reliable and cheaper suppliers selected for $\lambda = 0.25, 0.5, 0.75$, through the single, cheapest supplier, $i = 7$, for $\lambda = 0.99, 1$. While CVaR^{sl} and expected service level, E^{sl}, are significantly decreasing with the trade-off parameter λ, the associated expected fraction of fulfilled (on time or delayed) demand decreases at the much lower rate. This indicates that as λ approaches 1, the fulfilled demand becomes more delayed and the resulting service level decreases.

Examples of nondominated schedules for the confidence level $\alpha = 0.9$ and the trade-off parameter $\lambda = 0.5$ are shown in Fig. 5.11. Figure 5.11 compares the expected cumulative production with the expected worst-case cumulative production for model **SPSm_E(c)CV(sl)** maximizing either demand fulfillment rate or order fulfillment rate. The corresponding production schedules are similar for both objectives with a slightly greater expected and expected worst-case cumulative production for maximization of demand fulfillment rate.

Comparison of optimal CVaR^{sl} (5.25), with the actual expected worst-case service level, (5.29) and (5.30), clearly demonstrates that the optimal value of the auxiliary objective function (5.25) used to maximize conditional service-at-risk, overestimates the actual value of CVaR^{sl} calculated using (5.29) and (5.30), which is typical for the scenario-based optimization under uncertainty, when the probability measure is concentrated in finitely many points.

The major managerial insights are summarized in Table 5.14. Overall, the obtained solution results for the service level objectives are in line with other research in the area of supply chain risk management. For example, *the higher the confidence level, the more diversified the supply portfolio and the smaller the probability atoms at the lowest service level, i.e., the risk of low service level is more efficiently mitigated. Furthermore, the two service level measures are related in magnitude, while the worst-case service level and the expected cost are in opposition.*

The computational experiments were performed using the AMPL programming language and Gurobi 6.0 solver on a laptop MacBookPro with Intel Core i7 processor running at 2.8GHz and with 16GB RAM. The solver was capable of finding proven optimal solutions for all examples with CPU time ranging from a few seconds to around one hour. CPU time increases with the confidence level α and approaches one hour for $\alpha = 0.99$, when the supply portfolio is most diversified to mitigate the impact of worst 1% of all scenarios outcomes. Given confidence level, the computational effort required to find proven optimal solution is smaller for maximization of demand fulfillment rate than the order fulfillment rate, for which the size of customer order is not considered and hence all orders are equally important. For model **SPSm_E(c)CV(sl)**, given confidence level, the greatest CPU time is required for medium values of the trade-off parameter $\lambda = 0.25, 0.5, 0.75$, when both criteria are significant.

5.7 Notes

As manufacturers expand their business globally, their supply chains involve more global partners. According to an empirical study conducted by Shin et al. (2000) and Sounderpandian et al. (2008) dual or multiple sourcing is a common business practice to mitigate the impact of various operational and disruption risks, e.g., Heese (2015). In view of the recent trend of outsourcing, offshoring and globalization, integrated selection of part suppliers and allocation of order quantities and scheduling of customer orders may significantly improve performance of a multi-tier supply chain network under disruption risks. However, the research on quantitative approaches to the integrated supplier selection and customer order scheduling in the presence of supply chain disruption risks has not been often reported in the literature. Most work on integrated supply chain scheduling focuses on coordinat-

Table 5.14 Managerial insights: supply portfolio selection and production scheduling

Decision-making issue	Decision-making impact
Single vs. dual sourcing	Supplier reliability is a key selection parameter for both risk-neutral and risk-averse decision-making. In order to maximize service level, the most reliable supplier is selected as the main one (for a dual sourcing) or the only one (for a single sourcing), whereas to minimize cost, the cheapest one is selected from among most reliable suppliers.
Single vs. multiple sourcing	Multiple sourcing better mitigates the risk of high costs or low service levels. Comparison of probability mass functions for the two sourcing strategies and the two objective functions indicates that a multiple sourcing better shapes the distribution of cost or customer service level. The multiple sourcing eliminates large probability atoms at the highest cost or at the lowest service level that may occur, if a single sourcing is applied. In the risk-averse decision-making with CVaR of cost and low confidence levels, the optimal solutions for a single and a multiple sourcing are identical.
Order fulfillment rate vs. demand fulfillment rate	Maximization of the expected worst-case order fulfillment rate may better mitigate the impact of disruption risks. In the risk-averse decision-making, the worst-case order fulfillment rate shows a higher service performance than the worst-case demand fulfillment rate. The supply portfolio is more diversified and for most confidence levels the expected worst-case fraction of fulfilled orders is greater than the expected worst-case fraction of fulfilled demand. In the risk-averse decision-making, the expected production approximately follows product demand pattern for both service level measures and both single and multiple sourcing. A relatively small fraction of unfulfilled customer demand for low confidence levels is observed and a greater fraction for the highest confidence level. The expected worst-case production depends on the optimized service level measure and the confidence level. The largest unfulfilled demand has been observed for low confidence levels and model **SPSm_CV(sl)** for maximization of demand fulfillment rate. Then, the supply portfolio is less diversified than the corresponding portfolio for maximization of order fulfillment rate.
Risk-neutral vs. risk-averse service level	The expected service level exceeds the corresponding expected worst-case service level (conditional service-at-risk), i.e., the average performance is better than the worst-case performance of a supply chain.
Risk-averse decision-making	For most confidence levels, the total demand for parts is allocated among the most reliable suppliers only. However, for the highest level $\alpha = 0.99$, the impact of disruption risks is usually mitigated by diversification of the supply portfolio, i.e., by selecting more suppliers. In particular, less reliable supporting suppliers may be added to the risk-averse supply portfolio, which deteriorates the associated average performance measure, i.e., the expected service level.

ing the flows of supply and demand over a supply chain network to minimize the inventory, transportation and shortage costs. For example, Chen and Vairaktarakis (2005), Chen and Pundoor (2006) and Pundoor and Chen (2009) studied simplified models for integrated scheduling of production and distribution operations. The authors have analyzed computational complexity of various cases of the problem and have developed heuristics for NP-hard cases. Lei et al. (2006) considered an integrated production, inventory, and distribution routing problem involving heterogeneous transporters with non-instantaneous traveling times and many capacitated customer demand centers. A MIP approach combined with a heuristic routing algorithm was proposed to coordinate the production, inventory, and transportation operations. Bard and Nananukul (2009) developed a MIP model and a reactive tabu search-based algorithm for a transportation scheduling problem that included a single production facility, a set of customers with time-varying demand and a fleet of vehicles. Wang and Lei (2012) considered the problem of operations scheduling for a capacitated multi-echelon shipping network with delivery deadlines, where semi-finished goods are shipped from suppliers to customers through processing centers, with the objective of minimizing the shipping and penalty cost. The three polynomial-time solvable cases of this problem were reported: with identical order quantities; with designated suppliers; and with divisible customer order sizes. Liu and Papageorgiou (2013) developed a multi-objective MIP approach to address production, distribution and capacity planning of global supply chains considering cost, responsiveness and customer service level simultaneously. An integrated approach to deterministic coordinated supply chain scheduling was proposed by Sawik (2007) to simultaneously schedule manufacturing and supply of parts and assembly of finished products. Given a set of part suppliers and a set of customer orders for finished products, the problem objective was to determine which orders were provided with parts by each supplier, to schedule manufacturing of parts at each supplier and delivery of parts from each supplier to the producer, and to schedule customer orders at the producer, such that a high customer service level was achieved and the total cost was minimized. The selection of part supplier for each customer order was combined with a due date setting for some orders to maximize the number of orders that can be completed by customer requested due dates. A monolithic MIP model was presented and compared with a hierarchy of mixed integer programs for a sequential selection of suppliers and scheduling of manufacturing and delivery of parts and assembly of products. Different enhancements of the above MIP approach for the coordinated scheduling in multi-tier supply chain networks were presented in Sawik (2009a). Various perspectives on supply chain coordination issues were reported and reviewed by Arshinder et al. (2008) and the gaps existing in the literature were identified. Li and Wang (2007) reviewed coordination mechanisms of supply chain systems in a framework that was based on supply chain decision structure and nature of demand. A review of methods and literature on supply chain coordination through contracts was provided by Hezarkhani and Kubiak (2010).

The customer service level in a make-to-order environment was studied by Altendorfer and Jodlbauer (2011). Shao and Dong (2012) compared and analyzed order fulfillment performance measures for two different production control systems:

make-to-order versus make-to-stock. They formulated service-maximization models with inventory cost budget constraints. For the make-to-stock production, Larsen and Thorstenson (2008, 2014) differentiated between an order fill rate and a volume fill rate and specified their performance for different inventory control systems. They showed how the order and volume fill rates are related in magnitude. Bijulal et al. (2011) analyzed the production-inventory system performance in terms of order fill rate and average system costs, affected by the control parameters.

The material presented in this chapter is based on the research results reported in Sawik (2013e, 2014a,b) where single, dual and multiple sourcing strategies were compared for risk-neutral and risk-averse decision-making. In addition, Sawik (2016b) compared the two risk-averse service level measures: expected worst-case order fulfillment rate and expected worst-case demand fulfillment rate. The former corresponds to order fill rate and the latter to volume fill rate in the make-to-stock production (Larsen and Thorstenson 2008, 2014). The future research should focus on comparison of the proposed integrated supplier selection and customer order scheduling with a common hierarchical approach, where first the supplier selection and order quantity allocation subject to disruption risks is accomplished and then, given a schedule of part supplies, the optimal schedule of customer orders is determined for each disruption scenario, subject to parts availability constraints.

Problems

5.1. Modify the SMIP models presented in this chapter for multiple part types with subsets of part types required for each customer order and subsets of suppliers available for each part type.

5.2. Modify the SMIP models presented in this chapter for finite capacity suppliers.

5.3. Mixed mean-risk decision-making
(a) Formulate model **SPSm_E(sl)CV(c)** for a mixed mean-risk selection of supply portfolio and scheduling of customer orders to trade-off expected service level and CVaR of cost.
(b) How should the values of the optimized objective functions be scaled into the interval [0,1] to avoid dimensional inconsistency among the two objectives and how should the trade-off parameter be selected?
(c) How would you interpret the mixed mean-risk solution?

5.4. Explain why order fulfillment rate shows a higher service performance than demand fulfillment rate.

5.5. Looking into the computational examples try to compare dual vs. multiple sourcing solution results.

Chapter 6
Integrated Selection of Supply Portfolio and Scheduling of Production and Distribution

6.1 Introduction

The key operational functions in a supply chain are supply, production, and distribution operations. To achieve a high performance of a cyber-physical system, such as contemporary supply chains, it is crucial to integrate these three functions and jointly schedule supply, production, and distribution in a coordinated manner. For example, in customer-driven supply chains, where customer orders are executed immediately or shortly after arrival of material supplies and the ordered products are delivered to customers immediately or shortly after their completion, the impact of disruption risks can be best mitigated when an integrated decision-making is applied. At the same time, the integrated decision-making allows reaching various conflicting objectives, such as reduction in total cost and increase in service level. The purpose of this chapter is to study the integrated decision-making to simultaneously select suppliers of parts, allocate order quantity, and schedule production and delivery of finished products to customers in a supply chain under disruption risks. In addition to supplier selection, order quantity allocation, and scheduling of customer orders, distribution of finished products to customers is simultaneously considered with different shipping methods to optimize the trade-off between cost and service level. The service level in this chapter denotes demand fulfillment rate (see, Chap. 5.3.2). The three different shipping methods will be modeled and compared for the distribution of products: batch shipping with a single shipment of different customer orders, batch shipping with multiple shipments of different customer orders, and individual shipping of each customer order immediately after its completion. In addition, the SMIP formulation based on the wait-and-see approach to optimize the trade-off between expected cost and expected service level will be compared with a deterministic MIP model based on the expected value approach, in which random parameters are replaced by their expected values.

T. Sawik, *Supply Chain Disruption Management*,
International Series in Operations Research & Management Science 291,
https://doi.org/10.1007/978-3-030-44814-1_6

The following time-indexed SMIP and MIP models are presented in this chapter:

SCS1_E for risk-neutral scheduling in supply chain with single batch shipping of products to distribution centers;
SCS2_E for risk-neutral scheduling in supply chain with individual shipping of products for each customer order to distribution centers;
SCS3_E for risk-neutral scheduling in supply chain with multiple batch shipping of products to distribution centers;
SCS1_CV for risk-averse scheduling in supply chain with single batch shipping of products to distribution centers;
SCS2_CV for risk-averse scheduling in supply chain with individual shipping of products for each customer order to distribution centers;
SCS3_CV for risk-averse scheduling in supply chain with multiple batch shipping of products to distribution centers;
ESCS1 for the expected value-based supply chain scheduling, corresponding to model **SCS1_E**.

The models incorporate supply-production, production-distribution, and supply-distribution coordinating constraints to efficiently coordinate supply, production, and distribution schedules.

Numerical examples and computational results are reported in Sect. 6.6 for risk-neutral, risk-averse, and expected value-based decision-making. The major managerial insights are provided at the end of this chapter.

6.2 Problem Description

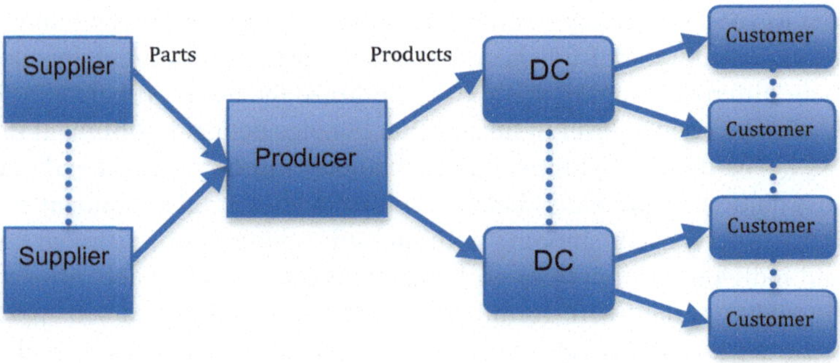

Fig. 6.1 A multi-tier supply chain network

A multi-tier supply chain network under disruption risk is studied with multiple suppliers of a critical part type, producer of one product type, and multiple distribution centers that provide with products a set of customers (see, Fig. 6.1).

Let $I = \{1,\ldots,\overline{I}\}$ be the set of $\overline{I}$ suppliers, $J = \{1,\ldots,\overline{J}\}$ the set of $\overline{J}$ customers, $K = \{1,\ldots,\overline{K}\}$ the set of $\overline{K}$ distribution centers, and $T = \{1,\ldots,\overline{T}\}$ the set of $\overline{T}$ planning periods (for notation used, see Table 6.1).

Table 6.1 Notation: supply portfolio selection and production and distribution scheduling

Indices	
i	= supplier, $i \in I$
j	= customer order, $j \in J$
k	= distribution center, $k \in K$
r	= region, $r \in R$
s	= disruption scenario, $s \in S$
t	= planning period, $t \in T$
Input Parameters	
a_k	= fixed transportation cost of each shipment to distribution center $k \in K$
b_j	= size (number of products) of customer order j
B	= $\sum_{j \in J} b_j$, total demand for products
C_P	= per period capacity of producer
C_V	= vehicle capacity
d_j	= due date for customer order j
e_i	= fixed cost of ordering parts from supplier i
g_j	= per unit penalty cost of unfulfilled customer order j
h_j	= per unit and per period penalty cost of delayed customer order j
I^r	= the subset of suppliers in region r
I_s	= the subset of suppliers non-disrupted under disruption scenario s
J_k	= the subset of customers (customer orders) served by distribution center k
o_i	= per unit price of parts purchased from supplier i
p_i	= local disruption probability for supplier i
p^r	= regional disruption probability for all suppliers in region r
τ_i	= delivery lead time from supplier i
σ_k	= transportation time to distribution center k

Denote by $J_k \subset J$ the subset of customers (or equivalently customer orders) served by the distribution center $k \in K$. The demand of each customer $j \in J_k$ is defined by the number of products, b_j and the latest period of their delivery to the distribution center k, d_j. The total demand for parts is identical with the total demand for products, $B = \sum_{j \in J} b_j$. For each customer order $j \in J$, denote by g_j and h_j, the unitary unfulfillment penalty cost and delay penalty cost, respectively. The producer aggregate capacity in each planning period is denoted by C_P. The capacity depends on machine configuration of production facility and the processing time available on each machine in a single planning period, e.g., Sawik (2011a, 2009c). (In the computational examples provided in Sect. 6.6, the production facility is a flexible flow shop with parallel single and batch machines.)

Each supplier $i \in I$ is characterized by the four parameters: the unit purchasing price, o_i; the fixed ordering cost, e_i; the constant lead time (manufacturing and transportation time to producer), τ_i; and the local disruption probability, p_i, i.e., the probability that parts ordered from supplier i are not delivered. The parts ordered from supplier i can be used by the producer not earlier than in period $\tau_i + 1$.

Three distribution strategies of customer orders will be considered:

- The finished products for all customers $j \in J_k$ are delivered to distribution center k in a single batch shipment, after completion of all scheduled customer orders $j \in J_k$.
- The finished products for each customer order $j \in J_k$ are individually shipped to distribution center k, immediately after its completion.
- The finished products for customers $j \in J_k$ are delivered to distribution center k in multiple batch shipments, subject to limited transportation capacity.

The transportation time to distribution center k is constant and equals to σ_k periods so that the products shipped in period t are delivered in period $t + \sigma_k - 1$.

For the multiple batch shipping strategy, additionally denote by C_V, vehicle limited capacity defined as the maximum number of finished products that can be shipped together, and by a_k, a fixed transportation cost of each shipment to distribution center $k \in K$.

The above shipping methods can be easily justified from a practical point of view. The individual and immediate shipping are commonly used for distribution of time-sensitive products to achieve a fast delivery. The single and multiple batch shipping methods are used for orders going to the same distribution center. The shipping methods with unlimited transportation capacity, e.g., infinitely many vehicles, are applicable when the delivery is handled by a third-party logistics provider that typically owns a large number of vehicles. On the other hand, the multiple batch shipping with limited transportation capacity accounts for limited capacity of a single vehicle. The shipping methods considered do not account for transportation time and costs to customers since routing (e.g., vehicle routing) is not a part of the decision-making, and the customers are assumed to be individually served by the corresponding distribution centers. Alternatively, for each distribution center $k \in K$, all customers $j \in J_k$ can be assumed to be co-located.

The suppliers are assumed to be located in $\overline{R}$ disjoint geographic regions. Denote by $I^r \subseteq I$ the subset of suppliers in region $r \in R = \{1, \ldots, \overline{R}\}$, where $\bigcup_{r \in R} I^r = I$. The supplies of parts are subject to random local disruptions that are uniquely associated with a particular supplier, which may arise from equipment breakdowns, local labor strike, fires, etc. Denote by p_i the local disruption probability for supplier i, i.e., the parts ordered from supplier i are delivered without disruptions with probability $(1 - p_i)$, or not at all with probability p_i. In addition to independent local disruptions of each supplier, the supplies of parts are also subject to correlated regional disruption of all suppliers in the same region simultaneously, with probability p^r for region $r \in R$.

Let $S = \{1, \ldots, \overline{S}\}$ be the index set of $\overline{S} = 2^{\overline{I}}$ disruption scenarios, where each scenario $s \in S$ defines a subset $I_s \subset I$ of non-disrupted suppliers. The supplies from every supplier, $i \in I \setminus I_s$, can be independently disrupted either by a local or by a regional disaster event. The probability P_s of each disruption scenario $s \in S$ is a product over all regions $r \in R$ of probabilities P_s^r of realizing disruption scenario s for suppliers in I^r,

$$P_s = \prod_{r \in R} P_s^r,$$

where P_s^r is (cf. Sect. 1.3)

$$P_s^r = \begin{cases} (1-p^r)\prod_{i \in I^r \bigcap I_s}(1-p_i)\prod_{i \in I^r \setminus I_s} p_i & \text{if } I^r \bigcap I_s \neq \emptyset \\ p^r + (1-p^r)\prod_{i \in I^r} p_i & \text{if } I^r \bigcap I_s = \emptyset. \end{cases}$$

The problem of the integrated supply, production, and distribution scheduling will be formulated under the following simplified assumptions:

- orders for parts are placed at the start of the planning horizon, when all customer orders for products are known;
- each supplier capacity is sufficient to meet total demand for parts and to complete and prepare orders for shipping in a single planning period;
- all parts ordered from a supplier are shipped together in a single delivery;
- customers are assigned to distribution centers ahead of time and each customer is served by exactly one distribution center;
- each customer order can be completed by the producer in a single planning period;
- customer requested due dates are replaced with delivery due dates for distribution centers;
- transportation times to distribution centers are constant, and the delivery dates are determined by the shipping dates to distribution centers;
- transportation times and costs to individual customers are not considered and vehicle routing is not a part of the decision-making;
- inventory of parts and products are not considered;
- three shipping methods: single batch shipping, multiple batch shipping, and individual and immediate shipping are considered separately.

Some of the above assumptions can be easily relaxed, while others need more advanced models to be developed.

6.3 Models for Risk-Neutral Decision-Making

In this section, three SMIP models **SCS1_E**, **SCS2_E**, and **SCS3_E** are presented for the integrated supplier selection, order quantity allocation, and scheduling production of finished products and distribution to customers to minimize expected cost per product and maximize expected service level (expected demand fulfillment rate). Model **SCS1_E** is formulated for supply chain scheduling with a single batch shipping of finished products to each distribution center for all customers served by that center. Model **SCS2_E** accounts for individual and immediate shipment of products to distribution centers for each customer. In models **SCS1_E** and **SCS2_E** transportation cost and capacity are not considered. Model **SCS3_E** is formulated for multiple batch shipping of finished products to each distribution center and accounts for transportation cost and capacity.

The problem variables are defined in Table 6.2.

Table 6.2 Problem variables: supply portfolio selection and production and distribution scheduling

	First stage variables
u_i	$= 1$, if supplier i is selected; otherwise $u_i = 0$ (supplier selection)
v_i	$\in [0,1]$, the fraction of total demand for parts ordered from supplier i (supply portfolio)
	Second stage variables
w^s_j	$= 1$, if under disruption scenario s customer order j is delivered by its due date; otherwise $w^s_j = 0$ (customer order non-delay delivery, single batch shipping)
x^s_{jt}	$= 1$, if under disruption scenario s customer order j is scheduled for period t; otherwise $x^s_{jt} = 0$ (production scheduling, single and multiple batch shipping)
	$= 1$, if under disruption scenario s customer order j is scheduled for period t and shipped to distribution center in period $t+1$; otherwise $x^s_{jt} = 0$ (production and distribution scheduling, individual shipping)
y^s_{kt}	$= 1$, if under disruption scenario s, batch shipment of products to distribution center k is scheduled for period t; otherwise $y^s_{kt} = 0$ (distribution scheduling, single and multiple batch shipping)
z^s_{jt}	$= 1$, if under disruption scenario s customer order j is shipped to distribution center in period t; otherwise $z^s_{jt} = 0$ (distribution scheduling, multiple batch shipping)
	Auxiliary variables
VaR^c	cost-at-risk, the targeted cost such that for a given confidence level α, for $100\alpha\%$ of the scenarios, the outcome is below VaR^c
VaR^{sl}	service-at-risk, the targeted service level such that for a given confidence level α, for $100\alpha\%$ of the scenarios, the outcome is above VaR^{sl}
$\mathscr{C}_s$	≥ 0, the tail cost for scenario s, i.e., the amount by which costs in scenario s exceed VaR^c
$\mathscr{S}_s$	≥ 0, the tail service level for scenario s, i.e., the amount by which VaR^{sl} exceeds service level in scenario s

6.3.1 Scheduling with Single Batch Shipping

The total customer demand, $\sum_{j\in J} b_j$, under each disruption scenario $s \in S$, can be split into the three satisfaction levels: non-delayed, delayed, or unscheduled (rejected). For a single batch shipping, these portions of the customer demand are defined below.

- Non-delayed, $\sum_{j\in J} b_j w^s_j$,
- Delayed, $\sum_{j\in J} b_j (\sum_{t\in T} x^s_{jt} - w^s_j)$,
- Unscheduled (rejected), $\sum_{j\in J} b_j (1 - \sum_{t\in T} x^s_{jt})$.

The expected cost per product, E^c, and service level, E^{sl}, as well their normalized values, respectively, f^c and f^{sl}, are defined below.

$$E^c = (\sum_{i\in I} e_i u_i + \sum_{s\in S} P_s(\sum_{i\in I_s} Bo_i v_i$$

$$+\sum_{j\in J} h_j b_j(\sum_{t\in T} x^s_{jt} - w^s_j) + \sum_{j\in J} g_j b_j(1-\sum_{t\in T} x^s_{jt})))/B. \quad (6.1)$$

$$E^{sl} = \sum_{s\in S}\sum_{j\in J} P_s b_j w^s_j/B. \quad (6.2)$$

$$f^c = \frac{E^c - \underline{E}^c}{\overline{E}^c - \underline{E}^c}, \quad (6.3)$$

where $\underline{E}^c$ and $\overline{E}^c$ are the minimum and the maximum values of E^c, respectively.

$$f^{sl} = \frac{\overline{E}^{sl} - E^{sl}}{\overline{E}^{sl} - \underline{E}^{sl}}, \quad (6.4)$$

where $\underline{E}^{sl}$ and $\overline{E}^{sl}$ are the minimum and the maximum values of E^{sl}, respectively.

A subset of nondominated solutions can be obtained by the parameterization on λ, the model presented below.

SCS1_E: *Risk-neutral* **S**upply **C**hain **S**cheduling: *single batch shipping*

Minimize

$$\lambda f^c + (1-\lambda) f^{sl}, \quad (6.5)$$

where $0 \leq \lambda \leq 1$,

subject to (6.1)–(6.4) and

Supply portfolio selection constraints:

$$\sum_{i\in I} v_i = 1 \quad (6.6)$$

$$v_i \leq u_i;\ i \in I \quad (6.7)$$

Customer order scheduling constraints:

$$\sum_{t\in T} x^s_{jt} \leq 1;\ j \in J, s \in S \quad (6.8)$$

$$\sum_{j\in J} b_j x^s_{jt} \leq C_P;\ t \in T, s \in S \quad (6.9)$$

(continued)

Supply-production coordinating constraints:

$$\sum_{j\in J}\sum_{t'\in T:t'\le t} b_j x^s_{jt'} \le B \sum_{i\in I_s:\tau_i\le t-1} v_i;\ t\in T, s\in S \tag{6.10}$$

Single batch shipping constraints:

$$\sum_{t\in TK} y^s_{kt} \le 1;\ k\in K, s\in S \tag{6.11}$$

$$\sum_{t\in TK} y^s_{kt} \ge \sum_{t\in T} x^s_{jt};\ k\in K, j\in J_k, s\in S \tag{6.12}$$

$$\sum_{t\in TK} y^s_{kt} \le \sum_{j\in J_k}\sum_{t\in T} x^s_{jt};\ k\in K, s\in S, \tag{6.13}$$

where $TK = \{\min_{i\in I}\tau_i + 2, \ldots, \overline{T}+1\}$ is the set of shipping periods.

Production-distribution coordinating constraints:

$$\sum_{t\in TK} t y^s_{kt} \ge \sum_{t\in T}(t+1)x^s_{jt};\ k\in K, j\in J_k, s\in S \tag{6.14}$$

Non-delayed delivery constraints:

$$w^s_j \le \sum_{t\in T:t\le d_j-\sigma_k} x^s_{jt};\ k\in K, j\in J_k, s\in S \tag{6.15}$$

$$w^s_j \le \sum_{t\in TK:t\le d_j-\sigma_k+1} y^s_{kt};\ k\in K, j\in J_k, s\in S \tag{6.16}$$

$$\sum_{t\in T:t\le d_j-\sigma_k} x^s_{jt} + \sum_{t\in TK:t\le d_j-\sigma_k+1} y^s_{kt} - 1 \le w^s_j;\ k\in K, j\in J_k, s\in S \tag{6.17}$$

Non-negativity and integrality conditions:

$$u_i \in \{0,1\};\ i\in I \tag{6.18}$$

$$v_i \in [0,1];\ i\in I \tag{6.19}$$

$$w^s_j \ge 0;\ j\in J, s\in S \tag{6.20}$$

$$x^s_{jt} \in \{0,1\};\ j\in J, t\in T, s\in S \tag{6.21}$$

$$y^s_{kt} \in \{0,1\};\ k\in K, t\in TK, s\in S. \tag{6.22}$$

Constraints (6.6)–(6.7) define a feasible supply portfolio, and (6.8)–(6.9), a feasible assignment of customer orders to planning periods for each disruption scenario. The supply-production coordinating constraints (6.10) ensure that for each disruption scenario s, the cumulative demand for required parts of all customer orders scheduled by period t is not greater than the cumulative supplies by period $t-1$ from the non-disrupted suppliers $i\in I_s$. Eq. (6.11) ensures that for each disruption scenario, at most one batch shipment can be scheduled to each distribution center.

Constraint (6.12) denotes that for each disruption scenario, shipment to distribution center k is scheduled, only if at least one customer order $j \in J_k$ is completed, and Eq. (6.13) that no shipment to distribution center k is scheduled, if no customer order $j \in J_k$ is completed, The production-distribution coordinating constraints (6.14) ensure that for each disruption scenario, a shipment to distribution center k can be scheduled only after the latest completion period of scheduled customer orders $j \in J_k$. Finally, constraints (6.15)–(6.17) denote that for each disruption scenario $s \in S$, customer order $j \in J_k$ can be delivered without delay (i.e., $w_j^s = 1$), unless it is scheduled not later than $d_j - \sigma_k$ and shipped to distribution center k not later than $d_j - \sigma_k + 1$; otherwise the customer order is delayed, or unscheduled (i.e., $w_j^s = 0$).

In the above formulation w_j^s is not restricted to being binary. However, for any feasible solution satisfying customer due date meeting constraints (6.15)–(6.17), $w_j^s \in \{0,1\};\ j \in J, s \in S$. The following proposition clarifies that issue.

Proposition 6.1. *In model* **SCS1_E**, *the non-delay delivery variables,* w_j^s, *may take on only value 0 or 1, even with no restriction to being binary. Thus, it is sufficient to consider non-negativity conditions only,* $w_j^s \geq 0;\ j \in J, s \in S$, *(6.20).*

Proof. Notice that $w_j^s = 1$ only, if

$$\sum_{t \in T: t \leq d_j - \sigma_k} x_{jt}^s = 1;\ j \in J_k, s \in S$$

and

$$\sum_{t \in TK: t \leq d_j - \sigma_k + 1} y_{kt}^s = 1;\ j \in J_k, s \in S.$$

Then both constraints (6.15) and (6.16) become $w_j^s \leq 1$, while (6.17) becomes $w_j^s \geq 1$, hence $w_j^s = 1$.

Otherwise; i.e., if right-hand side of either (6.15) or (6.16), or of both (6.15) and (6.16) is 0, then $w_j^s = 0$.

Since the cost and the service level objectives are in conflict, the minimum and maximum values of expected cost $\underline{E}^c$, $\overline{E}^c$, and expected customer service level, $\underline{E}^{sl}$, $\overline{E}^{sl}$, are obtained by solving the two mixed integer programs presented below.

Minimize E^c, (6.1), subject to (6.6)–(6.22).

Maximize E^{sl}, (6.2), subject to (6.6)–(6.22).

The values $\underline{E}^c$ and $\underline{E}^{sl}$ are determined by solving the first problem, and $\overline{E}^{sl}$ and $\overline{E}^c$, by solving the second problem.

6.3.2 Scheduling with Individual and Immediate Shipping

In model **SCS2_E** presented below the variables, $y^s_{kt} \in \{0,1\}$; $k \in K, t \in TK, s \in S$ and $w^s_j \in \{0,1\}$; $j \in J, s \in S$ are not required anymore; however, the production scheduling variable, $x^s_{jt} \in \{0,1\}$; $j \in J, t \in T, s \in S$, now becomes a joint production and distribution scheduling variable, redefined as follows:

$x^s_{jt} = 1$, if under disruption scenario s customer order j is processed in period t and shipped to the distribution center in period $t+1$; otherwise $x^s_{jt} = 0$.

Accordingly, the objective functions E^c, (6.1) and E^{sl}, (6.2) are rewritten as below.

$$E^c = \big(\sum_{i\in I} e_i u_i + \sum_{s\in S} P_s\big(\sum_{i\in I_s} Bo_i v_i$$

$$+ \sum_{j\in J} h_j b_j\big(\sum_{t\in T} x^s_{jt} - \sum_{t\in T: t\le d_j-\sigma_{k_j}} x^s_{jt}\big) + \sum_{j\in J} g_j b_j\big(1 - \sum_{t\in T} x^s_{jt}\big)\big)\big)/B \qquad (6.23)$$

$$E^{sl} = \sum_{s\in S}\sum_{j\in J}\sum_{t\in T: t\le d_j-\sigma_{k_j}} P_s b_j x^s_{jt}/B, \qquad (6.24)$$

where σ_{k_j} is the transportation time to the distribution center k that serves customer $j \in J_k$.

In model **SCS2_E** presented below the objective functions E^c, (6.1) and E^{sl}, (6.2) have been replaced with (6.23) and (6.24), respectively.

SCS2_E: *Risk-neutral supply chain scheduling: individual shipping*
Minimize (6.5)
subject to (6.3), (6.4), (6.6)–(6.10), (6.18)–(6.20), (6.23)–(6.24).

For model **SCS2_E**, the values of $\underline{E}^c$, $\overline{E}^c$, and $\underline{E}^{sl}$, $\overline{E}^{sl}$, are obtained by solving the two mixed integer programs:

Minimize E^c, (6.23), subject to (6.6)–(6.10), (6.18)–(6.20),

Maximize E^{sl}, (6.24), subject to (6.6)–(6.10), (6.18)–(6.20).

6.3.3 Scheduling with Multiple Batch Shipping and Limited Transportation Capacity

In this subsection a multiple batch shipping method is modeled, where each batch size is limited by vehicle capacity C_V and, in addition, a fixed transportation cost, a_k, is incurred for each shipment to distribution center $k \in K$.

In model **SCS3_E** presented below the variables, $w_j^s \in \{0,1\};\ j \in J, s \in S$, are not required any more. Instead a new customer order shipping variable is introduced, $z_{jt}^s \in \{0,1\};\ j \in J, t \in TK, s \in S$, defined as follows:

$z_{jt}^s = 1$, if under disruption scenario s customer order j is shipped to the distribution center in period t; otherwise $z_{jt}^s = 0$.

The objective functions E^c and E^{sl} are defined below.

$$E^c = (\sum_{i \in I} e_i u_i + \sum_{s \in S} P_s(\sum_{i \in I_s} Bo_i v_i + \sum_{k \in K} \sum_{t \in TK} a_k y_{kt}^s$$

$$+ \sum_{j \in J} h_j b_j (\sum_{t \in T} x_{jt}^s - \sum_{t \in TK: t \le d_j - \sigma_{k_j} + 1} z_{jt}^s) + \sum_{j \in J} g_j b_j (1 - \sum_{t \in T} x_{jt}^s)))/B \qquad (6.25)$$

$$E^{sl} = \sum_{s \in S} \sum_{j \in J} \sum_{t \in TK: t \le d_j - \sigma_{k_j} + 1} P_s b_j z_{jt}^s / B, \qquad (6.26)$$

where $\sum_{s \in S} \sum_{k \in K} \sum_{t \in TK} P_s a_k y_{kt}^s$ is the expected transportation cost of shipping to distribution centers and $\sum_{s \in S} \sum_{j \in J} P_s h_j b_j (\sum_{t \in T} x_{jt}^s - \sum_{t \in TK: t \le d_j - \sigma_{k_j} + 1} z_{jt}^s)$ is the expected penalty for delayed orders.

SCS3_E: *Risk-neutral supply chain scheduling: multiple batch shipping*

Minimize (6.5)

subject to (6.3), (6.4), (6.6)–(6.10), (6.18)–(6.21), (6.25)–(6.26) and

Multiple batch shipping constraints:

$$\sum_{t \in TK} z_{jt}^s = \sum_{t \in T} x_{jt}^s;\ j \in J, s \in S \qquad (6.27)$$

$$\sum_{j \in J_k} b_j z_{jt}^s \le C_V y_{kt}^s;\ k \in K, t \in TK, s \in S \qquad (6.28)$$

$$\sum_{t \in TK} y_{kt}^s \ge \sum_{j \in J_k} \sum_{t \in T} b_j x_{jt}^s / C_V;\ k \in K, s \in S \qquad (6.29)$$

$$\sum_{j \in J_k} z_{jt}^s \ge y_{kt}^s;\ k \in K, t \in TK, s \in S \qquad (6.30)$$

Production-distribution coordinating constraints:

$$\sum_{t \in TK} t z_{jt}^s \ge \sum_{t \in T} (t+1) x_{jt}^s;\ j \in J, s \in S \qquad (6.31)$$

(continued)

Supply-distribution coordinating constraints:

$$\sum_{j\in J}\sum_{t'\in TK:t'\leq t} b_j z^s_{jt'} \leq B \sum_{i\in I_s:\tau_i\leq t-2} v_i;\ t\in TK, s\in S \tag{6.32}$$

$$z^s_{jt}\in\{0,1\};\ j\in J, t\in TK, s\in S. \tag{6.33}$$

For each disruption scenario, Eq. (6.27) ensures that each completed order is shipped to distribution center. Constraint (6.28) ensures that each batch shipment size cannot exceed vehicle limited capacity, and Eq. (6.29) determines minimum number of batch shipments to each distribution center. Eq. (6.30) guarantees that at least one order is shipped in every batch. The production-distribution coordinating constraints (6.31) ensure that each order can be shipped only after its completion. Finally, the supply-distribution coordinating constraints (6.32) ensure that for each disruption scenario s, the cumulative distribution of products shipped to customers by period t is not greater than the cumulative supplies of required parts delivered by the non-disrupted suppliers $i\in I_s$ by period $t-2$. (Note that parts delivered by supplier $i\in I_s$ in period τ_i can be used for production not earlier than in period τ_i+1, and then the completed products can be shipped to distribution center not earlier than in period τ_i+2.)

For model **SCS3_E**, the values of $\underline{E}^c$, $\overline{E}^c$, and $\underline{E}^{sl}$, $\overline{E}^{sl}$, are obtained by solving the two mixed integer programs:

Minimize E^c, (6.25), subject to (6.6)–(6.10), (6.18)–(6.21), (6.27)–(6.33).

Maximize E^{sl}, (6.26), subject to (6.6)–(6.10), (6.18)–(6.21), (6.27)–(6.33).

6.4 Models for Risk-Averse Decision-Making

In this section, three time-indexed SMIP models **SCS1_CV**, **SCS2_CV**, and **SCS3_CV** are proposed for risk-averse supply chain scheduling to optimize the weighted sum of worst-case cost and worst-case service level under disruption risks. The models are based on the risk-neutral models **SCS1_E**, **SCS2_E**, and **SCS3_E**, respectively.

Let VaR^c be the value-at-risk of cost per product, i.e., the targeted cost such that for a given confidence level, α, for $100\alpha\%$ of disruption scenarios, the outcome is below VaR^c and let CVaR^c be conditional value-at-risk of cost per product, i.e., the expected cost in the worst $100(1-\alpha)\%$ of the scenarios with the cost above VaR^c

$$CVaR^c = VaR^c + (1-\alpha)^{-1} \sum_{s \in S} P_s \mathscr{C}_s. \tag{6.34}$$

In a similar way, define by VaR^{sl} the value-at-risk of service level, i.e., the targeted service level such that for a given confidence level α, for $100\alpha\%$ of disruption scenarios, the outcome is above VaR^{sl}, and by CVaR^{sl} the conditional value-at-risk of service level, i.e., the expected service level in the worst $100(1-\alpha)\%$ of scenarios with the service level below VaR^{sl}

$$CVaR^{sl} = VaR^{sl} - (1-\alpha)^{-1} \sum_{s \in S} P_s \mathscr{S}_s. \tag{6.35}$$

The risk-averse integrated supply, production, and distribution schedule will be optimized by simultaneously calculating VaR^c and VaR^{sl} and minimizing weighted difference of CVaR^c and CVaR^{sl}. Model **SCS1_CV** is presented below.

SCS1_CV: *Risk-averse supply chain scheduling to minimize trade-off between CVaR of cost and CVaR of service level: single batch shipping*
Minimize

$$\lambda CVaR^c - (1-\lambda) CVaR^{sl}, \tag{6.36}$$

where $0 \le \lambda \le 1$,
subject to (6.6)–(6.22), (6.34), (6.35) and
Risk constraints:

- the tail cost for scenario s is defined as the nonnegative amount by which cost in scenario s exceeds VaR^c,
- the tail service level for scenario s is defined as the nonnegative amount by which VaR^{sl} exceeds service level in scenario s,

$$\begin{aligned} \mathscr{C}_s \ge (\sum_{i \in I} e_i u_i + \sum_{i \in I_s} Bo_i v_i \\ + \sum_{j \in J} h_j b_j (\sum_{t \in T} x^s_{jt} - w^s_j) \\ + \sum_{j \in J} g_j b_j (1 - \sum_{t \in T} x^s_{jt}))/B - VaR^c; \; s \in S \end{aligned} \tag{6.37}$$

$$\mathscr{S}_s \ge VaR^{sl} - \sum_{j \in J} b_j w^s_j / B; \; s \in S \tag{6.38}$$

$$\mathscr{C}_s \ge 0; \; s \in S \tag{6.39}$$

$$\mathscr{S}_s \ge 0; \; s \in S, \tag{6.40}$$

where $\mathscr{C}_s$ is the tail cost and $\mathscr{S}_s$ is the tail service level, for scenario s.

In a similar way we can formulate the risk-averse versions **SCS2_CV** and **SCS3_CV** of the risk-neutral models **SCS2_E** and **SCS3_E**. The models are presented below.

SCS2_CV: *Risk-averse supply chain scheduling to minimize trade-off between CVaR of cost and CVaR of service level: individual shipping*
Minimize (6.36)
subject to (6.6)–(6.10), (6.18)–(6.20), (6.23), (6.24), (6.34), (6.35), (6.39), (6.40) and
Risk constraints:

$$\mathscr{C}_s \geq (\sum_{i\in I} e_i u_i + \sum_{i\in I_s} Bo_i v_i + \sum_{k\in K}\sum_{t\in TK} a_k y^s_{kt} + \sum_{j\in J} h_j b_j (\sum_{t\in T} x^s_{jt} - \sum_{t\in TK: t\leq d_j-\sigma_{k_j}+1} z^s_{jt}) + \sum_{j\in J} g_j b_j (1 - \sum_{t\in T} x^s_{jt}))/B - VaR^c;\ s\in S \quad (6.41)$$

$$\mathscr{S}_s \geq VaR^{sl} - \sum_{j\in J}\sum_{t\in T: t\leq d_j-\sigma_{k_j}} b_j x^s_{jt}/B;\ s\in S. \quad (6.42)$$

SCS3_CV: *Risk-averse supply chain scheduling to minimize trade-off between CVaR of cost and CVaR of service level: multiple batch shipping*
Minimize (6.36)
subject to (6.6)–(6.10), (6.18)–(6.21), (6.27)–(6.35), (6.39), (6.40) and
Risk constraints:

$$\mathscr{C}_s \geq (\sum_{i\in I} e_i u_i + \sum_{i\in I_s} Bo_i v_i + \sum_{k\in K}\sum_{t\in TK} a_k y^s_{kt} + \sum_{j\in J} h_j b_j (\sum_{t\in T} x^s_{jt} - \sum_{t\in TK: t\leq d_j-\sigma_{k_j}+1} z^s_{jt}) + \sum_{j\in J} g_j b_j (1 - \sum_{t\in T} x^s_{jt}))/B - VaR^c;\ s\in S \quad (6.43)$$

$$\mathscr{S}_s \geq VaR^{sl} - \sum_{j\in J}\sum_{t\in TK: t\leq d_j-\sigma_{k_j}+1} b_j z^s_{jt}/B;\ s\in S. \quad (6.44)$$

6.5 Expected Value Problem

Stochastic mixed integer programs are usually hard to solve because they are large-scale optimization problems when applied to real-world problems. A simplified approach, which may sometimes be useful in practice, is to consider a simpler deterministic program, known as expected value problem, in which the random parameters are replaced by their expected values (e.g., Birge and Louveaux 2011; Kall and Mayer 2011; Delage et al. 2014; Sawik 2017c). As an example of such approach this section presents the expected value problem corresponding to the risk-neutral scheduling in supply chain with single batch shipping of products to distribution centers, described by SMIP model **SCS1_E**. In model **SCS1_E**, where the randomness is characterized by a set of disruption scenarios, the only random parameters are suppliers all-or-nothing fulfillment rates, which appear both in the objective function (6.1) and in constraints (6.10). Denote by **ESCS1**, a deterministic MIP model of the expected value problem corresponding to model **SCS1_E**. In model **ESCS1**, suppliers probabilistic all-or-nothing fulfillment rates (1, for a non-disrupted supplier and 0, for a disrupted supplier) defined for each disruption scenario are replaced by the expected fulfillment rates of each supplier

$$1 - \pi_i = (1 - p^r)(1 - p_i);\ i \in I^r, r \in R,$$

where

$$\pi_i = p^r + (1 - p^r)p_i,\ i \in I^r, r \in R \tag{6.45}$$

is total disruption probability of supplier i.

Notice that the expected fulfillment rate of a supplier is identical with non-disruption probability.

Accordingly, stochastic binary decision variables, w_j^s, x_{jt}^s, y_{kt}^s, (6.20)–(6.22), defined for each disruption scenario $s \in S$ are replaced by their deterministic equivalents W_j, X_{jt}, Y_{kt} (see, Table 6.3)

Table 6.3 Notation: expected value problem

	Parameters
π_i	$= p^r + (1 - p^r)p_i$, total disruption probability of supplier $i \in I^r, r \in R$
$1 - \pi_i$	$= (1 - p^r)(1 - p_i)$, the expected fraction of an order delivered by supplier $i \in I^r, r \in R$ (supplier expected fulfillment rate)
	Variables
W_j	$= 1$, if customer order j is delivered by its due date; otherwise $W_j = 0$ (customer order non-delay delivery)
X_{jt}	$= 1$, if customer order j is scheduled for period t; otherwise $X_{jt} = 0$ (production scheduling)
Y_{kt}	$= 1$, if batch shipment of products to distribution center k is scheduled for period t; otherwise $Y_{kt} = 0$ (distribution scheduling)

Now, the expected cost per product, E^c, and the expected service level, E^{sl}, are defined as follows

$$E^c = (\sum_{i\in I} e_i u_i + \sum_{i\in I} Bo_i(1-\pi_i)v_i$$
$$+\sum_{j\in J} h_j b_j(\sum_{t\in T} X_{jt} - W_j) + \sum_{j\in J} g_j b_j(1-\sum_{t\in T} X_{jt}))/B \tag{6.46}$$
$$E^{sl} = \sum_{j\in J} b_j W_j/B. \tag{6.47}$$

Model **ESCS1** is presented below.

Model ESCS1

Minimize (6.5)
subject to (6.3), (6.4), (6.6), (6.7), (6.18), (6.19), (6.46), (6.47) and

$$\sum_{t\in T} X_{jt} \le 1;\ j\in J \tag{6.48}$$
$$\sum_{j\in J} b_j X_{jt} \le C;\ t\in T \tag{6.49}$$
$$\sum_{j\in J}\sum_{t'\in T:t'\le t} b_j X_{jt'} \le B \sum_{i\in I:\tau_i\le t-1} (1-\pi_i)v_i;\ t\in T \tag{6.50}$$
$$\sum_{t\in TK} tY_{kt} \ge \sum_{t\in T}(t+1)X_{jt};\ k\in K, j\in J_k \tag{6.51}$$
$$\sum_{t\in TK} Y_{kt} \le 1;\ k\in K \tag{6.52}$$
$$W_j \le \sum_{t\in T:t\le d_j-\sigma_k} X_{jt};\ k\in K, j\in J_k \tag{6.53}$$
$$W_j \le \sum_{t\in TK:t\le d_j-\sigma_k+1} Y_{kt};\ k\in K, j\in J_k \tag{6.54}$$
$$\sum_{t\in T:t\le d_j-\sigma_k} X_{jt} + \sum_{t\in TK:t\le d_j-\sigma_k+1} Y_{kt} - 1 \le W_j; k\in K, j\in J_k \tag{6.55}$$
$$W_j \ge 0;\ j\in J \tag{6.56}$$
$$X_{jt}\in\{0,1\};\ j\in J, t\in T \tag{6.57}$$
$$Y_{kt}\in\{0,1\};\ k\in K, t\in TK. \tag{6.58}$$

Notice that unlike SMIP model **SCS1_E** which is formulated to determine optimal schedules for all potential disruption scenarios, model **ESCS1** accounts for a single scenario only, representing the expected supplies. Except for the expected values of the random parameters, this model does not take into account any distribution information and the solution remains the same as long as the expectations do not change. In contrast to model **SCS1_E**, where the selection of supply portfolio is combined with supply chain scheduling for all disruption scenarios considered, now the portfolio is determined along with a single schedule.

6.6 Computational Examples

The examples presented in this section are modeled after a real-world electronics supply chain (e.g., Sawik 2011a). The supply chain consists of multiple manufacturers/suppliers of electronic components, a single producer where finished products (e.g., cellular phones) are assembled to meet customer orders and a set of distribution centers that deliver the products to customers (e.g., carriers) who gene rate final demand for products. The completed customer orders are shipped to the distribution centers either in batches of different customer orders or each customer order is shipped individually, immediately after its completion using a direct shipping line. The line consists of a surface mount technology line, where printed wiring boards are assembled, material preparation stage, where all materials required for each product are prepared, postponement stage, where products for some orders are customized, flashing/flexing stations, where required software is downloaded and packing stations, where products and required accessories are packed for shipping. The input data for the example problems were prepared considering monthly production.

The basic input parameters for computational examples are presented in Table 6.4.

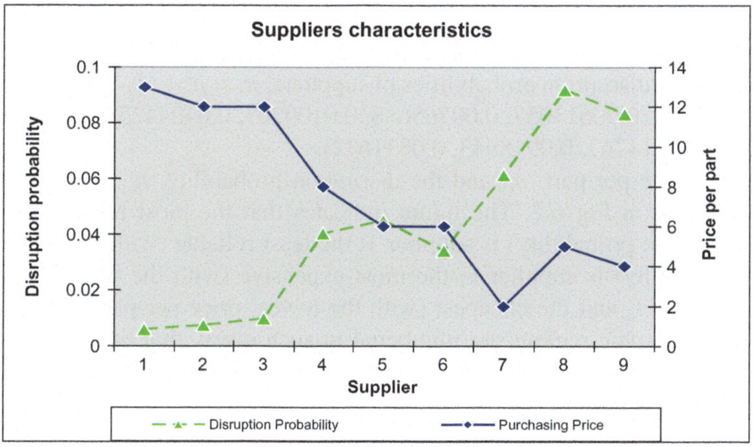

Fig. 6.2 Suppliers

Table 6.4 Input parameters: supply portfolio selection and production and distribution scheduling

$\overline{I}=9$ suppliers, $\overline{J}=25$ customer orders, $\overline{K}=3$ distribution centers,	
$\overline{R}=3$ three geographic regions, $\overline{T}=10$ planning periods, $\overline{S}=2^{\overline{I}}=512$ disruption scenarios	
$I^1=\{1,2,3\}$ three suppliers in region, $r=1$	
$I^2=\{4,5,6\}$ three suppliers in region, $r=2$	
$I^3=\{7,8,9\}$ three suppliers in region, $r=3$	
$J_1=\{1,\dots,10\}$ ten customers served by distribution center, $k=1$	
$J_2=\{11,\dots,20\}$ ten customers served by distribution center, $k=2$	
$J_3=\{21,\dots,25\}$ five customers served by distribution center, $k=3$	
$b=(1,2,9,7,8,5,1,7,5,4,7,4,10,6,8,1,4,2,4,8,6,3,8,7,3)\times 500$	customer order size
$B=66{,}000$	total demand for products
Production and distribution capacity:	
$C_P=20{,}000$	producer capacity
$C_V=\max_{j\in J} b_j=5000$	vehicle capacity
Local disruption probabilities of suppliers:	
$p=(0.00513571, 0.00666354, 0.00902974, 0.0356206, 0.040175, 0.0294692, 0.0519967, 0.0827215, 0.0739062)$	
Regional disruption probability:	
$p^1=0.001$, $p^2=0.005$, $p^3=0.01$	
Time parameters:	
$\sigma_k=1,2,3$ time periods for transportation to distribution center $k=1,2,3$, respectively	
$\tau_i=2,3,4$ time periods for delivery from suppliers $i\in I^1$, $i\in I^2$, $i\in I^3$, respectively	
$d_j\in\{2+\min_{i\in I}(\tau_i)+\min_{k\in K}(\sigma_k),\dots,\overline{T}+\max_{k\in K}(\sigma_k)\}$, $\forall j=1,\dots,25$	customer due-dates
Cost parameters:	
$a_k=C_V\sigma_k=5000, 10{,}000, 15{,}000$, for transportation to distribution center $k=1,2,3$, respectively	
$e=(8,7,10,13,14,11,19,17,19)\times 1000$, $o=(13,12,12,8,6,6,2,5,4)$	
$g_j=2\lceil a_j\max_{i\in I}(o_i)\rceil$, $h_j=1\ \forall j=1,\dots,25$ unit penalty for unfulfilled, delayed demand	
$\alpha\in\{0.50,0.75,0.90,0.95,0.99\}$	confidence level

Given local and regional disruption probabilities presented in Table 6.4, the corresponding total disruption probabilities of suppliers, $\pi_i=p^r+(1-p^r)p_i$, $i\in I^r, r\in R$, (6.45), are π=(0.00613057, 0.00765688, 0.0100207, 0.0404425, 0.0449741, 0.0343219, 0.0614767, 0.0918943, 0.0831672).

The unit price per part, o_i, and the disruption probability, π_i, of each supplier $i\in I$ are shown in Fig. 6.2. The figure indicates that the most reliable (with the lowest disruption probability) is supplier 1, the least reliable (with the highest disruption probability) is supplier 8, the most expensive (with the highest price per part) is supplier 1, and the cheapest (with the lowest price per part) is supplier 7. Note that geographic regions are numbered in such a way that the unit prices are nonincreasing with r, while the fixed ordering costs and the disruption probabilities are nondecreasing with r, i.e.,

$$o_{i_1}\geq o_{i_2}\geq o_{i_3},\quad e_{i_1}\leq e_{i_2}\leq e_{i_3}\quad\text{and}\quad \pi_{i_1}\leq\pi_{i_2}\leq\pi_{i_3};\forall i_1\in I^1, i_2\in I^2, i_3\in I^3.$$

6.6.1 Risk-Neutral Decision-Making

The solution results for the risk-neutral decision-making are presented in Table 6.5, 6.6, and 6.7, respectively, for single batch shipping, individual shipping, and multiple batch shipping with limited transportation capacity. For comparison, the expected cost per product in Table 6.7 does not include the expected transportation cost per product, which is shown separately. The results indicate that for $\lambda = 1$ (minimization of cost) the cheapest supplier $i = 7$ is selected only, while for $\lambda = 0$ (maximization of customer service level) the total demand for parts is allocated among all nine suppliers for batch shipping and eight suppliers, except of the least reliable supplier 8, for individual shipping. The highest proportion of demand is allotted among the three most reliable and most expensive suppliers $i = 1,2,3$. As λ increases from 0 to 1, i.e., the decision maker preference shifts from customer service level to cost, more demand is moved from expensive and reliable suppliers to low-cost, unreliable suppliers. The results for individual and multiple batch shipping, where more shipments are scheduled, are similar.

Table 6.5 Risk-neutral solutions: single batch shipping

λ	0	0.1	0.2	0.3	0.4	0.5	0.6	0.7	0.8	0.9	1
Var. = 124,207, Bin. = 112,723, Cons. = 84,406, Nonz. = 1,628,119[a]											
Exp. cost	16.42($\overline{E}^c$)	10.96	10.67	10.40	7.73	7.72	7.71	5.25	4.70	4.30	4.28($\underline{E}^c$)
Exp. service level[b]	78.65 ($\overline{E}^{sl}$)	78.36	78.30	78.04	74.30	74.29	74.17	56.67	51.71	41.96	41.94($\underline{E}^{sl}$)
Exp. fulfilled demand[c]	78.65	88.82	88.85	88.63	96.17	96.19	95.83	94.56	94.10	93.78	93.85
Suppliers selected	1(21)										
(% of total demand)	2(21)	2(37)	2(30)	2(30)	2(29)	2(29)	2(29)				
	3(18)										
	4(9)										
	5(5)	5(21)	5(20)	5(21)							
	6(8)	6(21)	6(20)	6(21)	6(30)	6(30)	6(30)	6(29)	6(11)		
	7(12)	7(21)	7(21)	7(28)	7(41)	7(41)	7(41)	7(71)	7(89)	7(100)	7(100)
	8(2)										
	9(4)		9(9)								
CPU [s][d]	15,285	74,074	21,819	60,464	31,004	42,468	14,903	37,542	26,009	69,686	113

[a] Var.: number of variables; Bin.: number of binary variables; Cons.: number of constraints; Nonz.: number of nonzero coefficients

[b] $\sum_{s\in S}\sum_{j\in J} P_s b_j w_j^s / B \times 100\%$

[c] $\sum_{s\in S}\sum_{j\in J}\sum_{t\in T} P_s b_j x_{jt}^s / B \times 100\%$

[d] CPU seconds for proven optimal solutions, except for $\lambda = 0.1, 0.9$ with GAP<1%

For all the shipping methods and most λ, the total demand for parts is allocated among four suppliers $i = 2,5,6,7$, where supplier $i = 2$ is the second most reliable and $i = 6$ is the most reliable in regions $r = 2,3$, while $i = 7$ is the cheapest among all suppliers. For small λ, however, more suppliers are selected to improve individual service level.

In general, for all shipping methods the service-oriented supply portfolio ($\lambda \leq 0.3$) is more diversified than the cost-oriented portfolio ($\lambda \geq 0.7$).

Table 6.6 Risk-neutral solutions: individual shipping

λ	0	0.1	0.2	0.3	0.4, 0.5, 0.6, 0.7	0.8, 0.9	1
Var. = 100,470, Bin. = 100,459, Cons. = 21,343, Nonz. = 909,247[a]							
Exp. cost	15.38($\overline{E}^c$)	10.59	7.87	6.19	5.57	4.77	4.15($\underline{E}^c$)
Exp. service level[b]	87.56($\overline{E}^{sl}$)	87.36	85.98	84.52	83.48	74.45	55.45($\underline{E}^{sl}$)
Exp. fulfilled demand[c]	87.56	97.36	95.74	95.06	94.86	94.37	93.85
Suppliers selected	1(52)						
(% of total demand)	2(12)	2(40)	2(9)	2(9)	2(9)		
	3(12)	3(12)					
	4(4)	4(12)					
	5(4)	5(12)	5(12)	5(12)			
	6(4)	6(12)	6(55)	6(20)	6(20)	6(20)	
	7(8)	7(12)	7(12)	7(59)	7(71)	7(80)	7(100)
	9(4)		9(12)				
CPU [s]	1	79	61	409	112, 8, 10, 12	11, 26	3

[a] Var.: number of variables; Bin.: number of binary variables; Cons.: number of constraints; Nonz.: number of nonzero coefficients

[b] $\sum_{s\in S}\sum_{j\in J}\sum_{t\in T:t\le d_j-\sigma_{k_j}} P_s b_j x^s_{jt}/B \times 100\%$

[c] $\sum_{s\in S}\sum_{j\in J}\sum_{t\in T} P_s b_j x^s_{jt}/B \times 100\%$

In addition to the expected customer service level, i.e., the expected fraction of customer demand fulfilled on time, Tables 6.5, 6.6, and 6.7 also show the expected fraction of fulfilled demand, $\sum_{s\in S}\sum_{j\in J}\sum_{t\in T} P_s b_j x^s_{jt}/B$, i.e., demand fulfilled on time or delayed. The solution results demonstrate that for the maximum service level objective, which is independent of any cost parameters, the largest expected fraction of non-delayed customer demand is associated with the smallest expected fraction of fulfilled demand. This indicates that to maximize expected service level, customer orders that cannot be fulfilled by requested due dates are rejected. Note that the difference between the expected fraction of fulfilled demand and the expected customer service level is the expected fraction of delayed customer demand. The latter measure is equal to zero for the maximum service level objective, increases with the trade-off parameter λ, and reaches its maximum for the minimum cost objective.

The results presented for the three shipping methods in Tables 6.5, 6.6, and 6.7 are illustrated and compared in Figs. 6.3 and 6.4. Figure 6.3 compares the expected values of cost per product, service level, and fulfilled demand for 11 levels of trade-off parameter λ, while Fig. 6.4 compares the obtained subsets of nondominated supply portfolios (the allocation of total demand for parts among selected suppliers), with the four major suppliers $i = 2,5,6,7$ indicated. The corresponding portfolios for different shipping methods are similar, with the cheapest suppliers dominating in the individual shipping. The results point out that individual and multiple batch shipping lead to better solution values, i.e., a higher expected service level and a lower expected ordering, purchasing, and shortage cost.

For single batch shipping, Figs. 6.5, 6.6, and 6.7 show the expected supply, production, and delivery schedules, respectively, for $\lambda = 0$ (i.e., for the maximum ex-

Table 6.7 Risk-neutral solutions: multiple batch shipping

λ	0	0.1	0.2	0.3	0.4	0.5	0.6	0.7, 0.8	0.9	1
Var. = 124,207, Bin. = 112,723, Cons. = 84,406, Nonz. = 1,628,119[a]										
Exp. cost[b]	17.85($\overline{E}^c$)	10.35	8.95	7.10	6.58	6.55	5.97	5.41	4.78	6.07($\underline{E}^c$)
Exp. transportation cost[c]	2.04	2.00	1.97	1.96	1.95	1.95	1.93	1.93	1.92	1.92
Exp. service level[d]	85.31($\overline{E}^{sl}$)	85.23	84.66	83.11	82.46	82.40	80.42	77.73	70.96	55.44($\underline{E}^{sl}$)
Exp. fulfilled demand[e]	85.31	97.04	95.78	95.56	95.44	95.43	95.07	94.69	94.36	93.85
Suppliers selected	1(42)	1(14)								
(% of total demand)	2(14)	2(14)	2(21)	2(21)	2(20)	2(20)	2(14)	2(7)		
	3(14)	3(14)								
	4(4)	4(15)	4(15)							
	5(5)	5(14)	5(14)	5(14)						
	6(6)	6(14)	6(21)	6(14)	6(20)	6(19)	6(19)	6(19)	6(19)	
	7(8)	7(15)	7(15)	7(51)	7(60)	7(61)	7(67)	7(74)	7(81)	7(100)
	8(2)									
	9(5)		9(14)							
CPU [s]	36	24,354	22,107	32,958	15,961	14,689	24,447	4560, 14,392	25,220	315

[a]Var.: number of variables; Bin.: number of binary variables; Cons.: number of constraints; Nonz.: number of nonzero coefficients

[b]Without expected transportation cost, except for $\lambda = 0$ and $\lambda = 1$

[c]$\sum_{s\in S}\sum_{k\in K}\sum_{t\in TK} P_s a_k y^s_{kt}/B$

[d]$\sum_{s\in S}\sum_{j\in J}\sum_{t\in TK:t\le d_j-\sigma_{k_j}+1} P_s b_j z^s_{jt}/B \times 100\%$

[e]$\sum_{s\in S}\sum_{j\in J}\sum_{t\in T} P_s b_j x^s_{jt}/B \times 100\%$

pected service), $\lambda = 0.5$ and $\lambda = 1$ (i.e., for the minimum expected cost). The expected schedules were computed using the formulae presented below:

- Expected schedule of supplies of parts to the producer

$$\sum_{s\in S}\sum_{i\in I_s:\tau_i=t} P_s B v_i;\ t\in T \tag{6.59}$$

- Expected production schedule

$$\sum_{s\in S}\sum_{j\in J} P_s b_j x^s_{jt};\ t\in T \tag{6.60}$$

- Expected schedule of shipping of products from the producer to the distribution centers

$$\sum_{s\in S}\sum_{j\in J_k} P_s b_j\Big(\sum_{t'\in T:t'<t} x^s_{jt'}\Big) y^s_{kt};\ k\in K, t\in TK, \tag{6.61}$$

  where $TK = \{\min_{i\in I}\tau_i + 2, \ldots, \overline{T}+1\}$ is the set of shipping periods.
  Note that the delivery schedule for the distribution center k is the shipping schedule delayed by σ_k periods.

The expected supply schedules shown in Figs. 6.5, 6.6, and 6.7 are marked with the supplier region, while the expected delivery schedules indicate the distribution center; DC1 ($k = 1$), DC2 ($k = 2$), and DC3 ($k = 3$).

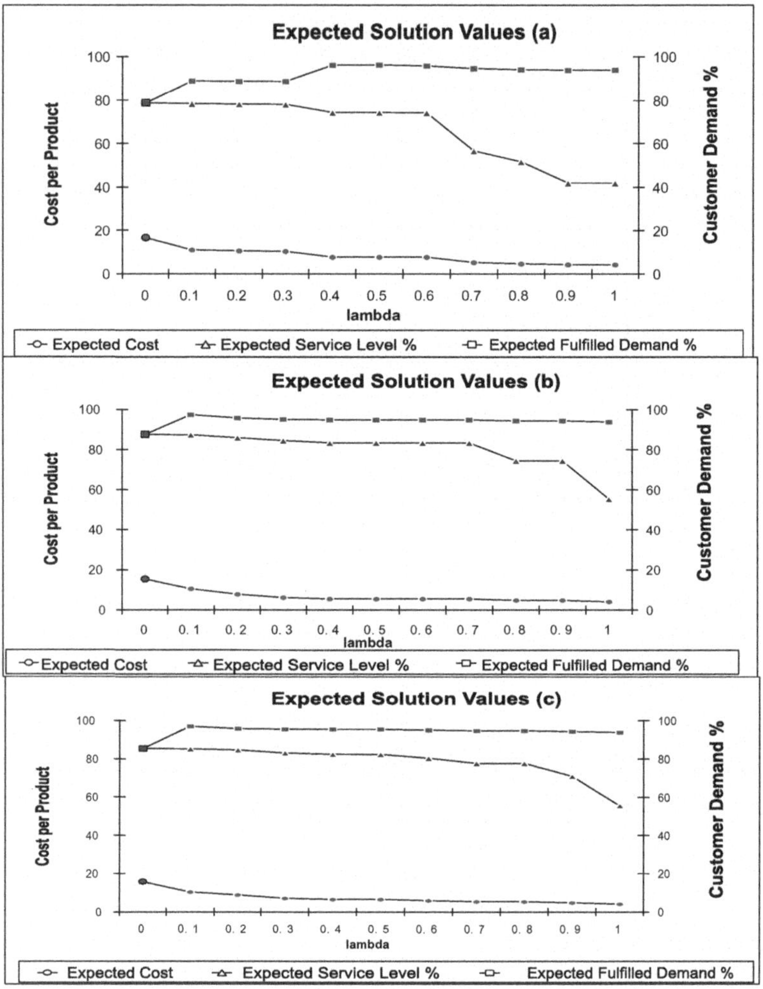

Fig. 6.3 Expected solution values:(**a**) single batch shipping, (**b**) individual shipping, (**c**) multiple batch shipping

As λ increases, i.e., the decision maker priority shifts from the maximum service level to minimum cost and more parts are ordered from less reliable and lower cost suppliers, the expected supply schedules and the corresponding production schedules are more delayed as well as the delivery of products to the customers. Note that despite constraint (6.11) for single batch shipping that ensures at most one shipment of products to each distribution center for each disruption scenario, the expected shipping schedule (6.61) may be split into more smaller size shipments, (cf. Figs. 6.6, 6.7, and 6.8).

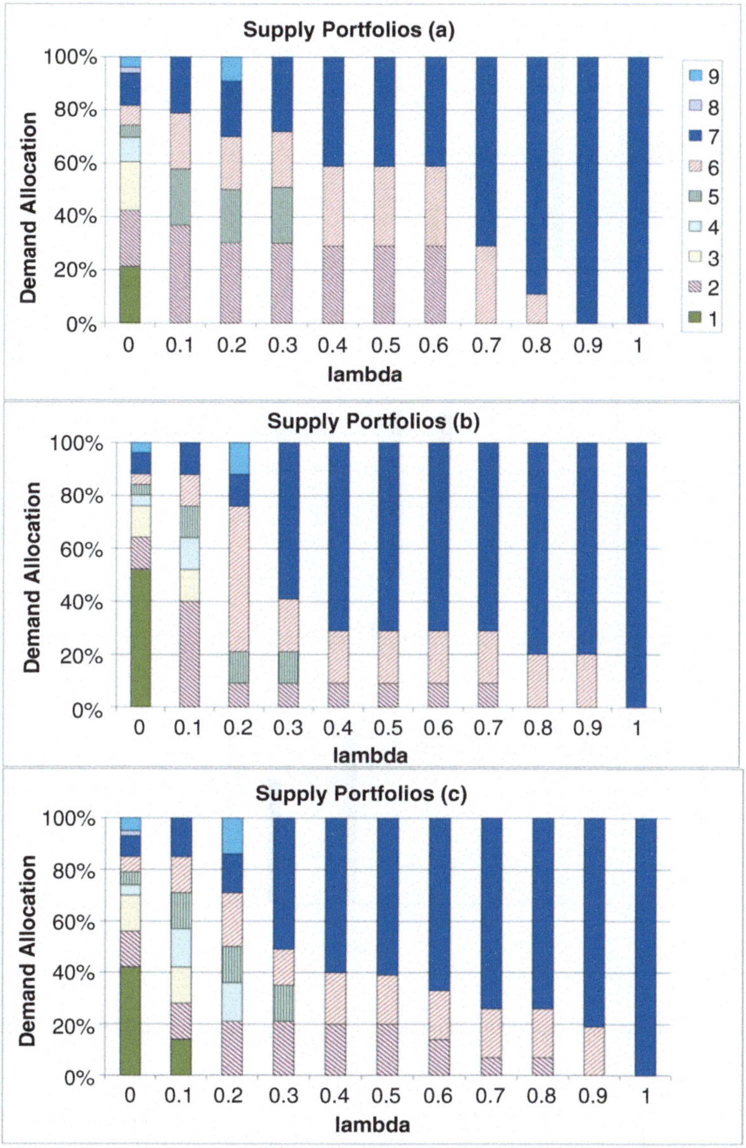

Fig. 6.4 Nondominated supply portfolios: (**a**) single batch shipping, (**b**) individual shipping, (**c**) multiple batch shipping

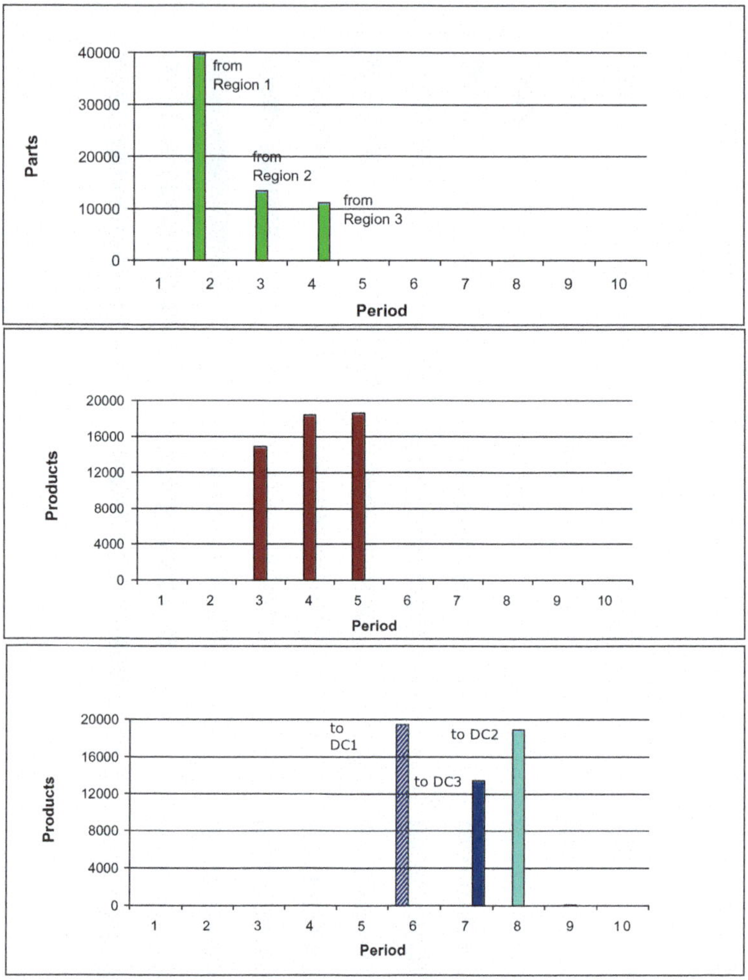

Fig. 6.5 Expected supplies, production, and deliveries for maximum service level ($\lambda = 0$): single batch shipping

For individual and multiple batch shipping, Figs. 6.8 and 6.9 show examples of expected supply, production, and delivery schedule for $\lambda = 0.5$. Comparison of Fig. 6.6 for single batch shipping with Fig. 6.8 and with Fig. 6.9 shows that the expected supplies from different regions are more diversified, and the expected production as well as the shipping schedule are distributed among more periods. The results indicate that for individual and multiple batch shipping, where more shipments are scheduled, supply, production, and delivery schedules are more unlevelled. On the other hand, comparison of the solution results shown in Tables 6.5, 6.6, and 6.7 indicates that individual and multiple batch shipping may lead to higher expected service level and lower expected ordering, purchasing, and shortage cost.

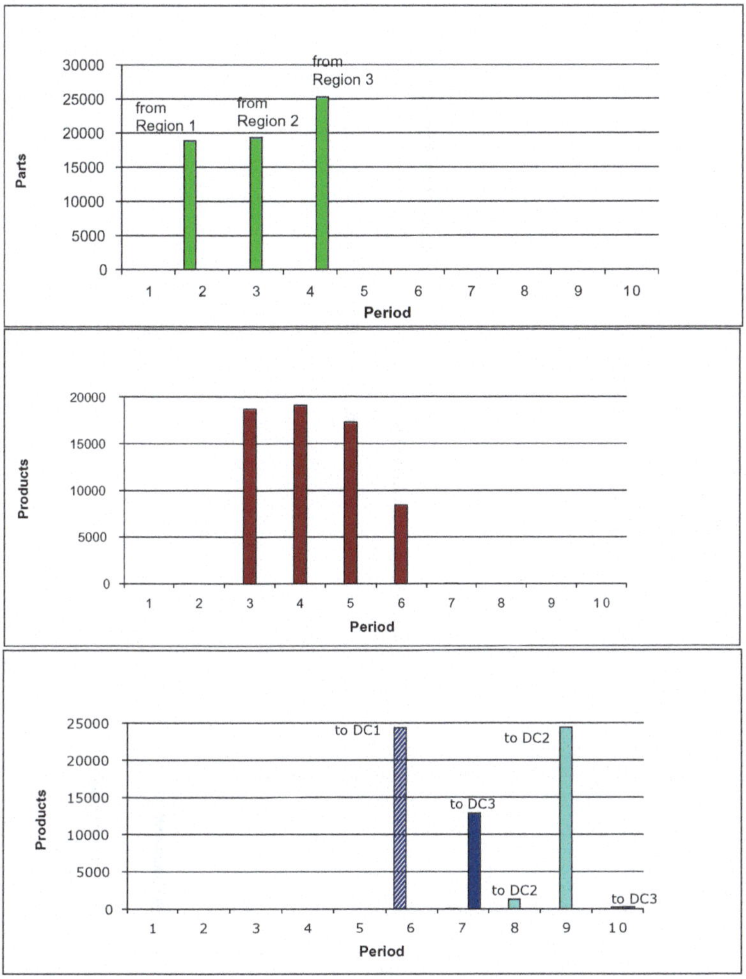

Fig. 6.6 Expected supplies, production, and deliveries for $\lambda = 0.5$: single batch shipping

The solution results obtained for numerical examples modeled after realistic problems in the electronics supply chain are in line with other approaches used in the area of supply chain risk management. The main findings are listed below.

- *The supply portfolios for different shipping methods are similar. For all shipping methods, the service-oriented supply portfolio ($\lambda \leq 0.3$) is more diversified than the cost-oriented portfolio ($\lambda \geq 0.7$).*
- *For the minimum cost objective the cheapest supplier is usually selected, while for the maximum service level objective the most reliable and most expensive suppliers are dominating in the supply portfolio.*

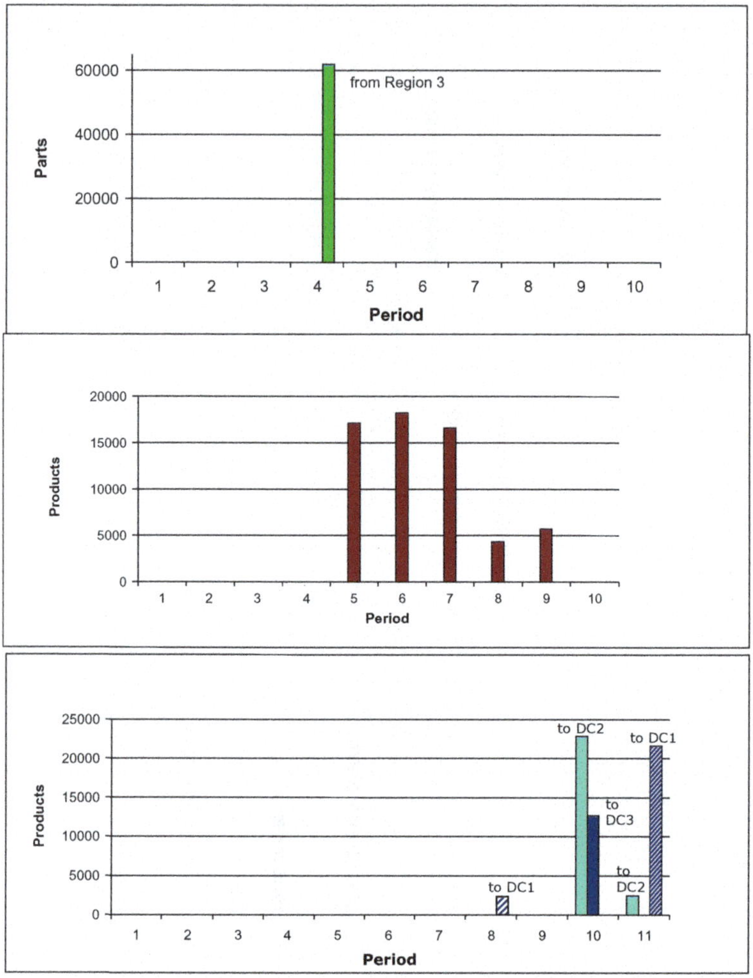

Fig. 6.7 Expected supplies, production, and deliveries for minimum cost ($\lambda = 1$): single batch shipping

- *For the maximum service level objective, which is independent of any cost parameters, the largest expected fraction of non-delayed customer demand simultaneously leads to the largest expected fraction of rejected demand. This indicates that in order to maximize expected service level, customer orders that cannot be fulfilled by requested due dates are rejected.*
- *The more cost-oriented decision-making, the more delayed the expected supply, production, and distribution schedules.*
- *The individual and multiple batch shipping lead to higher expected service level and lower expected ordering, purchasing, and shortage cost.*

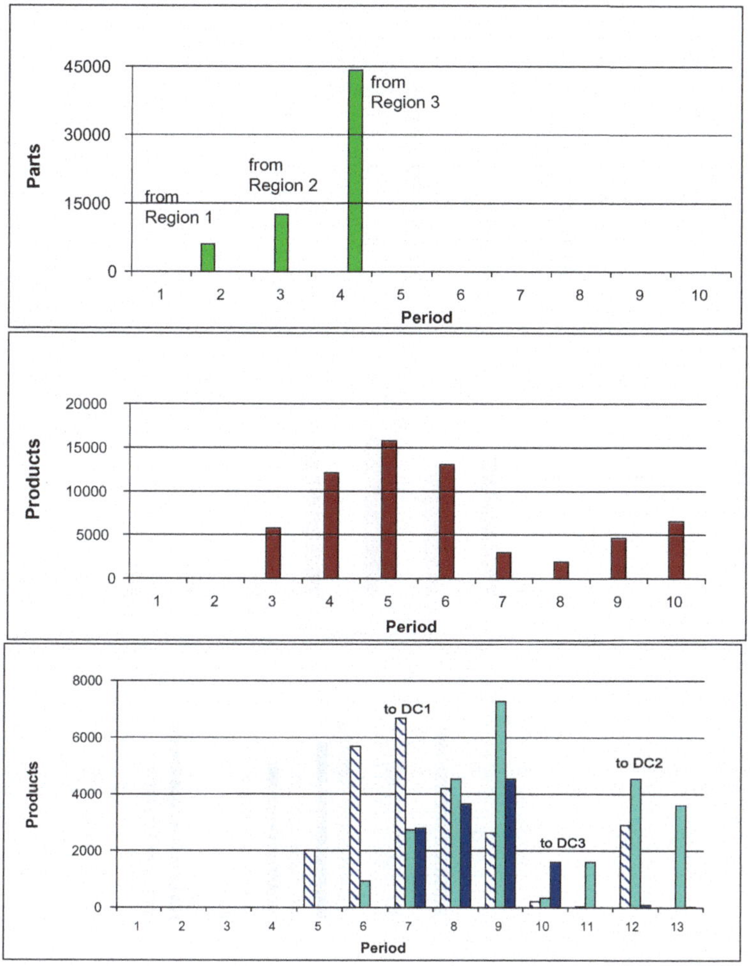

Fig. 6.8 Expected supplies, production, and deliveries for $\lambda = 0.5$: individual shipping

The computational experiments were performed using the AMPL programming language and the XPRESS 27.01 solver on a MacBookPro laptop with Intel Core i7 processor running at 2.8 GHz and with 16 GB RAM. (For comparison, the computational experiments have been also performed using CPLEX 12.6.2 and Gurobi 6.0.2 solvers, however they were outperformed by the XPRESS 27.01 solver.) The MIP problem sizes (the numbers of variables, constraints, and nonzero coefficients) and CPU seconds required to find optimal solutions (or with GAP less than 1%) for the examples are presented in Tables 6.5, 6.6, and 6.7. The solver was capable of finding proven optimal solution for all examples for models **SCS2_E** and **SCS3_E**, with CPU time ranging from several seconds to several hours. Proven optimal solutions were also found for most examples for single batch shipping model **SCS1_E**, except for $\lambda = 0.1, 0.9$, where the solution with GAP less than 1% was found and

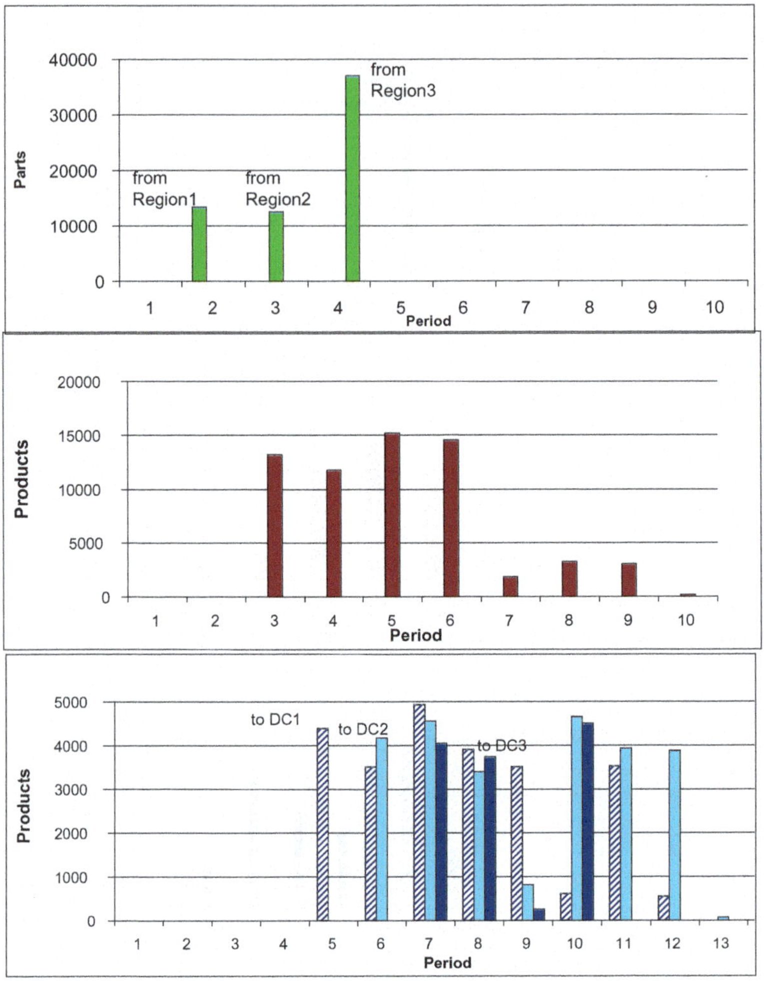

Fig. 6.9 Expected supplies, production, and deliveries for $\lambda = 0.5$: multiple batch shipping

no improvements were observed within next 3600 CPU seconds. The smallest CPU time was required for the individual shipping model **SCS2_E**. Since model **SCS2_E** is much simpler than **SCS1_E** and **SCS3_E**, the latter result is obvious.

6.6.2 Risk-Averse Decision-Making

In this subsection some computational examples are presented to illustrate the risk-averse scheduling of supply, production, and distribution using model **SCS1_CV**. In the computational experiments data sets provided at the beginning of this section were used. Table 6.8 shows solution results for the objective function (6.36)

Table 6.8 Risk-averse solutions for model **SCS1_CV**

Confidence level α	0.50	0.75	0.90	0.95	0.99
$\lambda = 0$ (maximization of $CVaR^{sl}$)					
Var. = 125,233, Bin. = 112,723, Cons. = 85,430, Nonz. = 1,638,087[a]					
$CVaR^{sl}$ 100%	78.58	78.40	77.84	76.89	68.83
VaR^{sl} 100%		78.79			75.76
E^{sl} 100%		78.69			75.88
E^{c}	16.54	16.56	16.72	16.64	17.40
Supply portfolio	1(21)	1(21)	1(21)	1(21)	1(21)
Supplier (% of total demand) [b]	2(21)	2(21)	2(21)	2(21)	2(21)
	3(21)	3(21)	3(20)	3(20)	3(18)
	4(8)	4(7)	4(7)	4(8)	4(8)
	5(7)	5(7)	5(7)	5(7)	5(7)
	6(7)	6(8)	6(8)	6(6)	6(8)
	7(14)	7(11)	7(14)	7(13)	7(11)
					8(3)
	9(1)	9(4)	9(2)	9(4)	9(3)
$\lambda = 1$ (minimization of $CVaR^{c}$)					
$CVaR^{c}$	5.66	8.56	12.78	13.82	16.10
VaR^{c}	2.77	2.77	11.25	12.32	15.30
E^{c}	4.22	4.22	10.17	11.27	13.05
E^{sl} 100%	48.35	48.35	15.57	11.26	5.75
Supply portfolio					
Supplier (% of total demand)			2(6)	2(26)	2(24)
					3(21)
			4(14)	4(19)	
			5(22)	5(18)	5(15)
			6(21)	6(20)	6(15)
	7(100)	7(100)	7(18)	7(17)	7(12)
			9(19)		9(13)
$\lambda = 0.5$					
$CVaR^{c}$	5.66	8.74	12.88	13.89	16.17
VaR^{c}	2.77	3.12	10.79	12.47	15.36
E^{c}	4.22	4.53	10.73	11.72	14.89
$CVaR^{sl}$ 100%	45.18	42.36	55.40	72.28	59.06
VaR^{sl} 100%	51.52	54.55	56.82	78.79	62.12
E^{sl} 100%	48.35	51.50	57.75	78.46	67.85
Supply portfolio					
Supplier (% of total demand)				2(23)	2(22)
				3(13)	3(20)
			4(15)	4(17)	4(17)
			5(20)	5(17)	
		6(5)	6(20)	6(17)	6(15)
	7(100)	7(95)	7(16)	7(13)	7(13)
			8(15)		
			9(14)		9(13)

[a] Var.: number of variables; Bin.: number of binary variables; Cons.: number of constraints; Nonz.: number of nonzero coefficients

with $\lambda = 0$ (maximization of CVaR of service level), $\lambda = 1$ (minimization of CVaR of cost), and $\lambda = 0.5$ (balanced trade-off between CVaR of cost and CVaR of service level). The risk-averse supply portfolios and solution values of VaR^c, $CVaR^c$ and VaR^{sl}, $CVaR^{sl}$ and the associated expected values E^c and E^{sl}, respectively, of cost and service level are presented for a subset of confidence levels $\alpha = 0.5, 0.75, 0.9, 0.95, 0.99$.

As α increases, the $CVaR^c$ of cost and the associated expected cost, E^c, increase, while expected service level, E^{sl}, decreases. Similarly, the $CVaR^{sl}$ of service level decreases and the associated expected cost, E^c, increases with confidence level α. However, VaR^{sl}, E^{sl} and risk-averse supply portfolios are nearly independent of confidence level, except for the highest $\alpha = 0.99$. At the same time, $CVaR^{sl}$ and E^{sl} are very close to the corresponding VaR^{sl}. For the $CVaR^c$ objective function and increasing confidence level, a greater diversification of supply portfolios is observed, with more demand shifted from the cheapest supplier 7 to more reliable and more expensive suppliers 2, 3, 5, 6, and 9. A low-cost and most unreliable supplier 8 was never selected (cf. Fig. 6.2). In contrast, for the $CVaR^{sl}$ objective function the supply portfolios are diversified, even for small confidence levels.

Comparison of solution results for models **SCS1_E** and **SCS1_CV** (cf. Tables 6.5 and 6.8) demonstrates that *solution values for the risk-neutral decision-making to maximize expected service level (that is, for* $\lambda = 0$*) and to minimize expected cost (that is, for* $\lambda = 1$*) are very close to the corresponding solution values for the risk-averse decision-makings with confidence level* $\alpha = 0.5$*. Moreover, for the cost objective function, the corresponding risk-neutral and the risk-averse supply portfolios are identical, while for the service level objective they are very close to each other.*

The computational experiments were performed using the AMPL programming language and the Gurobi 7.0.0 solver on a MacBookPro laptop with Intel Core i7 processor running at 2.8 GHz and with 16 GB RAM. The solver was incapable of finding proven optimal solutions within 3600 CPU seconds. However, for all examples with $\alpha = 0.5$ and $\alpha = 0.75$, the solution with GAP less than 1% was found much earlier, while for the higher confidence levels, more than 3600 CPU seconds were required to achieve GAP less than 1%.

6.6.3 Expected Value—Based Decision-Making

For the example input data, the suppliers expected fulfillment rates, $1 - \pi_i; i \in I^r, r \in R$, are (0.993869, 0.992343, 0.989979, 0.959558, 0.955026, 0.965 678, 0.938523, 0.908106, 0.916833), that is, supplier 1 is most reliable and supplier 8 most unreliable.

Table 6.9 presents a subset of nondominated solutions obtained for the expected value problem **ESCS1**, and Fig. 6.10 shows supply, production, and shipping schedules. Unlike the stochastic programming approach which accounts for all potential disruption scenarios to optimize an expected performance of the supply chain, the

Table 6.9 Solution results for the expected value model **ESCS1**

λ	0	0.1 and 0.2 and 0.3	0.4 and 0.5 and 0.6 and 0.7	0.8 and 0.9	1
	Var. = 265, Bin. = 233, Cons. = 157, Nonz. = 2249[a]				
Exp. cost	17.88($\overline{E}^c$)	9.33	7.65	4.49	4.39($\underline{E}^c$)
Exp. service level[b]	78.78 ($\overline{E}^{sl}$)	78.03	75.76	54.55	47.73($\underline{E}^{sl}$)
Exp. fulfilled demand[c]	83.33	90.15	95.45	93.94	93.18
Suppliers selected	1(97.6)				
(% of total demand)	2(0.8)	2(30)	2(26)		
	5(0.8)				
		6(31)	6(31)	6(5)	
	7(0.8)	7(39)	7(43)	7(95)	7(100)

[a]Var.: number of variables; Bin.: number of binary variables; Cons.: number of constraints; Nonz.: number of nonzero coefficients

[b]$\sum_{j\in J} b_j W_j / B \times 100\%$

[c]$\sum_{j\in J} \sum_{t\in T} b_j X_{jt} / B \times 100\%$

solution obtained using the deterministic approach is based on an aggregate information on suppliers expected fulfillments.

Comparison of results presented in Tables 6.5 and 6.9 indicates that for both **SCS1_E** and **ESCS1** models, the solution results are similar and the corresponding optimal solution values are close to each other.

For the deterministic, expected value approach, most of nondominated supply portfolios consist of three suppliers only: $i = 2,6,7$, while suppliers $i = 1,2,5,7$ are selected only for the maximum service level objective (that is, for $\lambda = 0$). In addition, the expected schedules are more delayed for the stochastic approach. Comparison of Figs. 6.5, 6.6, 6.7, and 6.10 indicates that the expected schedules for model **SCS1_E**, computed as expectations over all schedules for all potential disruption scenarios, (6.59)–(6.61), are similar to the corresponding schedules determined by model **ESCS1**. The main difference is a more delayed expected production and shipping schedule for model **SCS1_E** when minimization of cost is considered (that is, for $\lambda = 1$).

The major managerial insights are summarized in Table 6.10. Overall, the results of computational experiments are in line with other research.

In the risk-neutral decision-making, the service-oriented supply portfolio is more diversified than the cost-oriented portfolio, for all shipping methods. In the risk-averse decision-making, diversification of supply portfolios increases with the confidence level, with more demand shifted from the cheapest to more reliable and expensive suppliers. In contrast, the service-oriented supply portfolios are more diversified, even for small confidence levels.

The computational experiments were performed using the AMPL programming language and the Gurobi 7.0.0 solver on a MacBookPro laptop with Intel Core i7 processor running at 2.8 GHz and with 16 GB RAM. The solver was capable of finding proven optimal solution for all examples with CPU time ranging from fraction of a second to several seconds. The lowest computational effort is required for model **ESCS1**, which deals with a single scenario only.

Table 6.10 Managerial insights: supply portfolio selection and production and distribution scheduling

Decision-making issue	**Decision-making impact**
Risk-neutral decision-making.	The service-oriented supply portfolio is more diversified than the cost-oriented portfolio, for all shipping methods For the minimum cost objective, the cheapest supplier is usually selected, while for the maximum service level objective, the most reliable and most expensive suppliers are dominating in the supply portfolio For the service level objective, the largest expected fraction of non-delayed customer demand simultaneously leads to the largest expected fraction of rejected demand. This indicates that to maximize expected service level, customer orders that cannot be fulfilled by requested due dates are rejected The more cost-oriented decision-making, the more delayed the expected supply, production, and distribution schedules The individual and multiple batch shipping lead to higher expected service level and lower expected ordering, purchasing, and shortage cost
SCS1_E vs. **SCS1_CV**	The solution values for the risk-neutral decision-making (model **SCS1_E**) to maximize expected service level and to minimize expected cost are very close to the corresponding solution values for the risk-averse decision-making (model **SCS1_CV**) with low confidence levels. Moreover, for the cost objective function, the corresponding risk-neutral and risk-averse supply portfolios are identical, while for the service level objective, are very close to each other
SCS1_E vs. **ESCS1**	The optimal solution values are close to each other, which indicates that the expected value problem (**ESCS1**) can be used in practice, when the SMIP model is hard to solve. However, the stochastic wait-and-see approach (**SCS1_E**) leads to a more diversified supply portfolio that will hedge against a variety of scenarios. In particular, the service-oriented supply portfolio is more diversified and may combine both high-cost, reliable suppliers and low-cost, unreliable suppliers, while the cost-oriented portfolio depends mainly on low-cost, less reliable suppliers

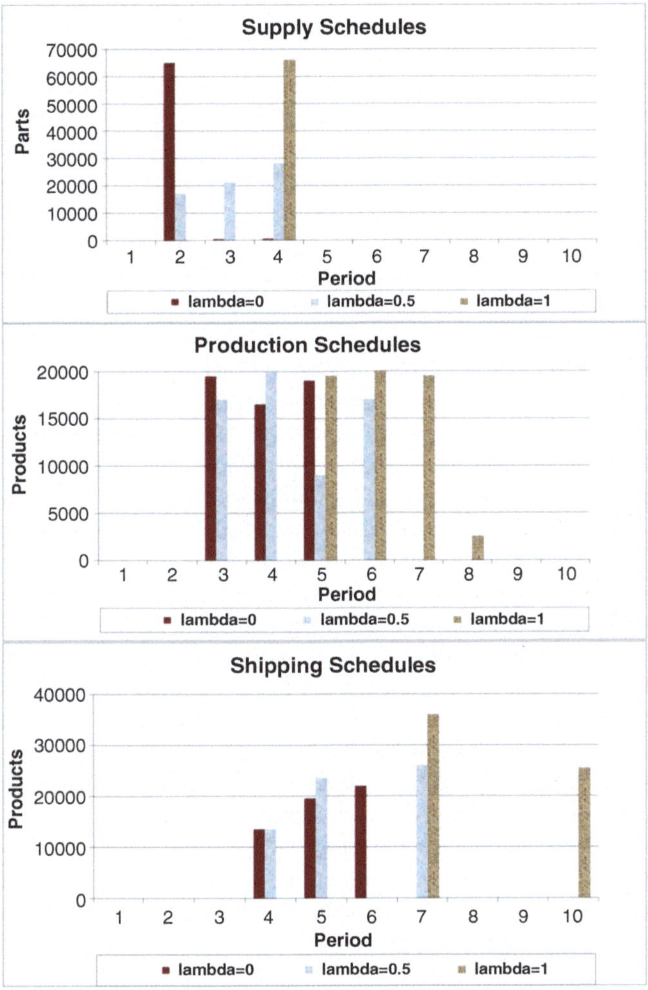

Fig. 6.10 Supplies, production, and deliveries for the expected value problem

6.7 Notes

There is a growing body of literature on deterministic models on integrated production-distribution planning and scheduling, however, without the supply operations, which are mostly considered separately, e.g., Erenguc et al. (1999). For example, Li et al. (2005) investigated an integrated scheduling of assembly and multi-destination air-transportation in a consumer electronics supply chain. The problem was divided into two sub-problems. The air transportation allocation was formulated and solved using a MIP approach, and two heuristics were proposed for the assembly scheduling problem. Lei et al. (2006) studied an integrated production, inventory, and distribution routing problem and proposed a MIP approach combined with a heuristic routing algorithm to coordinate the production, inventory, and transportation operations. A comprehensive review and classification of existing deterministic models that integrate production and outbound distribution operations at the detailed scheduling level was presented in Chen (2010). The variety of models were classified by shipping and delivery methods. In particular, models with individual and immediate delivery of each order and models with batch delivery and routing of orders to different customers delivered together in the same shipment were considered. Such models attempt to optimize detailed order-by-order production and delivery scheduling jointly by taking into account relevant revenues, costs, and customer service levels at the individual order level. In a more recent work on deterministic approaches, Cakici et al. (2012) proposed a MIP approach for multi-objective scheduling of customer orders in an integrated production and distribution system. The problem objective was to optimize the trade-off between total weighted tardiness as a service level measure and total distribution costs. Liu and Papageorgiou (2013) and Liu et al. (2014) developed a multi-objective MIP approach to address production, distribution, and capacity planning of global supply chains considering cost, responsiveness, and customer service level simultaneously to achieve an equitably efficient or Pareto optimal solution. An integrated production and distribution scheduling problem in a make-to-order supply chain with limited production and distribution capacity was considered by Viergutz and Knust (2014). The problem consists in finding a selection of customers to be supplied such that the total satisfied demand is maximized, see also Sawik (2018). There are also some reported studies on joint supplier selection and production and distribution planning. Sawik (2009a) proposed a MIP approach for integrated scheduling of material manufacturing, supply, and product assembly in a customer-driven supply chain. A monolithic approach, where the manufacturing, supply, and assembly schedules are determined simultaneously, was compared with a hierarchical approach. Numerical examples modeled after a real-world scheduling in the electronics supply chain were reported. A monolithic and a hierarchical approach to multi-objective, integrated supply chain scheduling were also compared in Sawik (2006, 2009b). A decomposition of the complex multi-objective production, manufacturing, and supply scheduling into a hierarchy of much simpler decision-making problems was proposed and simple MIP formulations were provided. The objective functions integrated both the total cost and the customer service level and the scheduling was combined with the selection of part suppliers for each customer order and due date setting for some orders. In Cui (2016) a MIP model was proposed for joint optimization problem of production planning and supplier selection. The objective was to maximize the manufacturer's total profit subject to various operating constraints of the supply chain. Gao et al. (2015) developed a MIP formulation and a heuristic with a guaranteed worst-case bound for integrated production and distribution problem in which orders are processed and delivered in batches with limited vehicle capacity. Cheng et al. (2015) considered an integrated scheduling of production and distribution to minimize total cost of production and distribution for the manufacturer. A MIP model was developed and an improved ant colony method was proposed to solve the production scheduling and the First-Fit-Decreasing heuristic used in the bin-packing problem, for the distribution scheduling. In a recent review, Panetto et al. (2019) discussed the impact of Industry 4.0 technology that require highly customized supply network control and developments in management decision-support systems for the design, scheduling, and control of resilient and digital manufacturing networks. The authors presented an analysis of the challenges facing the so-called Industry of the Future, performed on the enterprise-level topics with an interface to supply chains.

The material presented in this chapter is based on research reported in Sawik (2016a), where bi-objective SMIP models were proposed for an integrated selection of supply portfolio and scheduling of production and distribution in the presence of supply chain disruption risks. The models incorporate supply-production, production-distribution, and supply-distribution coordinating constraints to efficiently coordinate supply, production, and distribution schedules. However, in Sawik (2016a) only a risk-neutral trade-off between expected cost and expected service level has been considered to optimize an overall performance of a supply chain. In this chapter, the SMIP approach and the risk-neutral models have been enhanced for the risk-averse decision-making, using CVaR of cost and CVaR of service level as risk measures. Moreover, the risk-neutral SMIP formulation based on the wait-and-see approach is compared with a deterministic MIP model based on the expected value approach, in which random parameters are replaced by their expected values. The expected value problem is often used in practice as the related stochastic mixed integer program is in general much harder to solve since it considers multiple scenarios, e.g., Durbach and Stewart (2009), Maggioni and Wallace (2012).

The future research should concentrate on relaxations of the various simplified assumptions used to formulate the problem (see, Sect. 6.2). For example, multiple supplies of parts to the producer, e.g., partially disrupted supplies (e.g., Sawik 2015b), and multiple deliveries of finished products from distribution centers to customers, along with an assignment of customers to distribution centers should be considered. In addition, more shipping methods can be modeled such as batch shipping with vehicle routing, methods with fixed shipping or delivery dates or with shipping or delivery time windows, e.g., Chen (2010).

Problems

6.1. Modify the SMIP models presented in this chapter for multiple part types with subsets of part types required for each customer order and subsets of suppliers available for each part type.

6.2. Modify the SMIP models with single and multiple batch shipping of products to distribution centers to account for a limited storage space for products waiting for shipments.

6.3. Mean-risk scheduling in a supply chain

(a) Modify the SMIP models presented in this chapter to minimize expected cost and CVaR of cost or maximize expected service level and CVaR of service level.
(b) How should the values of the optimized objective functions be scaled into the interval [0,1] to avoid dimensional inconsistency among the two objectives and how should the trade-off parameter be selected?
(c) How would you interpret the mean-risk supply chain schedules?

6.4. Mixed mean-risk scheduling in a supply chain

(a) Modify the SMIP models presented in this chapter to optimize expected cost and CVaR of service level or optimize expected service level and CVaR of cost.
(b) How should the values of the optimized objective functions be scaled into the interval [0,1] to avoid dimensional inconsistency among the two objectives and how should the trade-off parameter be selected?
(c) How would you interpret the mixed mean-risk supply chain schedules?

6.5. Explain why the supply portfolios for different shipping methods of products to distribution centers are similar and the service-oriented portfolios are more diversified than the cost-oriented portfolios.

Part III
Equitably Efficient Selection of Supply Portfolio and Scheduling

Chapter 7
A Fair Decision-Making under Disruption Risks

7.1 Introduction

In global supply chains the optimization of material flows subject to unexpected disruptive events, focuses on a variety of different optimality criteria. The most commonly used criteria are minimization of cost and maximization of service level, where the latter measures either fraction of customer orders or fraction of customer demand satisfied on time. The cost and service level are in conflict and, in addition, the decision makers often do not have preference to any objective, i.e., the two objectives are equally important. Then, an equitably efficient solution should be generated, in which the two normalized objective function values are as much close to each other as possible. In this chapter, the two risk-neutral conflicting criteria: expected cost and expected service level are fairly optimized to achieve an equitably efficient supply portfolio and production schedule in the presence of supply chain disruption risks. The supplies of parts are subject to independent random local and correlated regional disruptions. The cost includes the cost of ordering, purchasing, and shortage of parts, while the service level is independent of any cost parameters. The two alternative service level measures are compared: the expected order fulfillment rate and the expected demand fulfillment rate (see, Chap. 5). In order to obtain an equitably efficient solution to the combinatorial stochastic optimization problem, the Ordered Weighted Averaging (OWA) aggregation of the two conflicting objective functions is applied, e.g., Yager (1988). The equitably efficient solutions obtained for the ordered weighted averaging aggregation of the two conflicting objective functions will be compared with nondominated solutions obtained using the weighted-sum aggregation approach.

T. Sawik, *Supply Chain Disruption Management*,
International Series in Operations Research & Management Science 291,
https://doi.org/10.1007/978-3-030-44814-1_7

The following time-indexed SMIP models are presented in this chapter:

ESPS_E for risk-neutral selection of equitably efficient supply portfolio and scheduling of customer orders to minimize expected cost and maximize expected service level;
SPS_E(c) for risk-neutral selection of supply portfolio and scheduling of customer orders to minimize expected cost;
SPS_E(sl) for risk-neutral selection of supply portfolio and scheduling of customer orders to maximize expected service level;
SPSW_E for risk-neutral selection of supply portfolio and scheduling of customer orders to minimize weighted-sum of normalized expected cost and service level.

Numerical examples and some computational results are provided in Sect. 7.5, in particular, comparison of OWA aggregation with weighted-sum aggregation of the two objective functions is discussed. The major managerial insights are provided at the end of this chapter.

7.2 The Lexicographic Minimax Optimization and Equitable Aggregation

This section provides an overview of the lexicographic minimax optimization and equitable aggregation.

Consider an optimization problem with m objective functions $f_k(x), k = 1, \ldots m$, that are to be minimized. The problem can be formulated as follows:

$$\min\{f_1(x), f_2(x), \ldots, f_m(x) : x \in Q\}, \tag{7.1}$$

where $f(x) = (f_1(x), f_2(x), \ldots, f_m(x))$ is a vector-function that maps the decision space $X = \mathscr{R}^n$ into the criterion space $Y = \mathscr{R}^m$, $Q \subset X$ denotes the feasible set, and $x \in X$ denotes the vector of decision variables. The elements of the criterion space are referred to as outcome vectors.

Model (7.1) specifies that the objective is to minimize of all outcomes $y_k, k = 1, \ldots m$; however, the solution concepts are defined by properties of the corresponding preference model within the outcome space. If all the objective functions are equally important to the decision makers, an equitable solution is sought, in which all normalized objective function values should be as much close to each other as possible.

The relation of equitable dominance can be expressed as a vector inequality on the cumulative ordered outcomes. Let $\Theta : \mathscr{R}^m \to \mathscr{R}^m$ be the ordering map such that $\Theta(y) = (\theta_1(y), \theta_2(y), \ldots \theta_m(y))$, where $\theta_1(y) \geq \theta_2(y) \geq \ldots \geq \theta_m(y))$ and there exists a permutation ϖ of set $\{1,\ldots,m\}$ such that $\theta_k(y) = y_{\varpi(k)}$ for $k = 1,\ldots m$. Apply to ordered outcomes $\Theta(y)$, a linear cumulative map that results in the cumulative ordering map $\overline{\Theta}(y) = (\overline{\theta}_1(y), \overline{\theta}_2(y), \ldots \overline{\theta}_m(y))$, defined as $\overline{\theta}_l(y) = \sum_{k=1}^{l} \theta_k(y)$ for $l = 1,\ldots,m$. The coefficients of vector $\overline{\Theta}(y)$ express the largest outcome, the total of the two largest outcomes,. . . ,the total of the m outcomes, respectively.

The outcome vector $y' \in Y$ equitably dominates $y'' \in Y$, if and only if $\overline{\theta}_l(y') \leq \overline{\theta}_l(y'')$ for all $l = 1,\ldots,m$ where at least one strict inequality holds.

The following relationship between equitable efficiency and Pareto-optimality can be provided (e.g., Kostreva et al. 2004).

Theorem 1: A feasible solution $x \in Q$ is an equitably efficient solution of the multiple criteria problem (7.1), iff it is a Pareto-optimal solution of the multiple criteria problem with objectives $\overline{\Theta}(f(x))$:

$$\min\{(\overline{\theta}_1(f(x)), \overline{\theta}_2(f(x)), \ldots \overline{\theta}_m(f(x))) : x \in Q\}. \tag{7.2}$$

Note that minimization of the first objective in (7.2) corresponds to minimization of the worst outcome, while minimization of the last (mth) objective in (2) aims at minimizing the sum of outcomes.

The quantities $\overline{\theta}_l(y)$ can be modeled with simple auxiliary variables and constraints. For a given outcome vector y, the total of the l largest outcomes, $\overline{\theta}_l(y)$, may be found by solving the following linear program (e.g., Ogryczak and Tamir 2003):

$$\overline{\theta}_l(y) = \min\{l\lambda + \sum_{k=1}^{m} \delta_k : \lambda + \delta_k \geq y_k,\ \delta_k \geq 0;\ k = 1,\ldots m\}, \tag{7.3}$$

where λ is an unrestricted variable while nonnegative variables δ_k represent, for several outcome values y_k, their upside deviations from the value of λ.

Using the linear program (7.3), problem (7.2) can be formulated as the following multiple criteria optimization problem:

$$\begin{aligned}\min\{(\lambda_1 + \sum_{k=1}^{m} \delta_{k1}, 2\lambda_2 + \sum_{k=1}^{m} \delta_{k2}, \ldots, m\lambda_m + \sum_{k=1}^{m} \delta_{km}) :&\\ x \in Q,\ \lambda_l + \delta_{kl} \geq f_k(x),\ \delta_{kl} \geq 0;\ k, l = 1,\ldots m\}.&\end{aligned} \tag{7.4}$$

In order to find an equitable efficient solution, the following lexicographic minimax problem can be solved:

$$\text{lex}\min\{\Theta(f(x)) : x \in Q\}. \tag{7.5}$$

Unlike in the standard minimax problem where only the worst objective value is minimized, in the lexicographic minimax problem (7.5), first the worst objective value is minimized, then the second worst objective value is minimized (provided that the worst one is as small as possible), the third worst objective value is minimized (provided that the two worst remain as small as possible), and so on. The lexicographic minimax solution can be considered in some sense the "most equitable solution" (Kostreva et al. 2004).

The multiple criteria problem (7.2) can be replaced with the minimization problem of an equitably aggregated objective function. An equitable aggregation can be based on the use of the cumulative ordered outcomes $\overline{\theta}_k(y)$, for example, the weighted sum of $\overline{\theta}_k(y)$,

$$\sum_{k=1}^{m} v_k \overline{\theta}_k(y). \tag{7.6}$$

The above aggregation is the so-called ordered weighted averaging (OWA) aggregation, was introduced by Yager (1988). If the weights v_k are positive, then applying OWA aggregation to the multiple criteria problem (7.1) yields an equitably efficient solution of the latter problem.

Note that in the OWA aggregation the weights are assigned to the ordered values (i.e., to the largest value, the two largest values, etc.) rather than to the specific criteria, like in the commonly used weighted-sum aggregation.

By applying the OWA aggregation (7.6) to (7.4), the following OWA problem can be formulated:

$$\min\{\sum_{l=1}^{m} v_l(l\lambda_l + \sum_{k=1}^{m} \delta_{kl}) : x \in Q,\ \lambda_l + \delta_{kl} \geq f_k(x),\ \delta_{kl} \geq 0;\ k,l = 1,\ldots m\}. \tag{7.7}$$

Liu and Papageorgiou (2013) considered the optimization problem (7.7) with equal weights $v_l = 1$ for all $l = 1,\ldots,m$ and $m = 2$ objective functions

$$\min\{\sum_{l=1}^{2} (l\lambda_l + \sum_{k=1}^{2} \delta_{kl}) : x \in Q,\ \lambda_l + \delta_{kl} \geq f_k(x),\ \delta_{kl} \geq 0;\ k,l = 1,2\}, \tag{7.8}$$

and in a recent paper Liu et al. (2014) proved the following new theorem.

Theorem 2: If there exists an optimal solution $x^* \in Q$ of the optimization problem (7.7) with equal weights $v_l = 1$ for all $l = 1,\ldots,m$, which also satisfies perfect equity (i.e., $f_1(x^*) = f_2(x^*) = \ldots = f_m(x^*)$), then it is the optimal solution of the lexicographic minimax problem (7.5).

In addition, the following theorem was introduced in Liu et al. (2014), according to Theorem 1 (Kostreva et al. 2004).

Theorem 3: If $x^* \in Q$ is an optimal solution of the optimization problem (7.7), with equal weights $v_l = 1$ for all $l = 1,\ldots,m$, then it is an equitably efficient solution, as well as a Pareto-optimal solution, of the multi-objective optimization problem (7.1).

The results of this section will be used to formulate and solve the fair, risk-neutral optimization problem in this chapter and the fair, mean-risk (robust) optimization problem in Chap. 8, for a bi-objective selection of supply portfolio and scheduling of customer orders in the presence of supply chain disruption risks.

7.3 Problem Description

Consider a two-tier customer-driven supply chain in which various types of products are assembled by a single producer to meet customer orders, using the same critical part type can be manufactured and provided by many suppliers (for notation used, see Table 7.1).

Table 7.1 Notation: equitably efficient supply portfolio selection and production scheduling

Indices	
i	= supplier, $i \in I$
j	= customer order, $j \in J$
r	= region, $r \in R$
s	= disruption scenario, $s \in S$
t	= planning period, $t \in T$
Input Parameters	
a_j	= per unit requirement for parts of each product in customer order j
b_j	= size (number of products) of customer order j
A	= $\sum_{j \in J} a_j b_j$, total demand for parts
B	= $\sum_{j \in J} b_j$, total demand for products
c_j	= per unit capacity consumption of producer for customer order j
C_t	= capacity of producer in period t
d_j	= due date for customer order j
e_i	= fixed cost of ordering parts from supplier i
g_j	= per unit penalty cost of unfulfilled customer order j
h_j	= per unit and per period penalty cost of delayed customer order j
I^r	= the subset of suppliers in region r
I_s	= the subset of suppliers non-disrupted under disruption scenario s
o_i	= per unit price of parts purchased from supplier i
p_i	= local disruption probability for supplier i
p^r	= regional disruption probability for all suppliers in region r
τ_i	= delivery lead time from supplier i

Let $I = \{1,\ldots,\overline{I}\}$ be the set of $\overline{I}$ suppliers, $J = \{1,\ldots,\overline{J}\}$ the set of $\overline{J}$ customer orders for products, and $T = \{1,\ldots,\overline{T}\}$ the set of $\overline{T}$ planning periods (for notation used, see Table 7.1).

Denote by b_j and d_j, respectively, the size and the due date of customer order $j \in J$, i.e., the number of units of ordered product type and the latest period of their completion required to deliver the products to the customer by requested date. Let a_j be the unit requirement for the critical part of each product in customer order $j \in J$. The total demand for all parts is $A = \sum_{j \in J} a_j b_j$ and the total demand for all products is $B = \sum_{j \in J} b_j$.

The customer orders are single-period orders such that each order must be completed in one planning period, e.g., Sawik (2011a). Assume that the producer has limited time-varying capacity, and denote by C_t the producer capacity available in planning period $t \in T$, and by c_j the unit capacity consumption for each product in customer order $j \in J$.

The orders for parts are assumed to be placed at the start of the planning horizon, when all customer orders for products are known. Let o_i be the unit purchasing price of parts from supplier $i \in I$ and denote by e_i the fixed ordering cost of creating contracts and maintaining relationships with supplier $i \in I$. Each supplier has sufficient capacity to meet total demand for parts. The order preparation and transportation time of a shipment from supplier $i \in I$ to the producer is constant and equals to τ_i periods so that the parts ordered from supplier $i \in I$ are delivered in period τ_i and then can be used for the assembly of products in period $\tau_i + 1$, at the earliest.

Assume that the suppliers are located in a number of disjoint geographic regions and denote by $I^r \subseteq I$ the subset of suppliers in region $r \in R$, where $\bigcup_{r \in R} I^r = I$.

The supplies of parts are subject to random independent local disruptions of each supplier individually, which may arise from equipment breakdowns, local labor strike, fires, etc. Denote by p_i the local disruption probability for supplier i.

In addition to independent local disruptions of each supplier, there are potential correlated regional disasters that may result in disruption of all suppliers in the same region simultaneously. For example, such regional disaster events may include an earthquake, flooding, etc. Denote by p^r the probability of correlated regional disruption of all suppliers $i \in I^r$ in region $r \in R$. The regional disasters in each region and the local disasters at each supplier are assumed to be independent events. Let π_i be the disruption probability of every supplier $i \in I^r, r \in R$

$$\pi_i = p^r + (1 - p^r)p_i; \; i \in I^r, r \in R. \tag{7.9}$$

Denote by $S = \{1, \ldots, \overline{S}\}$ the index set of all disruption scenarios, where each scenario $s \in S$ is comprised of a unique subset $I_s \subset I$ of suppliers who deliver parts without disruptions. All potential disruption scenarios will be considered, that is $\overline{S} = 2^{\overline{I}}$. For each scenario $s \in S$, the supplies from every supplier, $i \in I \setminus I_s$, can be disrupted either by a local or a regional disaster event. The probability P_s for disruption scenario $s \in S$ with the subset I_s of non-disrupted suppliers, and with all possible combinations of different disaster events considered, is (cf. Sect. 1.3)

$$P_s = \prod_{r \in R} P_s^r, \tag{7.10}$$

where P_s^r is the probability of realizing of disruption scenario s for suppliers in I^r

$$P_s^r = \begin{cases} (1 - p^r)\prod_{i \in I^r \bigcap I_s}(1 - p_i)\prod_{i \in I^r \setminus I_s} p_i & \text{if } I^r \bigcap I_s \neq \emptyset \\ p^r + (1 - p^r)\prod_{i \in I^r} p_i & \text{if } I^r \bigcap I_s = \emptyset. \end{cases} \tag{7.11}$$

The producer can be charged with a contractual, order specific penalty cost for delayed or unfulfilled customer orders, caused by the shortage of parts, that are delivered late or not at all due to supply disruptions. Let g_j and h_j be, respectively, the per unit total penalty cost of unfulfilled customer order $j \in J$ and the per unit and per period penalty cost of delayed customer order $j \in J$.

The objective of the equitable optimization of a supply chain under disruption risks is to allocate the total demand for parts among a subset of selected suppliers and to schedule the customer orders for products over the planning horizon to equitably minimize expected cost of ordering, purchasing, and shortage of parts and maximize expected customer service level, i.e., the fraction of customer orders (order fulfillment rate) or customer demand (demand fulfillment rate) filled on or before their due dates. The resulting equitably efficient supply portfolio (the allocation of total demand for parts among the selected suppliers) is determined ahead of time as well as the equitably efficient schedule of customer orders for every potential disruption scenario.

7.4 Problem Formulation

In this section the time-indexed SMIP model **ESPS_E** is presented for selection of equitably efficient supply portfolio and scheduling of customer orders to fairly minimize expected cost per product and maximize expected service level, i.e., the expected fraction of customer orders filled on or before their due dates (expected order fulfillment rate). The decision variables that are jointly determined using the proposed model are defined in Table 7.2.

Table 7.2 Problem variables: equitably efficient supply portfolio selection and production scheduling

First stage variables	
u_i	$=1$, if supplier i is selected; otherwise $u_i = 0$ (supplier selection)
v_i	$\in [0,1]$, the fraction of total demand for parts ordered from supplier i (supply portfolio)
Auxiliary variables	
λ_l	unrestricted auxiliary variable, $l = 1,2$
δ_{kl}	≥ 0, upside deviation of outcome value $f_k, k = 1,2$ from the value of $\lambda_l, l = 1,2$
Second stage variables	
x^s_{jt}	$=1$, if under disruption scenario s customer order j is scheduled for period t; otherwise $x^s_{jt} = 0$ (production scheduling)

Let E^c be the expected cost per product to be minimized and E^{sl}, the expected service level to be maximized

$$E^c = (\sum_{i\in I} e_i u_i + \sum_{s\in S} P_s(\sum_{i\in I_s} Ao_i v_i$$
$$+\sum_{j\in J}\sum_{t\in T:t>d_j} h_j b_j (t-d_j)x^s_{jt} + \sum_{j\in J} g_j b_j (1-\sum_{t\in T} x^s_{jt})))/B \tag{7.12}$$

$$E^{sl} = \sum_{j\in J}\sum_{t\in T:t\leq d_j}\sum_{s\in S} P_s x^s_{jt}/\bar{J}. \tag{7.13}$$

In order to avoid dimensional inconsistency among the two objectives, the values of the optimized objective functions are scaled into the interval [0,1]. Denote by $f_1 = \frac{E^c - \underline{E}^c}{\overline{E}^c - \underline{E}^c}$, the normalized expected cost per product ($\underline{E}^c, \overline{E}^c$ are the minimum and the maximum values of E^c, respectively), and by $f_2 = \frac{\overline{E}^{sl} - E^{sl}}{\overline{E}^{sl} - \underline{E}^{sl}}$, the normalized expected customer service level ($\underline{E}^{sl}, \overline{E}^{sl}$ are the minimum and the maximum values of E^{sl}, respectively).

The normalized objective functions f_1 and f_2 are defined below

$$f_1 = (\sum_{i \in I} e_i u_i + \sum_{s \in S} P_s (\sum_{i \in I_s} A o_i v_i + \sum_{j \in J} \sum_{t \in T: t > d_j} h_j b_j (t - d_j) x^s_{jt}$$
$$+ \sum_{j \in J} g_j b_j (1 - \sum_{t \in T} x^s_{jt}))) / B - \underline{E}^c) / (\overline{E}^c - \underline{E}^c) \tag{7.14}$$

$$f_2 = \frac{\overline{E}^{sl} - \sum_{j \in J} \sum_{t \in T: t \le d_j} \sum_{s \in S} P_s x^s_{jt} / J}{(\overline{E}^{sl} - \underline{E}^{sl})}. \tag{7.15}$$

The SMIP model **ESPS_E** is formulated below. The model is based on the SMIP formulations **SPSm_E(c)** and **SPSm_E(sl)** presented in Chap. 5. The objective function (7.16) subject to constraints (7.17) represent the so-called ordered weighted averaging aggregation of the two conflicting criteria with equal weights assigned to each criterion, see, OWA aggregation in Sect. 7.2. Applying OWA aggregation to the bi-objective problem yields an equitably efficient solution to the problem. In the model presented below, λ_l are unrestricted variables, while nonnegative variables, δ_{kl} represent, for outcome values f_k, their upside deviations from the value of λ_l (see, Sect. 7.2).

ESPS_E: *Risk-neutral selection of* **E***quitably efficient* **S***upply* **P***ortfolio and* **S***cheduling of customer orders to minimize expected cost and maximize expected service level*

Minimize

$$\sum_{l=1}^{2} (l \lambda_l + \sum_{k=1}^{2} \delta_{kl}) \tag{7.16}$$

subject to (7.14), (7.15) and

$$\lambda_l + \delta_{kl} \ge f_k; k, l = 1, 2. \tag{7.17}$$

Demand allocation constraints:

- the total demand for parts must be fully allocated among the selected suppliers (7.18),

(continued)

– demand for parts cannot be assigned to non-selected suppliers (7.19),

$$\sum_{i \in I} v_i = 1 \tag{7.18}$$

$$v_i \le u_i;\ i \in I. \tag{7.19}$$

Order-to-period assignment constraints:

– for each disruption scenario s, each customer order j is either scheduled during the planning horizon ($\sum_{t \in T} x^s_{jt} = 1$) or unscheduled and rejected ($\sum_{t \in T} x^s_{jt} = 0$) (7.20),
– for each disruption scenario s and each planning period t, the cumulative demand for parts of all customer orders scheduled in periods 1 through t cannot exceed the cumulative deliveries of parts in periods 1 through $t-1$, from the non-disrupted suppliers $i \in I_s$ (7.21),
– for each disruption scenario s, the total requirement for parts of scheduled customer orders is not greater than the total supplies from the non-disrupted suppliers $i \in I_s$ (7.22),

$$\sum_{t \in T} x^s_{jt} \le 1;\ j \in J, s \in S \tag{7.20}$$

$$\sum_{j \in J} \sum_{t' \in T: t' \le t} a_j b_j x^s_{jt'} \le A \sum_{i \in I_s: \tau_i \le t-1} v_i;\ t \in T, s \in S \tag{7.21}$$

$$\sum_{j \in J} \sum_{t \in T} a_j b_j x^s_{jt} \le A \sum_{i \in I_s} v_i;\ s \in S. \tag{7.22}$$

Producer capacity constraints:

– for any period t and each disruption scenario s, the total demand on capacity of all customer orders scheduled in period t must not exceed the producer capacity available in this period,

$$\sum_{j \in J} b_j c_j x^s_{jt} \le C_t;\ t \in T, s \in S. \tag{7.23}$$

Non-negativity and integrality conditions:

$$\delta_{kl} \ge 0; k, l = 1, 2 \tag{7.24}$$

$$u_i \in \{0, 1\};\ i \in I \tag{7.25}$$

$$v_i \in [0, 1];\ i \in I \tag{7.26}$$

$$x^s_{jt} \in \{0, 1\};\ j \in J, t \in T, s \in S. \tag{7.27}$$

In the above model, each supplier is assumed to have sufficient capacity to meet total demand for parts. Such an assumption allows the decision maker to select a single sourcing solution, if such a portfolio is an equitably efficient supply portfolio.

However, the assumption can be easily relaxed to account for multiple capacitated suppliers, see Sect. 7.4.2.

7.4.1 Minimum and Maximum Values of the Objective Functions

In this subsection the minimum and maximum values for all objective functions are calculated to determine the normalized values of the objective functions, f_1, (7.14), f_2, (7.15), that is, the values of the optimized objective functions scaled into the interval [0,1]. Note that the cost and the service level objectives are in conflict. Therefore, the minimum and maximum values of expected cost $\underline{E}^c$, $\overline{E}^c$, and expected customer service level, $\underline{E}^{sl}$, $\overline{E}^{sl}$, are obtained by solving the following stochastic mixed integer programs:

SPS_E(c): *Risk-neutral selection of* **S**upply **P**ortfolio *and* **S**cheduling *of customer orders to minimize expected cost per product*
Minimize E^c, (7.12),
subject to (7.18)–(7.23), (7.25)–(7.27).

SPS_E(sl): *Risk-neutral selection of* **S**upply **P**ortfolio *and* **S**cheduling *of customer orders to maximize expected service level*
Maximize E^{sl}, (7.13)
subject to (7.18)–(7.23), (7.25)–(7.27).

In problem **SPS_E(c)**, E^c is the minimized objective function, while E^{sl} is not considered. In problem **SPS_E(sl)**, E^{sl} is the maximized objective function, while E^c is not considered. Thus, by solving problem **SPS_E(c)**, the minimum value $\underline{E}^c$ of E^c and the minimum value $\underline{E}^{sl}$ of E^{sl} are determined. Similarly, by solving problem **SPS_E(sl)**, the maximum value $\overline{E}^{sl}$ of E^{sl} and the maximum value $\overline{E}^c$ of E^c are determined.

So far, in the proposed models the customer service level is measured by order fulfillment rate, i.e., the number of customer orders fulfilled by their due dates, with no account for the size of each customer order. For example, a high customer service level can be achieved by fulfilling a large number of small-size orders, while leaving the unfulfilled demand relatively high. To avoid such a solution, in particular when the customer orders of different sizes are simultaneously considered, the service level can be measured by demand fulfillment rate, i.e., the fraction of total customer demand fulfilled by the requested due dates.

If the customer service level is defined as demand fulfillment rate, then E^{sl}, (7.13) and f_2, (7.15) should be replaced with the following formulae, (7.28) and (7.29), respectively:

$$E^{sl} = \sum_{j \in J} \sum_{t \in T: t \le d_j} \sum_{s \in S} P_s b_j x^s_{jt} / B \tag{7.28}$$

$$f_2 = \frac{\overline{E}^{sl} - \sum_{j \in J} \sum_{t \in T: t \le d_j} \sum_{s \in S} P_s b_j x^s_{jt} / B}{(\overline{E}^{sl} - \underline{E}^{sl})}, \tag{7.29}$$

where $\underline{E}^{sl}$, $\overline{E}^{sl}$ are the minimum and the maximum values of E^{sl}, (7.28), respectively. The values of $\underline{E}^{sl}$, $\overline{E}^{sl}$ can be determined using the models **SPS_E(c)** and **SPS_E(sl)**.

In the computational examples presented in the next section, the two metrics of the customer service level will be considered and compared against each other.

7.4.2 Model Limitations and Possible Enhancements

The proposed model has been developed to support decision-making in a make-to-order environment under disruption risks. The model, however, has been formulated under various simplified assumptions that may limit its practical usefulness. The basic assumptions are listed below.

1. A single critical part type is required to fulfill all customer orders for products.
2. The orders for parts are placed at the start of the planning horizon, when all customer orders for products are known.
3. Each supplier has sufficient capacity to meet the total demand for parts.
4. The order preparation time at each supplier is constant, independent of order size, and all parts ordered from a supplier are delivered during a fixed transportation time.
5. Transportation costs are not explicitly considered and the unit purchasing price from each supplier is constant, independent of total volume or value of order for parts, i.e., no quantity or business volume discounts are considered.
6. The customer orders are single-period orders such that each order must be completed in one planning period.
7. The penalty costs for delayed or unfulfilled customer demand are linear.
8. The inventory of parts and products are not considered.
9. The two conflicting objectives: reduction of expected cost and increase of expected customer service level are equally important for the decision maker.

Some of the above assumptions can be easily relaxed, while the other needs a more advanced model to be developed. Possible relaxations of the corresponding assumptions and the model enhancements are listed below.

1. The model can be easily enhanced for multiple part types required to fulfill all customer orders with different subsets of part types needed for different product types and different subsets of capacitated suppliers capable of providing different subsets of part types, e.g., Sawik (2013c).
2. A rolling planning horizon approach can be used to account for a dynamic arrival and scheduling of customer orders as well as the corresponding supply portfolio.

In practice, however, the supply portfolio needs to be decided at the start of the planning horizon, based on a forecast of the customer demand.

3. The model can be easily enhanced for multiple capacitated suppliers by the addition of suppliers capacity constraints.
4. The more advanced model can be developed to consider order-dependent processing and transportation times to better coordinate manufacturing and transportation of parts and production of finished products.
5. The model can be easily enhanced to account for quantity or business volume discounts (e.g., Sawik 2010a) and the unit purchasing price from each supplier can include unit transportation cost.
6. The model can be easily enhanced to account for large, multi-period customer orders that cannot be completed in one period and must be split into single-period portions to be processed in consecutive planning periods, e.g., Sawik (2011a).
7. The introduction of non-linear penalty costs may lead to a non-linear SMIP model. The model, however, can be linearized in some cases, e.g., by using a piecewise linear representation of the non-linear penalty cost.
8. The model can be enhanced to account for the output inventory of parts at suppliers, the input inventory of parts at the producer and the output inventory of products waiting for shipment to customers. The inventory balance constraints can be added to the model and the producer inventory holding costs, to the cost-based objective function.
9. The number of conflicting and equally important objectives can be increased, for example, by the addition of responsiveness (e.g., Liu and Papageorgiou 2013) as another objective function.

7.5 Computational Examples

In this section the proposed SMIP approach for selection of equitably efficient supply portfolio and scheduling of customer orders in a supply chain under disruption risks is compared with the weighted-sum approach and illustrated with computational examples. The following parameters have been used for the example problems (cf. Sect. 6.6):

- $\overline{I} = 9, \overline{J} = 25, \overline{T} = 10$, and $\overline{S} = 2^{\overline{I}} = 512$;
- $R = \{1,2,3\}$, and $I^1 = \{1,2,3\}, I^2 = \{4,5,6\}, I^3 = \{7,8,9\}$;
- $a_j \in \{1,2,3\}, b_j \in \{500, 1000, \ldots, 5000\}, c_j \in \{1,2,3\}, d_j \in \{1+\min_{i\in I}(\tau_i), \ldots, \overline{T}\}$;
- C_t, the capacity of producer in each period t, was integer drawn from $1000\lceil(2\sum_{j\in J} b_j c_j/(\overline{T} - \max_{i\in I}\tau_i))U[0.75;1.25]/1000\rceil$ distribution, i.e., in each period the producer capacity was from 75% to 125% of the double capacity required to complete all customer orders during the planning horizon, after the latest delivery of parts;
- o_i, the unit price of parts purchased from supplier i, was uniformly distributed over [11,16], [6,11], and [1,6], respectively, for suppliers $i \in I^1$, $i \in I^2$, and $i \in I^3$;

- $g_j = 2\lceil a_j \max_{i\in I}(o_i)\rceil$, $j \in J$, i.e., the unit penalty cost of each unfulfilled customer order j was approximately twice as large as the maximum unit price of required parts;
- $h_j = \lceil a_j \max_{i\in I}(o_i)/350\rceil$, $j \in J$, i.e., the unit penalty cost per period of each delayed customer order j was approximately 0.28% of the maximum unit price of required parts;
- p_i, the local disruption probability was uniformly distributed over [0.005,0.01], [0.01,0.05], and [0.05;0.10], respectively, for suppliers $i \in I^1$, $i \in I^2$, and $i \in I^3$.

The detailed data set (see Table 7.3) is based on the example presented in Sawik (2013c).

Table 7.3 Input parameters: equitably efficient supply portfolio selection and production scheduling

$\overline{I} = 9$ suppliers, $\overline{J} = 25$ customer orders, $\overline{R} = 3$ three geographic regions, $\overline{T} = 10$ planning periods, $\overline{S} = 2^{\overline{I}} = 512$ disruption scenarios	
$I^1 = \{1,2,3\}$	three suppliers in region $r = 1$
$I^2 = \{4,5,6\}$	three suppliers in region $r = 2$
$I^3 = \{7,8,9\}$	three suppliers in region $r = 3$
$a = (2,1,3,3,1,3,2,1,2,2,2,2,3,2,1,3,2,1,3,3,2,1,1,2,1)$	per unit requirement for parts
$b = (1,2,9,7,8,5,1,7,5,4,7,4,10,6,8,1,4,2,4,8,6,3,8,7,3) \times 500$	customer order size
$A = 132500$	total demand for parts
$B = 66{,}000$	total demand for products
$c = (2,1,1,2,3,3,1,3,2,1,2,1,3,1,1,3,2,3,1,1,3,2,2,1,2)$	per unit capacity consumption
$C_t = C = 38{,}000, \forall t = 1,\dots,10$	producer capacity in period t
Local disruption probabilities of suppliers:	
p = (0.00513571, 0.00666354, 0.00902974, 0.0356206, 0.040175, 0.0294692, 0.0519967, 0.0827215, 0.0739062)	
Regional disruption probability: $p^1 = 0.001$, $p^2 = 0.005$, $p^3 = 0.01$	
Cost parameters:	
$e = (8,7,10,13,14,11,19,17,19) \times 1000$, $o = (13,12,12,8,6,6,2,5,4)$	
$g = (52,26,78,78,26,78,52,26,52,52,52,52,78,52,26,78,52,26,78,78,52,26,26,52,26)$ unit penalty for unfulfilled demand	
$h_j = 1 \forall j = 1,\dots,25$ unit penalty for delayed demand	
Time parameters:	
$d = (7,4,10,10,3,8,7,5,5,6,8,7,10,5,3,9,6,4,9,9,7,4,5,7,4)$ customer due-dates	
τ_i = 2, 3 and 4 time periods, respectively, for suppliers $i \in I^1$, $i \in I^2$ and $i \in I^3$ delivery times	

Given local and regional disruption probabilities presented in Table 7.3, the corresponding total disruption probabilities of suppliers, $\pi_i = p^r + (1-p^r)p_i$, $i \in I^r, r \in R$, (7.9), are: π=(0.00613057, 0.00765688, 0.0100207, 0.0404425, 0.0449741, 0.0343219, 0.0614767, 0.0918943, 0.0831672).

Note that the constant producer capacity, $C = 38{,}000$, allows for completing all customer orders in at most $\lceil \sum_{j\in J} b_j c_j / C \rceil = \lceil 3.26 \rceil = 4$ planning periods, that is, in less than $(\overline{T} - \max_{i\in I} \tau_i) = 6$ periods remaining in the planning horizon after the latest delivery of parts.

In the computational experiments all potential disruption scenarios, $\overline{S} = 2^{\overline{I}} = 512$, and all possible combinations of local and regional disaster events were considered. Each scenario $s \in S$ with the subset I_s of non-disrupted suppliers is represented by an $\overline{I}$-dimensional 0–1 vector with 1, if $i \in I_s$, i.e., if supplier i is not disrupted, and 0; otherwise. The corresponding disruption probability, P_s, for each scenario $s \in S$ was calculated using formulae (7.10) and (7.11).

The unit price per part o_i and the disruption probability π_i, (7.9), of each supplier $i \in I$ are shown in Fig. 6.2. The figure indicates that the most reliable (with the lowest disruption probability, $\pi_1 = 0.00613057$) is supplier 1, the least reliable (with the highest disruption probability, $\pi_8 = 0.0918943$) is supplier 8, the most expensive (with the highest price per part, $o_1 = 13$) is supplier 1, and the cheapest (with the lowest price per part, $o_7 = 2$) is supplier 7. Note that the geographic regions are numbered in such a way that the unit prices are nonincreasing with r, while the fixed ordering costs and the disruption probabilities are nondecreasing with r.

The solution results are presented in Table 7.4. In addition to the optimal absolute and normalized solution values for the primary objective functions and the allocation of demand among the selected suppliers, Table 7.4 presents the expected values of the associated objective function, i.e., the minimum expected service level, $\underline{E}^{sl}$, for model **SPS_E(c)** and the maximum expected cost per product, $\overline{E}^{c}$, for model **SPS_E(sl)**. Table 7.4 indicates that for the cost-based objective (model **SPS_E(c)**) the cheapest supplier $i = 7$ is selected only, while for the customer service level (model **SPS_E(sl)**), the total demand for parts is allocated among the three most reliable and most expensive suppliers $i = 1,2,3$ for objective (7.13) and among the two suppliers $i = 1,2$ for objective (7.28). For the equitable solution (model **ESPS_E**), the supply portfolio contains one reliable and expensive supplier $i = 2$ and two low-cost and unreliable suppliers $i = 6,7$ for both (7.15) and (7.29), service level objectives.

As an illustrative example, Fig. 7.1 presents the demand for products, $\sum_{j \in J: d_j = t} b_j$, $t \in T$, and the expected production schedules, $\sum_{s \in S} P_s \sum_{j \in J} b_j x^s_{jt}, t \in T$ for the optimal cost, optimal customer service level and for the equitably efficient solution. Figure 7.1 compares the expected production schedules for the two different metrics of customer service level: (a) order fulfillment rate (i.e., fraction of customer orders fulfilled on time), (7.13), and (b) demand fulfillment rate (i.e., fraction of customer demand fulfilled on time), (7.28). While for the two different service level metrics, the corresponding supply portfolios are very similar (cf. Table 7.4), and the expected production schedules with respect to the two service level objectives are also very similar, the corresponding schedules for the equitably efficient solutions are different.

In general, the service level-based solution, when no cost components are included in the objective function, better meets the customer demand, with the smallest fraction of unfulfilled demand. The total customer demand is met with only a small fraction of the expected unfulfilled demand: 0.0615 for the cost-based solution, 0.0055 for the service level-based solution (a), 0.0051 for the service level-based solution (b), 0.0468 for the equitably efficient solution (a), and 0.0470 for the equitably efficient solution (b). In addition, *for the service level-based objective*

Table 7.4 Solution results

Model **SPS_E(c)**: Var. = 100,468, Bin. = 100,459, Cons. = 21,341, Nonz. = 765,902[a]	
Expected cost ($\underline{E}^c$)	7.66
Suppliers selected(% of total demand)	7(100)
Expected service level ($\underline{E}^{sl}$)	67.60[b], 66.32[c]
Model **SPS_E(sl)**: Var.=100,468, Bin.=100,459, Cons.=21,341, Nonz.=765,902[a]	
Expected service level ($\overline{E}^{sl}$)	99.62[b]
Suppliers selected (% of total demand)	1(48), 2(31), 3(21)
Expected cost ($\overline{E}^c$)	25.64
Model **ESPS_E**: Var. = 100,476, Bin. = 100,459, Cons. = 21,347, Nonz.=921,880[a]	
Expected cost	9.31
Expected service level	96.06[b]
Normalized expected cost	0.098
Normalized expected service level	0.111
Suppliers selected (% of total demand)	2(6), 6(13), 7(81)
Model **SPS_E(sl)** with (7.13) replaced by (7.28)	
Expected service level ($\overline{E}^{sl}$)	99.49[c]
Suppliers selected (% of total demand)	1(55), 2(45)
Expected cost ($\overline{E}^c$)	25.61
Model **ESPS_E** with (7.15) replaced by (7.29)	
Expected cost	9.25
Expected service level	95.29[c]
Normalized expected cost	0.088
Normalized expected service level	0.127
Suppliers selected (% of total demand)	2(6), 6(10), 7(84)

[a] Var.: number of variables; Bin.: number of binary variables; Cons.: number of constraints; Nonz.: number of nonzero coefficients

[b] $(\sum_{s\in S} P_s \sum_{j\in J} \sum_{t\in T:t\le d_j} x^s_{jt}/J)100\%$

[c] $(\sum_{s\in S} P_s \sum_{j\in J} \sum_{t\in T:t\le d_j} b_j x^s_{jt}/B)100\%$

functions, the expected production schedule approximately follows the customer demand pattern, while for the minimum cost objective function, production schedule hardly meets demand pattern.

In order to compare the solution results for the two different service level-based objective functions (7.13) and (7.28), the computational experiments were repeated for another example with more diversified size of customer orders, $b_j \in \{500, 1000, \ldots, 12{,}000\}$, i.e., from 500 to 12,000 products, for the same total demand for parts and products, $A = 132{,}500$ and $B = 66{,}000$. The solution results are presented in Table 7.5, and Fig. 7.2 compares the expected production schedules for the two different metrics of customer service level. In general, the results presented in Table 7.5 are similar to those in Table 7.4, in particular the supply portfolios are very similar. However, Fig. 7.2 shows that for the more diversified customer orders, the less smoothed expected production schedules are obtained for the equitably efficient solutions. On the other hand, Table 7.5 shows that the equitably efficient solutions with a perfect equity were found (i.e., with identical values of normalized expected cost and normalized expected customer service level), which indicates that the obtained supply portfolios and schedules of customer orders are also the lexico-

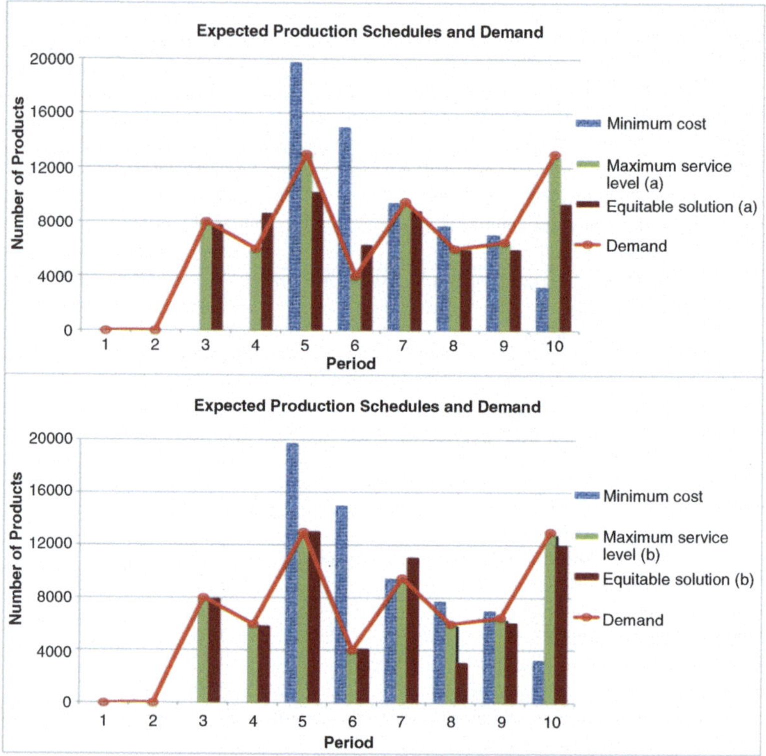

Fig. 7.1 Expected production schedules: (**a**) order fulfillment rate; (**b**) demand fulfillment rate

graphic minimax optimal solutions as well as the Pareto-optimal solutions (see, Liu and Papageorgiou 2013; Liu et al. 2014).

7.5.1 Weighted-Sum Approach

The equitably efficient solutions obtained using model **ESPS_E** have been compared with the nondominated solutions obtained by minimizing the weighted-sum aggregation of the two normalized objective functions, f_1, (7.14) and f_2, (7.29), i.e., the weighted-sum of expected cost per product and expected fraction of customer demand fulfilled on time. The weighted-sum model **SPSW_E** is shown below and the solution results, for the example, with similar and diversified customer orders are presented in Tables 7.6 and 7.7, respectively.

Table 7.5 Solution results: diversified customer orders

Model **SPS_E(c)**	
Expected cost ($\underline{E}^c$)	7.44
Suppliers selected (% of total demand)	7(100)
Expected service level ($\underline{E}^{sl}$)	71.32[a], 78.83[b]
Model **SPS_E(sl)**	
Expected service level ($\overline{E}^{sl}$)	99.77[a]
Suppliers selected (% of total demand)	1(42), 2(41), 3(17)
Expected cost ($\overline{E}^c$)	25.15
Model **ESPS_E**	
Expected cost	9.38
Expected service level	97.48[a]
Normalized expected cost	0.110
Normalized expected service level	0.110
Suppliers selected (% of total demand)	2(3), 6(22), 7(75)
Model **SPS_E(sl)** with (7.13) replaced by (7.28)	
Expected service level ($\overline{E}^{sl}$)	99.47[b]
Suppliers selected (% of total demand)	1(65), 2(35)
Expected cost ($\overline{E}^c$)	25.43
Model **ESPS_E** with (7.15) replaced by (7.29)	
Expected cost	10.44
Expected service level	95.99[b]
Normalized expected cost	0.168
Normalized expected service level	0.168
Suppliers selected (% of total demand)	2(3), 5(5), 6(30), 7(62)

[a] $(\sum_{s\in S} P_s \sum_{j\in J} \sum_{t\in T: t\leq d_j} x^s_{jt}/n)100\%$

[a] $(\sum_{s\in S} P_s \sum_{j\in J} \sum_{t\in T: t\leq d_j} b_j x^s_{jt}/B)100\%$

SPSW_E: *Risk-neutral selection of* **S**upply **P**ortfolio *and* **S**cheduling *of customer orders to minimize* **W**eighted-sum *of normalized expected cost and service level*

Minimize

$$\lambda f_1 + (1-\lambda) f_2, \tag{7.30}$$

where $0 \leq \lambda \leq 1$,

subject to (7.14), (7.18)–(7.23), (7.25)–(7.27), (7.29).

Tables 7.6 and 7.7 indicate that for $\lambda = 1$ (minimization of cost) the cheapest supplier $i = 7$ is selected only, while for $\lambda = 0$ (maximization of customer service level) the total demand for parts is allocated among the two most reliable and most expensive suppliers $i = 1, 2$. As λ increases from 0 to 1, i.e., the decision maker preference shifts from customer service level to cost, more demand is moved from expensive and reliable suppliers to low-cost, unreliable suppliers. For the example with similar customer orders, the subset of nondominated solutions contains (for $\lambda = 0.3$, 0.4)

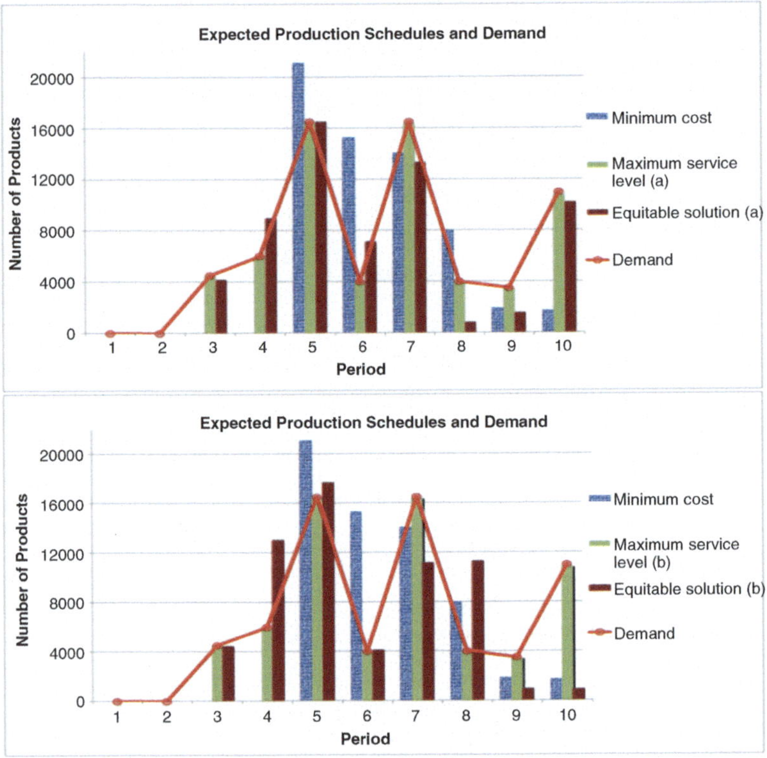

Fig. 7.2 Expected production schedules for diversified customer orders: (**a**) order fulfillment rate; (**b**) demand fulfillment rate

the optimal solution obtained for model **ESPS_E** with constraint (7.15) replaced by (7.29) (cf., Tables 7.4 and 7.6). In contrast to the example with diversified customer orders (cf., Tables 7.5 and 7.7). For diversified orders, Fig. 7.3 shows the nondominated supply portfolios (the allocation of total demand for parts among selected suppliers) for 11 levels of trade-off parameter λ. The subset of selected suppliers consists of four suppliers $i = 1,2,6,7$ of which suppliers $i = 1,2$ are most reliable and suppliers $i = 6,7$ are the cheapest suppliers in region $r = 2,3$, respectively.

For the example with diversified customer orders, the trade-off between the expected cost and the expected customer service level is clearly shown in Fig. 7.4, where the efficient frontier is presented. The results emphasize the effect of varying service level/cost preference of the decision maker; the higher the trade-off parameter λ, the more cost-oriented the decision-making.

The major managerial insights are summarized in Table 7.8. Overall, the solution results demonstrate that *for the minimum cost objective function, the cheapest supplier is usually selected, for the maximum service level objective, a subset of most*

Table 7.6 Nondominated solutions

λ	0	0.1	0.2	0.3	0.4	0.5	0.6	0.7	0.8	0.9	1
Var. = 100,470, Bin. = 100,459, Cons. = 21,343, Nonz. = 921,868[a]											
Exp. cost	25.61	19.40	11.93	9.25	9.25	8.88	8.88	8.88	8.88	8.01	7.66
Exp. service level[b]	99.48	98.57	96.86	95.29	95.29	94.73	94.73	94.73	94.73	82.69	66.32
Normalized exp. cost	1	0.654	0.238	0.088	0.088	0.068	0.068	0.068	0.068	0.020	0
Normalized exp. service level	0	0.028	0.079	0.127	0.127	0.143	0.143	0.143	0.143	0.506	1
Suppliers selected	1(55)	1(6)									
(% of total demand)	2(45)	2(44)	2(10)	2(6)	2(6)	2(6)	2(6)	2(6)	2(6)		
		6(50)	6(40)	6(10)	6(10)	6(5)	6(5)	6(5)	6(5)	6(5)	
			7(50)	7(84)	7(84)	7(89)	7(89)	7(89)	7(89)	7(95)	7(100)

[a]Var.: number of variables; Bin.: number of binary variables; Cons.: number of constraints
Nonz.: number of nonzero coefficients
[b]$(\sum_{s\in S} P_s \sum_{j\in J} \sum_{t\in T:t\le d_j} b_j x^s_{jt}/B)100\%$

Table 7.7 Nondominated solutions: diversified customer orders

λ	0	0.1	0.2	0.3	0.4	0.5	0.6	0.7	0.8	0.9	1
Var. = 100,470, Bin. = 100,459, Cons. = 21,343, Nonz. = 921,868[a]											
Exp. cost	25.43	24.81	12.95	11.00	8.61	8.61	8.31	8.31	8.31	8.31	7.44
Exp. service level[b]	99.47	99.42	97.07	96.30	94.90	94.90	94.45	94.45	94.45	94.45	78.83
Normalized exp. cost	1	0.9655	0.3062	0.1977	0.0650	0.0650	0.0486	0.0486	0.0486	0.0486	0
Normalized exp. service level	0	0.0025	0.1169	0.1532	0.2226	0.2224	0.2439	0.2440	0.2442	0.2446	1
Suppliers selected	1(65)	1(33)									
(% of total demand)	2(35)	2(67)	2(13)	2(8)	2(4)	2(4)	2(3)	2(3)	2(3)	2(3)	
			6(53)	6(35)	6(9)	6(9)	6(5)	6(5)	6(5)	6(5)	
			7(34)	7(57)	7(87)	7(87)	7(92)	7(92)	7(92)	7(92)	7(100)

[a]Var.: number of variables; Bin.: number of binary variables; Cons.: number of constraints;
Nonz.: number of nonzero coefficients
[b]$(\sum_{s\in S} P_s \sum_{j\in J} \sum_{t\in T:t\le d_j} b_j x^s_{jt}/B)100\%$

reliable and most expensive suppliers is usually chosen, whereas the equitably efficient supply portfolio usually combines the two types of suppliers. In addition, *for the service level-based objective functions, the expected production approximately follows the customer demand pattern, while for the minimum cost objective function, the expected production hardly meets the demand pattern.*

The computational experiments were performed using the AMPL programming language and the CPLEX 12.5 solver on a MacBookPro laptop with Intel Core i7 processor running at 2.8 GHz and with 16 GB RAM. The solver was capable of finding proven optimal solution for all examples with CPU time ranging from several seconds to a few hours.

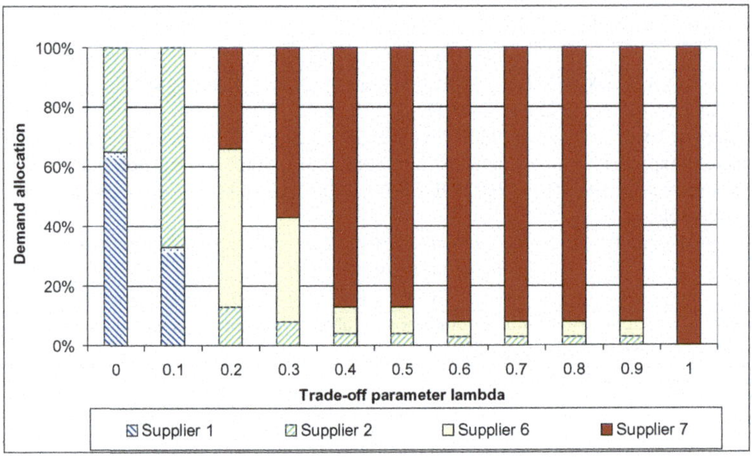

Fig. 7.3 Nondominated supply portfolios: diversified customer orders

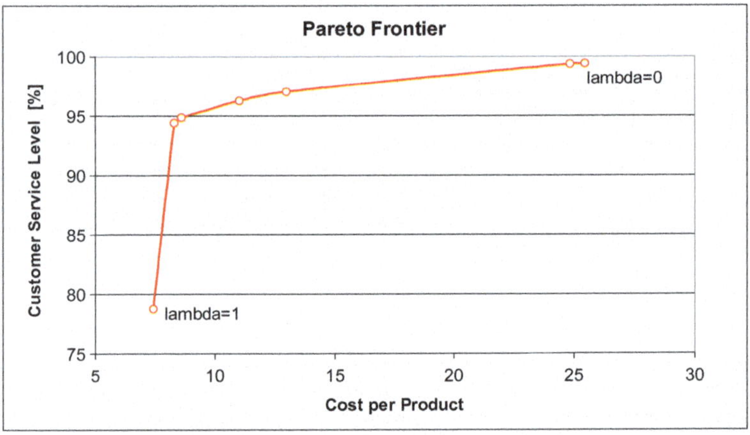

Fig. 7.4 Pareto front for model **SPSW_E**: diversified customer orders

7.6 Notes

The research on equitably efficient multiple objective optimization of a supply chain under disruption risks is rarely reported in the literature. In order to generate an equitably efficient solution with respect to multiple, equally important objective functions, the lexicographic minimax method, e.g., Kostreva et al. (2004), as a special case of the ordered weighted averaging aggregation introduced by Yager (1988), can be applied. The lexicographic minimax problem can be transferred to a lexicographic minimization problem, e.g., Ogryczak et al. (2008). Recently Liu and Papageorgiou (2013) developed an approach to transfer the lexicographic minimax

Table 7.8 Managerial insights: equitably efficient supply portfolio selection and production scheduling

Decision-making issue	Decision-making impact
Expected cost vs. expected service level	The service level-based solution better meets customer demand with the smallest fraction of unfulfilled demand. The expected production approximately follows the customer demand pattern, while the cost-related production hardly meets demand pattern
Weighted-sum approach vs. fair approach	The nondominated supply portfolio for the weighted-sum approach depends on the decision maker preference. When the preference shifts from service to cost, more demand is moved from expensive and reliable suppliers to low-cost, unreliable suppliers. In contrast, the fair supply portfolio usually combines the two types of suppliers
Order fulfillment rate vs. demand fulfillment rate	For the two service level measures, the service-oriented supply portfolios are similar, while the fair supply portfolio depends on the selected service level measure
Similar vs. diversified customer orders	The more diversified customer orders, the less smoothed expected production schedule for the fair decision-making

problem to a minimization optimization problem, instead of a lexicographic minimization problem, which needs to solve a sequence of optimization problems iteratively. The recent approach, however, is restricted to some special cases of a multiple objective problem (Liu et al. 2014).

The material presented in this chapter is based on research reported by Sawik (2014d, 2015a), where a SMIP approach was proposed for the integrated selection of supply portfolio and scheduling of customer orders in a supply chain under disruption risks. In order to equitably optimize expected cost and expected service level, the ordered weighted averaging aggregation of the two conflicting objective functions was applied. The equitably efficient solutions obtained for the ordered weighted averaging aggregation were compared with nondominated solutions obtained using the weighted-sum aggregation approach.

In the proposed model, each supplier is assumed to have sufficient capacity to meet total demand for parts, which allows the decision maker to select a single sourcing type of a supply portfolio, if it is an equitably efficient solution. In the future research, however, that assumption can be easily relaxed to account for multiple capacitated suppliers. Furthermore, the other assumptions can also be relaxed to develop a more advanced model (for possible model relaxations and enhancements, see Sect. 7.4.2). In particular, the future research should focus on equitably efficient decision-making with respect to more equally important and conflicting objectives such as responsiveness, robustness, competitiveness, etc.

Problems

7.1. Modify the SMIP models presented in this chapter for multiple part types with subsets of part types required for each customer order and subsets of capacitated suppliers available for each part type.

7.2. Enhance the SMIP models presented in this chapter for multi-period customer orders that cannot be completed in one period and must be split into single-period portions to be processed in consecutive planning periods (see, Sawik 2011a).

7.3. Limited storage space for parts and products
Enhance the SMIP models presented in this chapter:

(a) for the the limited input inventory of parts at the producer;
(b) for the limited output inventory of products at the producer. Formulate the inventory balance constraints that should be added to the SMIP models.

7.4. In addition to cost and service level, what are the other significant conflicting and equally important objective functions that can be considered when selecting supply portfolio and scheduling customer orders in the presence of supply chain disruption risks?

7.5. How would you select a nondominated solution obtained using the weighted-sum approach that is as close as possible to the equitably efficient solution?

Chapter 8
A Robust Decision-Making under Disruption Risks

8.1 Introduction

In this chapter we look for an equitably efficient solution with respect to both average-case and worst-case performance measures of a supply chain. Such an equitably efficient average and worst-case solution, or equivalently equitably efficient risk-neutral and risk-averse solution will be called a fair mean-risk solution. The solution will equitably focus on the two objective functions: the expected value (average-case performance measure) and the expected worst-case value (worst-case performance measure), i.e., conditional value-at-risk of the selected optimality criterion, cost or service level. The fair mean-risk decision-making aims at achieving the normalized expected and expected worst-case values of the selected objective function as much close to each other as possible, that is, the decision-making aims at equalizing the distance to optimality both under business-as-usual and under worst-case conditions. The mean-risk fairness reflects the decision makers common requirement to maintain an equally good performance of a supply chain under varying operating conditions, which is close to the idea of robustness of a supply chain. Therefore, the mean-risk fairness, i.e., the equitably efficient performance of a supply chain in the average-case as well as in the worst-case, in this chapter will be called robustness.

In a make-to-order environment, the choice between high-cost, reliable suppliers and low-cost, unreliable suppliers has a direct impact on both cost and service level. The more reliable and expensive suppliers selected, the higher the service level and the purchasing costs, and the lower penalties for delayed or unfulfilled customer orders. While the risk-neutral solution for business-as-usual conditions does not account for high-impact disruptions and hence may imply poor results when such disruptions occur, the risk-averse solution that focuses on potential worst-case losses may not perform well under business-as-usual conditions. It is often most desirable to equitably consider these conflicting objectives, that is, the expected value and the

T. Sawik, *Supply Chain Disruption Management*,
International Series in Operations Research & Management Science 291,
https://doi.org/10.1007/978-3-030-44814-1_8

expected worst-case value of cost or customer service level. The models presented in this chapter provide such a mean-risk fairness in decision-making to equitably optimize average and worst-case performance of a customer-driven supply chain.

The following time-indexed SMIP models are presented in this chapter:

RSPS_ECV(c) for selection of robust supply portfolio and scheduling of customer orders to equitably optimize expected cost and CVaR of cost;

RSPS_ECV(sl) for selection of robust supply portfolio and scheduling of customer orders to equitably optimize expected service level and CVaR of service level;

SPS_E(c) for risk-neutral selection of supply portfolio and scheduling of customer orders to minimize expected cost;

SPS_E(sl) for risk-neutral selection of supply portfolio and scheduling of customer orders to maximize expected service level;

SPS_CV(c) for risk-averse selection of supply portfolio and scheduling of customer orders to minimize CVaR of cost;

SPS_CV(sl) for risk-averse selection of supply portfolio and scheduling of customer orders to maximize CVaR of service level;

SPS_E(c,α), enhanced model **SPS_E(c)** for computing the maximum CVaR of cost, for a given confidence level α;

SPS_E(sl,α), enhanced model **SPS_E(sl)** for computing the minimum CVaR of service level, for a given confidence level α;

SPS_ECV(c) for mean-risk minimization of expected cost and CVaR of cost;

SPS_ECV(sl) for mean-risk maximization of expected service level and CVaR of service level.

Numerical examples and some computational results, in particular comparison of the mean-risk with the fair mean-risk approach, are provided in Sect. 8.4. The major managerial insights are provided at the end of this chapter.

8.2 Problem Description

In a supply chain under consideration various types of products are assembled over a planning horizon by a single producer to meet customer demand, using the same critical part type that can be manufactured and provided by different suppliers. For detailed problem description see Sect. 7.3, and the notation used is introduced in Table 8.1.

The objective of the equitably efficient optimization of average and worst-case performance of a supply chain under disruption risks is to allocate the total demand

Table 8.1 Notation: robust supply portfolio selection and production scheduling

Indices	
i	= supplier, $i \in I$
j	= customer order, $j \in J$
r	= region, $r \in R$
s	= disruption scenario, $s \in S$
t	= planning period, $t \in T$
Input parameters	
a_j	= per unit requirement for parts of each product in customer order j
b_j	= size (number of products) of customer order j
A	= $\sum_{j \in J} a_j b_j$, total demand for parts
B	= $\sum_{j \in J} b_j$, total demand for products
c_j	= per unit capacity consumption of producer for customer order j
C_t	= capacity of producer in period t
d_j	= due date for customer order j
e_i	= fixed cost of ordering parts from supplier i
g_j	= per unit penalty cost of unfulfilled customer order j
h_j	= per unit and per period penalty cost of delayed customer order j
I^r	= the subset of suppliers in region r
I_s	= the subset of suppliers non-disrupted under disruption scenario s
o_i	= per unit price of parts purchased from supplier i
p_i	= local disruption probability for supplier i
p^r	= regional disruption probability for all suppliers in region r
τ_i	= delivery lead time from supplier i

for parts among a subset of selected suppliers and to schedule the customer orders for products over the planning horizon to equitably minimize expected cost and expected worst-case cost or equitably maximize expected service level and expected worst-case service level. The cost includes the cost of ordering and purchasing of parts plus penalty cost of delayed and unfulfilled customer orders due to the parts shortages, while the customer service level is a performance measure independent of any cost parameters, defined as the fraction of customer orders fulfilled on or before their due dates (i.e., order fulfillment rate). The equitable solution means an equitably efficient risk-neutral and risk-averse solution. In this chapter such an equitably efficient solution will be called a robust solution. The robust solution is capable of equitably optimizing the performance of a supply chain in the average-case as well as in the worst-case. The robust solution (the supply portfolio and the schedule of customer orders) aims at achieving the normalized expected and expected worst-case values of the selected objective function as much close to each other as possible. To this end, the ordered weighted averaging aggregation of the expected value and conditional value-at-risk of the selected objective function will be applied. The resulting robust supply portfolio (the allocation of total demand for parts among the selected suppliers) is determined ahead of time along with the equitably efficient schedule of customer orders for every potential disruption scenario.

8.3 Problem Formulation

In this section the two time-indexed SMIP models **RSPS_ECV(c)** and **RSPS_ECV(sl)** are proposed for selection of robust supply portfolio and customer order scheduling to equitably optimize average and worst-case performance of a supply chain in the presence of disruption risks. The models are based on the bi-objective optimization problem formulation (7.8). The objective of model **RSPS_ECV(c)** is to equitably minimize expected cost per product and expected worst-case cost per product and the objective of model **RSPS_ECV(sl)** is to equitably maximize expected service level and expected worst-case service level. The problem variables are introduced in Table 8.2.

Table 8.2 Problem variables: robust supply portfolio selection and production scheduling

	First stage variables
u_i	$=$ 1, if supplier i is selected; otherwise $u_i = 0$ (supplier selection)
v_i	$\in [0,1]$, the fraction of total demand for parts ordered from supplier i (supply portfolio)
	Auxiliary variables
λ_l	unrestricted auxiliary variable, $l = 1,2$
δ_{kl}	≥ 0, upside deviation of outcome value $f_k, k = 1,2$ from the value of $\lambda_l, l = 1,2$
	Second stage variables
x^s_{jt}	$=$ 1, if under disruption scenario s customer order j is scheduled for period t; otherwise $x^s_{jt} = 0$ (production scheduling)
	Auxiliary variables
VaR^c	cost-at-Risk, the targeted cost such that for a given confidence level α, for $100\alpha\%$ of the scenarios, the outcome is below VaR^c
VaR^{sl}	service-at-Risk, the targeted service level such that for a given confidence level α, for $100\alpha\%$ of the scenarios, the outcome is above VaR^{sl}
$\mathscr{C}_s$	≥ 0, the tail cost for scenario s, i.e., the amount by which costs in scenario s exceed VaR^c
$\mathscr{S}_s$	≥ 0, the tail service level for scenario s, i.e., the amount by which VaR^{sl} exceeds service level in scenario s
z_s	$=$ 1, if for scenario s, cost per product is not less than VaR^c; otherwise $z_s = 0$
ζ_s	$=$ 1, if for scenario s, service level is not less than VaR^{sl}; otherwise $\zeta_s = 0$

To control the risk of supply disruptions, the two percentile measures of risks will be applied: Value-at-Risk (VaR) and Conditional Value-at-Risk (CVaR). In the selection of supply portfolio and scheduling of customer orders under disruption risks, the decision maker controls the risk of high losses due to supply disruptions by choosing the confidence level α. The greater the confidence level α, the more risk aversive is the decision maker and the smaller percent of the highest cost (or the lowest service level) outcomes is focused on.

8.3.1 *Equitable Minimization of Average and Worst-Case Costs*

Let VaR^c be the targeted cost such that for a given confidence level α, for $100\alpha\%$ of the scenarios, the outcome is below VaR^c, and let CVaR^c be the expected cost in the worst $100(1-\alpha)\%$ of the cases with the cost above VaR^c. Define $\mathscr{C}_s$ as the tail cost for scenario s, where tail cost is defined as the amount by which costs in scenario s exceed VaR^c. The risk-averse supply portfolio and the production schedule will be optimized by calculating VaR^c and minimizing CVaR^c simultaneously.

Let E^c be the minimized expected cost per product and $CVaR^c$, the minimized expected worst-case cost per product

$$E^c = (\sum_{i\in I} e_i u_i + \sum_{s\in S} P_s(\sum_{i\in I_s} Ao_i v_i$$

$$+\sum_{j\in J}\sum_{t\in T:t>d_j} h_j b_j (t-d_j) x^s_{jt} + \sum_{j\in J} g_j b_j (1-\sum_{t\in T} x^s_{jt})))/B \tag{8.1}$$

$$CVaR^c = VaR^c + (1-\alpha)^{-1}\sum_{s\in S} P_s\mathscr{C}_s. \tag{8.2}$$

In order to avoid dimensional inconsistency among various objectives, the values of the optimized objective functions are scaled into the interval [0,1]. Denote by $f_1^c = \frac{E^c-\underline{E}^c}{\overline{E}^c-\underline{E}^c}$, the normalized expected cost per product ($\underline{E}^c, \overline{E}^c$ are the minimum and the maximum values of E^c, respectively), and by $f_2^c = \frac{CVaR^c-\underline{CVaR}^c}{\overline{CVaR}^c-\underline{CVaR}^c}$, the normalized expected worst-case cost per product ($\underline{CVaR}^c, \overline{CVaR}^c$ are the minimum and the maximum values of $CVaR^c$ for a given confidence level α, respectively). The normalized cost objective functions f_1^c and f_2^c are defined below

$$f_1^c = (\sum_{i\in I} e_i u_i + \sum_{s\in S} P_s(\sum_{i\in I_s} Ao_i v_i + \sum_{j\in J}\sum_{t\in T:t>d_j} h_j b_j (t-d_j) x^s_{jt}$$

$$+\sum_{j\in J} g_j b_j (1-\sum_{t\in T} x^s_{jt})))/B - \underline{E}^c)/(\overline{E}^c - \underline{E}^c) \tag{8.3}$$

$$f_2^c = (VaR^c + (1-\alpha)^{-1}\sum_{s\in S} P_s\mathscr{C}_s - \underline{CVaR}^c)/(\overline{CVaR}^c - \underline{CVaR}^c). \tag{8.4}$$

The SMIP model **RSPS_ECV(c)** for selection of robust supply portfolio and customer order scheduling to equitably minimize expected and expected worst-case cost per product is formulated below.

RSPS_ECV(c): *Selection of* **R***obust* **S***upply* **P***ortfolio and* **S***cheduling of customer orders to equitably optimize* **E***xpected cost and* **CV***aR of* **c***ost*

Minimize

$$\sum_{l=1}^{2}(l\lambda_l + \sum_{k=1}^{2}\delta_{kl}) \tag{8.5}$$

(continued)

subject to (8.3), (8.4) and

$$\lambda_l + \delta_{kl} \geq f_k^c; k,l = 1,2. \tag{8.6}$$

Demand allocation constraints:

- the total demand for parts must be fully allocated among the suppliers (8.7),
- demand for parts cannot be assigned to non-selected suppliers (8.8),

$$\sum_{i \in I} v_i = 1 \tag{8.7}$$

$$v_i \leq u_i; \; i \in I. \tag{8.8}$$

Order-to-period assignment constraints:

- for each disruption scenario s, each customer order j is either scheduled during the planning horizon ($\sum_{t \in T} x_{jt}^s = 1$), or unscheduled and rejected ($\sum_{t \in T} x_{jt}^s = 0$) (8.9),
- for each disruption scenario s and each planning period t, the cumulative demand for parts of all customer orders scheduled in periods 1 through t cannot exceed the cumulative deliveries of parts in periods 1 through $t-1$, from the non-disrupted suppliers $i \in I_s$ (8.10),
- for each disruption scenario s, the total requirement for parts of scheduled customer orders is not greater than the total supplies from the non-disrupted suppliers $i \in I_s$ (8.11),

$$\sum_{t \in T} x_{jt}^s \leq 1; \; j \in J, s \in S \tag{8.9}$$

$$\sum_{j \in J} \sum_{\tau \in T: \tau \leq t} a_j b_j x_{j\tau}^s \leq A \sum_{i \in I_s: \tau_i \leq t-1} v_i; \; t \in T, s \in S \tag{8.10}$$

$$\sum_{j \in J} \sum_{t \in T} a_j b_j x_{jt}^s \leq A \sum_{i \in I_s} v_i; \; s \in S. \tag{8.11}$$

Producer capacity constraints:

- for any period t and each disruption scenario s, the total demand on capacity of all customer orders scheduled in period t must not exceed the producer capacity available in this period,

$$\sum_{j \in J} b_j c_j x_{jt}^s \leq C_t; \; t \in T, s \in S. \tag{8.12}$$

(continued)

Risk constraints:

– the tail cost for scenario s is defined as the nonnegative amount by which cost in scenario s exceeds VaR^c,

$$\mathscr{C}_s \geq \sum_{i\in I} e_i u_i/B + \sum_{i\in I_s} Ao_i v_i/B$$
$$+\sum_{j\in J}\sum_{t\in T:t>d_j} h_j b_j(t-d_j)x^s_{jt}/B + \sum_{j\in J} g_j b_j(1-\sum_{t\in T} x^s_{jt})/B$$
$$-VaR^c;\ s\in S. \quad (8.13)$$

Non-negativity and integrality conditions:

$$\delta_{kl} \geq 0; k,l = 1,2 \quad (8.14)$$
$$u_i \in \{0,1\};\ i\in I \quad (8.15)$$
$$v_i \in [0,1];\ i\in I \quad (8.16)$$
$$x^s_{jt} \in \{0,1\};\ j\in J, t\in T, s\in S \quad (8.17)$$
$$\mathscr{C}_s \geq 0;\ s\in S. \quad (8.18)$$

8.3.2 *Equitable Maximization of Average and Worst-Case Service Level*

In the next model, VaR^{sl} is the targeted customer service level such that for a given confidence level α, for $100\alpha\%$ of the scenarios, the outcome is above VaR^{sl}, while CVaR^{sl} is the expected service level in the worst $100(1-\alpha)\%$ of the cases with the service level below VaR^{sl}. Let $\mathscr{S}_s$ be the tail service level for scenario s, where tail service level is defined as the amount by which VaR^{sl} exceeds service level in scenario s. The aim of the following SMIP model is to maximize average customer service level as well as to reduce the outcomes below the service level represented by VaR^{sl}, and by this to equitably optimize average and worst-case customer service level. The solutions equitably maximize the expected and the expected worst-case fraction of customer orders filled on or before their due dates.

Let E^{sl} be the maximized expected service level and $CVaR^{sl}$, the maximized expected worst-case service level

$$E^{sl} = \sum_{j\in J}\sum_{t\in T:t\leq d_j}\sum_{s\in S} P_s x^s_{jt}/\bar{J} \quad (8.19)$$

$$CVaR^{sl} = VaR^{sl} - (1-\alpha)^{-1}\sum_{s\in S} P_s \mathscr{S}_s. \quad (8.20)$$

Denote by $f_1^{sl} = \frac{\overline{E}^{sl}-E^{sl}}{\overline{E}^{sl}-\underline{E}^{sl}}$, the normalized expected service level ($\underline{E}^{sl}, \overline{E}^{sl}$ are the minimum and the maximum values of E^{sl}, respectively), and by $f_2^{sl} = \frac{\overline{CVaR}^{sl}-CVaR^{sl}}{\overline{CVaR}^{sl}-\underline{CVaR}^{sl}}$, the normalized expected worst-case service level ($\underline{CVaR}^{sl}, \overline{CVaR}^{sl}$ are the minimum and the maximum values of $CVaR^{sl}$ for a given confidence level α, respectively).

The normalized service level objective functions f_1^{sl} and f_2^{sl} are defined below

$$f_1^{sl} = \frac{\overline{E}^{sl} - \sum_{j\in J}\sum_{t\in T:t\le d_j}\sum_{s\in S} P_s x_{jt}^s/\overline{J}}{(\overline{E}^{sl} - \underline{E}^{sl})} \tag{8.21}$$

$$f_2^{sl} = \frac{\overline{CVaR}^{sl} - (VaR^{sl} - (1-\alpha)^{-1}\sum_{s\in S} P_s \mathscr{S}_s)}{\overline{CVaR}^{sl} - \underline{CVaR}^{sl}}. \tag{8.22}$$

The SMIP model **RSPS_ECV(sl)** for selection of robust supply portfolio and customer order scheduling to equitably maximize expected and expected worst-case service level is formulated below.

RSPS_ECV(sl): *Selection of* **R***obust* **S***upply* **P***ortfolio and* **S***cheduling of customer orders to equitably optimize* **E***xpected service level and* **CV***aR of* **s***ervice* **l***evel*

Minimize (8.5)

subject to (8.6)–(8.12), (8.14)–(8.17), (8.21), (8.22) and

Risk constraints:

– the tail service level for scenario s is defined as the nonnegative amount by which VaR^{sl} exceeds service level in scenario s,

$$\mathscr{S}_s \ge VaR^{sl} - \sum_{j\in J}\sum_{t\in T:t\le d_j} x_{jt}^s/\overline{J};\ s\in S \tag{8.23}$$

$$\mathscr{S}_s \ge 0. \tag{8.24}$$

Note that the objective functions (8.19) and (8.20) (or normalized objective functions (8.21) and (8.22)) do not directly account for any cost components. As a result, the optimal solution to **RSPS_ECV(sl)** that aims at equitable optimization of average and worst-case service level depends mainly on the distribution of disruption probabilities among the suppliers. However, the optimization of service level implicitly reduces the penalty costs of delayed and unfulfilled customer orders, that represent an important part of the total cost structure.

8.3.3 Minimum and Maximum Values of the Objective Functions

In this subsection the minimum and maximum values for all objective functions are calculated to determine the normalized values of all objective functions, f_1^c (8.3), f_2^c (8.4), f_1^{sl} (8.21), and f_2^{sl} (8.22), that is, the values of the optimized objective functions scaled into the interval [0,1]. Note that the cost and the service level objectives are in conflict. Similarly, the expected and the expected worst-case values of a given objective function are also in conflict.

8.3.3.1 Minimum and Maximum Values of Expected Values

The minimum and maximum values of expected cost $\underline{E}^c$, $\overline{E}^c$, and expected service level, $\underline{E}^{sl}$, $\overline{E}^{sl}$, are obtained by solving the following stochastic mixed integer programs:

> **SPS_E(c)**: *Risk-neutral selection of* **S**upply **P**ortfolio *and* **S**cheduling *of customer orders to minimize* **E**xpected **c**ost
> Minimize E^c, (8.1),
> subject to (8.7)–(8.12), (8.15)–(8.17).

> **SPS_E(sl)**: *Risk-neutral selection of* **S**upply **P**ortfolio *and* **S**cheduling *of customer orders to maximize* **E**xpected **s**ervice **l**evel
> Maximize E^{sl}, (8.19)
> subject to (8.7)–(8.12), (8.15)–(8.17).

In problem **SPS_E(c)**, E^c is the minimized objective function, while E^{sl} is not considered. In problem **SPS_E(sl)**, E^{sl} is the maximized objective function, while E^c is not considered. Thus, by solving problem **SPS_E(c)**, the minimum value $\underline{E}^c$ of E^c and the minimum value $\underline{E}^{sl}$ of E^{sl} are determined. Similarly, by solving problem **SPS_E(sl)**, the maximum value $\overline{E}^{sl}$ of E^{sl} and the maximum value $\overline{E}^c$ of E^c are determined.

8.3.3.2 Minimum and Maximum Values of CVaR

The minimum value of expected worst-case cost, $\underline{CVaR}^c$, and the maximum value of expected worst-case service level, $\overline{CVaR}^{sl}$, for a given confidence level α, are obtained as the optimal solution of problem **SPS_CV(c)** and **SPS_CV(sl)**, respectively. The SMIP models **SPS_CV(c)** and **SPS_CV(sl)** are shown below.

> **SPS_CV(c)**: *Risk-averse selection of* **S**upply **P**ortfolio *and* **S**cheduling *of customer orders to minimize* **CV***aR of* **c**ost
> Minimize (8.2)
> subject to (8.7)–(8.13), (8.15)–(8.18).

SPS_CV(sl): *Risk-averse selection of* **S**upply **P**ortfolio *and* **S**cheduling *of customer orders to maximize* **CV***aR of* service **l**evel
Maximize (8.20)
subject to (8.7)–(8.12), (8.15)–(8.17), (8.23), (8.24).

Since the expected and the expected worst-case values of a given objective function are in conflict, the maximum value of expected worst-case cost $\overline{CVaR}^c$, and the minimum value of expected worst-case service level, $\underline{CVaR}^{sl}$ for a given confidence level α, are associated with the optimal solutions of SMIP problem **SPS_E(c)** and **SPS_E(sl)**, respectively. In order to compute the associated values of CVaR, the SMIP models **SPS_E(c)** and **SPS_E(sl)**, will be enhanced using the results of the following two propositions.

Proposition 8.1. *Given confidence level, α, the scenarios $s \in S$ with cost per product not less than VaR^c can be determined using the following set of constraints:*

$$z_s \geq \Big(\sum_{i\in I} e_i u_i/B + \sum_{i\in I_s} Ao_i v_i/B + \sum_{j\in J}\sum_{t\in T:t>d_j} h_j b_j(t-d_j)x^s_{jt}/B + \sum_{j\in J} g_j b_j\big(1-\sum_{t\in T} x^s_{jt}\big)/B - VaR^c\Big)/C_{\max};\ s\in S \quad (8.25)$$

$$z_s \leq 1 + \Big(\sum_{i\in I} e_i u_i/B + \sum_{i\in I_s} Ao_i v_i/B + \sum_{j\in J}\sum_{t\in T:t>d_j} h_j b_j(t-d_j)x^s_{jt}/B + \sum_{j\in J} g_j b_j\big(1-\sum_{t\in T} x^s_{jt}\big)/B - VaR^c\Big)/C_{\max};\ s\in S \quad (8.26)$$

$$\sum_{s\in S} P_s z_s \leq 1-\alpha \quad (8.27)$$

$$z_s \in \{0,1\};\ s\in S. \quad (8.28)$$

Proof. Eqs. (8.25) and (8.26) determine the scenarios $s \in S$ with cost per product not less than VaR^c, and Eq. (8.27) ensures that the total probability of all such scenarios is not greater than $1-\alpha$.
The additional binary variable, z_s, is defined as follows:
$z_s = 1$, if for scenario s, cost per product,

$$\sum_{i\in I} e_i u_i/B + \sum_{i\in I_s} Ao_i v_i/B + \sum_{j\in J}\sum_{t\in T:t>d_j} h_j b_j(t-d_j)x^s_{jt}/B + \sum_{j\in J} g_j b_j\big(1-\sum_{t\in T} x^s_{jt}\big)/B$$

(continued)

is not less than VaR^c; otherwise $z_s = 0$.
$C_{\max}$ is an upper bound on cost per product.

Proposition 8.2. *Given confidence level, α, the scenarios $s \in S$ with customer service level not less than VaR^{sl} can be determined using the following set of constraints:*

$$\zeta_s \geq \sum_{j \in J} \sum_{t \in T: t \leq d_j} x^s_{jt}/\overline{J} - VaR^{sl};\ s \in S \tag{8.29}$$

$$\zeta_s \leq 1 + \sum_{j \in J} \sum_{t \in T: t \leq d_j} x^s_{jt}/\overline{J} - VaR^{sl};\ s \in S \tag{8.30}$$

$$\sum_{s \in S} P_s \zeta_s \geq \alpha \tag{8.31}$$

$$\zeta_s \in \{0, 1\};\ s \in S. \tag{8.32}$$

Proof. Eqs. (8.29) and (8.30) determine the scenarios $s \in S$ with customer service level not less than VaR^{sl}, and Eq. (8.31) ensures that the total probability of all such scenarios is not less than the confidence level α.
The additional binary variable is defined as follows:
$\zeta_s = 1$, if for scenario s, customer service level, $\sum_{j \in J} \sum_{t \in T: t \leq d_j} x^s_{jt}/\overline{J}$, is not less than VaR^{sl}; otherwise $\zeta_s = 0$.

The enhanced SMIP models **SPS_E(c,α)** and **SPS_E(sl,α)** are presented below.

SPS_E(c,α)

Minimize (8.1),
subject to (8.2), (8.7)–(8.13), (8.15)–(8.18), (8.25)–(8.28).

SPS_E(sl,α)

Maximize (8.19)
subject to (8.7)–(8.12), (8.15)–(8.17), (8.20), (8.23), (8.24), (8.29)–(8.32).

For a given confidence level α, the maximum value of expected worst-case cost, $\overline{CVaR}^c$, is obtained by solving problem **SPS_E(c,α)**, and the minimum value of expected worst-case service level, $\underline{CVaR}^{sl}$ by solving problem **SPS_E(sl,α)**.

8.4 Computational Examples

In this section some computational examples are presented to illustrate possible applications of the proposed SMIP approach for the robust supplier selection, order quantity allocation, and customer orders scheduling in a supply chain under disruption risks. The basic input parameters for computational examples are presented in Table 8.3 (cf. Sect. 7.5).

Table 8.3 Input parameters: robust supply portfolio selection and production scheduling

$\overline{I} = 9$ suppliers, $\overline{J} = 25$ customer orders, $\overline{R} = 3$ three geographic regions, $\overline{T} = 10$ planning periods, $\overline{S} = 2^{\overline{I}} = 512$ disruption scenarios	
$I^1 = \{1,2,3\}$	three suppliers in region $r = 1$
$I^2 = \{4,5,6\}$	three suppliers in region $r = 2$
$I^3 = \{7,8,9\}$	three suppliers in region $r = 3$
$a = (2,1,3,3,1,3,2,1,2,2,2,2,3,2,1,3,2,1,3,3,2,1,1,2,1)$	per unit requirement for parts
$b = (1,2,9,7,8,5,1,7,5,4,7,4,10,6,8,1,4,2,4,8,6,3,8,7,3) \times 500$	customer order size
$A = 132500$	total demand for parts
$B = 66000$	total demand for products
$c = (2,1,1,2,3,3,1,3,2,1,2,1,3,1,1,3,2,3,1,1,3,2,2,1,2)$	per unit capacity consumption
$C_t = C = 38000, \forall t = 1,\dots,10$	producer capacity in period t
Local disruption probabilities of suppliers:	
p = (0.00513571, 0.00666354, 0.00902974, 0.0356206, 0.040175, 0.0294692, 0.0519967, 0.0827215, 0.0739062)	
Regional disruption probability:	
$p^1 = 0.001$, $p^2 = 0.005$, $p^3 = 0.01$	
Cost parameters:	
$e = (8,7,10,13,14,11,19,17,19) \times 1000$, $o = (13,12,12,8,6,6,2,5,4)$	
$g = (52,26,78,78,26,78,52,26,52,52,52,52,78,52,26,78,52,26,78,78,52,26,26,52,26)$ unit penalty for unfulfilled demand	
$h_j = 1 \forall j = 1,\dots,25$ unit penalty for delayed demand	
Time parameters:	
$d = (7,4,10,10,3,8,7,5,5,6,8,7,10,5,3,9,6,4,9,9,7,4,5,7,4)$ customer due-dates	
τ_i = 2, 3 and 4 time periods, respectively for suppliers $i \in I^1$, $i \in I^2$ and $i \in I^3$ delivery times	
Confidence level:	
$\alpha \in \{0.50, 0.75, 0.90, 0.95, 0.99\}$	

The computational experiments were performed for the same replication of the above input data set. All potential disruption scenarios were considered and to calculate the corresponding disruption probabilities, formulae (7.10) and (7.11) were applied. The unit price per part o_i and the disruption probability π_i, (7.9), of each supplier $i \in I$ are shown in Fig. 6.2.

The solution results for the risk-neutral, risk-averse, and robust decision-making are presented in Tables 8.4, 8.5 and 8.6, respectively. The minimum and maximum values of all considered objective functions are presented in Table 8.7. In addition to the optimal absolute and normalized solution values for the primary objective functions and the allocation of demand among the selected suppliers, Table 8.6 presents also the expected values of the associated objective function, i.e., the expected service level for model **RSPS_ECV(c)** and the expected cost per product for model **RSPS_ECV(sl)**.

Table 8.4 Risk-neutral solutions

Model **SPS_E(c)**: Var.=100468, Bin.=100459, Cons.=21341, Nonz.=765902[a]	
Expected Cost	7.66
Suppliers Selected(% of total demand)	7(100)
Model **SPS_E(sl)**: Var.=100468, Bin.=100459, Cons.=21341, Nonz.=765902[a]	
Expected Service Level(%)	99.62
Suppliers Selected(% of total demand)	1(48), 2(31), 3(21)

[a] Var.=number of variables, Bin.=number of binary variables
Cons.=number of constraints, Nonz.= number of nonzero coefficients

Table 8.5 Risk-averse solutions

Confidence level α	0.50	0.75	0.90	0.95	0.99
Model **SPS_CV(c)**: Var.=100981, Bin.=100459, Cons.=21853, Nonz.=874288[a]					
CVaR^{c}	10.60	16.47	23.75	26.66	30.90
Suppliers Selected(% of total demand)	7(100)	7(100)	4(23)	2(30)	2(22)
			5(21)	4(16)	3(22)
			6(21)	5(19)	5(15)
			7(17)	6(19)	6(16)
			9(18)	7(16)	7(12)
					9(13)
Model **SPS_CV(sl)**: Var.=100981, Bin.=100459, Cons.=21853, Nonz.=822422[a]					
CVaR^{sl}(%)	99.25	98.49	96.23	92.46	85.73
Suppliers Selected(% of total demand)	1(48)	1(48)	1(48)	1(48)	1(21)
	2(31)	2(31)	2(31)	2(31)	2(21)
	3(21)	3(21)	3(21)	3(21)	3(14)
					4(7)
					5(7)
					6(8)
					7(8)
					8(7)
					9(7)

[a] Var.=number of variables, Bin.=number of binary variables
Cons.=number of constraints, Nonz.= number of nonzero coefficients

For the risk-neutral minimization of cost, Table 8.4 indicates that the cheapest supplier $i = 7$ is selected only, while for the risk-neutral maximization of service level, the total demand is allocated among the three most reliable and most expensive suppliers $i = 1,2,3$.

For the risk-averse minimization of cost, Table 8.5 indicates that for the lowest confidence levels $\alpha = 0.5, 0.75$, the cheapest supplier $i = 7$ is selected only, while for a higher α, more suppliers are selected, including the reliable but expensive suppliers: $i = 2$ for $\alpha = 0.95$ and $i = 2,3$ for $\alpha = 0.99$.

Table 8.6 Robust solutions

Confidence level α	0.50	0.75	0.90	0.95	0.99
Model **RSPS_ECV(c)**: Var.=100989, Bin.=100459, Cons.=21859, Nonz.=975283[a]					
E^c	7.66	7.66	10.67	11.75	11.04
CVaR^c	10.60	16.47	28.55	29.05	35.44
VaR^c	4.73	4.73	18.51	23.70	30.83
Normalized E^c	0	0	0.167	0.227	0.188
Normalized CVaR^c	0	0	0.167	0.093	0.210
Suppliers Selected(% of total demand)	7(100)	7(100)	6(26)	6(37)	6(54)
			7(50)	7(30)	7(46)
			9(24)	9(33)	
E^{sl} (%)	67.46	67.56	85.37	84.79	87.18
Model **RSPS_ECV(sl)**: Var.=100989, Bin.=100459, Cons.=21859, Nonz.=878445[a]					
E^{sl} (%)	99.62	99.62	99.62	99.62	98.91
CVaR^{sl}(%)	99.25	98.49	96.23	92.46	85.61
VaR^{sl}(%)	100	100	100	100	92.00
Normalized E^{sl}	0	0	0	0	0.022
Normalized CVaR^{sl}	0	0	0	0	0.002
Suppliers Selected(% of total demand)	1(48)	1(48)	1(48)	1(48)	1(22)
	2(31)	2(31)	2(31)	2(31)	2(22)
	3(21)	3(21)	3(21)	3(21)	3(19)
					4(11)
					6(10)
					7(9)
					9(7)
E^c	25.64	25.64	25.64	25.64	21.84

[a] Var.=number of variables, Bin.=number of binary variables, Cons.=number of constraints, Nonz.= number of nonzero coefficients

Table 8.7 Minimum and maximum values of objective functions

Expected values	
$\underline{E}^c$	7.66
$\overline{E}^c$ (associated with Maximum Expected Service Level)	25.64
$\overline{E}^{sl}$(%)	99.62
$\underline{E}^{sl}$(%) (associated with Minimum Expected Cost)	67.60

Expected worst-case values, CVaR					
Confidence level α	0.50	0.75	0.90	0.95	0.99
$\underline{CVaR}^c$	10.60	16.47	23.75	26.66	30.91
$\overline{CVaR}^c$ (associated with Minimum Expected Cost)	52.48 for all confidence levels				
$\overline{CVaR}^{sl}$(%)	99.25	98.49	96.23	92.46	85.73
$\underline{CVaR}^{sl}$(%) (associated with Maximum Expected Service Level)	67.86	67.72	71.26	74.43	39.58

For the risk-averse maximization of service level, where the supplier selection is independent of any cost parameters and the solution mainly depends on the distribution of disruption probabilities, Table 8.5 demonstrates that the most reliable (and most expensive) suppliers $i = 1, 2, 3$ are selected for all confidence levels, except for the highest confidence level, $\alpha = 0.99$, for which all nine suppliers are selected.

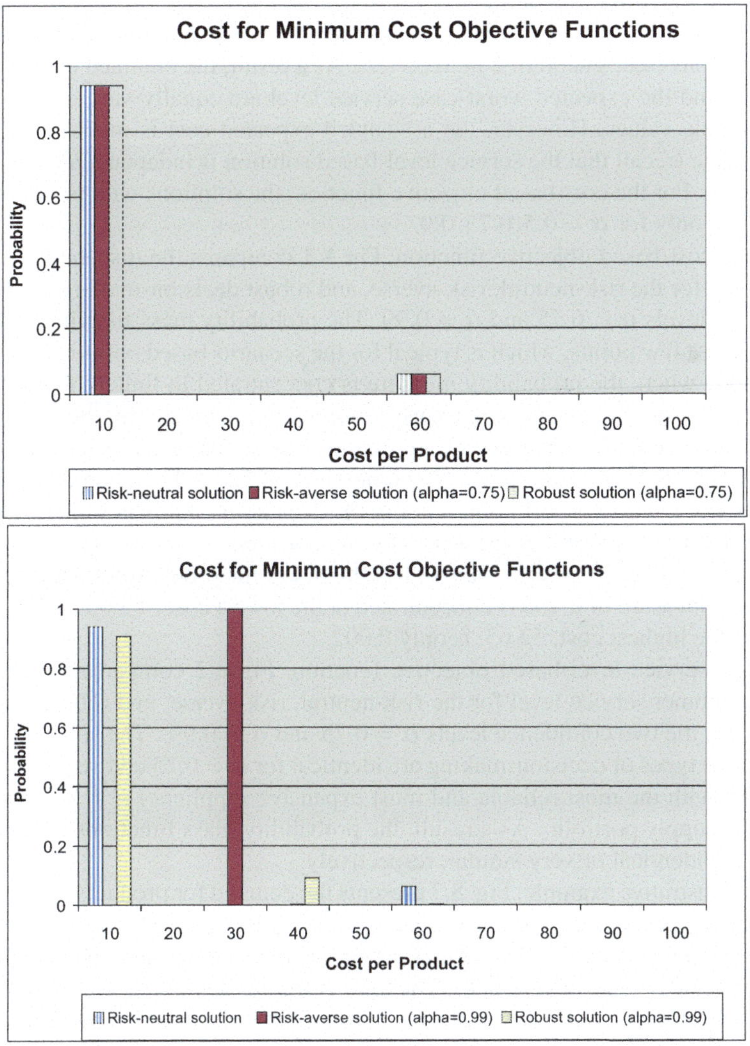

Fig. 8.1 Distribution of cost per product for α=0.75 and 0.99

Table 8.6 indicates that for minimization of cost and the low confidence levels, $\alpha = 0.5, 0.75$, the robust supply portfolio is identical with the risk-neutral portfolio. Similarly, for maximization of service level and the confidence levels, $\alpha = 0.5, 0.75, 0.9, 0.95$, the robust supply portfolio is identical with the risk-neutral portfolio, while for $\alpha = 0.99$, the robust supply portfolio is similar to the risk-averse portfolio. For the service level objective function, the robust solutions with a perfect equity were found for all confidence levels, except for the highest $\alpha = 0.99$, which indicates that the obtained robust solutions are also the lexicographic minimax optimal solutions (see, Theorem 2 in Sect. 7.2). As a result, the obtained expected service level and the expected worst-case service level are equally very close to their best available values. However, the associated expected cost is much higher than its best value (recall that the service level-based solution is independent of any cost parameters). For the cost-based objective function, the solutions with perfect equity were found only for $\alpha = 0.5, 0.75, 0.9$.

For the cost-based objective function, Fig. 8.1 compares the distribution of cost per product for the risk-neutral, risk-averse, and robust decision-making for the two confidence levels $\alpha = 0.75$ and $\alpha = 0.99$. The probability mass functions are concentrated in a few points, which is typical for the scenario-based optimization under uncertainty, where the probability measure is concentrated in finitely many points. For $\alpha = 0.75$, the risk-averse and robust solution are identical with the risk-neutral solution, and so are the corresponding probability mass functions. A large probability atom 0.06148 is concentrated at the highest cost of 52.49. For $\alpha = 0.99$, different solutions are obtained for different types of the decision-making. For the risk-averse solution, the total probability measure of costs between 24 and 29 is 0.9977, while the probability of costs greater than 40 is 0.0023. For the robust solution, the largest probability measure of 0.906 is concentrated at the lowest cost of 8.95 and the probability of the highest cost, 52.65, is only 0.002.

For the service level-based objective function, Fig. 8.2 compares the distribution of customer service level for the risk-neutral, risk-averse, and robust decision-making, for the two confidence levels $\alpha = 0.75$ and $\alpha = 0.99$. The solution results for the three types of decision-making are identical for $\alpha = 0.75$ and very similar for $\alpha = 0.99$, with the most reliable and most expensive suppliers $i = 1, 2, 3$ predominating the supply portfolio. As a result, the probability mass functions presented in Fig. 8.2 are identical or very similar, respectively.

As an illustrative example, Fig. 8.3 presents the demand for products, $\sum_{j \in J: d_j = t} b_j$, $t \in T$, and the expected production schedules, $\sum_{s \in S} P_s \sum_{j \in J} b_j x_{jt}^s, t \in T$ for the optimal cost- and service level-based, risk-neutral, risk-averse, and robust solutions with the confidence level $\alpha = 0.99$. The total customer demand is met with a small fraction of the expected rejected demand (from 0.0038 for the risk-neutral, service level-based solution up to 0.2407 for the risk-averse, cost-based solution). In general, the service level-based solution, when no cost components are included in the objective function, better meets the customer demand, with a smaller fraction of unfulfilled demand. For the service level objective, the expected production approximately follows the demand pattern. In addition, the expected production schedules for the risk-neutral and the robust decision-making are very close to each other,

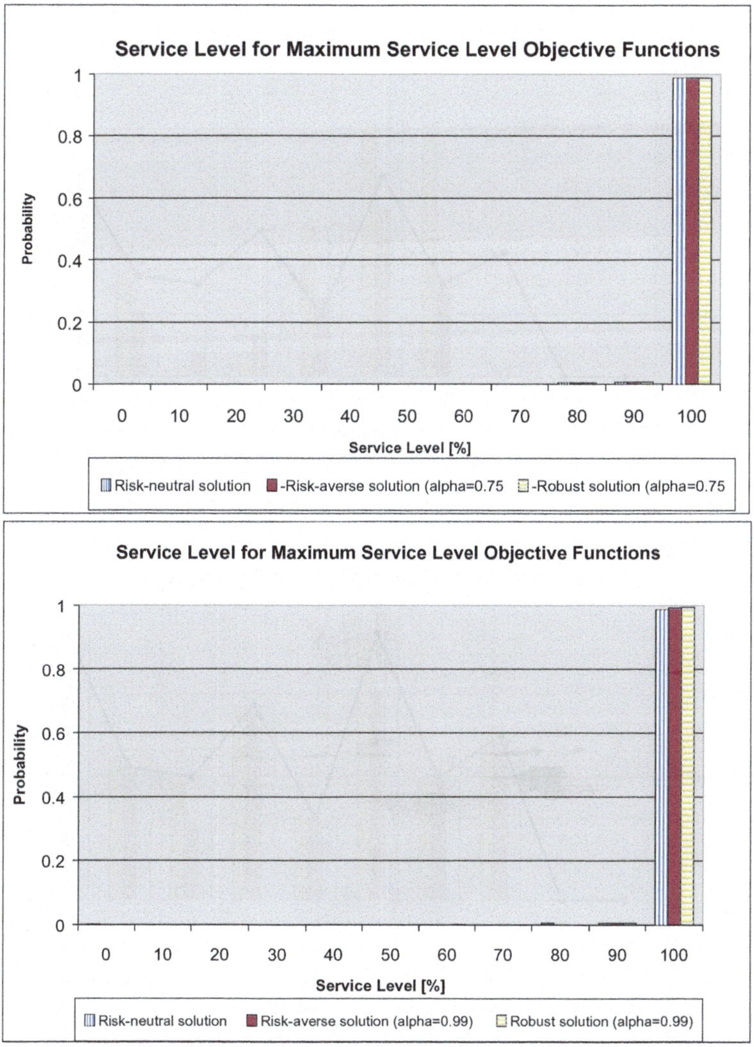

Fig. 8.2 Distribution of service level for $\alpha = 0.75$ and 0.99

which indicates that in the average-case, the robust solution is nearly as good as the risk-neutral solution.

Finally, Figs. 8.4 and 8.5 present the expected worst-case production schedules for the optimal risk-averse and the robust solutions with the confidence level $\alpha = 0.99$, for maximization of service level (Fig. 8.4),

$$\frac{\sum_{s \in S: \mathscr{S}_s > 0} (P_s \sum_{j \in J} b_j x^s_{jt})}{\sum_{s \in S: \mathscr{S}_s > 0} P_s}, t \in T,$$

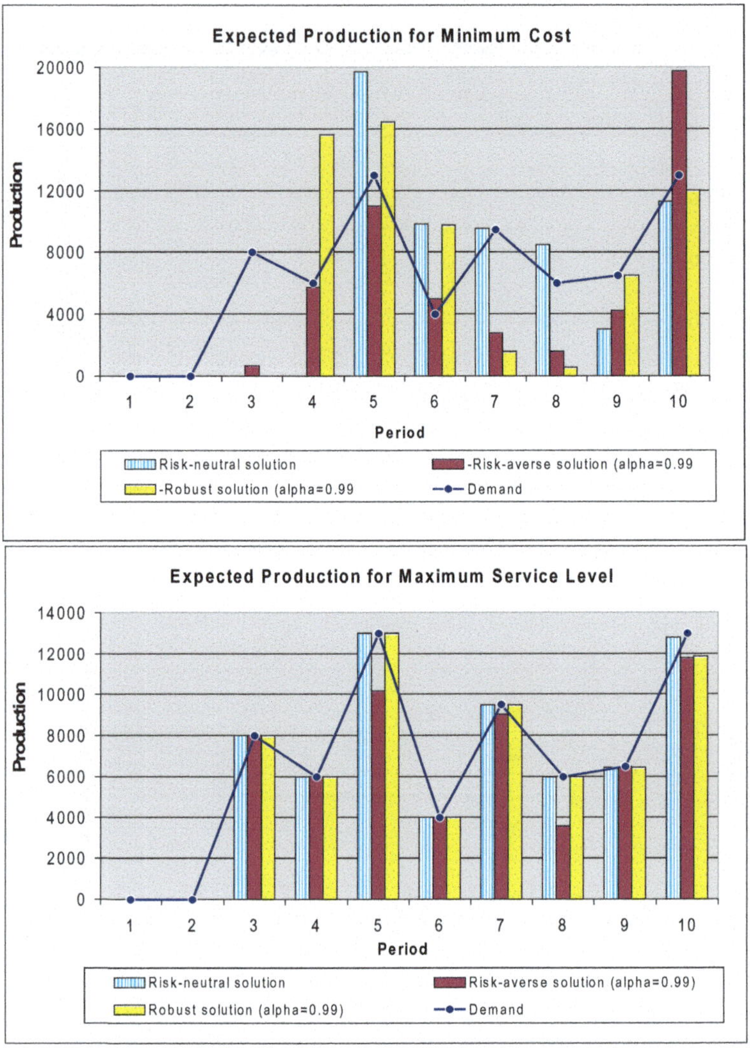

Fig. 8.3 Expected production schedules for $\alpha = 0.99$

and for minimization of cost (Fig. 8.5),

$$\frac{\sum_{s \in S:\mathscr{C}_s > 0}(P_s \sum_{j \in J} b_j x_{jt}^s)}{\sum_{s \in S:\mathscr{C}_s > 0} P_s}, t \in T,$$

respectively.

The risk-averse and the robust expected worst-case schedules for maximization of service level are very close to each other (cf. similar supply portfolios in Ta-

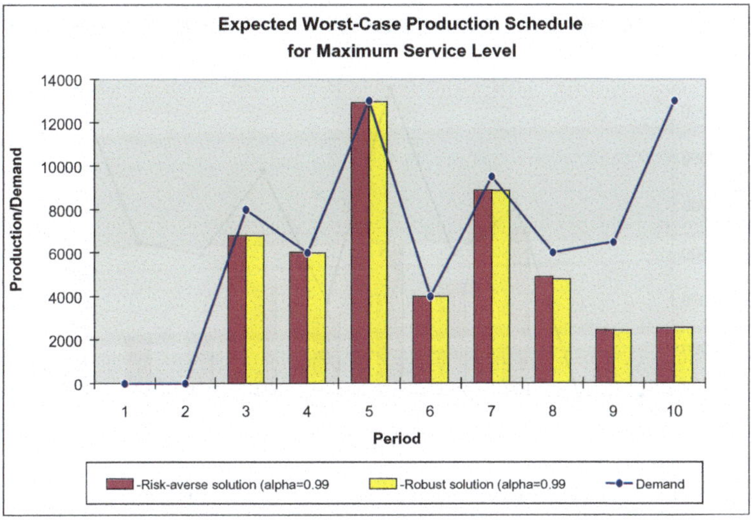

Fig. 8.4 Expected worst-case production schedules for maximum service level, $\alpha = 0.99$

bles 8.5 and 8.6), which indicates that in the worst-case the robust solution is nearly as good as the risk-averse solution. The expected worst-case fractions of fulfilled customer demand for products are similar and close to 75%.

For the minimization of cost, however, the expected worst-case schedules for the risk-averse and the robust solutions are different (cf. different supply portfolios in Tables 8.5 and 8.6). For the risk-averse solution, above 55% of the customer demand is fulfilled in the expected worst-case. However, the expected worst-case production for the robust solution is virtually negligible (see, Fig. 8.5a). The corresponding expected worst-case fraction of fulfilled customer demand for products is nearly equal to zero. Such a result may be due to the too low unit penalty cost, ($g_j = 2\lceil a_j \max_{i\in I}(o_i)\rceil$, $j \in J$), for unfulfilled customer demand, with respect to the purchasing cost of required parts, i.e., the unit penalty cost is approximately twice as large as the maximum unit price of required parts. The resulting robust supply portfolio is based on the low-cost, unreliable suppliers $i = 6,7$, similarly to the risk-neutral portfolio. In contrast, the pure risk-averse portfolio contains the most reliable suppliers $i = 2,3$.

In order to emphasize the impact of higher penalty costs, the computational experiments were repeated with a double unit penalty cost, $g_j = 4\lceil a_j \max_{i\in I}(o_i)\rceil$, $j \in J$, i.e., with the unit penalty cost for unfulfilled customer demand, approximately four times as large as the maximum unit price of required parts. The solution results for the minimization of cost with a double penalty are presented in Table 8.8. The table shows that for the low confidence levels, $\alpha = 0.5, 0.75$, the risk-averse and the robust supply portfolios are identical with the risk-neutral portfolio that assigns total demand for parts to a single, cheapest supplier $i = 7$. For a higher confidence level, the risk-averse portfolio contains a single, reliable supplier ($i = 2$, for $\alpha = 0.9$,

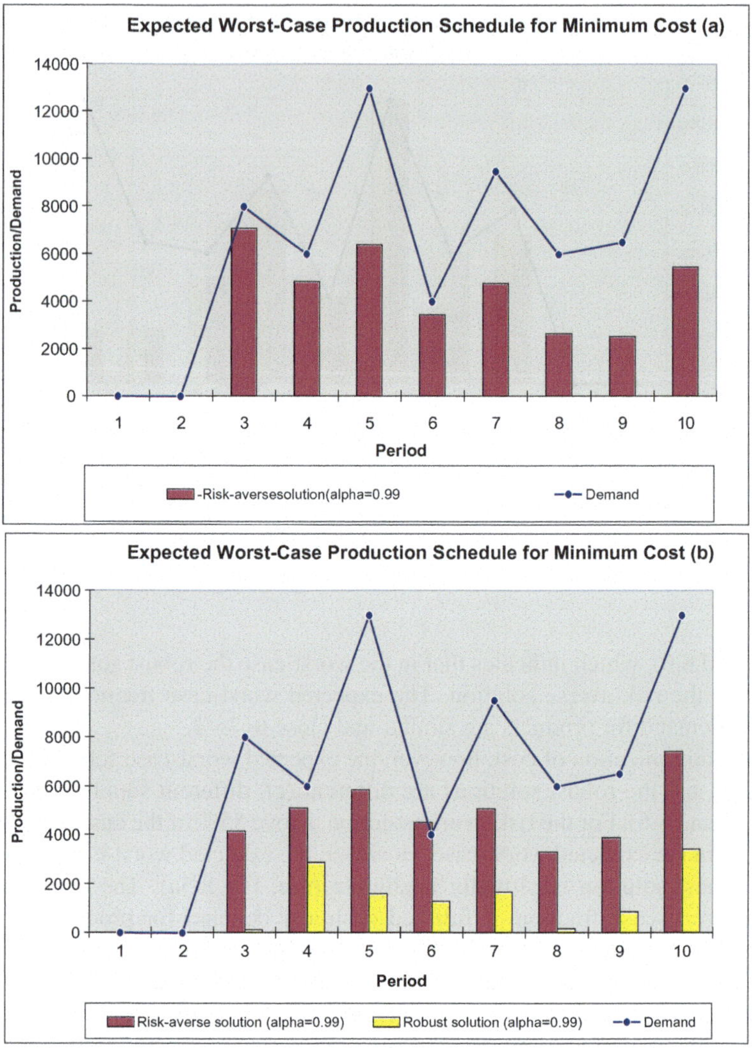

Fig. 8.5 Expected worst-case production schedules for minimum cost, $\alpha = 0.99$: (**a**) $g_j = 2\lceil a_j \max_{i \in I}(o_i) \rceil$, (**b**) $g_j = 4\lceil a_j \max_{i \in I}(o_i) \rceil$

and $i = 1$, for $\alpha = 0.95$), whereas the robust portfolio contains the reliable supplier, $i = 2$, only for the highest confidence level, $\alpha = 0.99$. This simultaneously leads to the highest expected service level, 94.03%. Table 8.8 shows that for most confidence levels, the robust solution with a perfect equity was obtained, which indicates that the obtained robust solutions are also the lexicographic minimax optimal solutions (see, Theorem 2 in Sect. 7.2).

Table 8.8 Solution results for a double unit penalty cost, $g_j = 4\lceil a_j \max_{i \in I}(o_i) \rceil$

Risk-neutral solution: model **SPS_E(c)**					
E^c	10.87				
Suppliers Selected(% of total demand)	7(100)				
Confidence level α	0.50	0.75	0.90	0.95	0.99
Risk-averse solutions: model **SPS_CV(c)**					
CVaR^c	17.02	24.99	30.35	35.82	46.97
Suppliers Selected(% of total demand)	7(100)	7(100)	2(100)	1(100)	1(19)
					2(19)
					3(20)
					4(6)
					5(7)
					6(9)
					7(9)
					8(5)
					9(6)
Minimum and maximum values of objective functions.					
$\underline{E}^c$	10.87				
$\overline{E}^c$	26.03				
$\underline{CVaR}^c$	17.02	24.99	30.35	35.82	46.97
$\overline{CVaR}^c$	104.68 for all confidence levels				
Robust solutions: model **RSPS_ECV(c)**					
E^c	10.87	10.87	14.35	14.41	14.92
CVaR^c	17.02	29.31	47.38	51.56	62.38
VaR^c	4.73	4.73	37.85	46.17	52.88
Normalized E^c	0	0	0.23	0.23	0.27
Normalized CVaR^c	0	0.05	0.23	0.23	0.27
Suppliers Selected(% of total demand)	7(100)	7(100)	5(30)	5(25)	2(13)
			6(31)	6(39)	6(45)
			7(39)	7(36)	7(42)
E^{sl} (%)	67.38	67.46	86.55	87.12	94.03

The expected worst-case schedules for a double unit penalty cost, $g_j = 4\lceil a_j \max_{i \in I}(o_i) \rceil$, $j \in J$, are shown in Fig. 8.5b. The obtained expected worst-case production for the robust solution is no longer negligible; however, the production is still smaller than that for a pure risk-averse solution. Now, the robust supply portfolio has been enforced with a reliable supplier $i = 2$, while the pure risk-averse portfolio contains all nine suppliers (see, Table 8.8). For the robust solution only 18% of the total customer demand is fulfilled in the expected worst-case, while for the risk-averse solution, over 60% of the customer demand is fulfilled. On the other hand, the expected percentage of fulfilled customer demand is over 95% for the robust solution and only 75% for the risk-averse solution. Thus, the risk-averse solution outperforms the robust solution in the worst-case, and vice versa the robust solution outperforms the risk-averse solution in the average-case. The above results demonstrate that the robust solution for the cost-based objective functions is strongly dependent on the cost parameters.

Table 8.9 Mean-risk solutions

Confidence level α	0.50	0.75	0.90	0.95	0.99
Model **SPS_ECV(c)** (8.33)					
E^c	7.67	7.67	12.74	12.74	11.04
CVaR^c	10.61	16.49	24.17	27.36	35.44
VaR^c	4.73	4.73	20.97	21.00	30.83
Normalized E^c	0.0005	0.0004	0.2827	0.2828	0.1880
Normalized CVaR^c	0.0002	0.0005	0.0146	0.0269	0.2100
Suppliers Selected(% of total demand)	7(100)	7(100)	5(27)	5(27)	6(54)
			6(27)	6(27)	7(46)
			7(22)	7(22)	
			9(24)	9(24)	
$E^{sl}(\%)$	67.01	67.80	84.32	85.15	87.18
Model **SPS_ECV(c)** (8.33) for a double unit penalty cost, $g_j = 4\lceil a_j \max_{i\in I}(o_i)\rceil$.					
E^c	10.87	10.87	15.53	14.73	13.37
CVaR^c	17.02	29.31	41.13	48.70	66.94
VaR^c	4.73	4.73	33.74	41.98	56.76
Normalized E^c	0	0	0.307	0.255	0.165
Normalized CVaR^c	0	0.05	0.145	0.187	0.346
Suppliers Selected(% of total demand)	7(100)	7(100)	5(25)	5(34)	
			6(26)	6(34)	6(52)
			7(24)	7(32)	7(48)
			9(25)		
$E^{sl}(\%)$	67.05	67.50	85.06	86.59	87.13
Model **SPS_ECV(sl)** (8.34)					
$E^{sl}(\%)$	99.62	99.62	99.62	99.62	99.07
CVaR$^{sl}(\%)$	99.25	98.49	96.20	92.40	84.92
VaR$^{sl}(\%)$	100	100	100	100	92.00
Normalized E^{sl}	0	0	0	0	0.0172
Normalized CVaR^{sl}	0	0	0.001	0.003	0.0176
Suppliers Selected(% of total demand)	1(48)	1(48)	1(48)	1(48)	1(21)
	2(31)	2(31)	2(31)	2(31)	2(21)
	3(21)	3(21)	3(21)	3(21)	3(18)
					4(10)
					5(10)
					6(10)
					7(10)
E^c	25.64	25.64	26.03	25.64	21.44

The equitably efficient solutions obtained using models **RSPS_ECV(c)** and **RSPS_ECV(sl)** have been compared with the nondominated solutions minimizing the weighted-sum of the normalized objective functions, respectively, f_1^c, f_2^c and f_1^{sl}, f_2^{sl}, with equal weights of each objective. In the literature this type of trade-off model (with a varying trade-off parameter) is known as the mean-risk model (e.g., Ogryczak and Ruszczynski 2002) with weighted-sum objective consisting of the expected value and the CVaR as a risk measure. The weighted-sum programs (with equal weights of each objective function), **SPS_ECV(c)** (8.33), and **SPS_ECV(sl)** (8.34), are presented below and the solution results are shown in Table 8.9.

$$\textbf{SPS\_ECV}(\mathbf{c}) \tag{8.33}$$

Minimize $f_1^c + f_2^c$
subject to (8.3), (8.4), (8.7)–(8.13), (8.15)–(8.18).

$$\textbf{SPS\_ECV}(\mathbf{sl}) \tag{8.34}$$

Maximize $f_1^{sl} + f_2^{sl}$
subject to (8.7)–(8.12), (8.15)–(8.17), (8.21)–(8.24).

Table 8.9 indicates that the mean-risk solutions are close to the robust solutions presented in Tables 8.6 and 8.8, while the perfect equity is very rarely achieved. In particular, for the service level objective function the robust and the mean-risk solutions are nearly identical for most confidence levels. However, for the cost objective function, the robust solutions based on the ordered weighted averaging aggregation approach, in most cases are different from the nondominated solutions obtained using the mean-risk approach.

The robust solution that equitably optimizes both the average and the worst-case performance of a supply chain has been compared with the risk-neutral solution that focuses on the average performance only, and with the risk-averse solution that focuses on the worst-case performance only. The major managerial insights are summarized in Table 8.10.

Overall, the computational experiments demonstrate that *the robust solution may be outperformed by an optimal risk-neutral solution in the average-case and by an optimal risk-averse solution in the worst-case. However, the robust solution outperforms the risk-averse solution in the average-case and the risk-neutral solution in the worst-case.*

The computational experiments were performed using the AMPL programming language and the CPLEX 12.5 solver on a MacBookPro laptop with Intel Core i7 processor running at 2.8 GHz and with 16 GB RAM. The solver was capable of finding proven optimal solutions for all examples with CPU time ranging from several seconds for the risk-neutral solutions to a few hours for the robust or mean-risk solutions and cost-based objectives.

Table 8.10 Managerial insights: robust supply portfolio selection and production scheduling

Decision-making issue	Decision-making impact
Robust vs. lexicographic minimax vs. Pareto-optimal	The robust solutions are frequently also the lexicographic minimax optimal solutions as well as the Pareto-optimal solutions. For both cost-based and service-based objective functions, the solutions with a perfect equity are frequently found (see, Theorems 2 and 3 in Sect. 7.2)
Robust decision-making vs. risk-neutral decision-making	In the average-case the robust solution may perform as well as the risk-neutral solution. For both cost-based and service-based objective functions, the robust solution for a low confidence level is identical with the risk-neutral solution, i.e., the supply portfolio is based on low-cost, unreliable suppliers and on high-cost, reliable suppliers, respectively. Moreover, the expected production schedules for the robust and for the risk-neutral solutions are very close to each other
Robust decision-making vs. risk-averse decision-making for the maximum service objective	In the worst-case, the robust solution for the maximum service level objective is nearly as good as the risk-averse solution. For the maximum service level objective and a high confidence level, the robust solution is very close to the risk-averse solution. Moreover, the expected worst-case schedules for the robust and for the risk-averse solutions are very close to each other
Robust decision-making for the minimum cost objective	For the minimum cost objective function and a high confidence level, the robust solution is strongly dependent on the cost parameters. If unit penalty cost for unfulfilled demand for products is comparable with purchasing cost of required parts, the robust solution for a high confidence level may more resemble the risk-neutral solution, while for a high unit penalty cost the robust solution may be closer to the risk-averse solution
Mean-risk vs. lexicographic minimax	The mean-risk solutions are rarely the lexicographic minimax optimal solutions. The mean-risk solutions that minimize the sum of the normalized expected value and the normalized CVaR are close to the robust solutions, while the perfect equity is very rarely achieved. For the service level objective function the robust and the mean-risk solutions are nearly identical for most confidence levels. However, for the cost objective function, the robust solution and the mean-risk solutions are different in most cases
Cost vs. service level.	The service level and the cost objective functions are in conflict. The best service level-based solutions may perform poorly with respect to the cost-based metrics, and vice versa.

8.5 Notes

In the literature on supply chain risk management, robustness implies that the supply chain is strong enough to be unaffected by disturbances, while retaining its original structure, e.g., Klibi et al. (2010); Gabrel et al. (2014). In contrast to resiliency (see, Chap. 4) that implies that the supply chain needs to adapt (reconfigure) its structure to survive and grow in the face of change and uncertainty (e.g., Fiksel 2006; Klibi and Martel 2012).

The major contribution of this chapter is that it proposes a simple approach for the robust decision-making associated with supplies of parts and deliveries of finished products in a customer-driven supply chain under disruption risks. Here, the robustness is defined as the mean-risk fairness and refers to an equitably efficient performance of a supply chain in the average-case as well as in the worst-case. The fair mean-risk decision-making aims at equalizing the distance to optimality both under business-as-usual and under worst-case conditions, which reflects the decision makers common requirement to maintain an equally good performance of a supply chain under varying operating conditions. The robust decision-making equitably focuses on the two objective functions: the expected value and the expected worst-case value (i.e., conditional value-at-risk) of the selected criterion, cost or service level.

The material presented in this chapter is based on results presented in Sawik (2014c, 2016c), where SMIP models were developed. The models were the enhancements of formulations proposed in Sawik (2013e; 2014a; 2014b, see also Chap. 5) for a single-objective decision-making. In contrast, the stochastic optimization problems considered in this chapter are formulated as bi-objective SMIP models with the two conflicting objective functions: expected cost and expected worst-case cost (conditional cost-at-risk) or expected service level and expected worst-case service level (conditional service-at-risk). In order to obtain an equitably efficient solution to the combinatorial stochastic optimization problem, the ordered weighted averaging aggregation (Yager 1988) of the expected and the expected worst-case value of the selected objective function has been applied. The equitable optimization of the supply chain network under disruption risks and the associated coordinated scheduling of the disrupted material flows are rarely considered. However, another type of a trade-off model is well known in the literature: the mean-risk model (e.g., Ogryczak and Ruszczynski 2002). The mean-risk model is formulated as the optimization of weighted-sum objective consisting of the expected value and the CVaR as a risk measure. The mean-risk approach aims at balancing the expected value with the risk tolerance, however, the mean-risk model is not designed to achieve an equitably efficient solution.

Problems

8.1. Modify the SMIP models presented in this chapter for multiple part types with subsets of part types required for each customer order and subsets of suppliers available for each part type.

8.2. How would you determine an upper bound, C_{max}, on the cost per product for model **SPS_E(c,**α**)**?

8.3. Mixed mean-risk decision-making

(a) Modify models **SPS_ECV(sl)** and **SPS_ECV(c)** to optimize expected cost and CVaR of service level and optimize expected service level and CVaR of cost, respectively.
(b) How should the values of the optimized objective functions be scaled into the interval [0,1] to avoid dimensional inconsistency among the two objectives and how should the trade-off parameter be selected?
(c) How would you interpret the mixed mean-risk solutions?

8.4. Explain why the robust supply portfolios for the cost-based objective is strongly dependent on the cost parameters and which parameters are the most influential?

8.5. Explain why the expected worst-case production for the robust solution in Fig. 8.5a is nearly negligible?

Part IV
Selection of Primary and Recovery Portfolios and Scheduling

Chapter 9
Selection of Primary and Recovery Supply Portfolios and Scheduling

9.1 Introduction

In this chapter the portfolio approach presented in previous chapters for the selection of primary suppliers and order quantity allocation to mitigate the impact of disruption risks is enhanced also for the recovery process, i.e., for the selection of both primary and recovery suppliers and order quantity allocation to mitigate the impact of disruption risks and optimize the recovery process. Unlike most of reported research on the supply chain risk management which focuses on the risk mitigation decisions taken prior to a disruption, this chapter combines decisions made before, during, and after the disruption. The two decision-making approaches will be considered: an integrated approach with the perfect information about the future disruption scenarios, and a hierarchical approach with no such information available ahead of time. In the integrated approach, which accounts for all potential disruption scenarios, the primary supply portfolio that will hedge against all scenarios is determined along with the recovery supply portfolio and production schedule of finished products for each scenario, to minimize expected cost or expected worst-case cost over all scenarios. In the hierarchical approach first the primary supply portfolio is selected, and then, when a primary supplier is hit by a disruption, the recovery supply portfolio is selected to optimize the process of recovery from the disruption.

The following time-indexed SMIP and MIP models are presented in this chapter:

Support_E for risk-neutral selection of primary and recovery supply portfolios and production scheduling;
SupportMP_E model **Support_E** for multiple part types and product types;

(continued)

T. Sawik, *Supply Chain Disruption Management*,
International Series in Operations Research & Management Science 291,
https://doi.org/10.1007/978-3-030-44814-1_9

PSupport for selection of primary supply portfolio and production scheduling under deterministic conditions;
RSupport(s) for selection of recovery supply portfolio and production scheduling, for predetermined primary supply portfolio and realized disruption scenario;
RSupport_E model **Support_E** for predetermined primary supply portfolio;
Support_CV(c) for risk-averse selection of primary and recovery supply portfolios and production scheduling to minimize CVaR of cost;
Support_CV(sl) for risk-averse selection of primary and recovery supply portfolios and production scheduling to maximize CVaR of service level.

Numerical examples and computational results are reported in Sects. 9.5.1 and 9.5.2, respectively, for risk-neutral and risk-averse decision-making, and the major managerial insights are provided at the end of this chapter.

In the next chapter, the portfolio approach will be further enhanced to simultaneously select supply and demand portfolios, when a disruption impacts both a primary supplier of parts and the manufacturer's primary assembly plant. Then, in addition to determining the primary and recovery supply portfolios, the manufacturer may also choose to move production to alternate (recovery) plants along with transshipment of parts from the impacted primary assembly plant to the recovery plants. The resulting allocation of unfulfilled demand for products among recovery assembly plants will determine a recovery, demand, or capacity portfolio.

9.2 Problem Description

Consider a supply chain in which a single producer of one product type assembles products to meet customer demand, using a critical part type that can be manufactured and provided by several suppliers.

Let $I = \{1,\ldots,\overline{I}\}$ be the set of $\overline{I}$ suppliers, $T = \{1,\ldots,\overline{T}\}$ the set of $\overline{T}$ planning periods, and let d_t be the demand for products in period $t \in T$ (for notation, see Table 9.1).

The orders for parts are assumed to be placed at the beginning of the planning horizon, and under normal conditions the parts ordered from supplier i are delivered in period τ_i, where τ_i represents total of manufacturing lead time and transportation time. If production does not meet the demand, the producer is charged with penalty cost for unfulfilled demand for products.

The suppliers of parts are located in $\overline{R}$ geographic regions, subject to potential regional disasters that may result in complete shutdown of all suppliers in the same region simultaneously. In addition to correlated regional disruptions, each supplier $i \in I$ is subject to random independent local disruptions of different levels, $l \in L_i =$

Table 9.1 Notation: primary and recovery supply portfolio selection and production scheduling

Indices	
i	= supplier, $i \in I$
l	= disruption level, $l \in L_i, i \in I$
r	= region, $r \in R$
s	= disruption scenario, $s \in S$
t	= planning period, $t \in T$
Input parameters	
c	= per period capacity of producer
C	= total available capacity of producer
d_t	= demand for products in period t
D	= total demand for parts/products
e_i	= fixed ordering cost of creating contracts and maintaining relationships with supplier i
g	= per unit penalty cost for unfulfilled demand for products
o_i	= per unit price of parts purchased from supplier i
p_{il}	= probability of disruption level l for supplier i
p^r	= regional disruption probability for region r
t_s	= start time period of disruptive event s
γ_{il}	= fraction of an order delivered by supplier i under disruption level l (supplier fulfillment rate)
γ_i^s	= fraction of an order delivered by supplier i under scenario s
τ_i	= delivery lead time from supplier i
ρ_{is}	= manufacturer's portion of supplier i cost-to-recover from disruption under scenario s
θ_{is}	= time-to-recover of supplier i from disruption under scenario s
$CTR(i,l)$	= manufacturer's portion of cost-to-recover for supplier i hit by disruption at level l
$TTR(i,l)$	= time-to-recover for supplier i hit by disruption at level l

$\{0,\ldots,\overline{L_i}\}$, where the disruption level refers to the fraction of an order that can be delivered (supplier fulfillment rate). Level $l = 0$ represents complete shutdown of a supplier, i.e., no order delivery, while level $l = \overline{L_i}$ represents normal conditions with no disruption, i.e., full order delivery. The fraction of an order that can be delivered by supplier i under disruption level l is described by the associated supplier fulfillment rate, γ_{il}

$$\gamma_{il} = \begin{cases} 0 & \text{if } l = 0 \\ \in (0,1) & \text{if } l = 1,\ldots,\overline{L_i}-1 \\ 1 & \text{if } l = \overline{L_i}. \end{cases} \tag{9.1}$$

Denote by $S = \{1,\ldots,\overline{S}\}$ the index set of all disruption scenarios, and by P_s the probability of disruption scenario $s \in S$. Each scenario $s \in S$ can be represented by an integer-valued vector $\lambda_s = \{\lambda_{1s},\ldots,\lambda_{\overline{I}s}\}$, where $\lambda_{is} \in L_i$ is the disruption level of an order delivery from supplier $i \in I$ under scenario $s \in S$. When all potential disruption scenarios are considered, then $\overline{S} = \prod_{i \in I}(\overline{L_i}+1)$.

For each scenario $s \in S$, the supplies from every supplier can be disrupted either by a local or a regional disaster event. Denote by $I_s \subset I$ the subset of non-shutdown suppliers, who can deliver parts under scenario s. The probability P_s for disruption scenario $s \in S$ with the subset I_s of non-shutdown suppliers is

$$P_s = \prod_{r \in R} P_s^r,$$

where P_s^r is the probability of realizing of disruption scenario s for suppliers in I^r (cf. Sect. 1.4)

$$P_s^r = \begin{cases} (1-p^r)\prod_{i \in I^r}(p_{i,\lambda_{is}}) & \text{if } I^r \bigcap I_s \neq \emptyset \\ p^r + (1-p^r)\prod_{i \in I^r} p_{i0} & \text{if } I^r \bigcap I_s = \emptyset. \end{cases}$$

$p_{i,\lambda_{is}}$ is the probability of occurrence the disruption at level $l = \lambda_{is}$, of an order delivery from supplier i under scenario s.

A disruptive event under scenario $s \in S$ is assumed to occur in period t_s. When supplier i is hit by disruption at level l, its recovery process to normal conditions starts in period $t_s + 1$ and takes $TTR(i,l)$ time periods (Time-To-Recover), so that the supplier i recovers to its full pre-disruption capacity in period $t = t_s + TTR(i,l) + 1$. The cost of recovery process can be shared between the supplier of parts and the manufacturer of products. Let $CTR(i,l)$ be the manufacturer's portion of Cost-To-Recover. Note that $CTR(i,\overline{L}) = TTR(i,\overline{L}) = 0$, i.e., no recovery is required for a non-disrupted supplier.

For each supplier i, denote by θ_{is} and ρ_{is}, respectively, time-to-recover and the manufacturer's portion of cost-to-recover from disruption under scenario s

$$\theta_{is} = TTR(i,l);\ i \in I, s \in S : l = \lambda_{is} \tag{9.2}$$

$$\rho_{is} = CTR(i,l);\ i \in I, s \in S : l = \lambda_{is}. \tag{9.3}$$

The following assumptions are made to formulate the problem:

- Each supplier has sufficient capacity to meet total demand for parts.
- A single recovery mode is considered for each supplier and each disruption level.
- Disruption start times are constant parameters.
- Time-to-recover and the associated cost-to-recover are constant parameters that represent recovery of a disrupted supplier to its full capacity.
- The manufacturer of finished products may participate in supplier's cost-to-recover.
- Disruption to primary supplier under scenario $s \in S$ occurs in period t_s and recovery process starts in period $t_s + 1$, so that the disrupted supplier $i \in I$ returns to its full capacity in period $t = t_s + \theta_{is} + 1$.
- Multiple disruptions, one after the other in a series, during the recovery process are not considered.
- A recovery supplier can be a disrupted primary supplier after its recovery to full capacity or a new supplier.
- Transition time required for switching to recovery supplier different from the primary suppliers is negligible.
- Each product requires one unit of a critical part type.

- A penalty cost is charged for the demand for products unfulfilled by the end of the planning horizon.

Many of the above assumptions can be easily relaxed. For example, the assumption that each supplier has sufficient capacity to meet total demand for parts, which allows for a single sourcing, is often met in practice. Some assumptions, however, would require development of more advanced models to be relaxed, e.g., random disruption start times, durations, and suppliers cost-to-recover.

Notice that no recovery supply portfolio needs to be selected if a disruptive event occurs after the latest delivery lead time from primary suppliers, i.e., when $t_s > \max_{i \in I}(\tau_i)$.

In the sequel two decision-making approaches will be considered: an integrated approach with the perfect information about the future disruption scenarios, and a hierarchical approach with no such information available ahead of time. In the integrated approach, which accounts for all potential disruption scenarios, decisions are made prior to a disruption. The primary supply portfolio that will hedge against all scenarios is determined along with the recovery supply portfolio and production schedule for each scenario. The problem objective is to minimize expected cost or maximize expected service level over all scenarios, for risk-neutral models, and minimize CVaR of cost or maximize CVaR of service level for risk-averse models. In the hierarchical approach, first prior to a disruption, the primary supply portfolio is determined to minimize cost or maximize service level, and then, when a primary supplier is hit by a disruption, the recovery supply portfolio is selected to optimize the process of recovery from the disruption.

9.3 Models for Risk-Neutral Decision-Making

9.3.1 Integrated Selection of Primary and Recovery Supply Portfolios

In this subsection a time-indexed SMIP model **Support_E** is presented for the integrated risk-neutral decision-making in the presence of supply chain disruption risks.

The objective of the integrated decision-making (see, Fig. 9.1) is to jointly

- determine the primary supply portfolio, i.e., to allocate the total demand for parts among a subset of selected primary suppliers,
- determine the recovery supply portfolio for each disruption scenario, i.e., to allocate the unfulfilled demand for parts among a subset of selected recovery suppliers, when primary suppliers are disrupted,
- schedule production over the planning horizon for each disruption scenario,

to optimize the expected cost. The cost includes ordering and purchasing cost of parts, recovery cost of disrupted suppliers, and penalty cost for unfulfilled demand for products. The variables used to formulate the SMIP model are defined in Table 9.2.

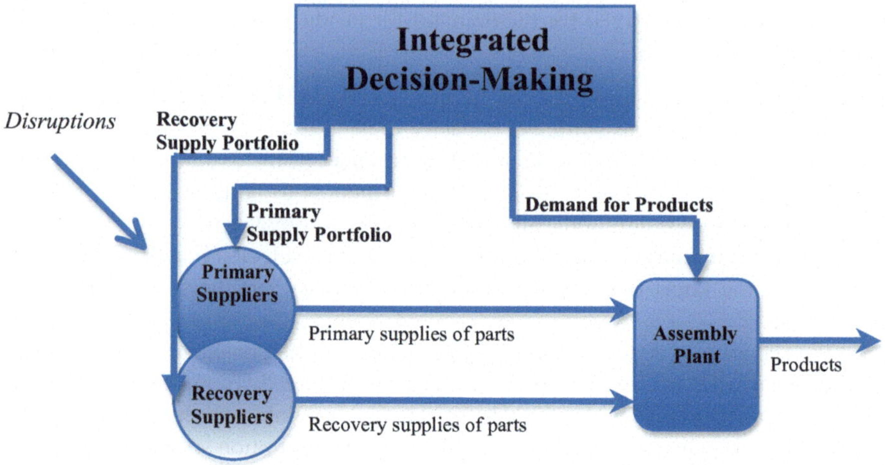

Fig. 9.1 Primary and recovery supply portfolio selection under disruption risks

Table 9.2 Problem variables: primary and recovery supply portfolio selection and production scheduling

	First stage variables
u_i	$=$ 1, if supplier i is selected as a primary supplier; otherwise $u_i = 0$ (primary supplier selection)
v_i	$\in [0,1]$, the fraction of total demand for parts ordered from primary supplier i (primary supply portfolio)
	Second stage variables
U_i^s	$=$ 1, if supplier i is selected as a recovery supplier under scenario s; otherwise $U_i^s = 0$ (recovery supplier selection)
V_i^s	$\in [0,1]$, the fraction of total demand for parts ordered from recovery supplier i under scenario s (recovery supply portfolio)
x_t^s	≥ 0, production in period t under scenario s (production scheduling)
	Auxiliary variables
VaR^c	cost-at-risk, the targeted cost such that for a given confidence level α, for $100\alpha\%$ of the scenarios, the outcome is below VaR^c
VaR^{sl}	service-at-risk, the targeted service level such that for a given confidence level α, for $100\alpha\%$ of the scenarios, the outcome is above VaR^{sl}
$\mathscr{C}_s$	≥ 0, the tail cost for scenario s, i.e., the amount by which costs in scenario s exceed VaR^c
$\mathscr{S}_s$	≥ 0, the tail service level for scenario s, i.e., the amount by which VaR^{sl} exceeds service level in scenario s
μ_i^s	$=$ 1, if $u_i = U_i^s = 1$; otherwise $\mu_i^s = 0$ (elimination of double fixed ordering costs)

The primary supply portfolio, $(v_1, \ldots, v_{\overline{I}})$, where $\sum_{i\in I} v_i = 1$ and $0 \leq v_i \leq 1, i \in I$, is the initial allocation of total demand for parts among primary suppliers.

The recovery supply portfolio for each disruption scenario s, $(V_1^s, \ldots, V_{\overline{I}}^s)$, where $\sum_{i\in I}(\gamma_i^s v_i + V_i^s) = 1$ and $0 \leq V_i^s \leq 1, i \in I$, is the allocation among recovery suppliers of unfulfilled demand for parts caused by supply disruptions under scenario s, and $\sum_{i\in I} \gamma_i^s v_i$, is total delivery of parts from primary suppliers under scenario s.

$\gamma_i^s = \gamma_{i,\lambda_{is}}$, denotes the fulfillment rate of supplier i under disruption scenario s, i.e., the fraction of an order delivered by supplier i under disruption at level $l = \lambda_{is}$.

Notice that the actual quantity ordered from each supplier i can be determined by multiplying total demand for parts ($D = \sum_{t \in T} d_t$), by fractional variable v_i or V_i^s, i.e., by the fraction of total demand allotted to supplier i.

In addition, the following auxiliary binary variable, μ_i^s, is introduced in model **Support_E**: $\mu_i^s = 1$, if $u_i = U_i^s = 1$; otherwise $\mu_i^s = 0$. This variable eliminates double charging with fixed ordering cost e_i of each supplier i, who is selected both as a primary and recovery supplier.

Let E^c be the expected cost per product to be minimized and E^{sl}, the expected service level to be maximized, i.e., the expected demand fulfillment rate (expected fraction of the total fulfilled demand for products)

$$E^c = \sum_{s \in S} P_s (\sum_{i \in I} e_i(u_i + U_i^s - \mu_i^s)/D$$
$$+ \sum_{i \in I} \rho_{is} U_i^s/D + \sum_{i \in I} o_i(\gamma_i^s v_i + V_i^s)) + g(1 - E^{sl}), \tag{9.4}$$

$$E^{sl} = \sum_{s \in S} \sum_{t \in T} P_s x_t^s / D. \tag{9.5}$$

The expected cost per product, E^c (9.4), includes
expected ordering cost per product,
$\sum_{s \in S} P_s \sum_{i \in I} e_i(u_i + U_i^s - \mu_i^s)/D = \sum_{i \in I} e_i u_i/D + \sum_{s \in S} P_s \sum_{i \in I} e_i(U_i^s - \mu_i^s)/D$,
manufacturer's expected portion of suppliers per product cost-to-recover from disruptions,
$\sum_{s \in S} P_s \sum_{i \in I} \rho_{is} U_i^s/D$,
expected purchasing cost per product, for delivered parts,
$\sum_{s \in S} P_s \sum_{i \in I} o_i(\gamma_i^s v_i + V_i^s) = \sum_{i \in I} o_i \Gamma_i v_i + \sum_{s \in S} P_s \sum_{i \in I} o_i V_i^s$,
where, $\Gamma_i = \sum_{s \in S} P_s \gamma_i^s$, is the expected fulfillment rate of supplier i,
and expected penalty cost per product for the unfulfilled demand for products,
$g(1 - \sum_{s \in S} \sum_{t \in T} P_s x_t^s/D) = g(1 - E^{sl})$.

Support_E: *Risk-neutral selection of primary and recovery* **Supp***ly* **port***folios and production scheduling*

Minimize (9.4)
subject to
Primary supply portfolio selection constraints

– the total demand for parts must be fully allocated among the selected primary suppliers (9.6),

(continued)

– demand for parts cannot be assigned to non-selected primary suppliers (9.7),

$$\sum_{i\in I} v_i = 1 \tag{9.6}$$

$$v_i \le u_i;\ i \in I \tag{9.7}$$

Recovery supply portfolio selection constraints

– the unfulfilled demand for parts caused by supply disruptions under scenario s must be fully allocated among the selected recovery suppliers (9.8),
– the unfulfilled demand for parts caused by supply disruptions under scenario s cannot be assigned to non-selected recovery suppliers (9.9),
– each supplier selected to both primary and recovery portfolio is charged exactly once, (9.10), with fixed ordering cost in the objective function (9.4),

$$\sum_{i\in I} (\gamma_i^s v_i + V_i^s) = 1;\ s \in S \tag{9.8}$$

$$V_i^s \le U_i^s;\ i \in I, s \in S \tag{9.9}$$

$$\mu_i^s \le (u_i + U_i^s)/2;\ i \in I, s \in S \tag{9.10}$$

Supply-production coordinating constraints

– for each disruption scenario s and each period t, the cumulative demand for parts of production scheduled in periods 1 through t cannot exceed the initial inventory of parts and the cumulative deliveries by period $t-1$ (delivery in period, $\tau_i \le t-1$, from each primary supplier i and in period, $t_s + \theta_{is} + \tau_i \le t-1$, from each recovery supplier i),

$$\sum_{t'\in T: t' \le t} x_{t'}^s / D \le$$
$$V_0 + \sum_{i\in I: \tau_i \le t-1} \gamma_i^s v_i + \sum_{i\in I: t_s+\theta_{is}+\tau_i \le t-1} V_i^s;\ t \in T, s \in S, \tag{9.11}$$

where DV_0 is the initial inventory of parts

Production capacity constraints

– for each disruption scenario s and each period t, production cannot exceed the producer capacity,

$$x_t^s \le c;\ t \in T, s \in S \tag{9.12}$$

Non-negativity and integrality conditions

$$u_i \in \{0, 1\};\ i \in I \tag{9.13}$$

$$v_i \in [0, 1];\ i \in I \tag{9.14}$$

(continued)

$$U_i^s \in \{0,1\};\ i \in I, s \in S \tag{9.15}$$

$$V_i^s \in [0,1];\ i \in I, s \in S \tag{9.16}$$

$$x_t^s \geq 0;\ t \in T, s \in S \tag{9.17}$$

$$\mu_i^s \in \{0,1\};\ i \in I, s \in S. \tag{9.18}$$

Best-case Versus Worst-Case Service Level

A simple upper bound on the expected service level (9.5) derived below is based on production capacity-to-demand ratio, C/D,

Proposition 9.1.

$$E^{sl} \leq \min\{1, C/D\}, \tag{9.19}$$

where $C = \min\{DV_0, c\tau_{\min}\} + c(\overline{T} - \tau_{\min})$, *is the total available capacity of producer,* $D = \sum_{t \in T} d_t$, *is the total demand for products, and* $\tau_{\min} = \min_{i \in I}(\tau_i)$ *is the minimum delivery lead time from suppliers.*

Proof. Supply-production coordinating constraints (9.11) and production capacity constraints (9.12) imply that

$$\sum_{t \in T: t \leq \tau_{\min}} x_t^s \leq \min\{DV_0, c\tau_{\min}\};\ s \in S$$

$$\sum_{t \in T: t > \tau_{\min}} x_t^s \leq c(\overline{T} - \tau_{\min});\ s \in S.$$

Thus

$$E^{sl} = \sum_{s \in S} P_s \sum_{t \in T} x_t^s / D \leq \sum_{s \in S} P_s C/D = C/D.$$

Since E^{sl} cannot be greater than 1, its upper bound is 1, if capacity-to-demand ratio, $C/D > 1$.

The next proposition defines the lowest service level, $\underline{SL}$, that can be applied to identify worst-case disruption scenarios with respect to service.

Proposition 9.2. *If there is no initial inventory of parts (i.e., $V_0 = 0$), the lowest service level, $\underline{SL}$, can be calculated as below.*

$$\underline{SL} = (\overline{T} - \max_{s \in S}\{t_s + \min_{i \in I}\{\theta_{is} + \tau_i\}\})c/D. \tag{9.20}$$

Proof. The lowest service level is associated with worst-case disruption scenario, $s \in S$, for which all primary suppliers are hit by disruption at time t_s and then the supplier that leads to smallest remaining capacity of final manufacturer is selected as a recovery supplier. The manufacturer smallest remaining capacity is determined by disruption scenario with the smallest number of periods remaining for production, after the delivery of parts by recovery supplier with the shortest recovery and delivery lead time, i.e., after, $\max_{s \in S}\{t_s + \min_{i \in I}\{\theta_{is} + \tau_i\}\}$, periods.

The results of Propositions 9.1 and 9.2 imply the following constraints on service level

$$\underline{SL} \leq \sum_{t \in T} x_t^s / D \leq 1; \; s \in S.$$

The implied constraints can be added to model **Support_E** to strengthen its formulation.

Model **Support_E** is a deterministic equivalent mixed integer program of a two-stage stochastic mixed integer program with recourse. The primary portfolio selection variables, u_i, v_i, are referred to as first stage decisions, and the recovery portfolio selection variables, U_i^s, V_i^s and production scheduling variables, x_t^s, are referred to as recourse or second stage decisions (cf. Table 9.2). Unlike the first stage decisions, the latter variables are dependent on disruption scenario $s \in S$. Model **Support_E** illustrates the wait-and-see approach. Basically, this approach is based on the perfect information about the future. In contrast to the two-stage stochastic program with recourse, in the wait-and-see approach, both the first stage and the second stage decisions are made simultaneously only when disruption scenario is known. Model **Support_E** is capable of simultaneously determining both stage variables to minimize expected cost over all disruption scenarios. When this problem is solved a recommendation is obtained for selection of primary supply portfolio (u_i, v_i) that will hedge against a variety of disruption scenarios in which fulfillment rates of certain suppliers are not sufficient to satisfy demand for parts. The recovery supply portfolio selection and production scheduling (U_i^s, V_i^s and x_t^s) are decisions that will be implemented in the future, when scenario $s \in S$ is finally realized.

Even though the recovery portfolio selection, U_i^s, V_i^s, and production scheduling, x_t^s, decisions are associated with each disruption scenario, they can be viewed as recourse decisions during disruption: each scenario $s \in S$ is associated with time period of disruption occurrence, t_s, as well as disruption duration, θ_{is} (i.e., time-to-recover from disruption of each supplier i). Moreover, a recovery supplier selection decision is interrelated with production scheduling during supply disruption.

9.3.2 Multiple Part Types and Product Types

In this subsection an enhancement of model **Support_E** is described for multiple types of parts and products. Let H and K be, respectively, the set of part types and set of product types, and denote by a_{hk}, $h \in H, k \in K$ the number of parts type h required to produce one unit of product type k.

If we denote by D_k the total demand for products type k, then $D = \sum_{k \in K} D_k$ is the total demand for all products and $A_h = \sum_{k \in K} a_{hk} D_k$ is the total demand for parts type h.

Now, the portfolio decision variables v_i, V_i^s, $i \in I, s \in S$ are replaced by v_{ih}, V_{ih}^s; $i \in I, h \in H, s \in S$, defined as fractions of total demand A_h for parts type h, ordered from primary supplier i, recovery supplier i under disruption scenario s, respectively. In addition, the production scheduling variable x_t^s, $t \in T, s \in S$ is replaced by x_{kt}^s, $k \in K, t \in T, s \in S$—the production of product type k in period t under disruption scenario s.

Model **SupportMP_E** for multiple part and product types is presented below.

SupportMP

Minimize

$$E^c = \sum_{s \in S} P_s \Big(\sum_{i \in I} e_i (u_i + U_i^s - \mu_i^s)$$

$$+ \sum_{i \in I} \rho_{is} U_i^s + \sum_{i \in I} \sum_{h \in H} o_{ih} A_h (\gamma_i^s v_{ih} + V_{ih}^s)$$

$$+ \sum_{k \in K} g_k (D_k - \sum_{t \in T} x_{kt}^s)\Big)/D,$$

where o_{ih} is unit purchasing price of part type h from supplier i, and g_k is unit penalty cost for unfulfilled demand for product type k,

subject to

$$\sum_{i \in I} v_{ih} = 1; \; h \in H$$

$$v_{ih} \le u_i; \; i \in I, h \in H$$

$$\sum_{i \in I} (\gamma_i^s v_{ih} + V_{ih}^s) = 1; \; h \in H, s \in S$$

$$V_{ih}^s \le U_i^s; \; i \in I, h \in H, s \in S$$

$$\mu_i^s \le (u_i + U_i^s)/2; \; i \in I, s \in S$$

$$\sum_{k \in K} \sum_{t' \in T: t' \le t} a_{hk} x_{kt'}^s / A_h \le V_{h0} + \sum_{i \in I: \tau_i \le t-1} \gamma_i^s v_{ih}$$

$$+ \sum_{i \in I: t_s + \theta_{is} + \tau_i \le t-1} V_{ih}^s; \; h \in H, t \in T, s \in S$$

(continued)

$$\sum_{k\in K} x_{kt}^s \le c;\ t\in T, s\in S$$
$$u_i \in \{0,1\};\ i\in I v_{ih} \in [0,1];\ i\in I, h\in H$$
$$U_i^s \in \{0,1\};\ i\in I, s\in S$$
$$V_{ih}^s \in [0,1];\ i\in I, h\in H, s\in S$$
$$x_{kt}^s \ge 0;\ k\in K, t\in T, s\in S,$$
$$\mu_i^s \in \{0,1\};\ i\in I, s\in S,$$

where $A_h V_{h0}$ is the initial inventory of parts type h.

Notice that the supply-production coordinating constraints (9.11) are also bill-of-material constraints in model **SupportMP_E**.

The above model can be further enhanced, for example, by introducing subsets $I_h \subset I$ of suppliers for each part type h, unit capacity consumption for each product type k in the left-hand side of production capacity constraints, etc.

9.3.3 Hierarchical Selection of Primary and Recovery Supply Portfolios

In this subsection two deterministic MIP models **PSupport** and **RSupport(s)** are presented for the hierarchical decision-making in the presence of supply chain disruption risks. The two-stage decision-making is described below (Fig. 9.2).

1. Selection of primary supply portfolio for deterministic environment.
   The primary suppliers are determined ahead of time with no disruption scenarios considered, using deterministic MIP model **PSupport**.
2. Selection of recovery supply portfolio, after disruption of primary suppliers.
   The recovery suppliers are determined after a primary supplier was hit by a disruption, to optimize the process of recovery from the disruption, using MIP model **RSupport(s)**.

The hierarchical approach is a purely top-down approach.

In the deterministic MIP model **PSupport**, stochastic variable, x_t^s (9.18), defined in model **Support_E** for each disruption scenario $s \in S$ has been replaced by its deterministic equivalent X_t.

PSupport: **P***rimary* **Supp***ly* **port***folio selection and production scheduling*

Minimize

$$P^c = \sum_{i\in I}(e_i u_i/D + o_i v_i) + g(1 - \sum_{t\in T} X_t/D) \qquad (9.21)$$

(continued)

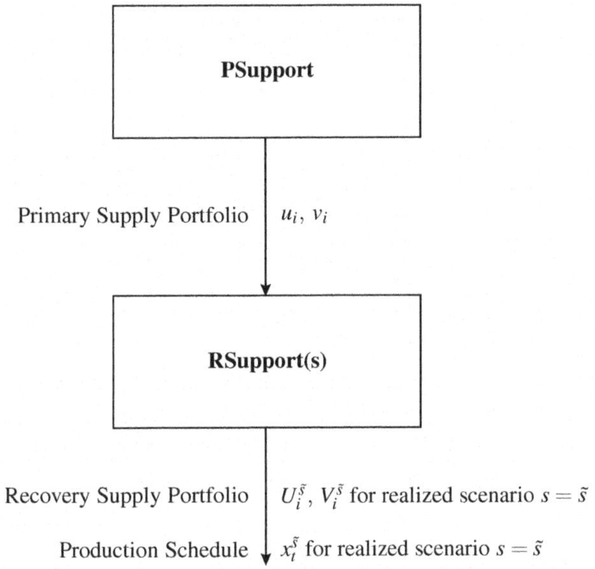

Fig. 9.2 Hierarchical selection of supply portfolios

subject to

$$\sum_{i \in I} v_i = 1 \tag{9.22}$$

$$v_i \leq u_i;\ i \in I \tag{9.23}$$

$$\sum_{t' \in T: t' \leq t} X_{t'}/D \leq V_0 + \sum_{i \in I: \tau_i \leq t-1} v_i;\ t \in T \tag{9.24}$$

$$X_t \leq c;\ t \in T \tag{9.25}$$

$$u_i \in \{0,1\};\ i \in I \tag{9.26}$$

$$v_i \in [0,1];\ i \in I \tag{9.27}$$

$$X_t \geq 0;\ t \in T. \tag{9.28}$$

The solution to **PSupport** is primary supply portfolio: u_i^*, v_i^*; $i \in I$, and the associated production schedule, X_t^*.

A simple lower bound on the objective function P^c (9.21), of the top level problem **PSupport** is derived below.

Proposition 9.3.

$$P^c \geq \min_{i \in I}(e_i/D + o_i) + g \max\{0, 1 - C/D\}. \tag{9.29}$$

(continued)

Proof. Primary supply portfolio selection constraints (9.22),(9.23) and **Proposition 9.1** (9.19), imply that

$$P^c = \sum_{i\in I}(e_i u_i/D + o_i v_i) + g(1 - \sum_{t\in T} X_t/D) \geq$$
$$\sum_{i\in I}(e_i/D + o_i)v_i + g\max\{0, 1 - C/D\} \geq$$
$$\min_{i\in I}(e_i/D + o_i) + g\max\{0, 1 - C/D\}.$$

In the deterministic MIP model **RSupport(s)**, stochastic variables, U_i^s, V_i^s, x_t^s (9.15), (9.16), (9.18), defined in model **Support_E** for each disruption scenario $s \in S$ have been replaced by their deterministic equivalents, $U_i^{\tilde{s}}$, $V_i^{\tilde{s}}$, $x_t^{\tilde{s}}$ for the realized disruption scenario $s = \tilde{s}$.

RSupport(*s*): **R***ecovery* **Supp***ly* **port***folio selection and production scheduling for predetermined primary supply portfolio and realized disruption scenario s*

Minimize

$$R_{\tilde{s}}^c = \sum_{i\in I}(e_i(1-u_i^*)U_i^{\tilde{s}}/D + \rho_{i\tilde{s}}U_i^{\tilde{s}}/D + o_i V_i^{\tilde{s}}) + g(1 - \sum_{t\in T} x_t^{\tilde{s}}/D) + \sum_{i\in I}(e_i u_i^*/D + o_i \gamma_i^{\tilde{s}} v_i^*) \quad (9.30)$$

subject to

$$\sum_{i\in I} V_i^{\tilde{s}} = 1 - \sum_{i\in I}\gamma_i^{\tilde{s}} v_i^* \quad (9.31)$$

$$V_i^{\tilde{s}} \leq U_i^{\tilde{s}};\ i \in I \quad (9.32)$$

$$\sum_{t'\in T:t'\leq t} x_{t'}^{\tilde{s}}/D - \sum_{i\in I:t_{\tilde{s}}+\theta_{i\tilde{s}}+\tau_i\leq t-1} V_i^{\tilde{s}} \leq V_0 + \sum_{i\in I:\tau_i\leq t-1}\gamma_i^{\tilde{s}} v_i^*;\ t \in T \quad (9.33)$$

$$x_t^{\tilde{s}} \leq c;\ t \in T \quad (9.34)$$

$$U_i^{\tilde{s}} \in \{0,1\};\ i \in I \quad (9.35)$$

$$V_i^{\tilde{s}} \in [0,1];\ i \in I \quad (9.36)$$

$$x_t^{\tilde{s}} \geq 0;\ t \in T. \quad (9.37)$$

Since the primary supply portfolio has been predetermined, the last summation term in Eq. (9.30) as well as the right-hand sides of Eqs. (9.31) and (9.33) are constant. The solution to **RSupport(s)** is the recovery supply portfolio, $U_i^{\tilde{s}}$, $V_i^{\tilde{s}}$; $i \in I$, and the production schedule, $x_t^{\tilde{s}}$; $t \in T$, for the realized disruption scenario $s = \tilde{s}$:

Notice that no recovery supply portfolios, $V_i^s = 0, \forall i \in I$, will be determined for scenarios s with non-disrupted primary suppliers, i.e., for scenarios $s \in S$, such that $\lambda_{is} = \overline{L_i}$ if $v_i^* > 0$. Since $\gamma_{i,\overline{L_i}} = 1$, $\sum_{i \in I} \gamma_{i,\overline{L_i}} v_i^* = 1$, and hence the right-hand side of (9.31) is equal to zero.

Model **Support_E** for the predetermined primary supply portfolio is separable with respect to disruption scenarios $s \in S$, since the objective function (9.4) is additive and separable with respect to s and all constraints (9.8)–(9.12) are also separable with respect to s. Thus, given the primary supply portfolio, the recovery supply portfolios, U_i^s, V_i^s; $i \in I$, and production schedules, x_t^s, can be found simultaneously for all potential disruption scenarios $s \in S$, by solving

RSupport_E = **Support_E** for predetermined primary supply portfolio: $u_i = u_i^*$, $v_i = v_i^*$; $i \in I$.

RSupport_E: **R***ecovery* **Sup***ply* **port***folio selection and production scheduling for predetermined primary supply portfolio*

Minimize (9.4)

subject to (9.8)–(9.12) (9.15)–(9.18) and

$$u_i = u_i^*; \; i \in I \tag{9.38}$$

$$v_i = v_i^*; \; i \in I. \tag{9.39}$$

9.4 Models for Risk-Averse Decision-Making

In this section the two time-indexed SMIP models **Support_CV(c)** and **Support_CV(sl)** are proposed for the integrated, risk-averse selection of primary and recovery supply portfolios and production scheduling to optimize, respectively, CVaR of cost and CVaR of service level under disruption risks. The models are based on the risk-neutral model **Support_E**.

Let VaR^c be the Value-at-Risk of cost per product, i.e., the targeted cost such that for a given confidence level α, for $100\alpha\%$ of disruption scenarios, the outcome is below VaR^c. Accordingly, let CVaR^c be conditional value-at-risk of cost per product, i.e., the expected cost in the worst $100(1-\alpha)\%$ of the scenarios with the cost above VaR^c.

$$CVaR^c = VaR^c + (1-\alpha)^{-1} \sum_{s \in S} P_s \mathscr{C}_s. \tag{9.40}$$

The risk-averse primary and recovery supply portfolio and the production schedule will be optimized by calculating VaR^c and minimizing CVaR^c simultaneously. Model **Support_CV(c)** is presented below.

Support_CV(c): *Risk-averse selection of primary and recovery* **Supp***ly* **port***folios and production scheduling to minimize* **CV***aR of* **c***ost*

Minimize (9.40)

subject to (9.6)–(9.18) and

Risk constraints:

- the tail cost for scenario s is defined as the nonnegative amount by which cost in scenario s exceeds VaR^c,

$$\mathscr{C}_s \geq \sum_{i \in I} e_i(u_i + U_i^s - \mu_i^s)/D + \sum_{i \in I} \rho_{is} U_i^s / D$$
$$+ \sum_{i \in I} o_i(\gamma_i^s v_i + V_i^s) + g(1 - \sum_{t \in T} x_t^s / D) - VaR^c; \ s \in S \qquad (9.41)$$
$$\mathscr{C}_s \geq 0; \ s \in S, \qquad (9.42)$$

where $\mathscr{C}_s$ is the tail cost for scenario s.

In the next model, VaR^{sl} is the Value-at-Risk of service level, i.e., the targeted service level such that for a given confidence level α, for $100\alpha\%$ of disruption scenarios, the outcome is above VaR^{sl}, while CVaR^{sl} is the conditional value-at-risk of service level, i.e., the expected service level in the worst $100(1-\alpha)\%$ of scenarios with the service level below VaR^{sl}.

$$CVaR^{sl} = VaR^{sl} - (1-\alpha)^{-1} \sum_{s \in S} P_s \mathscr{S}_s. \qquad (9.43)$$

The risk-averse primary and recovery supply portfolio and the production schedule will be optimized by calculating VaR^{sl} and maximizing CVaR^{sl} simultaneously. Model **Support_CV(sl)** is presented below.

Support_CV(sl): *Risk-averse selection of primary and recovery* **Supp***ly* **port***folios and production scheduling to maximize* **CV***aR of* **s***ervice* **l***evel*

Maximize (9.43)

subject to (9.6)–(9.18) and

Risk constraints:

- the tail service level for scenario s is defined as the nonnegative amount by which VaR^{sl} exceeds service level in scenario s,

$$\mathscr{S}_s \geq VaR^{sl} - \sum_{t \in T} x_t^s / D; \ s \in S \qquad (9.44)$$
$$\mathscr{S}_s \geq 0; \ s \in S, \qquad (9.45)$$

where $\mathscr{S}_s$ is the tail service level for scenario s.

9.5 Computational Examples

In this section some computational examples are presented to illustrate the proposed portfolio approach for selection of primary and recovery supply portfolios. The problem of joint selection of primary and recovery supply portfolios and production scheduling under disruption risks considered here is different from the existing literature in many different ways. Although the input data for the examples are hypothetical, their relations to each other are real and in part they have been taken from real case studies. In particular, the case studies of Toyota supply chain disruption and recovery after the Great East Japan earthquake and tsunami of March 11, 2011 (Fujimoto and Park 2013; Park et al. 2013; MacKenzie et al. 2014, Matsuo 2015) and a case study of Thailand's floods in 2011 (Haraguchi and Lall 2015) have been analyzed.

The computational experiments were performed using the AMPL programming language and the Gurobi 7.0.0 solver on a MacBookPro laptop with Intel Core i7 processor running at 2.8 GHz and with 16 GB RAM. The basic input parameters for computational examples are presented in Table 9.3. Notice that three variants of producer capacity per period, c, will be compared in the computational study.

The total number of all potential scenarios is $\overline{S} = (\overline{L}+1)^{\overline{I}} = 4^4 = 256$ scenarios, where each scenario $s \in S$ is represented by vector $\lambda_s = \{\lambda_{1s}, \ldots, \lambda_{4s}\}$, where $\lambda_{is} \in \{0,1,2,3\}$, $i \in I$, see Table 9.4.

The probability of realizing of disruption scenario $s \in S$ for suppliers in region $r = 1,2$ is calculated as follows (see, Sect. 9.2)

$$P_s^r = \begin{cases} (1-p^r)(\prod_{i\in I^r:\lambda_{is}=0} 0.1(1-p_{i3}))(\prod_{i\in I^r:\lambda_{is}=1} 0.3(1-p_{i3})) & \\ \times(\prod_{i\in I^r:\lambda_{is}=2} 0.6(1-p_{i3}))(\prod_{i\in I^r:\lambda_{is}=3} p_{i3}) & \text{if } \sum_{i\in I^r}\lambda_{is} > 0 \\ p^r + (1-p^r)\prod_{i\in I^r} 0.1(1-p_{i3}) & \text{if } \sum_{i\in I^r}\lambda_{is} = 0, \end{cases}$$

and the probability for disruption scenario $s \in S$ is given by $P_s = P_s^1 P_s^2$.

Notice that, $p_{i0} \le p_{i1} \le p_{i2} \le p_{i3}$, i.e., the probability of disruption level of a supplier increases with the level such that the smallest probability was assigned to complete shutdown (level 0) and the largest to non-disruptive operation (level 3).

The regional disruption probabilities are chosen to be lower than local shutdown probabilities of suppliers in that region. In addition, the two regional disruption probabilities are significantly different to represent two geographic regions differently exposed to disruptive events.

Figure 9.3 presents basic characteristics of all suppliers: purchasing price, o_i, $i \in I$, probability of complete shutdown, $p^r + (1-p^r)p_{i0}$, $i \in I^r$, $r \in R$, and expected fulfillment rate, $\Gamma_i = \sum_{l=1,2,3}(1-p^r)p_{il}\gamma_{il}$, $i \in I^r$, $r \in R$.

Table 9.3 Input parameters: primary and recovery supply portfolio selection and production scheduling

$\overline{I}=4$ suppliers, $\overline{R}=2$ two geographic regions, $\overline{T}=30$ planning periods	
$I^1=\{1,2\}$	two suppliers in region $r=1$
$I^2=\{3,4\}$	two suppliers in region $r=2$
$V_0=0$	initial inventories of parts
$d_t=10000,\ t\in T$	demand per period for parts/products
$D=300000$	total demand for parts/products
Disruption levels:	
$\overline{L}_i=3$	four disruption levels for each supplier $i\in I$
Local disruption probabilities of suppliers:	
$p_{i0}=0.1(1-p_{i3})$	complete shutdown (level $l=0$)
$p_{i1}=0.3(1-p_{i3})$	major disruption (level $l=1$)
$p_{i2}=0.6(1-p_{i3})$	minor disruption (level $l=2$)
$p_{i3}\in[0.89,0.99];\ i\in I^1,\ p_{i3}\in[0.79,0.89];\ i\in I^2$	non disruption (level $l=3$)
Regional disruption probability:	$p^1=0.001,\ p^2=0.01$
Fulfillment rates of suppliers:	
$\gamma_{i0}=0\ \forall i\in I,\ \gamma_{i3}=1\ \forall i\in I$	
$\gamma_{i1}\in[0.01,0.50]\ \forall i\in I^1\ \gamma_{i1}\in[0.01,0.30]\ \forall i\in I^2,\ \gamma_{i2}\in[0,51,0.99]\ \forall i\in I^1\ \gamma_{i2}\in[0,31,0.99]\ \forall i\in I^2$	
Plant capacity:	
$c\in\{5000,10000,15000\}$	Three variants of producer capacity per period
$C\in\{140000,\ 280000,420000\}$	Three variants of total producer capacity, $C=c(\overline{T}-\tau_{\min})$
$C/D\in\{0.47,0.93,1.4\}$	Three variants of capacity-to-demand ratio
Cost parameters:	
$o=(14,12,8,9),\ e=(8000,6000,12,000,13,000)$	
$g\in\{1,10,100,1000,10,000,100,000,\infty\}$	unit penalties for unfulfilled demand for products, where $g=\infty$ denotes maximization of service level
$CTR(i,l)$=if $l=0$ then $100e_i$; if $l=1$ then $10e_i$; if $l=2$ then $e_i\ \forall i\in I$	
Time parameters:	
$TTR(i,l)$=if $l=0$ then 12; if $l=1$ then 10; if $l=2$ then 8 $\forall i\in I$	
$\tau_{1j}=\tau_{2j}=2,\ \tau_{3j}=\tau_{4j}=4,\forall j\in J,\ \sigma_2=2$	

9.5.1 Risk-neutral Decision-Making

9.5.1.1 Scenarios with a Common Disruption Start Time

In this subsection each disruption $s\in S$ to primary suppliers is assumed to occur in the same period, $t_s=1$, before the earliest delivery lead time, $\min_{i\in I}(\tau_i)=2$, and the recovery process starts in period, $t=2$, so that the disrupted supplier i returns to its full capacity in period, $t=TTR(i,l)+1$.

The optimal solutions for both the integrated and the hierarchical approach are summarized in Tables 9.5, 9.6, and 9.7, respectively, for capacity-to-demand ratio, $C/D=0.47$, 0.93, and 1.4.

Table 9.4 Multi-level disruption scenarios: $\lambda_{1s}, \lambda_{2s}, \lambda_{3s}, \lambda_{4s}; s = 1, \ldots, 256$

s	i=1	2	3	4	s	i=1	2	3	4	s	i=1	2	3	4	s	i=1	2	3	4
1	0	0	0	0	65	1	0	0	0	129	2	0	0	0	193	3	0	0	0
2	0	0	0	1	66	1	0	0	1	130	2	0	0	1	194	3	0	0	1
3	0	0	0	2	67	1	0	0	2	131	2	0	0	2	195	3	0	0	2
4	0	0	0	3	68	1	0	0	3	132	2	0	0	3	196	3	0	0	3
5	0	0	1	0	69	1	0	1	0	133	2	0	1	0	197	3	0	1	0
6	0	0	1	1	70	1	0	1	1	134	2	0	1	1	198	3	0	1	1
7	0	0	1	2	71	1	0	1	2	135	2	0	1	2	199	3	0	1	2
8	0	0	1	3	72	1	0	1	3	136	2	0	1	3	200	3	0	1	3
9	0	0	2	0	73	1	0	2	0	137	2	0	2	0	201	3	0	2	0
10	0	0	2	1	74	1	0	2	1	138	2	0	2	1	202	3	0	2	1
11	0	0	2	2	75	1	0	2	2	139	2	0	2	2	203	3	0	2	2
12	0	0	2	3	76	1	0	2	3	140	2	0	2	3	204	3	0	2	3
13	0	0	3	0	77	1	0	3	0	141	2	0	3	0	205	3	0	3	0
14	0	0	3	1	78	1	0	3	1	142	2	0	3	1	206	3	0	3	1
15	0	0	3	2	79	1	0	3	2	143	2	0	3	2	207	3	0	3	2
16	0	0	3	3	80	1	0	3	3	144	2	0	3	3	208	3	0	3	3
17	0	1	0	0	81	1	1	0	0	145	2	1	0	0	209	3	1	0	0
18	0	1	0	1	82	1	1	0	1	146	2	1	0	1	210	3	1	0	1
19	0	1	0	2	83	1	1	0	2	147	2	1	0	2	211	3	1	0	2
20	0	1	0	3	84	1	1	0	3	148	2	1	0	3	212	3	1	0	3
21	0	1	1	0	85	1	1	1	0	149	2	1	1	0	213	3	1	1	0
22	0	1	1	1	86	1	1	1	1	150	2	1	1	1	214	3	1	1	1
23	0	1	1	2	87	1	1	1	2	151	2	1	1	2	215	3	1	1	2
24	0	1	1	3	88	1	1	1	3	152	2	1	1	3	216	3	1	1	3
25	0	1	2	0	89	1	1	2	0	153	2	1	2	0	217	3	1	2	0
26	0	1	2	1	90	1	1	2	1	154	2	1	2	1	218	3	1	2	1
27	0	1	2	2	91	1	1	2	2	155	2	1	2	2	219	3	1	2	2
28	0	1	2	3	92	1	1	2	3	156	2	1	2	3	220	3	1	2	3
29	0	1	3	0	93	1	1	3	0	157	2	1	3	0	221	3	1	3	0
30	0	1	3	1	94	1	1	3	1	158	2	1	3	1	222	3	1	3	1
31	0	1	3	2	95	1	1	3	2	159	2	1	3	2	223	3	1	3	2
32	0	1	3	3	96	1	1	3	3	160	2	1	3	3	224	3	1	3	3
33	0	2	0	0	97	1	2	0	0	161	2	2	0	0	225	3	2	0	0
34	0	2	0	1	98	1	2	0	1	162	2	2	0	1	226	3	2	0	1
35	0	2	0	2	99	1	2	0	2	163	2	2	0	2	227	3	2	0	2
36	0	2	0	3	100	1	2	0	3	164	2	2	0	3	228	3	2	0	3
37	0	2	1	0	101	1	2	1	0	165	2	2	1	0	229	3	2	1	0
38	0	2	1	1	102	1	2	1	1	166	2	2	1	1	230	3	2	1	1
39	0	2	1	2	103	1	2	1	2	167	2	2	1	2	231	3	2	1	2
40	0	2	1	3	104	1	2	1	3	168	2	2	1	3	232	3	2	1	3
41	0	2	2	0	105	1	2	2	0	169	2	2	2	0	233	3	2	2	0
42	0	2	2	1	106	1	2	2	1	170	2	2	2	1	234	3	2	2	1
43	0	2	2	2	107	1	2	2	2	171	2	2	2	2	235	3	2	2	2
44	0	2	2	3	108	1	2	2	3	172	2	2	2	3	236	3	2	2	3
45	0	2	3	0	109	1	2	3	0	173	2	2	3	0	237	3	2	3	0
46	0	2	3	1	110	1	2	3	1	174	2	2	3	1	238	3	2	3	1
47	0	2	3	2	111	1	2	3	2	175	2	2	3	2	239	3	2	3	2
48	0	2	3	3	112	1	2	3	3	176	2	2	3	3	240	3	2	3	3
49	0	3	0	0	113	1	3	0	0	177	2	3	0	0	241	3	3	0	0
50	0	3	0	1	114	1	3	0	1	178	2	3	0	1	242	3	3	0	1
51	0	3	0	2	115	1	3	0	2	179	2	3	0	2	243	3	3	0	2
52	0	3	0	3	116	1	3	0	3	180	2	3	0	3	244	3	3	0	3
53	0	3	1	0	117	1	3	1	0	181	2	3	1	0	245	3	3	1	0
54	0	3	1	1	118	1	3	1	1	182	2	3	1	1	246	3	3	1	1
55	0	3	1	2	119	1	3	1	2	183	2	3	1	2	247	3	3	1	2
56	0	3	1	3	120	1	3	1	3	184	2	3	1	3	248	3	3	1	3
57	0	3	2	0	121	1	3	2	0	185	2	3	2	0	249	3	3	2	0
58	0	3	2	1	122	1	3	2	1	186	2	3	2	1	250	3	3	2	1
59	0	3	2	2	123	1	3	2	2	187	2	3	2	2	251	3	3	2	2
60	0	3	2	3	124	1	3	2	3	188	2	3	2	3	252	3	3	2	3
61	0	3	3	0	125	1	3	3	0	189	2	3	3	0	253	3	3	3	0
62	0	3	3	1	126	1	3	3	1	190	2	3	3	1	254	3	3	3	1
63	0	3	3	2	127	1	3	3	2	191	2	3	3	2	255	3	3	3	2
64	0	3	3	3	128	1	3	3	3	192	2	3	3	3	256	3	3	3	3

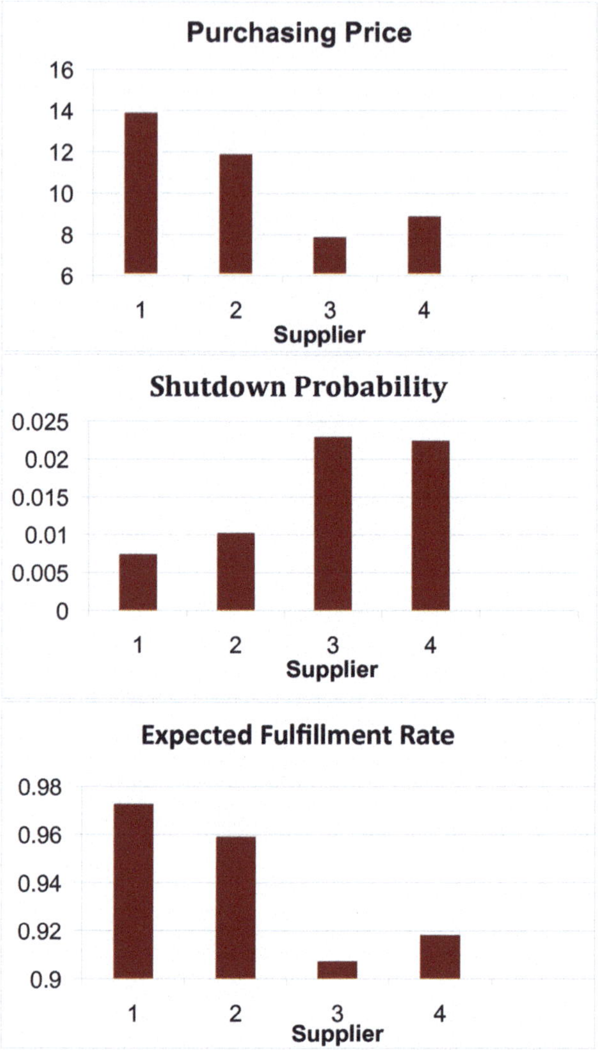

Fig. 9.3 Suppliers

For different values of unit penalty cost g, the tables show:

- optimal primary supply portfolio, $(v_1, v_2, v_3.v_4)$, indicating percent of total demand for parts ordered from each supplier;
- optimal expected recovery supply portfolio, $(\sum_{s\in S} P_s V_1^s, \sum_{s\in S} P_s V_2^s, \sum_{s\in S} P_s V_3^s, \sum_{s\in S} P_s V_4^s)$, indicating expected percent of total demand for parts ordered from each supplier.

For the integrated approach and the bottom level problem **RSupport_E** of the hierarchical approach, expected cost E^c (9.4), is shown along with the associated

Table 9.5 Solution results for a common disruption start time: $C/D = 0.47$

g	1	10	10^2	10^3	10^4	10^5	∞[a]
Integrated approach: model **Support_E**							
Var.=10205, Bin.=2052, Cons.=10462, Nonz.=10205[e]							
Exp.Cost E^c (9.4)	8.80	13.72	61.76	542	5343	53,337	–
Exp.Service $E^{sl} \times 100\%$ (9.5)	43	47	47	47	47	47	47
Primary Portfolio:	1(0)	1(0)		1(2)	1(4)	1(9)	1(31)
Supplier(% of total demand)[b]	2(0)	2(3)		2(2)	2(4)	2(10)	2(19)
	3(100)	3(97)		3(96)	3(92)	3(79)	3(26)
	4(0)	4(0)		4(0)	4(0)	4(2)	4(24)
Exp.Recovery Portfolio:	1(0.18)	1(0.27)		1(0.22)	1(0.17)	1(0.16)	1(1.82)
Supplier(% of total demand)[c]	2(1.86)	2(1.81)		2(1.82)	2(1.71)	2(1.51)	2(1.46)
	3(0)	3(0.04)		3(0.06)	3(0.25)	3(0.7)	3(1.25)
	4(7.21)	4(6.95)		4(6.93)	4(6.63)	4(5.80)	4(1.43)
Hierarchical approach: models **PSupport** and **RSupport**							
PSupport: Var.=36, Bin.=4, Cons.=37, Nonz.=534							
Primary Portfolio:	1(0)	1(0)					1(0)
Supplier(% of total demand)[b]	2(0)	2(3)					2(100)
	3(100)	3(97)					3(0)
	4(0)	4(0)					4(0)
Cost P_c (9.21)	8.61	13.53	61.53	542	5342	53,342	–
Service[d]	43	47	47	47	47	47	47
RSupport_E: Var.=8050, Bin.=768, Cons.=6250, Nonz.=8050							
Exp.Recovery Portfolio:	1(0.18)	1(0.22)					1(0.83)
Supplier(% of total demand)[c]	2(1.86)	2(1.73)					2(3.14)
	3(0)	3(0.20)					3(0.07)
	4(7.21)	4(6.75)					4(0.04)
Exp.Cost E^c (9.4)	8.80	13.84	61.85	542	5344	53,365	–
Exp.Service $E^{sl} \times 100\%$ (9.5)	43	47	47	47	47	47	47

[a]Maximization of service level

[b]$1(v_1 \times 100), 2(v_2 \times 100), 3(v_3 \times 100\%), 4(v_4 \times 100)$

[c]$1(\sum_{s\in S} P_s V_1^s \times 100), 2(\sum_{s\in S} P_s V_2^s \times 100), 3(\sum_{s\in S} P_s V_3^s \times 100), 4(\sum_{s\in S} P_s V_4^s \times 100)$

[d]$\sum_{t\in T} X_t / D \times 100$

[e]Var.=number of variables, Bin.=number of binary variables, Cons.=number of constraints, Nonz.= number of nonzero coefficients

expected service level E^{sl} (9.5). Similarly, for the top level problem **PSupport** of the hierarchical approach, cost P^c (9.21), is shown along with the associated service level, $\sum_{t\in T} X_t/D$. In addition, Table 9.5 presents the size of each MIP model, **Support_E**, **PSupport**, and **RSupport_E**. The results indicate that for most examples the expected service level E^{sl} (9.5), and the service level for **PSupport** problem, $\sum_{t\in T} X_t/D$, attain their upper bounds C/D (9.19), whereas the cost P^c (9.21), remains at its lower bound 8.04 (9.29), only for $C/D = 1.4$.

Tables 9.5 and 9.6 indicate that for $C/D < 1$ and small unit penalty cost g, the cheapest supply portfolio is selected and the achieved service level is less than C/D. The same solution results are obtained for both the integrated and the hierarchical approach. As g increases to reduce the unfulfilled demand, more expensive and diversified supply portfolios are selected and the highest service level C/D (9.19), is attained. In particular, for the integrated approach, more diversified primary sup-

Table 9.6 Solution results for a common disruption start time: $C/D = 0.93$

g	1	10	10^2	10^3	10^4	10^5	∞ [a]
Integrated approach: model **Support_E**							
Exp.Cost E^c (9.4)	8.36	9.19	15.26	75.53	677	6685	–
Exp.Service $E^{sl} \times 100\%$ (9.5)	87	93	93	93	93	93	93
Primary Portfolio:	1(0)	1(0)		1(3)	1(8)	1(18)	1(27)
Supplier(% of total demand)[b]	2(0)	2(7)		2(4)	2(9)	2(20)	2(25)
	3(100)	3(93)		3(93)	3(83)	3(57)	3(7)
	4(0)	4(0)		4(0)	4(0)	4(5)	4(41)
Exp.Recovery Portfolio:	1(0.18)	1(0.36)		1(0.25)	1(0.16)	1(0.13)	1(1.99)
Supplier(% of total demand)[c]	2(1.86)	2(1.77)		2(1.78)	2(1.55)	2(1.16)	2(1.99)
	3(0)	3(0.07)		3(0.13)	3(0.49)	3(1.44)	3(1.37)
	4(7.21)	4(6.70)		4(6.65)	4(6.06)	4(4.26)	4(0.40)
Hierarchical approach: models **PSupport** and **RSupport_E**							
Primary Portfolio:	1(0)	1(0)					1(0)
Supplier(% of total demand)[b]	2(0)	2(7)					2(100)
	3(100)	3(93)					3(0)
	4(0)	4(0)					4(0)
Cost P^c (9.21)	8.17	9	15	75	675	6675	–
Service[d]	87	93					93
Exp.Recovery Portfolio:	1(0.18)	1(0.36)					1(1.18)
Supplier(% of total demand)[c]	2(1.86)	2(1.77)					2(1.27)
	3(0)	3(0.07)					3(1.55)
	4(7.21)	4(6.70)					4(0.07)
Exp.Cost E^c (9.4)	8.36	9.19	15.26	76.02	684	6759	–
Exp.Service $E^{sl} \times 100\%$ (9.5)	87	93					93

[a] Maximization of service level
[b] $1(v_1 \times 100), 2(v_2 \times 100), 3(v_3 \times 100), 4(v_4 \times 100)$
[c] $1(\sum_{s\in S} P_s V_1^s \times 100), 2(\sum_{s\in S} P_s V_2^s \times 100), 3(\sum_{s\in S} P_s V_3^s \times 100), 4(\sum_{s\in S} P_s V_4^s \times 100)$
[d] $\sum_{t\in T} X_t / D \times 100$

ply portfolios are selected for, $g > 1$, to hedge against all disruption scenarios. In contrast, for the hierarchical approach the results for, $g > 1$, are independent of g. When the objective is to maximize service level ($g \to \infty$), the hierarchical approach selects the most reliable supplier as the only primary supplier. In addition, Table 9.7 demonstrates that for $C/D > 1$ and both approaches, the primary portfolio and the expected recovery portfolio are independent of g, since for $C/D = 1.4$, the total demand for products is fulfilled for all g and hence no penalty cost appears.

Table 9.8 shows an example of optimal primary supply portfolio v_1, v_2, v_3, v_4 and optimal recovery supply portfolios $V_1^s, V_2^s, V_3^s, V_4^s, \forall s \in S$ for $C/D = 1.4$ and unit penalty $g = 1, 10, 100, 1000, 10{,}000$. Both the integrated approach and the hierarchical approach yield the same solution results (cf. Table 9.7). The table shows the optimal primary supply portfolio, $v_1 = v_2 = 0, v_3 = 1, v_4 = 0$, and the associated cost P^c (9.21), as well as the optimal recovery supply portfolios and costs R_s^c (9.30) for each disruption scenarios $s \in S$. The results demonstrate that no recovery supply portfolio is selected, i.e., $V_1^s = V_2^s = V_3^s = V_4^s = 0$, for scenarios $s \in S$ with non-disrupted primary supplier, $i = 3$, i.e., for $\{s \in S : \lambda_{3s} = 3\} =$

Table 9.7 Solution results for a common disruption start time: $C/D = 1.4$

g	1	10	10^2	10^3	10^4	10^5	∞[a]
Integrated approach: model **Support_E**							
Exp.Cost E^c (9.4)	8.23	8.23	8.23	8.24	8.31	9.05	–
Exp.Service $E^{sl} \times 100\%$ (9.5)	100						100
Primary Portfolio:	1(0)					1(0)	1(40)
Supplier(% of total demand)[b]	2(0)					2(6)	2(10)
	3(100)					3(94)	3(26)
	4(0)					4(0)	4(24)
Exp.Recovery Portfolio:	1(0.18)					1(0.17)	1(0.43)
Supplier(% of total demand)[c]	2(1.86)					2(1.74)	2(2.10)
	3(0)					3(0.22)	3(1.73)
	4(7.21)					4(6.78)	4(1.62)
Hierarchical approach: models **PSupport** and **RSupport_E**:							
Primary Portfolio:	1(0)						1(100)
Supplier(% of total demand)[b]	2(0)						2(0)
	3(100)						3(0)
	4(0)						4(0)
Cost P^c (9.21)	8.04						–
Service[d]	100						100
Exp.Recovery Portfolio:	1(0.18)						1(1.45)
Supplier(% of total demand)[c]	2(1.86)						2(1.04)
	3(0)						3(0.13)
	4(7.21)						4(0.08)
Exp.Cost E^c (9.4)	8.23	8.23	8.23	8.24	8.31	9.05	–
Exp.Service $E^{sl} \times 100\%$ (9.5)	100						100

[a] Maximization of service level
[b] $1(v_1 \times 100), 2(v_2 \times 100), 3(v_3 \times 100), 4(v_4 \times 100)$
[c] $1(\Sigma_{s \in S} P_s V_1^s \times 100), 2(\Sigma_{s \in S} P_s V_2^s \times 100), 3(\Sigma_{s \in S} P_s V_3^s \times 100), 4(\Sigma_{s \in S} P_s V_4^s \times 100)$
[d] $\Sigma_{t \in T} X_t / D \times 100$

$\{13-16, 29-32, 45-48, 61-64, 77-80, 93-96, 109-112, 125-128, 141-144, 157-160, 173-176, 189-192, 205-208, 221-224, 237-240, 253-256\}$.

9.5.1.2 Scenarios with Different Disruption Start Times

In real-life cases, supply disruptions may occur at any time and for any supplier. In this subsection, disruption scenario is defined as a combination of disruptive event and its start time. In order to consider non-trivial disruptions, the start time t_s of each disruptive event, $s \in S$, is assumed to be not greater than the maximum delivery lead time, $\max_{i \in I}(\tau_i) = 4$. In the computational examples, $t_s \in \{1,2,3\}$, and the total number of all potential scenarios to be considered is $256 \times 3 = 768$, i.e., $S = \{1, \ldots, 768\}$. Now, each disruption scenario $s \in S$ is represented by vector $\lambda_s = \{\lambda_{1s}, \ldots, \lambda_{4s}\}$ (see Table 9.4), where $\lambda_s = \lambda_{s+256} = \lambda_{s+512}$; $s \leq 256$, and its start time $t_s = 1$ for $s \leq 256$, $t_s = 2$ for $257 \leq s \leq 512$ and $t_s = 3$ for $513 \leq s \leq 768$.

Table 9.8 Optimal recovery supply portfolios for a common disruption start time, $C/D = 1.4$, $g = 1, 10, 10^2, 10^3, 10^4$: integrated approach/hierarchical approach[a]

Scenario: s Optimal recovery supply portfolio: $V_1^s\ V_2^s\ V_3^s\ V_4^s$ Optimal expected cost: R_s^c (9.30)											
s	i=1	2	3	4	Cost	s	i=1	2	3	4	Cost
1	0	1	0	0	2037.06	51	0	1	0	0	12.06
2	0	0	0	1	467.42	52	0	0	0	1	9.08
3	0	0	0	1	67.42	53	0	0.9	0	0	11.65
4	0	0	0	1	9.08	54	0	0.9	0	0	11.65
5	0	0	0.9	0	422.72	55	0	0.9	0	0	11.65
6	0	0	0.9	0	422.72	56	0	0	0	0.9	8.98
7	0	0	0	0.9	56.99	57	0	0.44	0	0	9.83
8	0	0	0	0.9	8.98	58	0	0.44	0	0	9.83
9	0	0	0.44	0	48.04	59	0	0.44	0	0	9.83
10	0	0	0.44	0	48.04	60	0	0	0	0.44	8.53
11	0	0	0.44	0	48.04	61	0	0	0	0	8.04
12	0	0	0	0.44	8.53	62	0	0	0	0	8.04
13	0	0	0	0	8.04	63	0	0	0	0	8.04
14	0	0	0	0	8.04	64	0	0	0	0	8.04
15	0	0	0	0	8.04	65	1	0	0	0	295.73
16	0	0	0	0	8.04	66	1	0	0	0	295.73
17	0	1	0	0	227.06	67	0	0	0	1	67.42
18	0	1	0	0	227.06	68	0	0	0	1	9.08
19	0	0	0	1	67.42	69	0.9	0	0	0	284.8
20	0	0	0	1	9.08	70	0.9	0	0	0	284.8
21	0	0.9	0	0	216.33	71	0	0	0	0.9	56.99
22	0	0.9	0	0	216.33	72	0	0	0	0.9	8.98
23	0	0	0	0.9	56.99	73	0	0	0.44	0	48.04
24	0	0	0	0.9	8.98	74	0	0	0.44	0	48.04
25	0	0	0.44	0	48.04	75	0	0	0.44	0	48.04
26	0	0	0.44	0	48.04	76	0	0	0	0.44	8.53
27	0	0	0.44	0	48.04	77	0	0	0	0	8.04
28	0	0	0	0.44	8.53	78	0	0	0	0	8.04
29	0	0	0	0	8.04	79	0	0	0	0	8.04
30	0	0	0	0	8.04	80	0	0	0	0	8.04
31	0	0	0	0	8.04	81	0	1	0	0	227.06
32	0	0	0	0	8.04	82	0	1	0	0	227.06
33	0	1	0	0	37.06	83	0	0	0	1	67.42
34	0	1	0	0	37.06	84	0	0	0	1	9.08
35	0	1	0	0	37.06	85	0	0.9	0	0	216.33
36	0	0	0	1	9.08	86	0	0.9	0	0	216.33
37	0	0.9	0	0	31.65	87	0	0	0	0.9	56.99
38	0	0.9	0	0	31.65	88	0	0	0	0.9	8.98
39	0	0.9	0	0	31.65	89	0	0	0.44	0	48.04
40	0	0	0	0.9	8.98	90	0	0	0.44	0	48.04
41	0	0.44	0	0	29.83	91	0	0	0.44	0	48.04
42	0	0.44	0	0	29.83	92	0	0	0	0.44	8.53
43	0	0.44	0	0	29.83	93	0	0	0	0	8.04
44	0	0	0	0.44	8.53	94	0	0	0	0	8.04
45	0	0	0	0	8.04	95	0	0	0	0	8.04
46	0	0	0	0	8.04	96	0	0	0	0	8.04
47	0	0	0	0	8.04	97	0	1	0	0	37.06
48	0	0	0	0	8.04	98	0	1	0	0	37.06
49	0	1	0	0	12.06	99	0	1	0	0	37.06
50	0	1	0	0	12.06	100	0	0	0	1	9.08

(continued)

Table 9.8 (continued)

Scenario: *s* Optimal recovery supply portfolio: V_1^s V_2^s V_3^s V_4^s Optimal expected cost: R_s^c (9.30)											
s	i=1	2	3	4	Cost	s	i=1	2	3	4	Cost
101	0	0.9	0	0	31.65	151	0.9	0	0	0	40.11
102	0	0.9	0	0	31.65	152	0	0	0	0.9	8.98
103	0	0.9	0	0	31.65	153	0.44	0	0	0	37.39
104	0	0	0	0.9	8.98	154	0.44	0	0	0	37.39
105	0	0.44	0	0	29.83	155	0.44	0	0	0	37.39
106	0	0.44	0	0	29.83	156	0	0	0	0.44	8.53
107	0	0.44	0	0	29.83	157	0	0	0	0	8.04
108	0	0	0	0.44	8.53	158	0	0	0	0	8.04
109	0	0	0	0	8.04	159	0	0	0	0	8.04
110	0	0	0	0	8.04	160	0	0	0	0	8.04
111	0	0	0	0	8.04	161	0	1	0	0	37.06
112	0	0	0	0	8.04	162	0	1	0	0	37.06
113	0	1	0	0	12.06	163	0	1	0	0	37.06
114	0	1	0	0	12.06	164	0	0	0	1	9.08
115	0	1	0	0	12.06	165	0	0.9	0	0	31.65
116	0	0	0	1	9.08	166	0	0.9	0	0	31.65
117	0	0.9	0	0	11.65	167	0	0.9	0	0	31.65
118	0	0.9	0	0	11.65	168	0	0	0	0.9	8.98
119	0	0.9	0	0	11.65	169	0	0.44	0	0	29.83
120	0	0	0	0.9	8.98	170	0	0.44	0	0	29.83
121	0	0.44	0	0	9.83	171	0	0.44	0	0	29.83
122	0	0.44	0	0	9.83	172	0	0	0	0.44	8.53
123	0	0.44	0	0	9.83	173	0	0	0	0	8.04
124	0	0	0	0.44	8.53	174	0	0	0	0	8.04
125	0	0	0	0	8.04	175	0	0	0	0	8.04
126	0	0	0	0	8.04	176	0	0	0	0	8.04
127	0	0	0	0	8.04	177	0	1	0	0	12.06
128	0	0	0	0	8.04	178	0	1	0	0	12.06
129	1	0	0	0	45.73	179	0	1	0	0	12.06
130	1	0	0	0	45.73	180	0	0	0	1	9.08
131	1	0	0	0	45.73	181	0	0.9	0	0	11.65
132	0	0	0	1	9.08	182	0	0.9	0	0	11.65
133	0.9	0	0	0	40.11	183	0	0.9	0	0	11.65
134	0.9	0	0	0	40.11	184	0	0	0	0.9	8.98
135	0.9	0	0	0	40.11	185	0	0.44	0	0	9.83
136	0	0	0	0.9	8.98	186	0	0.44	0	0	9.83
137	0.44	0	0	0	37.39	187	0	0.44	0	0	9.83
138	0.44	0	0	0	37.39	188	0	0	0	0.44	8.53
139	0.44	0	0	0	37.39	189	0	0	0	0	8.04
140	0	0	0	0.44	8.53	190	0	0	0	0	8.04
141	0	0	0	0	8.04	191	0	0	0	0	8.04
142	0	0	0	0	8.04	192	0	0	0	0	8.04
143	0	0	0	0	8.04	193	1	0	0	0	14.07
144	0	0	0	0	8.04	194	1	0	0	0	14.07
145	1	0	0	0	45.73	195	1	0	0	0	14.07
146	1	0	0	0	45.73	196	0	0	0	1	9.08
147	1	0	0	0	45.73	197	0.9	0	0	0	13.45
148	0	0	0	1	9.08	198	0.9	0	0	0	13.45
149	0.9	0	0	0	40.11	199	0.9	0	0	0	13.45
150	0.9	0	0	0	40.11	200	0	0	0	0.9	8.98

(continued)

Table 9.8 (continued)

Scenario: s Optimal recovery supply portfolio: V_1^s V_2^s V_3^s V_4^s Optimal expected cost: R_s^c (9.30)

s	i=1	2	3	4	Cost	s	i=1	2	3	4	Cost
201	0.44	0	0	0	10.72	229	0.9	0	0	0	13.45
202	0.44	0	0	0	10.72	230	0.9	0	0	0	13.45
203	0.44	0	0	0	10.72	231	0.9	0	0	0	13.45
204	0	0	0	0.44	8.53	232	0	0	0	0.9	8.98
205	0	0	0	0	8.04	233	0.44	0	0	0	10.72
206	0	0	0	0	8.04	234	0.44	0	0	0	10.72
207	0	0	0	0	8.04	235	0.44	0	0	0	10.72
208	0	0	0	0	8.04	236	0	0	0	0.44	8.53
209	1	0	0	0	14.07	237	0	0	0	0	8.04
210	1	0	0	0	14.07	238	0	0	0	0	8.04
211	1	0	0	0	14.07	239	0	0	0	0	8.04
212	0	0	0	1	9.08	240	0	0	0	0	8.04
213	0.9	0	0	0	13.45	241	0	1	0	0	12.06
214	0.9	0	0	0	13.45	242	0	1	0	0	12.06
215	0.9	0	0	0	13.45	243	0	1	0	0	12.06
216	0	0	0	0.9	8.98	244	0	0	0	1	9.08
217	0.44	0	0	0	10.72	245	0	0.9	0	0	11.65
218	0.44	0	0	0	10.72	246	0	0.9	0	0	11.65
219	0.44	0	0	0	10.72	247	0	0.9	0	0	11.65
220	0	0	0	0.44	8.53	248	0	0	0	0.9	8.98
221	0	0	0	0	8.04	249	0	0.44	0	0	9.83
222	0	0	0	0	8.04	250	0	0.44	0	0	9.83
223	0	0	0	0	8.04	251	0	0.44	0	0	9.83
224	0	0	0	0	8.04	252	0	0	0	0.44	8.53
225	1	0	0	0	14.07	253	0	0	0	0	8.04
226	1	0	0	0	14.07	254	0	0	0	0	8.04
227	1	0	0	0	14.07	255	0	0	0	0	8.04
228	0	0	0	1	9.08	256	0	0	0	0	8.04

[a] Optimal primary supply portfolio: $v_1 = v_2 = 0, v_3 = 1, v_4 = 0$, optimal expected cost: 8.04 (9.21).

The probability P_s of realizing each disruption scenario $s \in S$ is calculated as follows:
$P_s = \beta_1 P_s^1 P_s^2$ for $s \leq 256$,
$P_s = \beta_2 P_{s-256}^1 P_{s-256}^2$ for $257 \leq s \leq 512$,
$P_s = \beta_3 P_{s-512}^1 P_{s-512}^2$ for $513 \leq s \leq 768$,
where probabilities P_s^r; $r = 1, 2$, $s \leq 256$ are defined at the beginning of this section, and β_1, β_2, and β_3 are nonnegative constants such that: $\beta_1 + \beta_2 + \beta_3 = 1$.

Notice that for $\beta_1 = 1$, P_s reduces to probability of realizing scenario s with a common disruption start time $t_s = 1$, from Sect. 9.5.1.

Table 9.9 Solution results for different disruption start times: $C/D = 0.93$

g	1	10	10^2	10^3	10^4	10^5	∞[a]
Integrated approach: model **Support_E**							
Var.=30596, Bin.=6148, Cons.=31365, Nonz.=451299[e]							
Exp.Cost E^c (9.4)	8.37	9.20	15.43	75.73	677	6686	–
Exp.Service $E^{sl} \times 100\%$ (9.5)	87	93	93	93	93	93	93
Primary Portfolio:	1(0)	1(0)	1(0)	1(7)	1(8)	1(17)	1(33)
Supplier(% of total demand)[b]	2(0)	2(7)	2(8)	2(7)	2(9)	2(20)	2(31)
	3(100)	3(93)	3(92)	3(86)	3(80)	3(51)	3(10)
	4(0)	4(0)	4(0)	4(0)	4(3)	4(12)	4(26)
Exp.Recovery Portfolio:	1(0.18)	1(0.20)	1(0.19)	1(0.17)	1(0.16)	1(1.13)	1(2.90)
Supplier(% of total demand)[c]	2(1.86)	2(1.81)	2(1.76)	2(1.64)	2(1.55)	2(1.16)	2(7.70)
	3(0)	3(0.22)	3(0.29)	3(0.39)	3(0.7)	3(1.84)	3(10.39)
	4(7.21)	4(6.67)	4(6.57)	4(6.27)	4(5.82)	4(3.80)	4(5.22)
Hierarchical approach: models **PSupport** and **RSupport_E**							
PSupport: Var.=36, Bin.=4, Cons.=37, Nonz.=534							
Primary Portfolio:	1(0)	1(0)					1(0)
Supplier(% of total demand)[b]	2(0)	2(7)					2(100)
	3(100)	3(93)					3(0)
	4(0)	4(0)					4(0)
Cost P^c, (21)	8.17	9	15	75	675	6675	–
Service[d]	87	93	93	93	93	93	93
RSupport_E: Var.=23846, Bin.=2304, Cons.=15838, Nonz.=203756							
Exp.Recovery Portfolio:	1(0.18)	1(0.20)					1(1.17)
Supplier(% of total demand)[c]	2(1.86)	2(1.81)					2(1.27)
	3(0)	3(0.22)					3(1.55)
	4(7.21)	4(6.67)					4(0.07)
Exp.Cost E^c (9.4)	8.37	9.20	15.48	78.26	706	6985	–
Exp.Service $E^{sl} \times 100\%$ (9.5)	87	93	93	93	93	93	93

[a] Maximization of service level
[b] $1(v_1 \times 100), 2(v_2 \times 100), 3(v_3 \times 100), 4(v_4 \times 100)$
[c] $1(\sum_{s\in S} P_s V_1^s \times 100), 2(\sum_{s\in S} P_s V_2^s \times 100), 3(\sum_{s\in S} P_s V_3^s \times 100), 4(\sum_{s\in S} P_s V_4^s \times 100)$
[d] $\sum_{t\in T} X_t / D \times 100$
[e] Var.=number of variables, Bin.=number of binary variables, Cons.=number of constraints, Nonz.= number of nonzero coefficients

As illustrative examples, the optimal solutions for $\beta_1 = 0.1$, $\beta_2 = 0.2$, $\beta_3 = 0.7$, and capacity-to-demand ratio, $C/D = 0.93$, are summarized in Tables 9.9, for both integrated and hierarchical approach. In contrast to examples in Sect. 9.5.1, now the probabilities for disruptions with later start times are higher.

Comparison of solution results in Tables 9.6 and 9.9 indicate for the two types of disruption scenarios, nearly identical optimal values of expected cost are achieved for the integrated approach and slightly worse for the hierarchical approach. However, the optimal supply portfolios for the corresponding penalty, g, for unfulfilled demand for products, are not identical (except for the smallest, $g = 1$). This indicates

Table 9.10 Worst-case scenarios: $C/D = 0.93,\ g = 100$

Disruption scenario $(\lambda_{1s}, \lambda_{2s}, \lambda_{3s}, \lambda_{4s}),\ t_s$	Recovery portfolio $(V_1^s, V_2^s, V_3^s, V_4^s)$	Cost[a]	Service Level[b]
Integrated approach: primary portfolio (v_1, v_2, v_3, v_4)= (0,0.08,0.92,0)			
Hierarchical approach: primary portfolio (v_1, v_2, v_3, v_4)= (0,0.07,0.93,0)			
(0,0,0,0), 1	(0,1,0,0)	2062	50
(0,0,0,0), 2	(0,1,0,0)	2065	47
(0,0,0,0), 3	(0,1,0,0)	2069	43[c]
(0,0,0,1), 1	(0,0,0,1)	492	50
(0,0,0,1), 2	(0,0,0,1)	496	47
(0,0,0,1), 3	(0,0,0,1)	499	43[c]
(1,0,0,0), 3	(1,0,0,0)	331	50
(1,0,0,1), 3	(1,0,0,0)	331	50
(0,0,0,2), 3	(0,0,0,1)	102	50
(1,0,0,2), 3	(0,0,0,1)	102	50

[a] $\sum_{i\in I} e_i(u_i + U_i^s - \mu_i^s)/D + \sum_{i\in I} p_{is}U_i^s/D + \sum_{i\in I} o_i(\gamma_i^s v_i + V_i^s) + g(1 - \sum_{t\in T} x_t^s/D)$

[b] $\sum_{t\in T} x_t^s/D \times 100$

[c] $\underline{SL} \times 100$ (9.20)

that for both types of scenarios the impact of disruption risks can be similarly mitigated by best selection of primary and recovery portfolios, using the developed approach.

9.5.1.3 Best-case and Worst-Case Analysis

In order to provide more insights, best-case and worst-case scenarios were identified such that service level associated with optimal solution is not less than capacity-to-demand ratio, C/D, and not greater than some fraction of C/D, respectively. For example, the total number of best-case scenarios for, $C/D = 0.93$, and unit penalty, $g = 100$, with service level, $\sum_{t\in T} x_t^s/D = 0.93$, is 298 and 192, respectively, for integrated and hierarchical approach. The corresponding cost per product is ranging, respectively, from 15 to 36.69 and from 15 to 20.65. The best-case scenarios for hierarchical approach are a subset of those for the integrated approach. In contrast, the number of worst-case scenarios is much lower. Table 9.10 presents all worst-case scenarios for unit penalty $g = 100$, with service level $\sum_{t\in T} x_t^s/D \leq 0.5$. For both integrated and hierarchical approach the same set of 10 worst-case scenarios were identified. The table shows disruption scenario, $(\lambda_{1s}, \lambda_{2s}, \lambda_{3s}, \lambda_{4s})$ along with its start time, t_s, and the corresponding optimal recovery supply portfolio $(V_1^s, V_2^s, V_3^s, V_4^s)$, cost per product, $\sum_{i\in I} e_i(u_i + U_i^s - \mu_i^s)/D + \sum_{i\in I} p_{is}U_i^s/D + \sum_{i\in I} o_i(\gamma_i^s v_i + V_i^s) + g(1 - \sum_{t\in T} x_t^s/D)$, and service level $\sum_{t\in T} x_t^s/D$. Notice that the lowest service level, $\underline{SL} = 0.43$ (9.20), is attained for two scenarios only.

Table 9.10 indicates that for worst-case scenarios all primary suppliers are shut down as well as the other suppliers are hit by disruptions. When all suppliers were shut down, then a single sourcing recovery portfolio is selected with one of the primary suppliers chosen as a recovery supplier. Otherwise, a single recovery supplier is selected from among those less severely disrupted.
Similar results were obtained, for example, with other ratios C/D and unit penalty g as well as for scenarios with a common disruption start time.

The Gurobi solver was capable of finding proven optimal solution for all examples with CPU time ranging from fraction of a second for MIP models **PSupport** and **RSupport(s)** to a few seconds for SMIP models **Support_E** and **RSupport_E**.

9.5.2 Risk-averse Decision-Making

In this subsection some computational examples are presented to illustrate the risk-averse selection of primary and recovery supply portfolios to minimize CVaR of cost or maximize CVaR of service level. In the computational experiments data sets provided at the beginning of this section were used with producer per period capacity fixed to $c = 10000$ and hence $C/D = 0.93$. In addition, for minimization of CVaR of cost, unit penalty was fixed to $g = 100$. The solution results are shown in Tables 9.11 and 9.12, respectively, for scenarios with a common disruption start time and with different disruption start times. For the risk-averse solutions, VaR, CVaR, and the associated expected values of cost and service level are presented for a subset of confidence levels $\alpha = 0.5, 0.75, 0.9, 0.95, 0.99$. The risk-averse portfolio selected using model **Support_CV(sl)** is more diversified than the risk-averse portfolio determined by model **Support_CV(c)**. As α increases, $CVaR^c$ of cost increases and $CVaR^{sl}$ of service level decreases. At the same time, a greater diversification of the primary supply portfolio is observed for model **SupportCV(c)**, with more demand shifted from the cheapest supplier 3 to more reliable supplier 2. For model **Support_CV(sl)**, however, VaR^{sl}, E^{sl}, and primary supply portfolio are independent of α. $CVaR^{sl}$ and E^{sl} are very close to VaR^{sl}, while the latter is equal to $C/D = 0.93$. Comparison of solution results in Tables 9.11 and 9.12 for scenarios with a common and different disruption start times indicates that while the latter scenarios lead to higher values of cost, the corresponding values of service level are nearly identical.

Comparison of solution results for models **Support_E** and **Support_CV(c)**, **Support_CV(sl)** (cf. Tables 9.6, 9.9, 9.11 and 9.12), demonstrates that the risk-neutral solutions for minimization of expected cost are very close to the corresponding risk-averse solutions with confidence level $\alpha = 0.5$. Moreover, for the service level objective, the corresponding risk-neutral and the risk-averse primary supply portfolios are identical.

For the above numerical examples, proven optimal solutions were found with CPU time ranging from fraction of a second for model **Support_CV(sl)** to a few seconds for model **Support_CV(c)**.

Table 9.11 Risk-averse solutions for a common disruption start time: $C/D = 0.93$

Confidence level α	0.50	0.75	0.90	0.95	0.99
Model **Support_CV(c)**, $g = 100$: Var.=10462, Bin.=2052, Cons.=10718, Nonz.=163986[a]					
$CVaR^c$	15.50	15.92	16.93	18.03	22.37
VaR^c	15.05	15.09	15.57	16.09	18.82
E^c	15.28	15.30	15.71	15.65	17.70
$E^{sl}100\%$	93.29	93.31	92.92	93.13	93.21
Primary Portfolio:	1(2)	1(3)	1(3)	1(3)	1(4)
Supplier(% of total demand)[b]	2(5)	2(4)	2(5)	2(7)	2(64)
	3(93)	3(93)	3(92)	3(90)	3(31)
	4(0)	4(0)	4(0)	4(0)	4(0)
Exp.Recovery Portfolio:	1(0.34)	1(0.39)	1(0.31)	1(0.76)	1(0.29)
Supplier(% of total demand)[c]	2(1.81)	2(1.85)	2(1.78)	2(1.82)	2(2.34)
	3(0.03)	3(0.02)	3(0.07)	3(0.14)	3(1.67)
	4(6.69)	4(6.69)	4(6.64)	4(5.93)	4(1.39)
Model **Support_CV(sl)**: Var.=10462, Bin.=2052, Cons.=10718, Nonz.=159122 [a]					
$CVaR^{sl}100\%$	93.32	93.30	93.25	93.17	92.53
$VaR^{sl}100\%$	93.33				
$E^{sl}100\%$	93.32				
E^c	19.04	19.06	19.06	19.02	19.06
Primary Portfolio:	1(27)				
Supplier(% of total demand)[b]	2(25)				
	3(7)				
	4(41)				
Exp.Recovery Portfolio:	1(1.81)	1(1.14)	1(1.14)	1(1.80)	1(1.52)
Supplier(% of total demand)[c]	2(0.86)	2(1.55)	2(1.55)	2(0.86)	2(1.16)
	3(2.00)	3(2.41)	3(2.41)	3(2.54)	3(2.25)
	4(1.07)	4(0.64)	4(0.64)	4(0.53)	4(0.81)

[a] Var.=number of variables, Bin.=number of binary variables, Cons.=number of constraints, Nonz.= number of nonzero coefficients

[b] $1(v_1 \times 100), 2(v_2 \times 100), 3(v_3 \times 100), 4(v_4 \times 100)$

[c] $1(\sum_{s\in S} P_s V_1^s \times 100), 2(\sum_{s\in S} P_s V_2^s \times 100), 3(\sum_{s\in S} P_s V_3^s \times 100), 4(\sum_{s\in S} P_s V_4^s \times 100)$

The major managerial insights are summarized in Table 9.13. Overall, the main results of computational study for the risk-neutral and risk-averse decision-making models are in line with other research. The results of computational experiments indicate that *the proposed portfolio approach and developed SMIP models are a flexible and efficient tool for supply chain disruption management.* The approach has proven to be computationally very efficient.

Table 9.12 Risk-averse solutions for different disruption start times: $C/D = 0.93$

Confidence level α	0.50	0.75	0.90	0.95	0.99
Model **Support_CV(c)**, $g = 100$: Var.=31365, Bin.=6148, Cons.=32133, Nonz.=488799[a]					
$CVaR^c$	15.68	16.11	17.07	18.13	22.49
VaR^c	15.25	15.28	15.81	16.22	18.90
E^c	15.47	15.49	15.64	15.80	17.43
$E^{sl}100\%$	93.28	93.29	93.29	93.22	93.08
Primary Portfolio: Supplier(% of total demand)[b]	1(5)	1(4)	1(6)	1(6)	1(8)
	2(5)	2(6)	2(6)	2(7)	2(46)
	3(90)	3(90)	3(83)	3(80)	3(46)
	4(0)	4(0)	4(5)	4(7)	4(0)
Exp.Recovery Portfolio: Supplier(% of total demand)[c]	1(0.21)	1(0.18)	1(0.31)	1(0.45)	1(0.62)
	2(1.76)	2(1.71)	2(1.80)	2(2.45)	2(1.19)
	3(0.23)	3(0.13)	3(0.26)	3(0.38)	3(1.09)
	4(6.46)	4(6.60)	4(6.12)	4(5.13)	4(2.44)
Model **Support_CV(sl)**: Var.=31365, Bin.=6148, Cons.=32133, Nonz.=474207[a]					
$CVaR^{sl}100\%$	93.32	93.30	93.25	93.17	92.50
$VaR^{sl}100\%$	93.33				
$E^{sl}100\%$	93.32				
E^c	20.52	20.48	20.50	20.45	20.47
Primary Portfolio: Supplier(% of total demand)[b]	1(33)				
	2(31)				
	3(10)				
	4(26)				
Exp.Recovery Portfolio: Supplier(% of total demand)[c]	1(2.23)	1(2.35)	1(2.26)	1(1.79)	1(2.31)
	2(0.89)	2(0.81)	2(0.75)	2(0.59)	2(0.54)
	3(1.27)	3(1.16)	3(1.30)	3(1.77)	3(0.97)
	4(0.84)	4(0.90)	4(0.92)	4(1.07)	4(1.41)

[a] Var.=number of variables, Bin.=number of binary variables, Cons.=number of constraints, Nonz.= number of nonzero coefficients

[b] $1(v_1 \times 100), 2(v_2 \times 100), 3(v_3 \times 100), 4(v_4 \times 100)$

[c] $1(\sum_{s \in S} P_s V_1^s \times 100), 2(\sum_{s \in S} P_s V_2^s \times 100), 3(\sum_{s \in S} P_s V_3^s \times 100), 4(\sum_{s \in S} P_s V_4^s \times 100)$

9.6 Notes

The low-probability and high-impact disruptions of material flows in global supply chains and the resulting losses may threaten financial state of firms. For example, the disruptive events that occurred in 2011 in the automative and electronics supply chains (the Great East Japan earthquake and tsunami in March and Thailand's floods in October) resulted in huge losses of major automakers and electronics manufacturers, e.g., Park et al. (2013) and Haraguchi and Lall (2015). In order to minimize losses caused by the shortage of material supplies, customer companies (firms) apply different disruption management strategies, such as maintaining inventory, buying from an alternate supplier or helping a primary supplier recover more quickly. When the latter strategy is applied, the firm participates in supplier's recovery process after disruption to reduce recovery time, which means that the firm

Table 9.13 Managerial insights: primary and recovery supply portfolio selection and production scheduling

Decision-making issue	Decision-making impact
Risk-neutral decision-making	
Hierarchical vs. integrated decision-making	For the hierarchical decision-making the primary supply portfolio is made up of cheapest suppliers or a single, most reliable primary supplier only, to minimize expected cost or maximize expected service level, respectively The integrated decision-making selects a more diversified primary supply portfolio for both cost and service level objective function, that will hedge against all potential disruption scenarios Best-case scenarios for the hierarchical approach are a subset of those for the integrated approach
Worst-case scenarios	All primary suppliers are shut down. If all suppliers are shut down, then a single sourcing recovery portfolio is selected with one primary supplier chosen as a recovery supplier. Otherwise, a single recovery supplier is selected from among the less severely impacted suppliers
Worst service level	The lowest service level, associated with worst-case disruption scenario, depends on the smallest capacity of final manufacturer available for production after delivery of recovery supplies
Best expected service level	A simple upper bound on expected service level depends on production capacity-to-demand ratio
Risk-averse decision-making	
Worst-case service level	The impact of worst-case disruption scenarios can be fully mitigated by the optimal risk-averse selection of primary and recovery supply portfolios

may participate in supplier's cost-to-recover. The real-world examples presented in Sect. 1.1 illustrate a typical disruption management strategy. Whenever a primary supplier is hit by disruption, the customer company needs to choose whether to support recovery of disrupted primary supplier or select an alternate (recovery) supplier, non-disrupted or disrupted less severely than the primary supplier. The recovery suppliers are selected in such a way that the recovery process is optimized with respect to recovery time and cost. However, the literature on mitigation the impact of disruption risks and optimization of a recovery process in supply chains is limited (e.g., Tomlin 2006; Ruiz-Torres et al. 2013; MacKenzie et al. 2014; Zeng and Xia 2015; Hamdi et al. 2018; Ivanov et al. 2017). The literature reviews on Operations Research/Management Science models for supply chain disruptions were presented, for example, by Snyder et al. (2016) or Rajagopal et al. (2017). In the former review the authors discussed 180 scholarly works on the topic, organized into six categories: evaluating supply disruptions; strategic decisions; sourcing decisions; contracts and incentives; inventory; and facility location.

The supply chain recovery may sometimes be connected with the problem of scheduling jobs in an area contaminated with chemical or radio-active materials, e.g., disaster of Fukushima Daiichi Nuclear Power Plant in 2011 (Fujimoto and Park 2013). The jobs may include rescue, de-activation, or cleaning works to be executed by a team of workers. The dynamics of the harmful factor and the norms of organism recovery imply that each work period for a job should be accompanied by a rest period for the worker executing this job and the length of the rest period depends on the start time of the corresponding work period. In Sawik (2010b) a time-indexed integer programming formulation was proposed to support a joint decision-making on assignment, sequencing, and timing of jobs in a contaminated area.

In another stream of research, Paul et al. (2016) reviewed literature on managing risk and disruption in production-inventory systems and supply chains. They considered four categories of disruptions: disruption in production, disruption in supply, disruption in transportation, and fluctuation in demand. The authors focused on reviewing the mathematical models and the solution approaches used in solving the models using both hypothetical and real-world problem scenarios. Paul et al. (2014a,b, 2015a,b) developed mathematical models and solution algorithms for disruption management in production–inventory systems, under single or multiple disruptions. The authors considered back orders, lost sales, and/or outsourcing options as recovery strategies.

In practice, the probability and magnitude of low-probability and high-impact disruptive events are difficult and often impossible to be estimated. The classical risk assessment and mitigation methods that require such knowledge at an early stage of the risk analysis may fail to prepare supply chains for such disruptive events. A novel risk-exposure model for analyzing operational-disruption risk with no need to estimate the probability of any specific disruptive events was proposed by Simchi-Levi et al. (2015). The model is capable of assessing the impact of a disruption originating anywhere in a supply chain. The model has been applied by Ford Motor Company to identify risk exposures, evaluate risk mitigation actions, and develop optimal contingency plans in the automotive supply chain.

The material presented in this chapter is based on results achieved by Sawik (2017b), where a computationally efficient portfolio approach developed in Sawik (2011a,b, 2013e) was enhanced for simultaneous or sequential selection of primary and recovery supply portfolios under local and regional disruptions risks. However, in Sawik (2017b) only a risk-neutral, expected cost (or expected service level) objective has been considered to optimize an overall performance of a supply chain. Minimizing expected cost (or maximizing expected service level) under disruption risks may sometimes be impractical in the long run, especially when large losses could threaten financial state of firms. Then, a downside risk measure such as expected worst-case cost may be more appropriate. In this chapter, the portfolio approach and the risk-neutral models developed by Sawik (2017b) have been enhanced to assess the effect of risk-averse decision-making on the selection of primary and recovery supply portfolios, using CVaR as a risk measure.

In the literature on supply uncertainty, the supply is either subject to complete disruptions or yield uncertainty. Yield uncertainty occurs when the quantity of supply delivered is a random variable, modeled as either a random additive or multiplicative quantity, whereas disruptions occur when supply is subject to partial or complete failure. Typically disruptions are modeled as events which occur randomly and may have a random length. Schmitt and Singh (2012) considered inventory systems subject to both supply disruptions and yield uncertainty. They compared single-period vs. multi-period models and showed that the former can lead to selecting the wrong strategy for mitigating supply risk.

The future research should concentrate on relaxations of the various simplified assumptions used to formulate the problem. For example, the assumption that each supplier has sufficient capacity to meet total demand for parts that allows for a single sourcing can be easily relaxed. The developed models can be enhanced for finite capacity suppliers. A single recovery mode defined by constant recovery time, $TTR(i,l)$ and the associated recovery cost, $CTR(i,l)$ for each supplier i and each disruption level l, can be replaced by a number of available recovery modes, each represented by different value of time, $TTR(i,l)$ and the associated cost, $CTR(i,l)$. Then, a new recovery mode selection variable can enhance the problem formulation. In a more general setting, disruption start time, recovery time, and recovery cost can be modeled as random parameters, e.g., Schmitt (2011), Schmitt and Singh (2012), and Paul et al. (2017, 2018).

Problems

9.1. In the SMIP models presented in this chapter introduce delay penalty when demand d_t for period t is not met by that period.

9.2. In the SMIP models presented in this chapter introduce transition time required for switching to recovery supplier different from the primary supplier.

9.3. Enhance the SMIP models presented in this chapter for the limited output inventory of products at the producer. Formulate the inventory balance constraints that should be added to the models.

9.4. Formulate the mean-risk models **Support_ECV(c)** and **Support_ECV(sl)** to minimize expected cost and CVaR of cost or maximize expected service level and CVaR of service level, respectively.

9.5. Enhance model **Support_E** for multiple recovery modes defined by different recovery time, $TTR(i,l)$ and the associated recovery cost, $CTR(i,l)$ for each supplier i and each disruption level l.

Chapter 10
Selection of Primary and Recovery Supply and Demand Portfolios and Scheduling

10.1 Introduction

In this chapter the portfolio approach proposed in Chap. 9 for the selection of primary and recovery suppliers and order quantity allocation to mitigate the impact of disruption risks is enhanced also for the recovery process of the manufacturer's assembly plants for finished products. Unlike most of reported research on supply chain disruption management a disruptive event is assumed to impact both a primary supplier of parts and a primary assembly plant of the finished products manufacturer. Then the manufacturer may choose alternate (recovery) suppliers and move production to alternate (recovery) plants along with transshipment of parts from the impacted primary plant to the recovery plants. The resulting allocation of unfulfilled demand for parts among recovery suppliers, the inventory of parts, and unfulfilled demand for products among recovery assembly plants determines recovery supply, transshipment, and demand portfolio, respectively. The enhanced portfolio approach and SMIP formulations with an embedded network flow problem developed in this chapter are capable of selecting primary suppliers, the decision to be implemented before a disruption and of selecting recovery suppliers and recovery assembly plants, the decision to be implemented during and after the disruption. The supply, transshipment, and demand portfolios are determined along with production scheduling in assembly plants. Multi-level disruptions of suppliers and assembly plants will be considered. The objective of selection of primary and recovery suppliers and assembly plants and allocation of order quantity for parts, inventory of parts, and demand for products is to mitigate the impact of disruption risks and optimize the recovery process. The two decision-making approaches will be considered: an integrated approach with the perfect information about the future disruption scenarios, and a hierarchical approach with no such information available ahead of time. In the integrated approach, which accounts for all potential disruption scenarios, the primary supply portfolio that will hedge against all scenarios is

T. Sawik, *Supply Chain Disruption Management*,
International Series in Operations Research & Management Science 291,
https://doi.org/10.1007/978-3-030-44814-1_10

determined along with the recovery supply, transshipment, and demand portfolios and production schedule of finished products for each scenario. The problem objective is to minimize expected cost or maximize expected service level and minimize conditional cost-at-risk or maximize conditional service-at-risk over all scenarios, respectively, for risk-neutral and risk-averse decision-making. In the hierarchical approach first the primary supply portfolio is selected, and then, when a primary supplier or primary assembly plant is hit by a disruption, the recovery supply, transshipment, and demand portfolios are selected to optimize the process of recovery from the disruption.

The following time-indexed SMIP and MIP models are presented in this chapter:

DSupport_E for risk-neutral selection of primary and recovery supply and demand portfolios and production scheduling to minimize expected cost;

DSupportMP_E model **DSupport_E** for multiple part types and product types;

PSupport for selection of primary supply portfolio and production scheduling to minimize cost under deterministic conditions;

RDSupport(s) for selection of recovery supply and demand portfolios and production scheduling to minimize cost, for predetermined primary portfolios and realized disruption scenario;

RDSupport_E model **DSupport_E** for predetermined primary portfolios.

DSupport_CV(c) for risk-averse selection of primary and recovery supply and demand portfolios and production scheduling to minimize CVaR of cost;

DSupport_CV(sl) for risk-averse selection of primary and recovery supply and demand portfolios and production scheduling to maximize CVaR of service level;

RDSupport_CV(c) model **DSupport_CV(c)** for predetermined primary portfolios;

RDSupport_CV(sl) model **DSupport_CV(sl)** for predetermined primary portfolios.

Numerical examples, computational results, best-case and worst-case analysis, and some comparison of the two approaches for selection of primary and recovery supply and demand portfolios are provided in Sects. 10.5.1 and 10.5.2, respectively, for risk-neutral and risk-averse decision-making. The major managerial insights are summarized at the end of this chapter.

10.2 Problem Description

In this section the problem of simultaneous selection of primary and recovery supply and demand portfolios and production scheduling is considered.

Table 10.1 Notation: supply and demand portfolio selection and production scheduling

Indices	
i	= supplier, $i \in I$
j	= assembly plant, $j \in J$
l	= disruption level, $l \in L_i, i \in I, l \in L^j, j \in J$
r	= region, $r \in R$
s	= disruption scenario, $s \in S$
t	= planning period, $t \in T$
Input parameters	
c_j	= per period capacity of plant j
D	= total demand for products
ε_j	= additional per unit production cost at recovery plant $j > 1$
e_i	= fixed ordering cost for supplier i
φ_j	= fixed production setup cost at recovery plant $j > 1$
ψ_j	= per unit transshipment cost from primary plant $j = 1$ to plant $j > 1$
g	= per unit penalty cost for unfulfilled demand for products
o_i	= per unit price of parts purchased from supplier i
p_{il}	= probability of disruption level l for supplier i
π_{jl}	= probability of disruption level l for plant j
p^r	= regional disruption probability for region r
t_s	= start time period of disruptive event s
γ_{il}	= fraction of an order delivered by supplier i under disruption level l (supplier fulfillment rate)
γ_i^s	= fraction of an order delivered by supplier i under scenario s
δ_{jl}	= fraction of capacity of plant j available under disruption level l
δ_j^s	= fraction of capacity of plant j available under scenario s
τ_{ij}	= delivery lead time from supplier i to plant j
σ_j	= transshipment time from primary plant $j = 1$ to plant $j > 1$
θ_{is}	= time-to-recover of supplier i from disruption under scenario s
ϑ_{js}	= time-to-recover of plant j from disruption under scenario s
ρ_{is}	= cost-to-recover of supplier i from disruption under scenario s
ρ_{js}	= cost-to-recover of plant j from disruption under scenario s

Consider a supply chain in which a single producer of one product type assembles products in several assembly plants to meet customer demand, using a critical part type that can be manufactured and provided by several suppliers (for notation used, see Table 10.1). For example, (see, Sawik 2011a, 2017a), a single printed wiring board of a specific design that can be manufactured in surface mount technology lines at multiple locations is required for each type of cellular phone assembled at multiple distribution centers.

Let $I = \{1,\ldots,\overline{I}\}$ be the set of $\overline{I}$ suppliers, $J = \{1,\ldots,\overline{J}\}$, the set of $\overline{J}$ assembly plants, and $T = \{1,\ldots,\overline{T}\}$, the set of $\overline{T}$ planning periods.

Without loss of generality, we assume that one unit of critical part is required to manufacture one unit of finished product and denote by D, total demand for finished products as well as total demand for critical parts. Let $j = 1$ be the primary plant, where the total demand for products, D, is initially assigned.

The suppliers of parts and assembly plants are located in $\overline{R}$ geographic regions, subject to potential regional disasters that may result in complete shutdown of all suppliers and plants in the same region simultaneously. Denote by I^r and J^r, respectively, the subsets of suppliers and plants in region $r \in R$, and by p^r, the regional disruption probability for region r.

In addition to correlated regional disruptions, each supplier $i \in I$ is subject to independent local disruptions of different levels, $l \in L_i = \{0,\ldots,\overline{L_i}\}$, where disruption level refers to the fraction of an order that can be delivered, see, Sect. 9.2. Level $l = 0$ represents complete shutdown of a supplier, i.e., no order delivery, while level $l = \overline{L_i}$ represents normal conditions with no disruption, i.e., full order delivery. Denote by p_{il}, the probability of disruption level l for supplier i, and by γ_{il}, the fraction of an order that can be delivered by supplier i under disruption level l (supplier fulfillment rate),

$$\gamma_{il} = \begin{cases} 0 & \text{if } l = 0 \\ \in (0,1) & \text{if } l = 1,\ldots,\overline{L_i} - 1 \\ 1 & \text{if } l = \overline{L_i}. \end{cases} \tag{10.1}$$

Similarly to suppliers, each plant $j \in J$ is subject to random local disruptions of different levels, $l \in L^j = \{0,\ldots,\overline{L^j}\}$, where disruption level refers to available fraction of full capacity, c_j, available per period under normal conditions. Level $l = 0$ represents complete shutdown of an assembly plant, while level $l = \overline{L^j}$ represents normal conditions, i.e., full capacity, c_j, available. Denote by π_{jl}, the probability of disruption level l for plant j, and by δ_{jl}, the fraction of available capacity of plant j under disruption level l.

$$\delta_{jl} = \begin{cases} 0 & \text{if } l = 0 \\ \in (0,1) & \text{if } l = 1,\ldots,\overline{L^j} - 1 \\ 1 & \text{if } l = \overline{L^j}. \end{cases} \tag{10.2}$$

The total number of all potential scenarios is $\overline{S} = \prod_{i \in I}(\overline{L_i}+1)\prod_{j \in J}(\overline{L^j}+1)$. Each scenario $s \in S$ is represented by an $(\overline{I}+\overline{J})$—dimensional vector $\lambda_s = \{\lambda_{1s},\ldots,\lambda_{\overline{I}s}, \lambda_{\overline{I}+1,s},\ldots,\lambda_{\overline{I}+\overline{J},s}\}$, where $\lambda_{is} \in L_i$ is the disruption level of supplier $i \in I$ and $\lambda_{\overline{I}+j,s} \in L^j$ is the disruption level of plant $j \in J$, under scenario $s \in S$.

The probability P_s of disruption scenario $s \in S$ is, $P_s = \prod_{r \in R} P_s^r$, where P_s^r is the probability of realizing disruption scenario s in region r

$$P_s^r = \begin{cases} (1-p^r)(\prod_{i \in I^r}\prod_{l \in L_i:\lambda_{is}=l} p_{il})(\prod_{j \in J^r}\prod_{l \in L^j:\lambda_{\overline{I}+j,s}=l} \pi_{jl}), \\ \text{if } \sum_{i \in I^r}\lambda_{is} + \sum_{j \in J^r}\lambda_{\overline{I}+j,s} > 0 \\ p^r + (1-p^r)(\prod_{i \in I^r} p_{i0})(\prod_{j \in J^r}\pi_{j0}), \\ \text{if } \sum_{i \in I^r}\lambda_{is} + \sum_{j \in J^r}\lambda_{\overline{I}+j,s} = 0. \end{cases} \tag{10.3}$$

When supplier i is hit by disruption at level l, its recovery process to normal conditions takes $TTR(i,l)$ time periods (time-to-recover) and let $CTR(i,l)$ be the manufacturer's portion of cost-to-recover. For each supplier i, denote by θ_{is} and ρ_{is}, respectively, time-to-recover and manufacturer's portion of cost-to recover from disruption under scenario s

$$\theta_{is} = TTR(i,l);\ i \in I, s \in S : l = \lambda_{is}, \tag{10.4}$$

$$\rho_{is} = CTR(i,l);\ i \in I, s \in S : l = \lambda_{is}. \tag{10.5}$$

Similarly, when plant j is hit by disruption at level l, its recovery process to normal conditions takes $PRT(j,l)$ time periods (Plant Recovery Time) and cost $PRC(j,l)$ (Plant Recovery Cost). For each plant j, denote by ϑ_{js} and ρ_{js}, respectively, time-to-recover and cost-to-recover from disruption under scenario s

$$\vartheta_{js} = PRT(j,l);\ j \in J, s \in S : l = \lambda_{\bar{I}+j,s}, \tag{10.6}$$

$$\rho_{js} = PRC(j,l);\ j \in J, s \in S : l = \lambda_{\bar{I}+j,s}. \tag{10.7}$$

The manufacturer who moves production to an alternate assembly plant $j \in J$ incurs a fixed production setup cost φ_j and encounters additional per unit cost of production ε_j, and per unit cost, ψ_j, of transshipment of parts from the primary plant, where $\varphi_1 = 0$, $\varepsilon_1 = 0$, and $\psi_1 = 0$. A recovery plant can be a disrupted primary plant $j = 1$ with reduced capacity during recovery process and then with its full capacity or a new plant, non-disrupted or disrupted less severely than the primary plant.

The following assumptions are made to formulate the problem:

- Each supplier has sufficient capacity to meet total demand for parts
- Orders for parts are placed at the beginning of the planning horizon and, under normal conditions, parts ordered from supplier i are delivered to assembly plant j in period τ_{ij}, where τ_{ij} is the total of manufacturing lead time and transportation time.
- Disruption to primary supplier (primary assembly plant) under scenario $s \in S$ occurs in period t_s and recovery process starts in period $t_s + 1$, so that the disrupted supplier $i \in I$ (disrupted plant $j \in J$) returns to its full capacity in period $t = t_s + \theta_{is}$ ($t = t_s + \vartheta_{js}$).
- If a disruptive event $s \in S$ may occur at different start times $t_s \in T_s$, then it is represented by equivalent $|T_s|$ scenarios (where $|T_s|$ denotes the power of a set T_s), one for each start time. As a result, the total number of all potential scenarios to be considered is $\sum_{s \in S} |T_s|$.
- A single disruption scenario is assumed to realize over the entire planning horizon. Multiple disruptions, one after the other in a series, during the recovery process are not considered.
- A single recovery mode is considered for each supplier, each assembly plant, and each disruption level.
- A recovery supplier can be a disrupted primary supplier after its recovery to full capacity or a new supplier.
- A recovery assembly plant can be a disrupted primary assembly plant during and after its recovery to full capacity or a new assembly plant.
- Partial recovery of a disrupted supplier to its partial capacity as well as partial recovery of a disrupted assembly plant to its partial capacity are not considered.
- Time-to recover and the associated cost-to-recover are constant parameters that represent recovery of a disrupted supplier or a disrupted assembly plant to its full capacity.

- Transition time required for switching to recovery supplier different from the primary suppliers is negligible.
- The manufacturer participates in supplier's cost-to-recover.

The two different types of demand satisfaction scenarios will be considered;

1. Period-dependent demand that should be satisfied in each period with no inventory allowed and a penalty cost charged for unfulfilled demand in every period;
2. Period-independent demand, not related to planning periods, that should be satisfied by the end the planning horizon with a penalty cost charged for total unfulfilled demand.

In both cases the assumption of a single disruption scenario realized over the entire planning horizon allows a two-stage stochastic setting to be applied with multi-period production decisions made in the second stage, even if the demand is required at the end of the planning horizon. The multi-period decisions are necessary to account for different delivery lead times from suppliers and different transshipment times to recovery plants. The time-varying availability of parts restricts the usage of available production capacity under each disruption scenario.

Notice that the two-stage stochastic model for a multi-period problem is not an approximation of the related multistage one, since the realization of the disruption scenario in a given period is fully conditioning the unique realizations of the stochastic parameters in the next period. Actually, the assumption of a single disruption over the entire planning horizon ensures that disruption, s, realized in period, t_s, fully conditions the unique realizations of all stochastic parameters (i.e., fulfillment rates of suppliers, capacity of plants, time-to recover, and cost-to-recover of suppliers and plants) in all remaining periods of the planning horizon, $t_s, \dots, \overline{T}$.

10.3 Models for Risk-Neutral Decision-Making

10.3.1 Integrated Selection of Primary and Recovery Supply and Demand Portfolios

In this subsection a SMIP model **DSupport_E** is presented for the integrated risk-neutral selection of supply and demand portfolios under disruption risks. Problem **DSupport_E** is a stochastic multi-portfolio selection problem, in which primary supply portfolio is selected along with recovery supply, transshipment, and demand portfolios for all potential disruption scenarios (see, Sect. 1.6).

The following primary and recovery portfolios are jointly determined using the proposed model (for definitions of first and second stage variables, see Table 10.2):

- Primary supply portfolio: $v_i \in [0, 1]$ is the fraction of total demand for parts ordered from selected primary supplier $i \in I$ (such that $u_i = 1$) to be delivered to primary plant $j = 1$.

Table 10.2 Problem variables: supply and demand portfolio selection and production scheduling

	First stage variables
u_i	$=1$, if supplier i is selected as a primary supplier; otherwise $u_i=0$ (primary supplier selection)
v_i	$\in[0,1]$, the fraction of total demand for parts ordered from primary supplier i to primary plant $j=1$ (primary supply portfolio).
	Second stage variables
q^s_j	$=1$, if assembly plant j is selected as a recovery plant under disruption scenario s; otherwise $q^s_j=0$ (recovery plant selection)
Q^s_j	$\in[0,1]$ the fraction of total demand for products to be completed by recovery plant j under disruption scenario s (recovery demand portfolio)
U^s_i	$=1$, if supplier i is selected as a recovery supplier under disruption scenario s; otherwise $U^s_i=0$ (recovery supplier selection)
V^s_{ij}	$\in[0,1]$, the fraction of total demand for parts ordered from recovery supplier i to recovery plant j, under disruption scenario s (recovery supply portfolio)
w^s_j	$\in[0,1]$, the fraction of total demand for parts, transshipped from the primary plant $j=1$ to recovery plant j, under disruption scenario s, where w^s_1 represents the parts that remain in the primary plant $j=1$ (recovery transshipment portfolio)
x^s_{jt}	≥ 0, production in plant j in period t under disruption scenario s (production scheduling).
	Auxiliary variables
VaR^c	cost-at-risk, the targeted cost such that for a given confidence level α, for $100\alpha\%$ of the scenarios, the outcome is below VaR^c
VaR^{sl}	service-at-risk, the targeted service level such that for a given confidence level α, for $100\alpha\%$ of the scenarios, the outcome is above VaR^{sl}
$\mathscr{C}_s$	≥ 0, the tail cost for scenario s, i.e., the amount by which costs in scenario s exceed VaR^c
$\mathscr{S}_s$	≥ 0, the tail service level for scenario s, i.e., the amount by which VaR^{sl} exceeds service level in scenario s
μ^s_i	$=1$, if $u_i=U^s_i=1$; otherwise $\mu^s_i=0$ (elimination of double fixed ordering costs)

- Recovery supply portfolio: $V^s_{ij}\in[0,1]$ is the fraction of total demand for parts ordered from selected recovery supplier $i\in I$ (such that $U^s_i=1$) to selected recovery assembly plant $j\in J$ (such that $q^s_j=1$), under disruption scenario s.
- Recovery transshipment portfolio: $w^s_j\in[0,1]$ is the fraction of total demand for parts transshipped from primary assembly plant $j=1$ to selected recovery plants, $j\in J$ (such that $q^s_j=1$), under disruption scenario s.
- Recovery demand portfolio: $Q^s_j\in[0,1]$ is the fraction of total demand for products to be completed by recovery plant $j\in J$ (such that $q^s_j=1$) under disruption scenario s.

Since the total demand for products has been assigned to the single primary assembly plant $j=1$ and all parts from the primary suppliers are ordered for that plant, the primary demand portfolio is not considered.

The primary supply portfolio, $(v_1,\ldots,v_{\bar{I}})$, which determines supply of parts to the primary assembly plant $j=1$ is selected along with the recovery supply, transshipment, and demand portfolios for each disruption scenario.

The recovery supply portfolio for scenario s, $V^s_{ij}, i\in I, j\in J$, specifies supplies of parts, DV^s_{ij}, from recovery suppliers, $i\in I$, to recovery assembly plants, $j\in J$, where $0\leq V^s_{ij}\leq 1, i\in I, j\in J$ and $\sum_{i\in I}(\gamma^s_i v_i+\sum_{j\in J}V^s_{ij})=1$.

The recovery transshipment portfolio for scenario s, $(w_1^s,\ldots,w_J^s)$, specifies supplies of parts, Dw_j^s, from primary assembly plant, $j=1$, to recovery assembly plants, $j\in J$, where $0\le w_j^s\le 1, j\in J$ and $\sum_{j\in J} w_j^s = V_0+\sum_{i\in I}\gamma_i^s v_i - \sum_{t\in T:t<t_s} x_{1t}^s/D$. $\sum_{t\in T:t<t_s} x_{1t}^s$, denotes the usage of parts for production at primary assembly plant $j=1$, before disruption and DV_0 is the initial inventory of parts at primary assembly plant $j=1$.

The recovery demand portfolio for scenario s, $(Q_1^s,\ldots,Q_J^s)$, specifies an allocation of unfulfilled demand for products, DQ_j^s, among selected recovery plants $j\in J$, where $0\le Q_j^s\le 1, j\in J$ and $\sum_{t\in T:t<t_s} x_{1t}^s/D+\sum_{j\in J} Q_j^s=1$.

Let E^c be the expected cost per product to be minimized, and E^{sl} the expected service level, i.e., the expected fraction of the total fulfilled demand for products (the expected demand fulfillment rate) to be maximized.

$$E^c = \sum_{s\in S} P_s \sum_{i\in I}(e_i(u_i+U_i^s-\mu_i^s)/D+\rho_{is}U_i^s/D+o_i(\gamma_i^s v_i+\sum_{j\in J} V_{ij}^s))$$
$$+\sum_{s\in S} P_s \sum_{j\in J}(\psi_j w_j^s+(\varphi_j+\rho_{js})q_j^s/D+\varepsilon_j\sum_{t\in T} x_{jt}^s/D)+g(1-E^{sl}), \tag{10.8}$$

$$E^{sl} = \sum_{s\in S}\sum_{j\in J}\sum_{t\in T} P_s x_{jt}^s/D. \tag{10.9}$$

The auxiliary variable, μ_i^s, is introduced to eliminate double charging with fixed ordering cost e_i of each supplier i, who is selected both as primary and recovery supplier: $\mu_i^s=1$, if $u_i=U_i^s=1$; otherwise $\mu_i^s=0$.

The expected cost per product, E^c (10.8), constitutes of different fixed and variable cost per product. The fixed cost per product includes expected cost: of ordering parts from primary suppliers and from recovery suppliers different from primary suppliers, $\sum_{i\in I} e_i u_i/D+\sum_{s\in S}\sum_{i\in I} P_s e_i(U_i^s-\mu_i^s)/D$, of recovery process for impacted suppliers selected as recovery suppliers, $\sum_{s\in S}\sum_{i\in I} P_s\rho_{is}U_i^s/D$, of moving production from primary assembly plant to recovery assembly plants, $\sum_{s\in S}\sum_{j\in J} P_s\varphi_j$ q_j^s/D, and of recovery process for impacted assembly plants selected as recovery plants, $\sum_{s\in S}\sum_{j\in J} P_s\rho_{js}q_j^s/D$. The variable cost per product includes cost: of purchasing parts for partially fulfilled deliveries from primary suppliers, $\sum_{s\in S}\sum_{i\in I} P_s o_i\gamma_i^s v_i$, of purchasing parts from recovery suppliers, $\sum_{s\in S}\sum_{i\in I}\sum_{j\in J} P_s o_i V_{ij}^s$, of transshipment of parts from primary plant to recovery plants, $\sum_{s\in S}\sum_{j\in J} P_s\psi_j w_j^s$, of additional production cost in recovery plants different from primary plant, $\sum_{s\in S}\sum_{j\in J}\sum_{t\in T}\varepsilon_j x_{jt}^s/D$, and of penalty for unfulfilled demand, $g(1-\sum_{s\in S}\sum_{j\in J}\sum_{t\in T} P_s x_{jt}^s/D)$.

DSupport_E: *Risk-neutral selection of primary and recovery* **D***emand and* **Supp***ly* **port***folios and production scheduling*

Minimize (10.8)

subject to

Primary supply portfolio selection constraints:

- the total demand for parts must be fully allocated among primary suppliers (i.e., the primary supply portfolio must be selected) (10.10),

(continued)

– demand for parts cannot be assigned to non-selected primary suppliers (10.11),

$$\sum_{i \in I} v_i = 1, \tag{10.10}$$

$$v_i \le u_i;\ i \in I, \tag{10.11}$$

Recovery supply, transshipment, and demand portfolio selection constraints:

– the unfulfilled demand for parts cannot be assigned to non-selected recovery supplier (10.12),
– the unfulfilled demand for parts cannot be ordered for non-selected recovery plant (10.13),
– the unfulfilled demand for products cannot be assigned to non-selected recovery plant (10.14),
– the parts from primary plant cannot be transshipped to non-selected recovery plants (10.15),
– the unfulfilled demand for parts must be fully allocated among recovery suppliers (i.e., the recovery supply portfolio must be selected) (10.16),
– the unfulfilled demand for products must be fully allocated among recovery plants (i.e., the recovery demand portfolio must be selected) (10.17),
– the supply and demand balance constraints ensure that for each recovery plant, the recovery supplies and transshipment of parts are in balance with the demand for products to be fulfilled by that plant (10.18),
– the transshipment balance constraints ensure that total transshipments of parts from primary supplier is equal to the initial inventory of parts and partially fulfilled supplies from primary suppliers less the usage of parts for production before a disruptive event (10.19),
– each supplier selected to both primary and recovery portfolio is charged exactly once with fixed ordering cost in the objective function (10.20),

$$V_{ij}^s \le U_i^s;\ i \in I, j \in J, s \in S, \tag{10.12}$$

$$V_{ij}^s \le q_j^s;\ i \in I, j \in J, s \in S, \tag{10.13}$$

$$Q_j^s \le q_j^s;\ j \in J, s \in S, \tag{10.14}$$

$$w_j^s \le q_j^s;\ j \in J, s \in S, \tag{10.15}$$

$$\sum_{i \in I} (\gamma_i^s v_i + \sum_{j \in J} V_{ij}^s) = 1 - V_0;\ s \in S, \tag{10.16}$$

$$\sum_{t \in T: t < t_s} x_{1t}^s / D + \sum_{j \in J} Q_j^s = 1;\ s \in S, \tag{10.17}$$

(continued)

$$\sum_{i\in I} V_{ij}^s + w_j^s = Q_j^s;\ j \in J, s \in S, \tag{10.18}$$

$$\sum_{j\in J} w_j^s = V_0 + \sum_{i\in I} \gamma_i^s v_i - \sum_{t\in T:t<t_s} x_{1t}^s/D;\ s \in S, \tag{10.19}$$

$$\mu_i^s \le (u_i + U_i^s)/2;\ i \in I, s \in S, \tag{10.20}$$

where DV_0 is the initial inventory of parts at primary assembly plant $j = 1$.

Production capacity constraints:

- before a disruptive event, the production at primary plant in every period cannot exceed its capacity (10.21),
- after a disruptive event, the production at each selected recovery plant in every period cannot exceed the plant available capacity (10.22),
- the total production at each recovery plant cannot exceed the assigned portion of total demand for products (10.23),

$$x_{1t}^s \le c_1;\ t \in T, s \in S : t < t_s, \tag{10.21}$$

$$x_{jt}^s \le c_{jt}^s q_j^s;\ j \in J, t \in T, s \in S : t \ge t_s, \tag{10.22}$$

$$\sum_{t\in T:t\ge t_s} x_{jt}^s/D \le Q_j^s;\ j \in J, s \in S. \tag{10.23}$$

Supply-transshipment-production coordinating constraints:

- for each disruption scenario s, the cumulative demand for parts of production scheduled in primary assembly plant $j = 1$ by period t cannot exceed the initial inventory of parts and the cumulative deliveries by period $t - 1$, i.e., delivery in period $\tau_{i1} \le t - 1$, from each primary supplier i and in period $t_s + \theta_{is} + \tau_{i1} \le t - 1$, from each recovery supplier i, less transshipment of parts to recovery assembly plants $j > 1$, if $t > \bar{t}_s$ (10.24),
- for each disruption scenario s, the cumulative demand for parts of production scheduled by period t in recovery assembly plant $j > 1$, cannot exceed the cumulative deliveries by period $t - 1$, i.e., delivery in period $t_s + \theta_{is} + \tau_{ij} \le t - 1$ from each recovery supplier i, that can be used for production in period $t_s + \theta_{is} + \tau_{ij} + 1$, at the earliest, plus transshipment of parts from primary assembly plant $j = 1$, if $t > \bar{t}_s + \sigma_j$ (10.25),

$$\sum_{t'\in T:t'\le t} x_{1t'}^s/D \le$$
$$V_0 + \sum_{i\in I:\tau_{i1}\le t-1} \gamma_i^s v_i + \sum_{i\in I:t_s+\theta_{is}+\tau_{i1}\le t-1} V_{i1}^s$$
$$-(\text{if } t \le \bar{t}_s \text{ then } 0 \text{ else } \sum_{j\in J:j>1} w_j^s);$$

(continued)

$$t \in T, s \in S \quad (10.24)$$

$$\sum_{t' \in T: t' \leq t} x^s_{jt'}/D \leq \sum_{i \in I: t_s + \theta_{is} + \tau_{ij} \leq t-1} V^s_{ij}$$
$$+(\text{if } t \leq \bar{t}_s + \sigma_j \text{ then } 0 \text{ else } w^s_j);$$
$$j \in J, t \in T, s \in S : j > 1, \quad (10.25)$$

where $\bar{t}_s = \max\{t_s, \max_{i \in I} \tau_{i1}\}$ and, $\bar{t}_s + 1$, is the start period of transshipment of parts to recovery plants.

Non-negativity and integrality conditions:

$$q^s_j \in \{0,1\};\ j \in J, s \in S, \quad (10.26)$$
$$Q^s_j \in [0,1];\ j \in J, s \in S, \quad (10.27)$$
$$u_i \in \{0,1\};\ i \in I, \quad (10.28)$$
$$v_i \in [0,1];\ i \in I, \quad (10.29)$$
$$U^s_i \in \{0,1\};\ i \in I, s \in S, \quad (10.30)$$
$$V^s_{ij} \in [0,1];\ i \in I, j \in J, s \in S, \quad (10.31)$$
$$w^s_j \in [0,1];\ j \in J, s \in S, \quad (10.32)$$
$$x^s_{jt} \geq 0;\ j \in J, t \in T, s \in S, \quad (10.33)$$
$$\mu^s_i \in \{0,1\};\ i \in I, s \in S, \quad (10.34)$$

where c^s_{jt} is the capacity in period t of plant j under disruption scenario s

$$c^s_{jt} = \begin{cases} \delta^s_j c_j, & \text{if } t_s \leq t \leq t_s + \vartheta_{js} - 1 \\ c_j, & \text{if } t \leq t_s - 1, t \geq t_s + \vartheta_{js}, \end{cases} \quad (10.35)$$

where $\delta^s_j = \delta_{j,\lambda_{\bar{I}+j,s}}$ is the fraction of capacity available at assembly plant j under scenario s, and $\lambda_{\bar{I}+j,s}$ is disruption level of plant j under scenario s.

Proposition 10.1. *Model* **DSupport_E** *contains an embedded network flow problem.*

Proof. In model **DSupport_E**, the supply and demand portfolio selection constraints (10.16)–(10.19), are flow conservation equations (see, Fig. 10.1), for primary assembly plant (node $j = 1$) and each recovery assembly plant (each node j), respectively. In Eqs. (10.16)–(10.19), primary portfolio variable, v_i, represents primary flow of parts, $\sum_{i \in I} \gamma^s_i v_i$, from primary supplier i (source node) to primary plant, $j = 1$ (sink/transshipment

(continued)

node), recovery portfolio variable, V_{ij}^s, represents recovery flow of parts from recovery supplier i (source node) to recovery assembly plant j (sink node), recovery transshipment portfolio variable, w_j^s, represents recovery flow of parts from primary plant, $j = 1$ (transshipment node) to recovery plant j (sink node), and recovery demand portfolio variable, Q_j^s, is the recovery outflow of products/parts from recovery plant j.

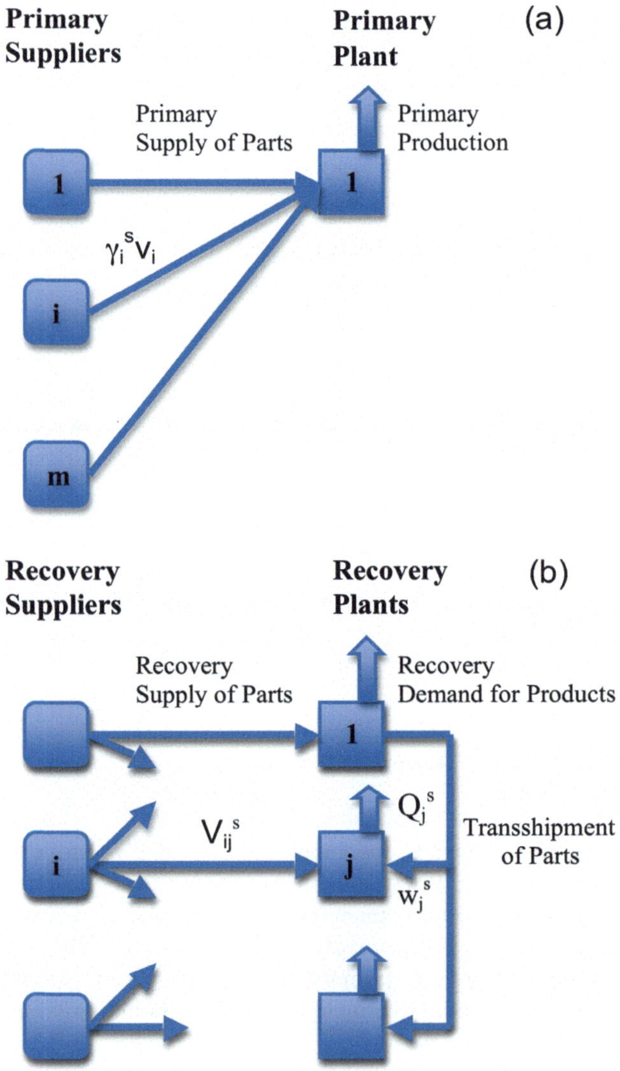

Fig. 10.1 Primary (**a**) and recovery (**b**) flow of parts under disruption risks

Model **DSupport_E** is a deterministic equivalent mixed integer program of a two-stage stochastic mixed integer program with recourse and illustrates the wait-and-see approach. The primary supply portfolio selection variables, u_i, v_i, are referred to as first stage decisions, and the recovery supply portfolio and recovery demand portfolio selection variables, U_i^s, V_{ij}^s and q_j^s, Q_j^s, as well as transshipment and production scheduling variables, w_j^s and x_{jt}^s, are referred to as recourse or second stage decisions (see, Table 10.2). Unlike the first stage decisions, the latter variables are dependent on disruption scenario $s \in S$.

Worst-Case Service Level

Proposition 10.2. *If there is no initial inventory of parts (i.e., $V_0 = 0$), the lowest service level, $\underline{SL}$, can be calculated as below.*

$$\underline{SL} = \max_{j \in J}\{c_j[\overline{T} - \max_{s \in S}\{t_s + \max(\min_{i \in I}\{\theta_{is} + \tau_{ij}\}, \vartheta_{js}, \sigma_j)]\}/D. \quad (10.36)$$

Proof. The lowest service level is associated with worst-case disruption scenario $s \in S$ for which primary suppliers and primary assembly plant are all hit by disruption at time t_s and then the plant j with maximum capacity available after its full recovery is selected as a recovery plant. The maximum available capacity of the recovery plant is based on the number of periods remaining for production after its full recovery (after $t_s + \vartheta_{js}$ periods), after the earliest delivery of parts by a recovery supplier (i.e., after $t_s + \min_{i \in I}\{\theta_{is} + \tau_{ij}\}$ periods) and after transshipment of parts from the primary plant (i.e., after $t_s + \sigma_j$ periods), whichever occurs later.

In order to strengthen MIP model **DSupport_E** the following constraint can be added:

$$\underline{SL} \le \sum_{t \in T: t < t_s} x_{1t}^s/D + \sum_{j \in J}\sum_{t \in T: t \ge t_s} x_{jt}^s/D \le 1; \; s \in S, \quad (10.37)$$

where the right-hand side of (10.37) is implied by Eqs. (10.17) and (10.23).

10.3.2 Multiple Part Types and Product Types

In this subsection an enhancement of model **DSupport_E** is described for multiple types of parts and products. Let H and K be, respectively, the set of part types and the set of product types, and denote by a_{hk}, $h \in H, k \in K$ the number of parts type

Table 10.3 Notation and variables: multiple part types and product types

	Indices	
h	=	part type, $h \in H$
k	=	product type, $k \in K$
	Input parameters	
a_{hk}	=	unit requirement for part type h of product type k
A_h	=	$\sum_{k\in K} a_{hk}D_k$ total demand for parts type h
D_k	=	total demand for products type k
ε_{jk}	=	additional per unit production cost for product type k at recovery plant $j > 1$
g_k	=	per unit penalty cost for unfulfilled demand for products type k
o_{ih}	=	per unit price of parts type h purchased from supplier i
	First stage variables	
v_{ih}	$\in$	$[0,1]$, the fraction of total demand for parts type h ordered from primary supplier i to primary plant $j = 1$ (primary supply portfolio)
	Second stage variables	
Q^s_{jk}	$\in$	$[0,1]$ the fraction of total demand for products type k to be completed by recovery plant j under disruption scenario s (recovery demand portfolio)
V^s_{ijh}	$\in$	$[0,1]$, the fraction of total demand for parts type h ordered from recovery supplier i to recovery plant j, under disruption scenario s (recovery supply portfolio)
w^s_{jh}	$\in$	$[0,1]$, the fraction of total demand for parts type h, transshipped from the primary plant $j = 1$ to recovery plant j, under disruption scenario s, where w^s_{1h} represents the parts type h that remain in the primary plant $j = 1$ (transshipment variables)
x^s_{jkt}	$\geq$	0, production in plant j of product type k in period t under disruption scenario s (production scheduling).

h required to produce one unit of product type k (for notation used and definition of problem variables, see Table 10.3).

If we denote by D_k the total demand for products type k, then $D = \sum_{k\in K} D_k$ is the total demand for all products and $A_h = \sum_{k\in K} a_{hk}D_k$ is the total demand for parts type h.

Now, the portfolio decision variables v_i, V^s_{ij}, $i \in I, j \in J, s \in S$ are replaced by v_{ih}, V^s_{ijh}; $h \in H, i \in I, j \in J, s \in S$, defined as fractions of total demand A_h for parts type h, ordered from primary supplier i, recovery supplier i for recovery plant j under disruption scenario s, respectively. Accordingly, the new transshipment variable w^s_{jh} denotes the fraction of total demand A_h for parts type h transshipped from the primary plant $j = 1$ to recovery plant j, under scenario s. In addition, the production scheduling variable x^s_{jt}, $j \in J, t \in T, s \in S$ is replaced by x^s_{jkt}, $j \in J, k \in K, t \in T, s \in S$—the production in plant j of product type k in period t under disruption scenario s, and Q^s_j, $j \in J, s \in S$ by Q^s_{jk}, $j \in J, k \in K, s \in S$—the fraction of demand for products type k to be completed in recovery plant j under disruption scenario s.

The portfolios are defined below.

The primary supply portfolio

$$(v_{1h}, \ldots, v_{\bar{I}h});\ h \in H,$$

specifies supplies of each part type h, $A_h v_{ih}$, from selected primary suppliers $i \in I$

(such that $u_i = 1$) to primary assembly plant $j = 1$, where

$$v_{ih} \in [0,1];\ i \in I, h \in H,$$
$$\sum_{i \in I} v_{ih} = 1;\ h \in H.$$

The recovery supply portfolio for scenario s

$$(V^s_{1jh}, \ldots, V^s_{\bar{I}jh});\ j \in J, h \in H,$$

specifies supplies of each part type h, $A_h V^s_{ijh}$ from selected recovery suppliers, $i \in I$ (such that $U^s_i = 1$), to selected recovery assembly plants, $j \in J$ (such that $q^s_j = 1$), where

$$V^s_{ijh} \in [0,1];\ i \in I, j \in J, h \in H,$$
$$\sum_{i \in I} (\gamma^s_i v_{ih} + \sum_{j \in J} V^s_{ijh}) = 1 - V_{h0};\ h \in H,$$

and $A_h V_{h0}$ is the initial inventory of parts type h at primary assembly plant $j = 1$, $\sum_{i \in I} \gamma^s_i v_{ih}$ denotes delivery of parts type h from primary suppliers to the primary plant.

The recovery transshipment portfolio for scenario s

$$(w^s_{1h}, \ldots, w^s_{\bar{J},h});\ h \in H,$$

specifies transshipment of each part type h, $A_h w^s_{jh}$ from primary assembly plant $j = 1$ to selected recovery plants, $j \in J$ (such that $q^s_j = 1$), where

$$w^s_{jh} \in [0,1];\ j \in J, h \in H,$$
$$\sum_{j \in J} w^s_{jh} = V_{h0} + \sum_{i \in I} \gamma^s_i v_{ih} - \sum_{k \in K} \sum_{t \in T: t < t_s} a_{hk} x^s_{1kt} / A_h;\ h \in H,$$

and $\sum_{k \in K} \sum_{t \in T: t < t_s} a_{hk} x^s_{1kt}$ denotes the usage of part type h for production at primary assembly plant $j = 1$, before disruption.

The recovery demand portfolio for scenario s

$$(Q^s_{1k}, \ldots, Q^s_{\bar{J},k});\ k \in K,$$

specifies an allocation of unfulfilled demand for each product type k, $D_k Q^s_{jk}$, among selected recovery plants $j \in J$ (such that $q^s_j = 1$), where

$$Q^s_{jk} \in [0,1];\ j \in J, k \in K,$$
$$\sum_{t \in T: t < t_s} x^s_{1kt} / D_k + \sum_{j \in J} Q^s_{jk} = 1;\ k \in K.$$

Model **DSupportMP_E** for multiple part and product types is presented below.

DSupportMP_E

Minimize

$$E^c = \sum_{s\in S} P_s \sum_{i\in I}(e_i(u_i + U_i^s - \mu_i^s) + \rho_{is}U_i^s + \sum_{h\in H} o_{ih}A_h(\gamma_i^s v_{ih} + \sum_{j\in J} V_{ijh}^s))/D$$
$$+ \sum_{s\in S} P_s \sum_{j\in J}(\sum_{h\in H} \psi_j A_h w_{jh}^s + (\varphi_j + \rho_{js})q_j^s + \sum_{k\in K}\sum_{t\in T} \varepsilon_{jk} x_{jkt}^s)/D$$
$$+ \sum_{k\in K} g_k(1 - \sum_{s\in S} P_s \sum_{j\in J}\sum_{t\in T} x_{jkt}^s / D_k),$$

subject to

Primary supply portfolio selection constraints:

- the total demand for each part type must be fully allocated among primary suppliers (i.e., the primary supply portfolio must be selected) (10.38),
- demand for parts cannot be assigned to non-selected primary suppliers (10.39),

$$\sum_{i\in I} v_{ih} = 1;\ h \in H, \tag{10.38}$$

$$v_{ih} \le u_i;\ h \in H, i \in I, \tag{10.39}$$

Recovery supply, transshipment, and demand portfolio selection constraints:

- the unfulfilled demand for each part type cannot be assigned to non-selected recovery supplier (10.40),
- the unfulfilled demand for each part type cannot be ordered for non-selected recovery plant (10.41),
- the unfulfilled demand for each product type cannot be assigned to non-selected recovery plant (10.42),
- the parts from primary plant cannot be transshipped to non-selected recovery plants (10.43),
- the unfulfilled demand for each part type must be fully allocated among recovery suppliers (i.e., the recovery supply portfolio must be selected) (10.44),
- the unfulfilled demand for each product type must be fully allocated among recovery plants (i.e., the recovery demand portfolio must be selected) (10.45),
- the supply and demand balance constraints ensure that for each recovery plant, the recovery supplies and transshipment of each part type

(continued)

are in balance with the demand for all products to be fulfilled by that plant (10.46),

- the transshipment balance constraints ensure that total transshipments of each part type from primary supplier is equal to the initial inventory of parts and partially fulfilled supplies from primary suppliers less the usage of parts for production before a disruptive event (10.47),
- each supplier selected to both primary and recovery portfolio is charged exactly once with fixed ordering cost in the objective function (10.48),

$$V^s_{ijh} \le U^s_i;\ h \in H, i \in I, j \in J, s \in S, \quad (10.40)$$

$$V^s_{ijh} \le q^s_j;\ h \in H, i \in I, j \in J, s \in S, \quad (10.41)$$

$$Q^s_{jk} \le q^s_j;\ j \in J, k \in K, s \in S, \quad (10.42)$$

$$w^s_{jh} \le q^s_j;\ j \in J, h \in H, s \in S, \quad (10.43)$$

$$\sum_{i\in I}(\gamma^s_i v_{ih} + \sum_{j\in J} V^s_{ijh}) = 1 - V_{h0};\ h \in H, s \in S, \quad (10.44)$$

$$\sum_{t\in T:t<t_s} x^s_{1kt}/D_k + \sum_{j\in J} Q^s_{jk} = 1;\ k \in K, s \in S, \quad (10.45)$$

$$\sum_{i\in I} V^s_{ijh} + w^s_{jh} = \sum_{k\in K} a_{hk} D_k Q^s_{jk}/A_h;\ h \in H, j \in J, s \in S, \quad (10.46)$$

$$\sum_{j\in J} w^s_{jh} = V_{h0} + \sum_{i\in I}\gamma^s_i v_{ih} - \sum_{k\in K}\sum_{t\in T:t<t_s} a_{hk} x^s_{1kt}/A_h;\ h \in H, s \in S, \quad (10.47)$$

$$\mu^s_i \le (u_i + U^s_i)/2;\ i \in I, s \in S, \quad (10.48)$$

where $A_h V_{h0}$ is the initial inventory of parts type h at primary assembly plant.

Production capacity constraints:

- before a disruptive event, the production at primary plant in every period cannot exceed its capacity (10.49),
- after a disruptive event, the production at each selected recovery plant in every period cannot exceed the plant available capacity (10.50),
- the total production at each recovery plant cannot exceed the assigned portion of total demand for products (10.51),

$$\sum_{k\in K} x^s_{1kt} \le c_1;\ t \in T, s \in S : t < t_s, \quad (10.49)$$

$$\sum_{k\in K} x^s_{jkt} \le c^s_{jt} q^s_j;\ j \in J, t \in T, s \in S : t \ge t_s, \quad (10.50)$$

$$\sum_{t\in T:t\ge t_s} x^s_{jkt}/D_k \le Q^s_{jk};\ j \in J, k \in K, s \in S. \quad (10.51)$$

Supply-transshipment-production coordinating constraints:

(continued)

– for each disruption scenario s, the cumulative demand for each part type of production scheduled in primary assembly plant $j = 1$ by period t cannot exceed the initial inventory of parts and the cumulative deliveries by period $t-1$, i.e., delivery in period $\tau_{i1} \le t-1$, from each primary supplier i and in period $t_s + \theta_{is} + \tau_{i1} \le t-1$, from each recovery supplier i, less transshipment of parts to recovery assembly plants $j > 1$, if $t > \bar{t}_s$ (10.52),
– for each disruption scenario s, the cumulative demand for each part type of production scheduled by period t in recovery assembly plant $j > 1$, cannot exceed the cumulative deliveries by period $t-1$, i.e., delivery in period $t_s + \theta_{is} + \tau_{ij} \le t-1$ from each recovery supplier i, that can be used for production in period $t_s + \theta_{is} + \tau_{ij} + 1$, at the earliest, plus transshipment of parts from primary assembly plant $j = 1$, if $t > \bar{t}_s + \sigma_j$ (10.53),

$$\sum_{k\in K}\sum_{t'\in T:t'\le t} a_{hk}x^s_{1kt'}/A_h \le V_{h0} + \sum_{i\in I:\tau_{i1}\le t-1} \gamma^s_i v_{ih}$$
$$+ \sum_{i\in I:t_s+\theta_{is}+\tau_{i1}\le t-1} V^s_{i1h}$$
$$-(\text{if } t \le \bar{t}_s \text{ then } 0 \text{ else } \sum_{j\in J:j>1} w^s_{jh});$$
$$h \in H, t \in T, s \in S \qquad (10.52)$$

$$\sum_{k\in K}\sum_{t'\in T:t'\le t} a_{hk}x^s_{jkt'}/A_h \le \sum_{i\in I:t_s+\theta_{is}+\tau_{ij}\le t-1} V^s_{ijh}$$
$$+(\text{if } t \le \bar{t}_s + \sigma_j \text{ then } 0 \text{ else } w^s_{jh});$$
$$h \in H, j \in J, t \in T, s \in S : j > 1, \qquad (10.53)$$

where $\bar{t}_s = \max\{t_s, \max_{i\in I}\tau_{i1}\}$ and, $\bar{t}_s + 1$, is the start period of transshipment of parts to recovery plants.

Non-negativity and integrality conditions:

$$q^s_j \in \{0,1\};\ j \in J, s \in S, \qquad (10.54)$$
$$Q^s_{jk} \in [0,1];\ j \in J, k \in K, s \in S, \qquad (10.55)$$
$$u_i \in \{0,1\};\ i \in I, \qquad (10.56)$$
$$v_{ih} \in [0,1];\ i \in I, h \in H, \qquad (10.57)$$
$$U^s_i \in \{0,1\};\ i \in I, s \in S, \qquad (10.58)$$
$$V^s_{ijh} \in [0,1];\ i \in I, j \in J, h \in H, s \in S, \qquad (10.59)$$
$$w^s_{jh} \in [0,1];\ j \in J, h \in H, s \in S, \qquad (10.60)$$
$$x^s_{jkt} \ge 0;\ j \in J, k \in K, t \in T, s \in S, \qquad (10.61)$$
$$\mu^s_i \in \{0,1\};\ i \in I, s \in S. \qquad (10.62)$$

Problem **DSupportMP_E** is a multi-portfolio selection problem, where supply, transshipment, and demand portfolios are simultaneously selected for each part type and each product type. Moreover, in model **DSupportMP_E**, the supply and demand balance constraints (10.46), and the supply-transshipment-production coordinating constraints (10.52)–(10.53) are also bill-of-material constraints.

Proposition 10.3. *Model* **DSupportMP_E** *contains an embedded multicommodity network flow problem.*

Proof. (cf. Proof of **Proposition 10.1**). In model **DSupportMP_E**, the supply and demand portfolio selection constraints (10.44)–(10.47), form an embedded multicommodity network flow problem. In particular, Eqs. (10.46) are recovery flow conservation equations.

Proposition 10.4. *The product demand fulfillment constraints (10.45) for each product type* $k \in K$*,*

$$\sum_{t \in T: t < t_s} x^s_{1kt}/D_k + \sum_{j \in J} Q^s_{jk} = 1;\ k \in K, s \in S,$$

imply part requirement satisfaction constraints for each part type $h \in H$*,*

$$\sum_{k \in K} \sum_{t \in T: t < t_s} a_{hk} x^s_{1kt}/A_h + \sum_{j \in J} \sum_{k \in K} a_{hk} D_k Q^s_{jk}/A_h = 1;\ h \in H, s \in S.$$

Proof. The part requirement satisfaction constraints can be obtained from the product demand fulfillment constraints by multiplying both sides of (10.45) by $a_{hk}D_k$ and summing over all $k \in K$, and then replacing the right-hand side, $\sum_{k \in K} a_{hk} D_k$, by A_h.

The part requirement satisfaction constraints for each part type can be added to model **DSupportMP_E** to strengthen MIP formulation.

Proposition 10.5. *The total usage of each part type h under each scenario s is the sum of usage before disruption (primary supplies of delivered parts less transshipment of parts) plus usage of parts required to satisfy unfulfilled demand for products after disruption, i.e.,*

$$\left(\sum_{i \in I} \gamma^s_i v_{ih} - \sum_{j \in J} w^s_{jh}\right) + \sum_{j \in J} \sum_{k \in K} a_{hk} D_k Q^s_{jk}/A_h = 1 - V_{h0};\ h \in H, s \in S.$$

(continued)

Proof. The latter constraints can be derived from (10.47) by using part requirement satisfaction constraints, defined in **Proposition 10.4**, to eliminate x^s_{1kt}.

Notice that when the demand is period-dependent and should be satisfied in each period, then the following constraints should be added to model: **DSupportMP_E**:

$$\sum_{j\in J} x^s_{jkt} \le d_{kt};\ k\in K, t\in T, s\in S. \tag{10.63}$$

The last constraints ensure that in each planning period, production of each product type cannot exceed the product demand and hence product inventory is not allowed. Thus, the penalty cost can be charged for unfulfilled demand in each period.

Proposition 10.6. *The optimal performance of supply chain with demand required at the end of the planning horizon may outperform that for period-dependent demand, in particular for a single part/product type model* **DSupport_E**.

Proof. When model **DSupport_E** with Eq. (10.63) for a single product type is applied, then capacity constraints (10.21)–(10.23) and no inventory constraints (10.63) imply

$$\sum_{j\in J}\sum_{t\in T} x^s_{jt} \le \sum_{t\in T:t<t_s} \min\{c_1, d_t\} + \sum_{t\in T:t\ge t_s} \min\{\sum_{j\in J} c^s_{jt} q^s_j, d_t\};\ s\in S,$$

which may result in unused production capacity

$$\sum_{t\in T:t<t_s} \max\{c_1 - d_t, 0\} + \sum_{t\in T:t\ge t_s} \max\{\sum_{j\in J} c^s_{jt} q^s_j - d_t, 0\};\ s\in S,$$

and by that the inferior supply chain performance.

The above formulation for the period-dependent demand will be further enhanced in Chap. 12 to allow for delayed satisfaction of demand using product inventory. In Chap. 12 production and inventory of finished products will be scheduled simultaneously, and additional cost of inventory holding and cost of penalty for delayed demand will be introduced in the objective function.

Model **DSupportMP_E** can also extended in other directions, for example, by introducing subsets $I_h \subset I$ of suppliers for each part type h, unit capacity consumption for each product type k and each plant j in the left-hand side of production capacity constraints (10.49) (10.50), etc.

Notice that an optimal solution to model **DSupportMP_E** may include recovery plant, $j > 1$, different from primary plant, $j = 1$, even if the primary plant is not hit by disruption. In order to ensure that no recovery plant $j > 1$ is selected if the primary plant $j = 1$ is not disabled by disruption s, i.e., when $\lambda_{\bar{l}+1,s} = \overline{L^1}$, the following constraint should be added to model **DSupportMP_E**:

$$q_j^s \leq \overline{L^1} - \lambda_{\bar{l}+1,s};\ j \in J, s \in S : j > 1.$$

10.3.3 Hierarchical Selection of Primary and Recovery Supply and Demand Portfolios

In this subsection two deterministic MIP models **PSupport** and **RDSupport(s)** are presented for the hierarchical decision-making in the presence of supply chain disruption risks. The hierarchical approach is based on a nonanticipative policy (e.g., Birge and Louveaux 2011), which means the decisions made at each stage cannot be dependent on the future realization of stochastic parameters. The first stage decisions are made in a deterministic environment: a primary supply portfolio and the corresponding production schedule are determined to optimize cost or service level in a supply chain with no disruptions. When a primary supplier or a primary assembly plant is hit by a disruptive event, then in the second stage the recovery supply, transshipment and, demand portfolios are found along with the associated production schedule, to optimize supply chain recovery under the realized disruption scenario.

The two-stage decision-making is described below (Fig. 10.2).

1. Selection of primary supply portfolio for deterministic environment.
   The primary suppliers are determined ahead of time with no disruption scenarios considered, using deterministic MIP model **PSupport**. The primary assembly plant is plant $j = 1$, by default.
2. Selection of recovery supply, transshipment, and demand portfolios, after disruption of a primary supplier and/or the primary assembly plant.
   The recovery portfolios are determined to optimize the process of recovery from the disruption, using MIP model **RDSupport(s)**.

In the deterministic MIP model **PSupport**, stochastic variable, x_{1t}^s (10.33), defined in model **DSupport_E** for each disruption scenario $s \in S$ has been replaced by its deterministic equivalent X_t.

PSupport: **P***rimary* **Sup***ply* **port***folio selection and production scheduling*

Minimize

$$P^c = \sum_{i \in I}(e_i u_i / D + o_i v_i) + g(1 - \sum_{t \in T} X_t / D), \tag{10.64}$$

(continued)

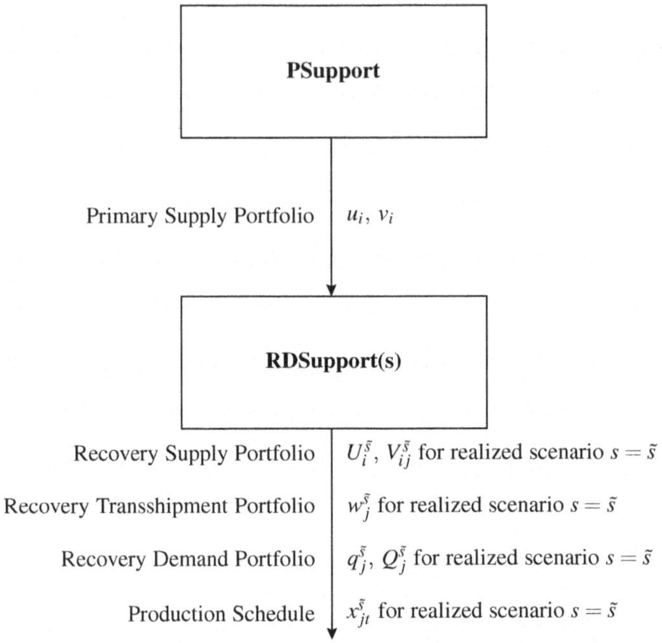

Fig. 10.2 Hierarchical selection of supply and demand portfolios

subject to

$$\sum_{i \in I} v_i = 1, \tag{10.65}$$

$$v_i \leq u_i;\ i \in I, \tag{10.66}$$

$$\sum_{t' \in T: t' \leq t} X_{t'}/D \leq V_0 + \sum_{i \in I: \tau_{i1} \leq t-1} v_i;\ t \in T, \tag{10.67}$$

$$X_t \leq c_1;\ t \in T, \tag{10.68}$$

$$u_i \in \{0,1\};\ i \in I, \tag{10.69}$$

$$v_i \in [0,1];\ i \in I, \tag{10.70}$$

$$X_t \geq 0;\ t \in T. \tag{10.71}$$

The solution to **PSupport** is primary supply portfolio: u_i^*, v_i^*; $i \in I$, and the associated production schedule, X_t^*.

A disruptive event under scenario $s \in S$ is assumed to occur in period t_s and it may impact a primary supplier (as well as the other suppliers) and the primary assembly plant $j = 1$ (as well as the other assembly plants). Then, the recovery process starts in period $t_s + 1$ so that the disrupted supplier $i \in I$ returns to its full capacity in period $t = t_s + \theta_{is}$ and disrupted plant $j \in J$, in period $t = t_s + \vartheta_{js}$.

Denote by D^s and V_0^s, respectively, the unfulfilled demand for products in period t_s and the inventory of parts at plant $j = 1$ in period t_s, expressed by the fraction of parts required to complete the unfulfilled demand D^s

$$D^s = D - \sum_{t \in T: t < t_s} x_{1t}^s; \ s \in S, \tag{10.72}$$

$$V_0^s = D(V_0 + \sum_{i \in I: \tau_i \le t_s} \gamma_i^s v_i^*)/D^s - \sum_{t \in T: t < t_s} x_{1t}^s / D^s; \ s \in S, \tag{10.73}$$

where $v_i^*, i \in I$, is a predetermined primary supply portfolio.

If $V_0^s > 0$, then selection of recovery portfolios may be combined with transshipment of parts from the primary plant to recovery plants.

When a disruption to primary assembly plant occurs after the latest delivery lead time from primary suppliers of parts, $t_s > \max_{i \in I}(\tau_{i1} u_i^*)$, i.e., after all required parts are delivered to the primary plant, then all parts can be transshipped to recovery plants and no recovery supply portfolio needs to be selected.

In model **RDSupport(s)** presented below, the recovery supply and demand portfolios are selected whenever a disruption occurs to a primary supplier or the primary assembly plant $j = 1$, given a primary supply portfolio and the realized disruption scenario $s = \tilde{s}$.

RDSupport(s): **R**_ecovery_ **D**_emand and_ **Sup**_ply_ **port**_folios selection and production scheduling for predetermined primary portfolios and realized disruption scenario s_

Minimize

$$R_{\tilde{s}}^c = \sum_{i \in I} e_i (1 - u_i^*) U_i^{\tilde{s}} / D^{\tilde{s}} + \sum_{i \in I} \rho_{i\tilde{s}} U_i^{\tilde{s}} / D^{\tilde{s}} + \sum_{j \in J} (\psi_j w_j^{\tilde{s}} + \sum_{i \in I} o_i V_{ij}^{\tilde{s}})$$
$$+ \sum_{j \in J} ((\varphi_j + \rho_{j\tilde{s}}) q_j^{\tilde{s}} + (\varepsilon_j - g) \sum_{t \in T^{\tilde{s}}} x_{jt}^{\tilde{s}}) / D^{\tilde{s}} + g$$
$$+ \sum_{i \in I} e_i u_i^* / D^{\tilde{s}} + \sum_{i \in I} o_i \gamma_i^{\tilde{s}} v_i^*, \tag{10.74}$$

subject to

$$V_{ij}^{\tilde{s}} \le U_i^{\tilde{s}}; \ i \in I, j \in J, \tag{10.75}$$

$$V_{ij}^{\tilde{s}} \le q_j^{\tilde{s}}; \ i \in I, j \in J, \tag{10.76}$$

$$Q_j^{\tilde{s}} \le q_j^{\tilde{s}}; \ j \in J, \tag{10.77}$$

$$\sum_{i \in I} V_{ij}^{\tilde{s}} + w_j^{\tilde{s}} = Q_j^{\tilde{s}}; \ j \in J, \tag{10.78}$$

$$\sum_{j \in J} w_j^{\tilde{s}} = V_0^{\tilde{s}}, \tag{10.79}$$

$$\sum_{j \in J} Q_j^{\tilde{s}} = 1, \tag{10.80}$$

(continued)

$$\sum_{t_{\tilde{s}} \le t' \le t} x^{\tilde{s}}_{jt'}/D^{\tilde{s}} \le \sum_{i \in I: t_{\tilde{s}} + \theta_{i\tilde{s}} + \tau_{ij} \le t-1} V^{\tilde{s}}_{ij};$$

$$j \in J, t \in T^{\tilde{s}} : t_{\tilde{s}} \le t \le \bar{t}_{\tilde{s}} + \sigma_j \qquad (10.81)$$

$$\sum_{t_{\tilde{s}} \le t' \le t} x^{\tilde{s}}_{jt'}/D^{\tilde{s}} \le \sum_{i \in I: t_{\tilde{s}} + \theta_{i\tilde{s}} + \tau_{ij} \le t-1} V^{\tilde{s}}_{ij} + w^{\tilde{s}}_j;$$

$$j \in J, t \in T^{\tilde{s}} : t > \bar{t}_{\tilde{s}} + \sigma_j, \qquad (10.82)$$

$$\sum_{t \in T^{\tilde{s}}} x^{\tilde{s}}_{jt}/D^{\tilde{s}} \le Q^{\tilde{s}}_j; \; j \in J, \qquad (10.83)$$

$$x^{\tilde{s}}_{jt} \le c^{\tilde{s}}_{jt} q^{\tilde{s}}_j; \; j \in J, t \in T^{\tilde{s}}, \qquad (10.84)$$

$$U^{\tilde{s}}_i \in \{0,1\}; \; i \in I, \qquad (10.85)$$

$$V^{\tilde{s}}_{ij} \in [0,1]; \; i \in I, j \in J, \qquad (10.86)$$

$$w^{\tilde{s}}_j \in [0,1]; \; j \in J, \qquad (10.87)$$

$$x^{\tilde{s}}_{jt} \ge 0; \; j \in J, t \in T^{\tilde{s}}, \qquad (10.88)$$

$$q^{\tilde{s}}_j \in \{0,1\}; \; j \in J, \qquad (10.89)$$

$$Q^{\tilde{s}}_j \in [0,1]; \; j \in J, \qquad (10.90)$$

where $T^s = \{t_s + 1, \ldots, \overline{T}\}$ and
$c^{\tilde{s}}_{jt}$ is the capacity in period t of plant j under disruption scenario $\tilde{s}$

$$c^{\tilde{s}}_{jt} = \begin{cases} \delta^{\tilde{s}}_j c_j, & \text{if } t_{\tilde{s}} \le t \le t_{\tilde{s}} + \vartheta_{j\tilde{s}} - 1 \\ c_j, & \text{if } t \ge t_{\tilde{s}} + \vartheta_{j\tilde{s}}. \end{cases} \qquad (10.91)$$

The objective function (10.74) constitutes of different fixed and variable cost per product (cf. Eq. (10.8)), and the last two constant components of (10.74) represent per product costs incurred by predetermined primary supply portfolio.

Notice that model **DSupport_E** for the predetermined primary supply portfolio is separable with respect to disruption scenarios $s \in S$, since the objective function (10.8) is additive and separable with respect to s as well as all constraints (10.12)–(10.25) are separable with respect to s. Thus, for a given primary supply portfolio and start time t_s of disruption s, the recovery supply portfolio, U^s_i, V^s_{ij}; $i \in I, j \in J$, the recovery transshipment portfolio, w^s_j, $j \in J$, the recovery demand portfolio, q^s_j, Q^s_j; $j \in J$, and the production schedules, x^s_{jt}, $j \in J$, can be found simultaneously for all potential disruption scenarios $s \in S$, by solving

RDSupport_E = **DSupport_E** for predetermined primary supply portfolio: $u_i = u^*_i$, $v_i = v^*_i$; $i \in I$.

RDSupport_E: *Risk-neutral* **R***ecovery* **D***emand and* **Sup***ply* **port***folios selection and production scheduling for predetermined primary portfolios*

Minimize (10.8)

subject to (10.12)–(10.27), (10.30)–(10.35) and

$$u_i = u_i^*;\ i \in I, \tag{10.92}$$

$$v_i = v_i^*;\ i \in I. \tag{10.93}$$

Worst-Case Cost Per Product

A simple upper bound on cost per product, R_s^c, for scenario s, of model **RDSupport_E** for predetermined primary supply and demand portfolios is derived below.

Proposition 10.7.

$$R_s^c \le \overline{R^c};\ s \in S, \tag{10.94}$$

where

$$\begin{aligned} R_s^c = {} & \sum_{i\in I} e_i(1-u_i^*)U_i^s/D + \sum_{i\in I} \rho_{is}U_i^s/D + \sum_{j\in J}(\varphi_j + \rho_{js})q_j^s/D \\ & + \sum_{j\in J}(\psi_j w_j^s + \sum_{i\in I} o_i V_{ij}^s) + \sum_{j\in J}\varepsilon_j \sum_{t\in T} x_{jt}^s/D + g(1 - \sum_{t\in T} x_{jt}^s/D) \\ & + \sum_{i\in I} e_i u_i^*/D + \sum_{i\in I} o_i \gamma_i^s v_i^*, \end{aligned} \tag{10.95}$$

and $\overline{R^c}$ *is the worst-case cost per product associated with worst-case disruption scenario with respect to cost*

$$\begin{aligned} \overline{R^c} = {} & \sum_{i\in I} e_i/D + g(1 - \underline{SL}) + \psi_{\max} + \varepsilon_{\max} \\ & + \max_{s\in S}\{\sum_{i\in I} \rho_{is}u_i^*/D + \max_{j\in J}(\varphi_j + \rho_{js})/D \\ & + (o_{\max} - \psi_{\min})(1 - \sum_{i\in I}\gamma_i^s v_i^*) + \sum_{i\in I} o_i\gamma_i^s v_i^*\}. \end{aligned} \tag{10.96}$$

(continued)

Proof. Without loss of generality, assume that $V_0 = 0$ and $\sum_{t\in T:t<t_s} x^s_{1t}/D = 0$. Recovery supply and demand portfolio selection constraints (10.16)–(10.18) and minimization of cost objective function imply

$$\sum_{i\in I}\rho_{is}U^s_i/D+\sum_{j\in J}(\varphi_j+\rho_{js})q^s_j/D+\sum_{j\in J}\psi_j w^s_j+\sum_{i\in I}\sum_{j\in J}o_iV^s_{ij}+\sum_{i\in I}o_i\gamma^s_i v^*_i =$$

$$\sum_{i\in I}\rho_{is}U^s_i/D+\sum_{j\in J}(\varphi_j+\rho_{js})q^s_j/D+\sum_{j\in J}\psi_j(Q^s_j-\sum_{i\in I}V^s_{ij})+\sum_{i\in I}\sum_{j\in J}o_iV^s_{ij}+\sum_{i\in I}o_i\gamma^s_i v^*_i =$$

$$\sum_{i\in I}\rho_{is}U^s_i/D+\sum_{j\in J}(\varphi_j+\rho_{js})q^s_j/D+\sum_{j\in J}\psi_jQ^s_j-\sum_{i\in I}\sum_{j\in J}\psi_jV^s_{ij}+\sum_{i\in I}\sum_{j\in J}o_iV^s_{ij}+\sum_{i\in I}o_i\gamma^s_i v^*_i \le$$

$$\sum_{i\in I}\rho_{is}U^s_i/D+\sum_{j\in J}(\varphi_j+\rho_{js})q^s_j/D+\psi_{\max}+(o_{\max}-\psi_{\min})\sum_{i\in I}\sum_{j\in J}V^s_{ij}+\sum_{i\in I}o_i\gamma^s_i v^*_i \le$$

$$\psi_{\max}+\max_{s\in S}\{\sum_{i\in I}\rho_{is}u^*_i/D+\max_{j\in J}(\varphi_j+\rho_{js})/D+(o_{\max}-\psi_{\min})(1-\sum_{i\in I}\gamma^s_i v^*_i)+\sum_{i\in I}o_i\gamma^s_i v^*_i\},$$

where $o_{\max} = \max_{i\in I} o_i$, $\psi_{\max} = \max_{j\in J}\psi_j$, $\psi_{\min} = \min_{j\in J}\psi_j$.

In addition, Eqs. (10.17), (10.23), and (10.37) imply

$$\sum_{i\in I}e_i(1-u^*_i)U^s_i/D+\sum_{i\in I}e_iu^*_i/D+\sum_{j\in J}\varepsilon_j\sum_{t\in T}x^s_{jt}/D+g(1-\sum_{t\in T}x^s_{jt}/D)\le$$

$$\sum_{i\in I}e_i/D+\varepsilon_{\max}\sum_{j\in J}\sum_{t\in T}x^s_{jt}/D+g(1-\underline{SL})\le$$

$$\sum_{i\in I}e_i/D+\varepsilon_{\max}\sum_{j\in J}Q^s_j+g(1-\underline{SL})\le$$

$$\sum_{i\in I}e_i/D+\varepsilon_{\max}+g(1-\underline{SL}),$$

where $\varepsilon_{\max} = \max_{j\in J}\varepsilon_j$.

By summing up the final parts of the above two expressions, formula (10.96) for $\overline{R^c}$ is achieved.

10.4 Models for Risk-Averse Decision-Making

10.4.1 Integrated Approach

In this subsection the two time-indexed SMIP models **DSupport_CV(c)** and **DSupport_CV(sl)** are proposed for the integrated, risk-averse selection of primary supply portfolio and recovery supply, transshipment, and demand portfolios along with production scheduling to optimize, respectively, expected worst-case cost and expected worst-case service level under disruption risks. The models are based on the risk-neutral model **DSupport_E**.

When the worst-case cost is focused on, the risk-averse primary and recovery supply and demand portfolios and the production schedule will be optimized by calculating VaR^c and minimizing CVaR^c simultaneously. Model **DSupport_CV(c)** is presented below.

DSupport_CV(c): *Risk-averse primary and recovery* **D***emand and* **Supp***ly* **port***folios selection and production scheduling to minimize* **CV***aR of* **c***ost*

Minimize

$$CVaR^c = VaR^c + (1-\alpha)^{-1} \sum_{s\in S} P_s \mathscr{C}_s, \tag{10.97}$$

subject to (10.10)–(10.34) and

Risk constraints:

- the tail cost for scenario s is defined as the nonnegative amount by which cost in scenario s exceeds VaR^c,

$$\begin{aligned}\mathscr{C}_s \geq \sum_{i\in I}(e_i(u_i + U_i^s - \mu_i^s)/D + \rho_{is}U_i^s/D + o_i(\gamma_i^s v_i + \sum_{j\in J} V_{ij}^s))\\ + \sum_{j\in J}(\psi_j w_j^s + (\varphi_j + \rho_{js})q_j^s/D + \varepsilon_j \sum_{t\in T} x_{jt}^s/D)\\ + g(1 - \sum_{j\in J}\sum_{t\in T} x_{jt}^s/D) - VaR^c;\ s\in S, \end{aligned} \tag{10.98}$$

$$\mathscr{C}_s \geq 0;\ s\in S \tag{10.99}$$

where $\mathscr{C}_s$ is the tail cost for scenario s.

If worst-case service level is to be maximized, the risk-averse primary and recovery supply portfolio and the production schedule will be optimized by calculating VaR^{sl} and maximizing CVaR^{sl} simultaneously. Model **DSupport_CV(sl)** is presented below.

DSupport_CV(sl): *Risk-averse* **D***emand and* **Supp***ly* **port***folios selection and production scheduling to maximize* **CV***aR of* **s***ervice* **l***evel*

Maximize

$$CVaR^{sl} = VaR^{sl} - (1-\alpha)^{-1} \sum_{s\in S} P_s \mathscr{S}_s, \tag{10.100}$$

subject to (10.10)–(10.34) and

Risk constraints:

(continued)

– the tail service level for scenario s is defined as the nonnegative amount by which VaR^{sl} exceeds service level in scenario s,

$$\mathscr{S}_s \geq VaR^{sl} - \sum_{j\in J}\sum_{t\in T} x^s_{jt}/D;\ s \in S, \tag{10.101}$$

$$\mathscr{S}_s \geq 0;\ s \in S, \tag{10.102}$$

where $\mathscr{S}_s$ is the tail service level for scenario s.

10.4.2 Hierarchical Approach

In this subsection the two time-indexed SMIP models **RDSupport_CV(c)** and **RDSupport_CV(sl)** are presented for the hierarchical, risk-averse selection of recovery supply, transshipment, and demand portfolios and production scheduling to optimize, respectively, expected worst-case cost and expected worst-case service level, given primary supply portfolio.

The models are based on the risk-averse decision-making models, **DSupport_CV(c)** and **DSupport_CV(sl)**, respectively:

RDSupport_CV(c) = **DSupport_CV(c)** for predetermined primary supply portfolio: $u_i = u^*_i,\ v_i = v^*_i;\ i \in I$;
RDSupport_CV(sl) = **DSupport_CV(sl)** for predetermined primary supply portfolio: $u_i = u^*_i,\ v_i = v^*_i;\ i \in I$.

For model **RDSupport_CV(c)**, the primary supply portfolio, $u^*_i, v^*_i; i \in I$, is determined using model **PSupport** with cost-based objective function (10.64). For model **RDSupport_CV(sl)**, however, the cost-based objective function (10.64) is replaced by the following service-based objective function:

$$Maximize \quad \sum_{t\in T} X_t/D. \tag{10.103}$$

RDSupport_CV(c): Risk-averse **R***ecovery* **D***emand and* **Supp***ly* **port***folios selection and production scheduling to minimize* **CV***aR of cost for predetermined primary portfolios.*

Minimize (10.97)
subject to (10.10)–(10.34), (10.92), (10.93), (10.98), (10.99).

RDSupport_CV(sl): *Risk-averse* **R**ecovery **D**emand *and* **Supp**ly **port**folios selection and production scheduling to maximize **CV**aR of service **l**evel for predetermined primary portfolios.

Maximize (10.100)

subject to (10.10)–(10.34), (10.92), (10.93), (10.101), (10.102).

10.5 Computational Examples

In this section some computational examples are presented to illustrate the proposed portfolio approach for selection of primary supply portfolio and recovery supply, transshipment, and demand portfolios along with production scheduling. The input data for the examples are in part hypothetical, however, their relations to each other are real and in part they have been taken from a real case study. In particular, the case studies of Toyota supply chain disruption and recovery after the Great East Japan earthquake and tsunami of March 11, 2011 and Thailand's floods in October 2011 (e.g., Fujimoto and Park 2013, Park et al. 2013, MacKenzie et al. 2014, Matsuo 2015, Haraguchi and Lall 2015) have been analyzed. The computational experiments were performed using the AMPL programming language and the Gurobi 7.5.0 solver on a MacBookPro laptop with Intel Core i7 processor running at 2.8 GHz and with 16 GB RAM.

The basic input parameters (see, Table 10.4) are taken from the computational examples presented in Sect. 9.5.

The total number of all potential scenarios is $\overline{S} = (\overline{L}+1)^{\overline{I}}(\overline{L^1}+1)(\overline{L^2}+1) = (4^4)(2^2) = 1024$ scenarios. Each scenario $s \in S$ is represented by a 6-dimensional vector $\lambda_s = (\lambda_{1s}, \dots, \lambda_{4s}, \lambda_{5s}, \lambda_{6s})$, where $\lambda_{is} \in L_i$, $i \in I$, and $\lambda_{5s}, \lambda_{6s} \in \{0,1\}$, where λ_{5s} and λ_{6s} represent disruption pattern under scenario s of assembly plant $j = 1$ and $j = 2$, respectively (for selected supply and production disruption scenarios, see Table 10.8).

The probability P_s^1 of realizing disruption scenario $s \in S$ for suppliers and assembly plants in region $r = 1$, and P_s^2 of realizing disruption scenario $s \in S$ for suppliers in region $r = 2$ are calculated as follows:

$$P_s^1 = \begin{cases} (1-p^1)(\prod_{i\in I^1:\lambda_{is}=0} 0.1(1-p_{i3}))(\prod_{i\in I^1:\lambda_{is}=1} 0.3(1-p_{i3})) \\ \times(\prod_{i\in I^1:\lambda_{is}=2} 0.6(1-p_{i3}))(\prod_{i\in I^1:\lambda_{is}=3} p_{i3}) \\ \times(\prod_{j\in J:\lambda_{\overline{I}+j,s}=0} \pi_{j1})(\prod_{j\in J:\lambda_{\overline{I}+j,s}=1} \pi_{j1}) \\ \text{if } \sum_{i\in I^1\cup\{5,6\}} \lambda_{is} > 0 \\ p^1 + (1-p^1)\prod_{i\in I^1} 0.1(1-p_{i3})\prod_{j\in J} \pi_{j0} \\ \text{if } \sum_{i\in I^1\cup\{5,6\}} \lambda_{is} = 0. \end{cases}$$

Table 10.4 Input parameters: supply and demand portfolio selection and production scheduling

$\overline{I}=4$ suppliers, $\overline{J}=2$ assembly plants, $\overline{T}=30$ planning periods, $\overline{R}=2$ two geographic regions	
$I^1=\{1,2\}$, $I^2=\{3,4\}$, $J^1=\{1,2\}$	two suppliers and two plants located in region $r=1$ and two suppliers in region $r=2$
$V_0=0$	initial inventories of parts
$d_t=10{,}000 \forall t \in T$	demand per period for parts/products
$D=300{,}000$	total demand for parts/products
Disruption levels:	
$\overline{L}_i=\overline{L}=3$	four disruption levels for each supplier $i \in I$
$\overline{L}^j=1$	two disruption levels for each plant $j \in J$
Local disruption probabilities of suppliers:	
$p_{i0}=0.1(1-p_{i3})$	complete shutdown (level $l=0$)
$p_{i1}=0.3(1-p_{i3})$	major disruption (level $l=1$)
$p_{i2}=0.6(1-p_{i3})$	minor disruption (level $l=2$)
$p_{i3}\in[0.89,0.99]$; $i\in I^1$, $p_{i3}\in[0.79,0.89]$; $i\in I^2$	non disruption (level $l=3$)
Local non disruption probabilities of plants:	$\pi_{11}=0.85$, $\pi_{21}=0.95$
Regional disruption probability:	$p^1=0.001$, $p^2=0.01$
Fulfillment rates of suppliers:	
$\gamma_{i0}=0\ \forall i\in I$, $\gamma_{i3}=1\ \forall i\in I$	
$\gamma_{i1}\in[0.01,0.50]\ \forall i\in I^1$ $\gamma_{i1}\in[0.01,0.30]\ \forall i\in I^2$, $\gamma_{i2}\in[0,51,0.99]\ \forall i\in I^1$ $\gamma_{i2}\in[0,31,0.99]\ \forall i\in I^2$	
Plant capacity:	
$c_1=10{,}000$, $c_2=5000$	
Cost parameters:	
$o=(14,12,8,9)$, $e=(8000,6000,12{,}000,13{,}000)$	
$g\in\{1,10,100,1000,10{,}000,100{,}000,\infty\}$	unit penalties for unfulfilled demand for products, where $g=\infty$ denotes maximization of service level
$\varepsilon_2=1$, $\varphi_2=100$, $\psi_2=0.1$	
$CTR(i,l)$=if $l=0$ then $100e_i$; if $l=1$ then $10e_i$; if $l=2$ then $e_i\ \forall i\in I$	
$PRC(1,0)=PRC(2,0)=10{,}000$	
Time parameters:	
$\tau_{1j}=\tau_{2j}=2$, $\tau_{3j}=\tau_{4j}=4, \forall j\in J$, $\sigma_2=2$	
$TTR(i,l)$=if $l=0$ then 12; if $l=1$ then 10; if $l=2$ then 8 $\forall i\in I$	
$PRT(1,0)=10$, $PRT(2,0)=5$	

$$P_s^2=\begin{cases}(1-p^2)(\prod_{i\in I^2:\lambda_{is}=0}0.1(1-p_{i3}))(\prod_{i\in I^2:\lambda_{is}=1}0.3(1-p_{i3}))\\ \times(\prod_{i\in I^2:\lambda_{is}=2}0.6(1-p_{i3}))(\prod_{i\in I^2:\lambda_{is}=3}p_{i3}) & \text{if } \sum_{i\in I^2}\lambda_{is}>0\\ p^2+(1-p^2)\prod_{i\in I^2}0.1(1-p_{i3}) & \text{if } \sum_{i\in I^2}\lambda_{is}=0.\end{cases}$$

The probability for disruption scenario $s\in S$ is given by $P_s=P_s^1P_s^2$.

10.5.1 Risk-neutral Decision-Making

In the computational examples presented in this subsection each disruption $s\in S$ to primary suppliers (and possibly other suppliers) and/or to primary plant $j=1$ (and possibly other plants) is assumed to occur in the same period $t_s=1$, before the earliest delivery lead time, $\min_{i\in I}(\tau_{i1})=2$, and the recovery process starts in period $t=2$, so that the disrupted supplier i and disrupted plant j return to its full capacity in period $t=\theta_{is}+1$ and $t=\vartheta_{js}+1$, respectively. The assumption of a common disruption start time of all disruptive events will be relaxed in Sect. 10.5.1.1 to provide more insights by best-case and worst-case analysis, and in Sect. 10.5.2 to illustrate the risk-averse decision-making.

The solution results for the integrated approach are summarized in Table 10.5 for period-independent demand and in Table 10.6 for period-dependent demand (model **DSupport_E** with Eq. (10.63) for a single product type). The tables show primary supply portfolio and expected recovery supply portfolio and recovery demand portfolio for different values of unit penalty cost g. The expected recovery supply portfolio refers to the expected supplies to meet demand that cannot be delivered by the primary suppliers, see constraint (10.16). Since the primary suppliers can meet demand for most of the scenarios (including the scenario without a disruption), the expected recovery supply is a small percentage of total demand. Thus, the numbers representing the expected fraction, over all scenarios, of total demand to be fulfilled by recovery suppliers do not need to add up to 100%. In contrast to expected recovery demand portfolio that sums up to 100%. All demand for products is assigned to a single primary plant, shutdown before any production can be started, if a disruptive event occurs at time $t_s = 1$, see (10.17).

Tables 10.5 and 10.6 show the expected cost E^c (10.8), along with the associated expected service level E^{sl} (10.9). In addition, the size of MIP models, respectively, **DSupport_E** and **DSupport_E** with Eq. (10.63) for a single product type, is presented.

As unit penalty g increases a higher service level is attained and the penalty for unfulfilled demand is reduced. As a result, for period-independent demand the expected cost, E^c, does not increase much with g, in contrast to period-dependent demand (cf. **Proposition 10.6**). For period-independent demand, a single sourcing primary supply portfolio with all parts ordered from the cheapest supplier, $i = 3$, is selected for most values of g. In contrast, for the period-dependent demand more expensive and diversified primary supply portfolios are selected to hedge against all disruption scenarios. Notice that even when $g \to \infty$, i.e., when the objective is to maximize service level, the expected service level is less than 100%, because the impact of disruptions cannot be fully mitigated for period-dependent demand.

The recovery demand is allocated among both assembly plants, which is indicated by expected recovery demand portfolio in Tables 10.5 and 10.6, in particular for period-independent demand.

The solution results for the hierarchical approach (with period-independent demand) are summarized in Table 10.7, in which the expected cost E^c (10.8), is shown along with the associated expected service level E^{sl} (10.9). Similarly, for the top level problem **PSupport**, cost P^c (10.64), is shown along with the associated service level, $\sum_{t \in T} X_t / D$. In addition, Table 10.7 presents the size of each MIP model: **PSupport** and **RDSupport**.

In contrast to the integrated approach, the results for the hierarchical approach are more independent of g, in particular the primary supply portfolio. When the objective is to maximize service level ($g \to \infty$), the hierarchical approach selects the most reliable supplier ($i = 2$) as the only primary supplier. Comparison of Ta-

Table 10.5 Supply and demand portfolios: model **DSupport_E**, period-independent demand

g	1	10	10^2	10^3	10^4	10^5	∞ [a]
Var.=72696, Bin.=10244, Cons.=132077, Nonz.=1082794[b]							
Exp.Cost E^c (10.8)	8.82	9.93	14.51	15.46	15.52	16.11	–
Exp.Service E^{sl} (10.9)×100%	74	91	99	100			100
Primary Supply Portfolio (% of total demand)	1(0)	1(0)	1(0)	1(0)			1(32)
	2(0)	2(0)	2(7)	2(0)			2(17)
	3(100)	3(100)	3(93)	3(100)			3(27)
	4(0)	4(0)	4(0)	4(0)			4(24)
Exp.Recovery Supply Portfolio (% of total demand) [c]	1(0.16)	1(0.17)	1(0.16)	1(0.16)			1(3.14)
	2(1.68)	2(1.74)	2(1.57)	2(1.68)			2(1.87)
	3(0)	3(0)	3(0.23)	3(0)			3(0.11)
	4(6.56)	4(6.49)	4(6.15)	4(6.56)			4(0.27)
Exp.Recovery Demand Portfolio	1(85.76)	1(75.04)	1(89.64)	1(83.68)			1(82.30)
Plant (% of total demand) [d]	2(14.24)	2(24.96)	2(10.36)	2(16.32)			2(17.70)
LP [e]	8.79	9.77	14.30	15.30	15.36	15.92	100[f]
CPU[g]	4	8	14	8	9	10	3

[a] Maximization of service level
[b] Var.=total number of variables, Bin.=number of binary variables, Cons.=number of constraints Nonz.= number of nonzero coefficients
[c] $1(\sum_{j\in J,s\in S} P_sV^s_{1j}\times 100), 2(\sum_{j\in J,s\in S} P_sV^s_{2j}\times 100), 3(\sum_{j\in J,s\in S} P_sV^s_{3j}\times 100), 4(\sum_{j\in J,s\in S} P_sV^s_{4j}\times 100)$
[d] $1(\sum_{s\in S} P_sQ^s_1\times 100), 2(\sum_{s\in S} P_sQ^s_2\times 100)$
[e] LP relaxation solution value of (10.8)
[f] LP relaxation solution value of (10.9)
[g] elapsed time in seconds for proven optimal solution

bles 10.5 and 10.7 demonstrates that for low values of g, the two approaches yield similar expected solution values. This is due to negligible penalty cost for unfulfilled demand for products with respect to other cost components. However, as g increases to reduce the unfulfilled demand, the integrated approach outperforms the purely deterministic, top-down hierarchical approach. In particular, more expensive and diversified primary supply portfolios are selected by the integrated approach to hedge against all disruption scenarios. Both, expected cost over all scenarios is lower and expected service level is higher. In contrast, for the hierarchical approach results for $g > 1$ are independent of g. When the objective is to maximize service level ($g \to \infty$), the hierarchical approach selects the most reliable supplier ($i = 2$) as the only primary supplier.

For the hierarchical approach (models **PSupport** and **RDSupport_E**) and selected supply and production disruption scenarios $s \in S$, Table 10.8 shows examples of recovery supply and demand portfolios $(V^s_{11}+V^s_{12}, V^s_{21}+V^s_{22}, V^s_{31}+V^s_{32}, V^s_{41}+V^s_{42}, Q^s_1, Q^s_2)$ along with the associated cost, and service level, for a unit penalty $g = 100$. The primary supply portfolio obtained to minimize purchasing cost for parts and penalty for unfulfilled demand for products in a deterministic operating environment (model **PSupport**) is (see, Table 10.7): $v_1 = 0$, $v_2 = 0.07$, $v_3 = 0.93$, $v_4 = 0$. Then, given the primary supply and demand portfolios and the realized disruption scenario, the recovery supply and demand portfolios are selected to minimize total

Table 10.6 Supply and demand portfolios: model **DSupport_E**, period-dependent demand

g	1	10	10^2	10^3	10^4	10^5	∞ [a]
Var.=72696, Bin.=10244, Cons.=157765, Nonz.=1135204[b]							
Exp.Cost E^c (10.8)	8.82	10.34	23.26	109	961	9461	–
Exp.Service E^{sl} (10.9)$\times$100%	74	86	90	90	90		91
Primary	1(0)	1(0)		1(3)	1(26)		1(37)
Supply Portfolio	2(0)	2(7)		2(8)	2(23)		2(35)
(% of total demand)	3(100)	3(93)		3(76)	3(38)		3(11)
	4(0)	4(0)		4(13)	4(13)		4(17)
Exp.Recovery	1(0.16)	1(0.33)		1(0.21)	1(0.14)		1(2.13)
Supply Portfolio	2(1.68)	2(1.64)		2(1.58)	2(0.96)		2(1.91)
(% of total demand) [c]	3(0)	3(0.08)		3(0.98)	3(1.94)		3(0.20)
	4(6.56)	4(6.05)		4(4.96)	4(2.57)		4(0.11)
Exp.Recovery Demand Portfolio	1(85.76)	1(85.76)	1(98.71)	1(98.53)	1(98.53)		1(92.24)
Plant (% of total demand) [d]	2(14.24)	2(14.24)	2(1.29)	2(1.47)	2(1.47)		2(7.76)
LP [e]	8.79	10.29	23.22	109	960	9461	91[f]
CPU [g]	4	12	11	10	10	5	3

[a] Maximization of service level
[b] Var.=total number of variables, Bin.=number of binary variables, Cons.=number of constraints Nonz.= number of nonzero coefficients
[c] $1(\sum_{j\in J,s\in S} P_s V_{1j}^s \times 100), 2(\sum_{j\in J,s\in S} P_s V_{2j}^s \times 100), 3(\sum_{j\in J,s\in S} P_s V_{3j}^s \times 100), 4(\sum_{j\in J,s\in S} P_s V_{4j}^s \times 100)$
[d] $1(\sum_{s\in S} P_s Q_1^s \times 100), 2(\sum_{s\in S} P_s Q_2^s \times 100)$
[e] LP relaxation solution value of (10.8)
[f] LP relaxation solution value of (10.9)
[g] elapsed time in seconds for proven optimal solution

Table 10.7 Supply and demand portfolios: hierarchical approach

g	1	10	10^2	10^3	10^4	10^5	∞ [a]
PSupport: Var.=64, Bin.=4, Cons.=65, Nonz.=968[b]							
Primary	1(0)	1(0)					1(0)
Supply Portfolio	2(0)	2(7)					2(100)
(% of total demand)	3(100)	3(93)					3(0)
	4(0)	4(0)					4(0)
Cost P^c (10.64)	8.17	9	15	75	675	6675	–
Service Level[c]	87	93					93
RDSupport: Var.=64758, Bin.=7168, Cons.=119276, Nonz.=935364							
Exp.Recovery	1(0.16)	1(0.18)	1(0.15)				1(3.42)
Supply Portfolio	2(1.68)	2(1.62)	2(1.57)				2(0)
(% of total demand) [d]	3(0)	3(0.21)	3(0.23)				3(0.41)
	4(6.56)	4(6.10)	4(6.15)				4(0.15)
Exp.Recovery Demand	1(85.76)	1(80.15)	1(89.64)	1(89.06)	1(89.06)	1(89.06)	1(61.90)
Portfolio (% of total demand) [e]	2(14.24)	2(19.85)	2(10.36)	2(10.94)	2(10.94)	2(10.94)	2(38.10)
Exp.Cost E^c (10.8)	8.82	10.11	14.51	15.66	15.71	16.14	–
Exp.Service E^{sl} (10.9)$\times$100%	74	91	99	100	100	100	100

[a] Maximization of service level
[b] Var.=total number of variables, Bin.=number of binary variables, Cons.=number of constraints Nonz.= number of nonzero coefficients
[c] $\sum_{t\in T} X_t / D \times 100\%$
[d] $1(\sum_{j\in J,s\in S} P_s V_{1j}^s \times 100), 2(\sum_{j\in J,s\in S} P_s V_{2j}^s \times 100), 3(\sum_{j\in J,s\in S} P_s V_{3j}^s \times 100), 4(\sum_{j\in J,s\in S} P_s V_{4j}^s \times 100)$
[e] $1(\sum_{s\in S} P_s Q_1^s \times 100), 2(\sum_{s\in S} P_s Q_2^s \times 100)$

Table 10.8 Supply and demand recovery portfolios, $(V^s_{11}+V^s_{12}, V^s_{21}+V^s_{22}, V^s_{31}+V^s_{32}, V^s_{41}+V^s_{42}, Q^s_1, Q^s_2)$ and associated cost and service, for selected supply and production disruption scenarios $s \in S$: hierarchical approach, $g = 100$

Disruption scenario							Supply portfolio				Demand portfolio		Solution values	
s	λ_{1s}	λ_{2s}	λ_{3s}	λ_{4s}	λ_{5s}	λ_{6s}	i=1	i=2	i=3	i=4	j=1	j=2	Cost	Service %
1	0	0	0	0	0	0	0	1	0	0	1	0	2095.39	50
7	0	0	1	2	0	0	0	0	0	0.9	1	0	119.38	66.3
17	0	1	0	0	0	0	0	0.98	0	0	1	0	286.49	58.91
19	0	1	0	2	0	0	0	0	0	0.98	1	0	126.93	58.91
26	0	1	2	1	0	0	0	0	0.46	0	1	0	114.82	66.67
28	0	1	2	3	0	0	0	0	0	0.46	1	0	75.32	66.67
35	0	2	0	2	0	0	0	0.95	0	0	1	0	98.73	66.67
36	0	2	0	3	0	0	0	0	0	0.95	1	0	75.93	66.67
42	0	2	2	1	0	0	0	0.43	0	0	1	0	96.65	66.67
44	0	2	2	3	0	0	0	0	0	0.43	1	0	75.41	66.67
49	0	3	0	0	0	0	0	0.93	0	0	1	0	78.73	66.67
50	0	3	0	1	0	0	0	0.93	0	0	1	0	78.73	66.67
321	1	0	0	0	0	1	1	0	0	0	0.57	0.43	329.7	85
322	1	0	0	1	0	1	1	0	0	0	0.72	0.28	329.7	85
323	1	0	0	2	0	1	0	0	0	1	0.57	0.43	101.39	85
324	1	0	0	3	0	1	0	0	0	1	0.67	0.33	43.1	100
325	1	0	1	0	0	1	0.9	0	0	0	0.66	0.34	319.49	94.63
326	1	0	1	1	0	1	0.9	0	0	0	0.66	0.34	319.49	94.63
328	1	0	1	3	0	1	0	0	0	0.9	0.67	0.33	43.01	100
329	1	0	2	0	0	1	0	0	0.48	0	0.72	0.28	87.01	95
332	1	0	2	3	0	1	0	0	0	0.48	0.67	0.33	42.58	100
339	1	1	0	2	0	1	0	0	0	0.98	0.72	0.28	99.21	87.24
340	1	1	0	3	0	1	0	0	0	0.98	0.67	0.33	43.17	100
341	1	1	1	0	0	1	0	0.88	0	0	0.67	0.33	250.62	95
344	1	1	1	3	0	1	0	0	0	0.88	0.67	0.33	43.07	100
353	1	2	0	0	0	1	0	0.95	0	0	0.68	0.32	67.71	98.33
356	1	2	0	3	0	1	0	0	0	0.95	0.67	0.33	43.26	100
357	1	2	1	0	0	1	0	0.85	0	0	0.68	0.32	67.32	98.33
360	1	2	1	3	0	1	0	0	0	0.85	0.67	0.33	43.16	100
361	1	2	2	0	0	1	0	0.43	0	0	0.67	0.33	65.63	98.33
381	1	3	3	0	0	1	0	0	0	0	1	0	74.99	66.67
382	1	3	3	1	0	1	0	0	0	0	1	0	74.99	66.67
383	1	3	3	2	0	1	0	0	0	0	1	0	74.99	66.67
384	1	3	3	3	0	1	0	0	0	0	1	0	74.99	66.67
640	1	3	3	3	1	0	0	0	0	0	1	0	14.99	93.33
646	2	0	1	1	1	0	0.9	0	0	0	1	0	67.21	72.96
647	2	0	1	2	1	0	0.9	0	0	0	1	0	67.21	72.96
648	2	0	1	3	1	0	0	0	0	0.9	1	0	22.34	86.67
657	2	1	0	0	1	0	0.98	0	0	0	1	0	75.14	65.57
658	2	1	0	1	1	0	0.98	0	0	0	1	0	75.14	65.57
659	2	1	0	2	1	0	0.98	0	0	0	1	0	75.14	65.57

(continued)

Table 10.8 (continued)

Disruption scenario							Supply portfolio				Demand portfolio		Solution values	
s	λ_{1s}	λ_{2s}	λ_{3s}	λ_{4s}	λ_{5s}	λ_{6s}	i=1	i=2	i=3	i=4	j=1	j=2	Cost	Service %
660	2	1	0	3	1	0	0	0	0	0.98	1	0	23.6	85.57
661	2	1	1	0	1	0	0.88	0	0	0	1	0	64.93	75.2
662	2	1	1	1	1	0	0.88	0	0	0	1	0	64.93	75.2
663	2	1	1	2	1	0	0.88	0	0	0	1	0	64.93	75.2
664	2	1	1	3	1	0	0	0	0	0.88	1	0	20.17	88.91
673	2	2	0	0	1	0	0	0.95	0	0	1	0	63.48	68.58
676	2	2	0	3	1	0	0	0	0	0.95	1	0	20.68	88.58
677	2	2	1	0	1	0	0	0.85	0	0	1	0	53.47	78.21
844	1	0	2	3	1	1	0	0	0	0.48	0.87	0.13	9.05	100
845	1	0	3	0	1	1	0	0	0.07	0	0.93	0.07	15.13	93.33
961	3	0	0	0	1	1	1	0	0	0	0.9	0.1	14.52	100
976	3	0	3	3	1	1	0	0	0.07	0	0.93	0.07	15.13	93.33
977	3	1	0	0	1	1	0.98	0	0	0	0.92	0.08	14.45	100
992	3	1	3	3	1	1	0	0	0.04	0	0.96	0.04	15.19	93.33
1009	3	3	0	0	1	1	0	0.93	0	0	0.93	0.07	12.46	100
1024	3	3	3	3	1	1	0	0	0	0	1	0	14.99	93.33

cost, including cost of recovery from the disruption. The results indicate that the cost is ranging from 9.05 for scenario $s = 844$ to 2095.39 for scenario $s = 1$ with complete shutdown of all suppliers and assembly plants. The corresponding service level is ranging from 100% to 50%.

Observe that no recovery supply or demand portfolio is selected, i.e., $V_{11}^s + V_{12}^s = V_{21}^s + V_{22}^s = V_{31}^s + V_{32}^s = V_{41}^s + V_{42}^s = 0, Q_1^s = 1, Q_2^s = 0$, for scenarios $s \in S$ with non-disrupted all primary suppliers or primary assembly plant, respectively, e.g., for scenarios $s = 640, 1024$.

More computational results for the integrated approach (with period-independent demand) with larger size problems are summarized in Table 10.9, where two-level (all-or-nothing) and four-level disruption scenarios are additionally compared. The basic parameters are randomly generated, in a similar way as in the above examples. The unit penalty is fixed to $g = 100$. The total number of disruption scenario is $\overline{S} \times |T_s|$, where $T_s = \{2\}$, $\{2,4\}$, $\{1,2,3,4\}$ is the set of possible start times, t_s, of each disruption $s \in S$. The results shown in the table clearly indicate that the integrated model **DSupport_E** has a very strong LP relaxation, which is due to an embedded network flow structure. The initial feasible solutions are very close to the proven optimal solution. The initial GAP is less than 1% for the multi-level disruption scenarios, and not greater than a few percent for the two-level scenarios. Comparison of the results for the two- and multi-level disruptions demonstrate that the impact of all-or-nothing disruptive events is more significant (greater optimal solution values) than that of multi-level disruptions. The partial disruptions do not necessarily mean full stoppage of supplies and production. Finally, similar solution values for different sets of scenarios indicate that the impact of disruption risks can

Table 10.9 Computational results: model **DSupport_E**

All-or-nothing disruptions of all suppliers and assembly plants						
$\overline{I}, \overline{J}, \overline{T}$ [a]	6,4,20			6,4,30		
$\overline{S} \times \vert T_s \vert$ [b]	1024	1024×2	1024×4	1024	1024×2	1024×4
Var. [c]	108,324	211,972	428,596	149,284	293,892	592,436
Bin. [c]	16,390	32,774	65,542	16,390	32,774	65,542
Cons. [c]	194,157	378,973	771,245	276,077	542,813	1,098,925
Nonz. [c]	1,204,222	2,299,598	4,726,862	2,471,662	4,787,838	9,749,982
LP [d]	12.71	12.81	12.79	12.11	12.17	12.16
MIP* [e]	15.09	15.18	15.17	13.72	13.78	13.77
CPU [f]	64	92	1064	20	108	963
Four-level disruptions of all suppliers and assembly plants						
$\overline{I}, \overline{J}, \overline{T}$ [a]	4,2,20			4,2,30		
$\overline{S} \times \vert T_s \vert$ [b]	4096	4096×2	4096×4	4096	4096×2	4096×4
Var. [c]	213,924	431,012	860,044	295,844	594,852	1,187,724
Bin. [c]	40,964	81,924	163,844	40,964	81,924	163,844
Cons. [c]	378,705	753,489	1,519,709	542,545	1,081,169	2,175,069
Nonz. [c]	2,235,612	4,518,108	9,035,204	4,661,492	9,401,588	18,782,444
LP [d]	9.64	9.65	9.69	8.97	8.96	8.96
MIP* [e]	9.79	9.81	9.81	9.14	9.17	9.16
CPU [f]	21	80	419	53	141	488

[a] $\overline{I}$ number of suppliers, $\overline{J}$ number of assembly plants, $\overline{T}$ planning horizon
[b] $\overline{S} \times |T_s|$ = total number of scenarios, where $|T_s|$ is number of possible start times of each disruptive event
[c] Var.=total number of variables, Bin.=number of binary variables, Cons.=number of constraints, Nonz.= number of nonzero coefficients
[d] LP relaxation solution value of (10.8)
[e] optimal solution value of (10.8)
[f] elapsed time in seconds for proven optimal solution

be similarly mitigated by best selection of primary and recovery portfolios, using the developed approach.

Overall, the computational results demonstrate that *the multi-portfolio approach forms an embedded network flow structure that leads to computationally efficient stochastic mixed integer program with a very strong LP relaxation.*

10.5.1.1 Best-case and Worst-Case Analysis

In order to provide more insights, in this subsection best-case and worst-case disruption scenarios with respect to cost and service level were identified. Since supply disruptions can happen at any time and for any supplier and assembly plant, from now on a disruption scenario is defined as a combination of disruptive event and its start time. In order to consider non-trivial supply disruptions, start time t_s of each disruptive event $s \in S$ is assumed to be not greater than the maximum delivery lead time, $\max_{i \in I, j \in J}(\tau_{ij}) = 4$. Thus $t_s \in \{1,2,3,4\}$, and the total number of all potential scenarios to be considered, is $1024 \times 4 = 4096$, i.e., $S = \{1, \ldots, 4096\}$. To simplify disruption modeling and analyses, disruption scenarios are numbered in nondecreasing order of their start times: the first 1024 scenarios are associated with $t_s = 1$; the second 1024 scenarios with $t_s = 2$; the third 1024 scenarios with $t_s = 3$; and the fourth 1024 scenarios with $t_s = 4$. Now, each disruption scenario $s \in S$ is represented by vector $\lambda_s = (\lambda_{1s}, \ldots, \lambda_{6s})$, where $\lambda_s = \lambda_{s+1024} = \lambda_{s+2048} = \lambda_{s+3072}$; $s \leq 1024$, and its start time $t_s = 1$ for $s \leq 1024$, $t_s = 2$ for $1025 \leq s \leq 2048$, $t_s = 3$ for $2049 \leq s \leq 3072$ and $t_s = 4$ for $3073 \leq s \leq 4096$.

The probability P_s of realizing each disruption scenario $s \in S$ is calculated as follows:

$$P_s = \beta_1 P_s^1 P_s^2 \text{ for } s \leq 1024,$$
$$P_s = \beta_2 P_{s-1024}^1 P_{s-1024}^2 \text{ for } 1025 \leq s \leq 2048,$$
$$P_s = \beta_3 P_{s-2048}^1 P_{s-2048}^2 \text{ for } 2049 \leq s \leq 3072,$$
$$P_s = \beta_4 P_{s-3072}^1 P_{s-3072}^2 \text{ for } 3073 \leq s \leq 4096,$$

where probabilities P_s^r; $r = 1, 2$, $s \leq 1024$ are defined at the beginning of this section, and $\beta_1, \beta_2, \beta_3$, and β_4 are nonnegative constants such that: $\beta_1 + \beta_2 + \beta_3 + \beta_4 = 1$. In the computational examples, optimal solutions were determined for $\beta_1 = 0.1$, $\beta_2 = 0.2$, $\beta_3 = 0.3$, $\beta_4 = 0.4$, i.e., disruptive events with later start times are modeled to be more likely.

In order to identify the cost of best-case and worst-case disruption scenarios, the cost per product associated with optimal solution of **DSupport_E** for the integrated approach (or **PSupport** and **RDSupport_E** for the hierarchical approach) is chosen to be not greater than 10 and not less than 2000, respectively. For example, for unit penalty $g = 1000$, the total number of cost best-case scenarios is 718 and 447, respectively, for the integrated and the hierarchical approach. The maximum service level, $\sum_{j \in J} \sum_{t \in T} x_{jt}^s / D = 1$, was achieved for all best-case scenarios with respect to cost, which clearly shows that for the maximum service level, no penalty cost for unfulfilled customer demand is incurred. The cost best-case scenarios for the hierarchical approach are a subset of those for the integrated approach.

In contrast, the number of cost worst-case disruption scenarios is much smaller, 12 and 16 scenarios, respectively, for the integrated and the hierarchical approach. The cost worst-case disruption scenarios for the integrated approach are a subset of those for the hierarchical approach.

In a similar way, to identify the service best-case and worst-case disruption scenarios, the service level associated with optimal solution of **DSupport_E** for the integrated approach (or **PSupport** and **RDSupport_E** for the hierarchical approach) is chosen to be not less than one and not greater than $\underline{SL} = 0.40$ (10.36), respectively. For example, for unit penalty $g = 1000$, the total number of best-case scenarios with service level, $\sum_{j \in J} \sum_{t \in T} x_{jt}^s / D = 1$, is 3220 and 2796, respectively, for the integrated and the hierarchical approach. The best-case scenarios for the hierarchical approach are a subset of those for the integrated approach. In contrast, for the integrated approach there are no worst-case scenarios with respect to service, while for the hierarchical approach there are only two such scenarios.

Table 10.10 Worst-case disruption scenarios: integrated/hierarchical approach, $g = 100$

Disruption scenario $(\lambda_{1s},\lambda_{2s},\lambda_{3s},\lambda_{4s},\lambda_{5s},\lambda_{6s}),\ t_s$	Recovery portfolio $(V^s_{11}+V^s_{12},V^s_{21}+V^s_{22},V^s_{31}+V^s_{32},V^s_{41}+V^s_{42},Q^s_1,Q^s_2)$	Cost [a]	Service Level [b]
Primary supply portfolio (v_1,v_2,v_3,v_4)= (0,0.07,0.93,0)			
Worst-case disruption scenarios with respect to cost			
(0,0,0,0,0,0), 1	(0,1,0,0,1,0)	2095	50
(0,0,0,0,0,1), 1	(0,1,0,0,0.75,0.25)	2071	75
(0,0,0,0,1,0), 1	(0,1,0,0,0.54,0.46)	2071	75
(0,0,0,0,1,1), 1	(0,1,0,0,0.75,0.25)	2038	75
(0,0,0,0,0,0), 2	(0,1,0,0,1,0)	2099	47
(0,0,0,0,0,1), 2	(0,1,0,0,0.47,0.53)	2076	70
(0,0,0,0,1,0), 2	(0,1,0,0,1,0)	2065	47
(0,0,0,0,1,1), 2	(0,1,0,0,0.77,0.23)	2043	70
(0,0,0,0,0,0), 3	(0,0.93,0,0,1,0)	2095	50
(0,0,0,0,0,1), 3	(0,0.93,0,0,0.78,0.22)	2074	72
(0,0,0,0,1,0), 3	(0,0.93,0,0,1,0)	2062	50
(0,0,0,0,1,1), 3	(0,0.93,0,0,0.50,0.50)	2041	72
(0,0,0,0,0,0), 4	(0,0.93,0,0,1,0)	2099	47
(0,0,0,0,0,1), 4	(0,0.93,0,0,0.47,0.53)	2079	67
(0,0,0,0,0,1), 4	(0,0.93,0,0,1,0)	2079	67
(0,0,0,0,1,0), 4	(0,0.93,0,0,1,0)	2065	47
(0,0,0,0,1,1), 4	(0,0.93,0,0,0.47,0.53)	2046	67
Worst-case disruption scenarios with respect to service level			
(0,0,1,0,0,1), 1	(0,0,0.9,0,0.4,0.6)	502	40[c]
(0,0,1,0,1,0), 2	(0,0,0.9,0,1,0)	486	40[c]

[a] $\sum_{i\in I}(f_i(u_i+U^s_i-\mu^s_i)/D+p_{is}U^s_i/D+e_i(\gamma^s_i v_i+\sum_{j\in J}V^s_{ij}))+\sum_{j\in J}(g_j w^s_j+(\varphi_j+\rho_{js})q^s_j/D$ $+\varepsilon_j\sum_{t\in T}x^s_{jt}/D)+b(1-\sum_{j\in J}\sum_{t\in T}x^s_{jt}/D)$
[b] $\sum_{j\in J}\sum_{t\in T}x^s_{jt}/D\times 100$
[c] $\underline{SL}\times 100$ (10.36)

Similar results are obtained for the other values of unit penalty g. In particular, for $g=100$, both integrated and hierarchical approach yield the same solution results. Table 10.10 presents all worst-case scenarios: 16 scenarios with cost per product not less than 2000, and two scenarios with service level $\sum_{j\in J}\sum_{t\in T}x^s_{jt}/D=\underline{SL}=0.4$.

Table 10.10 shows disruption scenario, $(\lambda_{1s},\lambda_{2s},\lambda_{3s},\lambda_{4s},\lambda_{5s},\lambda_{6s})$, along with its start time, t_s, and the corresponding optimal recovery supply and demand portfolio, $(V^s_{11}+V^s_{12},V^s_{21}+V^s_{22},V^s_{31}+V^s_{32},V^s_{41}+V^s_{42},Q^s_1,Q^s_2)$, cost per product,

$$\sum_{i\in I}(e_i(u_i+U^s_i-\mu^s_i)/D+\rho_{is}U^s_i/D+o_i(\gamma^s_i v_i+\sum_{j\in J}V^s_{ij}))$$
$$+\sum_{j\in J}(\psi_j w^s_j+(\varphi_j+\rho_{js})q^s_j/D+\varepsilon_j\sum_{t\in T}x^s_{jt}/D)$$
$$+g(1-\sum_{j\in J}\sum_{t\in T}x^s_{jt}/D),$$

and service level $\sum_{j\in J}\sum_{t\in T}x^s_{jt}/D$.

The results in Table 10.10 demonstrate that for the worst-case scenarios with respect to cost, all primary suppliers are completely shut down and then the one with shorter delivery lead time is selected as a single sourcing recovery portfolio.

When both assembly plants are shut down, the primary plant is selected as a single recovery plant, which saves both production delays due to transshipment time and additional transshipment and production costs. Otherwise, the recovery demand portfolio may include two plants.

The best-case and worst-case analysis demonstrate that:

- *when all primary suppliers are shut down, a single sourcing recovery supply portfolio is usually selected,*
- *if all assembly plants are shut down, the integrated approach selects the primary plant as a single recovery plant, whereas the hierarchical approach may choose multiple recovery plants,*
- *the best-case and worst-case disruption scenarios for the hierarchical approach are, respectively, subsets and supersets of the corresponding scenarios for the integrated approach.*

The computational experiments were performed using the AMPL programming language and the Gurobi 7.5.0 solver on a MacBookPro laptop with Intel Core i7 processor running at 2.8 GHz and with 16 GB RAM. The solver was capable of finding proven optimal solution for all examples with CPU time ranging from fraction of a second for deterministic models **PSupport** and **RDSupport(s)** to a few seconds for stochastic models **DSupport_E** and **RDSupport_E**.

10.5.2 Risk-averse Decision-Making

In this subsection some computational examples are presented to illustrate the risk-averse selection of primary and recovery supply and demand portfolios to minimize CVaR of cost. In the computational experiments the data sets provided at the beginning of this section were used, except for disruption scenarios that are described in Sect. 10.5.1.1. In addition, the unit penalty was fixed to $g = 100$.

The solution results are summarized in Table 10.11. The table shows primary supply portfolio and expected recovery supply and demand portfolios for different values of confidence level $\alpha = 0.5, 0.75, 0.9, 0.95, 0.99$. The solution values, $CVaR^c$, are presented along with the associated expected values of cost, E^c, and service level, E^{sl}, as well as with solution value of LP relaxation. While $CVaR^c$, VaR^c, and E^c increase with the confidence level α, the associated expected service level, E^{sl}, decreases as the decision maker becomes more risk aversive with respect to cost.

However, the primary supply portfolio does not vary much as the confidence level increases. For the presented examples, a dual primary sourcing is selected with most demand for parts allocated to the cheapest supplier $i = 3$ and the rest to a cheaper one, $i = 2$, among most reliable suppliers. The primary portfolio is very similar for both the lowest and the highest α and is similar to the risk-neutral portfolio. As a result the same portfolio is applied to mitigate the impact of disruptions in average-case as well as in worst-case, which may be due to relatively low unit penalty, $g = 100$, for unfulfilled demand and predominating purchasing cost.

Table 10.11 Risk-averse solutions for model **DSupport_CV**: $g = 100$

Confidence level α	0.50	0.75	0.90	0.95	0.99
Var.=290311, Bin.=40964, Cons.=530595, Nonz.=4698042 [a]					
$CVaR^c$ (10.97)	20.56	32.20	46.69	50.35	74.28
VaR^c	8.84	9.28	42.30	44.23	46.13
Exp.Cost E^c (10.8)	14.70	14.92	39.44	43.67	45.73
Exp.Service E^{sl} (10.9)×100%	99.48	99.35	75.61	70.83	69.50
LP [b]	20.36	32.03	46.54	50.06	72.97
CPU [c]	184	223	145	188	327
Primary supply portfolio:	2(8)	2(8)	2(3)	2(10)	2(10)
Supplier(% of total demand) [d]	3(92)	3(92)	3(97)	3(90)	3(90)
Exp.Recovery supply portfolio:	1(0.15)	1(0.15)	1(2.31)	1(2.41)	1(0.81)
Supplier(% of total demand) [e]	2(1.55)	2(1.55)	2(1.83)	2(3.42)	2(3.14)
	3(0.09)	3(0.09)	3(0.02)	3(0.05)	3(0.06)
	4(6.02)	4(6.02)	4(4.07)	4(1.78)	4(3.68)
Exp.Recovery demand portfolio:	1(87.52)	1(78.26)	1(91.28)	1(91.15)	1(91.66)
Plant (% of total demand) [f]	2(12.27)	2(21.53)	2(8.61)	2(8.67)	2(8.28)

[a] Var.=total number of variables, Bin.=number of binary variables, Cons.=number of constraints, Nonz.= number of nonzero coefficients
[b] LP relaxation solution value of (10.97)
[c] elapsed time in seconds for proven optimal solution
[d] $1(v_1 \times 100), 2(v_2 \times 100), 3(v_3 \times 100), 4(v_4 \times 100)$
[e] $1(\sum_{j\in J, s\in S} P_s V_{1j}^s \times 100), 2(\sum_{j\in J, s\in S} P_s V_{2j}^s \times 100), 3(\sum_{j\in J, s\in S} P_s V_{3j}^s \times 100), 4(\sum_{j\in J, s\in S} P_s V_{4j}^s \times 100)$
[f] $1(\sum_{s\in S} P_s Q_1^s \times 100), 2(\sum_{s\in S} P_s Q_2^s \times 100)$

Table 10.12 Best-case and worst-case results for risk-averse model **DSupport_CV**: $g = 100$

Confidence level α	0.50	0.75	0.90	0.95	0.99
Number of cost best-case scenarios (Cost≤ 10)	448	448	0	16	18
Minimum service level %	99.69	99.29	–	100	100
Maximum service level %	100	100	–	100	100
Number of service best-case scenarios (Service=100%)	1755	1751	918	818	668
Minimum cost	8.53	8.57	42.21	8.58	8.57
Maximum cost	116	116	116	116	244
Number of cost worst-case scenarios (Cost≥ 2000)	20	16	16	16	16
Minimum service %	40	47	43	47	47
Maximum service %			75		
Number of service worst-case scenarios (Service=$\underline{SL}$)	4	0	33	0	0
Minimum cost	6841	–	102	–	–
Maximum cost	7739	–	502	–	–

The best-case and worst-case scenario analysis for the risk-averse model **DSupport_CV** with unit penalty, $g = 100$, is summarized in Table 10.12. The table clearly indicates that the best-case scenarios with respect to cost are associated with the maximum or nearly maximum service level 100%, i.e., with virtually no penalty cost for unfulfilled demand. On the other hand, however, the best-case scenarios with respect to service, for which service level also attains 100%, can be associated with a higher cost, e.g., due to required recovery processes. The results indicate that

Table 10.13 Managerial insights: supply and demand portfolio selection and production scheduling

Decision-making issue	Decision-making impact
Risk-neutral decision-making	
Selection of recovery supply portfolio	If all primary suppliers are shut down, a single sourcing recovery supply portfolio is usually selected
Selection of recovery demand portfolio	If all assembly plants are shut down, the integrated approach selects primary plant as a single recovery plant, whereas the hierarchical approach may choose multiple recovery plants
Integrated vs. hierarchical approach	Integrated approach selects a more diversified primary supply portfolio that will hedge against all potential disruption scenarios Hierarchical approach selects primary supply portfolio that is made up of cheapest suppliers or a single, most reliable supplier only, to minimize expected cost or maximize expected service level, respectively
Best-case vs. worst-case disruption scenarios	Best-case and worst-case disruption scenarios for the hierarchical approach are, respectively, subsets and supersets of the corresponding scenarios for the integrated approach
Unit penalty for unfulfilled demand for products	The expected service level increases with unit penalty, whereas the expected cost significantly increases for the period-dependent demand and slightly increases for the period-independent demand
Period-independent demand vs. period-dependent demand	For the period-independent demand, cheapest supplier is selected only (single sourcing primary supply portfolio) for most values of unit penalty for unfulfilled demand for products. In contrast, for the period-dependent demand, more expensive and diversified primary supply portfolios are selected to hedge against all potential disruption scenarios
Two-level vs. multi-level disruptions.	The impact of all-or-nothing disruptions is more significant than that of multi-level disruptions
Risk-averse decision-making	
Selection of recovery supply portfolio	If all primary suppliers are shut down, a single sourcing recovery supply portfolio is usually selected
Selection of recovery demand portfolio	Multiple recovery plants may be selected even if all assembly plants are shut down, in particular for a low confidence level
Best-case vs. worst-case disruption scenarios	Under worst-case scenarios all suppliers and assembly plants are either completely shut down or hit by a major disruption
Best-case scenarios: cost vs. service level.	Best-case scenarios with respect to cost are a subset of best-case scenarios with respect to service, which attain 100% of service level. On the other hand, best-case scenarios with respect to service (with 100% of service level) can be associated with a higher cost, e.g., due to a high recovery cost.

the best-case scenarios with respect to cost are a subset of the best-case scenarios with respect to service.

The above results indicate that under the worst-case scenario, all suppliers, and assembly plants are either completely shut down or hit by a major disruption. When all primary suppliers are shut down, only one of them is selected for recovery to reduce recovery cost. In contrast to selection of recovery plants, where both shut-down plants may be selected for recovery. When both plants are selected as recovery plants, a larger fraction of demand for products is assigned to plant $j = 2$. While per period capacity, $c_2 = 5000$ of plant $j = 2$ is half of the capacity, $c_1 = 10000$ of plant $j = 1$, its recovery time $PRT(2,0) = 5$ is half of the recovery time $PRT(1,0) = 10$. In the examples, however, the largest portion of total cost is recovery cost of impacted suppliers, $\sum_{i \in I} p_{is} U_i^s / D$.

The major managerial insights are summarized in Table 10.13. Overall, the results of computational experiments indicate that *the proposed multi-portfolio approach and developed SMIP models with an embedded network flow problem and a very strong LP relaxation is a flexible and efficient tool for supply chain disruption management.* The approach has proven to be computationally very efficient.

10.6 Notes

Disruptions of material flows in global supply chains and the resulting losses may threaten financial state of firms. For example, the disruptive events that occurred in 2011 in the automotive and electronics supply chains (the Great East Japan earthquake and tsunami in March and Thailand's floods in October) resulted in huge losses of major automakers and electronics manufacturers, e.g., Fujimoto and Park (2013) and Marszewska (2016).

In order to minimize losses due to the shortage of material supplies, firms apply different disruption management strategies. Whenever a primary supplier is hit by disruption, the firm needs to choose whether to support recovery of disrupted primary supplier (e.g., the case of Toyota helping supplier of automotive semiconductors, Renesas Electronics after the earthquake in 2011 (Matsuo 2015), and after the Kumamoto earthquake in 2016 (Marszewska 2016)) or select an alternate (recovery) supplier, non-disrupted or disrupted less severely than the primary supplier. For example, after the Japanese earthquake disabled their primary supplier of a special pigment used in automobile paint, car manufacturers attempted to find alternate suppliers (Walsh 2011). Another example is the electronics supply chains (e.g., Sawik 2011a, 2017a), in which recovery suppliers need to be selected, whenever primary suppliers are hit by disruption. In a typical electronics supply chain, a wafer fabrication at multiple locations, where integrated circuits are manufactured, precedes printed wiring board assembly on surface mount technology lines at multiple locations, finally used for assembly of different types of electronic devices (e.g., cellular phones) at multiple assembly plants (e.g., cellular phone distribution centers). When a disruption hits a primary production facility, then a reasonable disruption management strategy would be to move production to alternate production facilities. For example, that strategy was applied by Texas Instruments and the Fujitsu Group after the Great East Japan earthquake (Godinez 2011; McGrath 2011). Intel replicated the manufacturing process at multiple plants to be prepared for a quick move of production in case one of its plant is damaged (Sheffi 2005).

The literature on mitigating the impact of disruption risks and optimization of a recovery process in supply chains is limited, e.g., Gurnani et al. (2012). For example, mitigation and contingency/recovery actions were studied by Tomlin (2006) in a dual sourcing setting, one unreliable supplier and another reliable and more expensive. A buyer that suffers a supply shortage can buy from a more expensive alternate supplier or produce less, and its decision depends on its inventory. If reliable supplier has volume flexibility, contingent rerouting by temporarily increasing its production may prove to be an effective way to speed up recovery process. The author established that, along with cost, percentage of supplier uptime, disruption length, capacity, and flexibility play an important role in determining a buyer's disruption management strategy. MacKenzie et al. (2014) proposed a model for severe disruptions in which a disruption simultaneously impacts several suppliers, while the suppliers' customers (firms) may face supply shortages. The model incorporated decisions made by both suppliers and firms during the disruption of random duration. The decisions may include whether or not suppliers move production to an alternate facility, hold parts inventory; a firm purchases parts from alternate suppliers that are not impacted, helps a primary supplier recover more quickly, or holds finished products inventory. The supplier and the firm optimal decisions are expressed in terms of model parameters, e.g., the cost of each strategy, the chances of losing business, and the probability of suppliers recovery. The impact of a disruption was evaluated based on the optimal response of an existing network once that disruption has occurred. The authors derived threshold parameters to support decisions whether and when a supplier moves production to an alternate facility and how much the firm should produce during the disruption to trade-off between two conflicting objectives: maximizing profit and service level. They applied model to a simulation based on the Great East Japan earthquake and tsunami in March 2011. Ivanov et al. (2016) analyzed seven proactive supply chain structures and computed recovery policies to re-direct material flows in the case of two disruption scenarios. They assessed the performance impact of the duration of disruptions and the costs of recovery for both service level and costs with the help of a supply chain (re)planning model containing elements of system dynamics and linear programming. A review of literature on Operations Research/Management Science models for supply chain disruptions was presented, for example, by Snyder et al. (2016) and Rajagopal et al. (2017). In the former review the authors discussed 180 scholarly works on the topic, organized into six categories: evaluating supply disruptions; strategic decisions; sourcing decisions; contracts and incentives; inventory; and facility location. In the latter review, quantitative decision-making models are presented and analyzed. A review of literature on disruption recovery in the supply chain can also be found in Ivanov et al. (2017).

The material presented in this chapter has been updated using results presented in Sawik (2019a), where the multi-portfolio approach and SMIP models were developed. The future research should concentrate on relaxations of the various simplified assumptions used to formulate the problem (see, Sect. 10.2). For example, a single recovery mode defined by constant recovery times, $TTR(i,l), PRT(j,l)$, and the associated recovery costs, $CTR(i,l), PRC(j,l)$, for each supplier i, each assembly plant j and each disruption level l, can be replaced by a number of available recovery modes, each represented by different values of recovery time and the associated recovery cost. Then, introduction of new recovery mode selection variables for each supplier and each assembly plant can enhance the problem formulation. Another assumption that can be relaxed is no partial recovery allowed. The recovery time is defined as the required time for recovering the whole capacity. In some practical cases, however, the fraction of capacity recovered in each period is assumed to be available, e.g., using a linear recovery function (Fattahi et al. 2017). In addition, the inventory of parts at the impacted primary assembly plant may be non available for transshipment or the transshipment itself can be non available or prohibitively expensive. Then, the recovery supply portfolio should be selected, even after full delivery of all required parts to the impacted primary assembly plant.

In a more general setting, both recovery time and recovery cost can be modeled as random parameters, e.g., Schmitt (2011) and Schmitt and Singh (2012).

Problems

10.1. In the SMIP models presented in this chapter introduce delay penalty when demand d_t for period t is not met by that period.

10.2. Enhance the SMIP models presented in this chapter for the case in which the inventory of parts at the impacted primary assembly plant is not available for transshipment.

10.3. Formulate the mean-risk models **DSupport_ECV(c)** and **DSupport_ECV(sl)** to minimize expected cost and CVaR of cost or maximize expected service level and CVaR of service level, respectively.

10.4. Enhance model **DSupport_E** for multiple recovery modes defined by different recovery time, $TTR(i,l), PRT(j,l)$, and the associated recovery cost, $CTR(i,l)$, $PRC(j,l)$, for each supplier i, each assembly plant j and each disruption level l.

10.5. Explain why the best-case and worst-case disruption scenarios for the hierarchical approach are, respectively, subsets and supersets of the corresponding scenarios for the integrated approach.

Chapter 11
Selection of Primary and Recovery Supply and Demand Portfolios and Scheduling: A Two-Period Approach

11.1 Introduction

In this chapter a multi-period SMIP decision-making model proposed in Chap. 10 for the integrated selection of primary and recovery portfolios and production scheduling is transformed into a simple two-period model. The planning horizon is divided into two aggregate periods: pre-disruption Period 1 and post-disruption Period 2. The corresponding mitigation and recovery decisions are: (1) primary supply and demand portfolios and production prior to disruption, and (2) recovery supply, transshipment and demand portfolios, and production after disruption. The two-period model supports the integrated decision-making to be implemented before a disruptive event to hedge against all potential disruption scenarios as well as the decisions for the aftermath period, for each scenario. The decisions to be made ahead of time include selection of primary suppliers and assembly plants, and allocation among them orders quantity and primary production volumes. Simultaneously, for each disruption scenario the aftermath decisions are made: recovery suppliers and assembly plants are selected, unfulfilled demands allocated among them and transshipment of parts inventory and recovery production volumes are determined. The two variants of the two-period modeling approach are proposed and compared: with best-case and with worst-case capacity constraints, where the available recovery production capacity may be overestimated or underestimated, respectively.

The following two-period SMIP models are presented in this chapter:

DSupport2bc_E two-period model with best-case capacity constraints for risk-neutral selection of primary and recovery supply and demand portfolios and production scheduling to minimize expected cost;

(continued)

T. Sawik, *Supply Chain Disruption Management*, International Series in Operations Research & Management Science 291, https://doi.org/10.1007/978-3-030-44814-1_11

DSupport2wc_E two-period model with worst-case capacity constraints for risk-neutral selection of primary and recovery supply and demand portfolios and production scheduling to minimize expected cost.

The proposed two-period modeling approach is compared with the multi-period approach, where both the approaches support integrated selection of primary suppliers, the decision to be implemented before a disruption, and recovery suppliers and assembly plants, the decision to be implemented during and after the disruption.

Numerical examples, computational results, and comparisons of the two two-period models with a multi-period model are provided in Sect. 11.4. The major managerial insights are summarized at the end of this chapter.

11.2 Problem Description

This section provides a brief description of the problem of simultaneous selection of primary and recovery supply and demand portfolios and production scheduling, which is presented in details in Sect. 10.2.

Consider a supply chain in which a single producer assembles one product type in several assembly plants to meet customer demand, using a critical part type that can be manufactured and provided by several suppliers. Let $I = \{1,\ldots,\overline{I}\}$ be the set of $\overline{I}$ suppliers, $J = \{1,\ldots,\overline{J}\}$, the set of $\overline{J}$ assembly plants, and $\overline{T}$ the length of the planning horizon. (for notation used, see Table 11.1).

Denote by D, the total demand for parts/products (each unit of product requires one unit of critical part) that should be met by the end of the planning horizon, i.e, period-independent demand is considered. Let $j = 1$ be the primary plant, where the total demand for products, D, is initially assigned.

The suppliers of parts and assembly plants are located in $\overline{R}$ geographic regions, subject to potential regional disasters that may result in complete shutdown of all suppliers and plants in the same region simultaneously. Denote by I^r and J^r, respectively, the subsets of suppliers and plants in region $r \in R$, and by p^r, the regional disruption probability for region r.

Both suppliers and plants are subject to random local disruptions of different levels, $l \in L_i = \{0,\ldots,\overline{L_i}\}$ and, $l \in L^j = \{0,\ldots,\overline{L^j}\}$, respectively. Supplier disruption level refers to the fraction of an order that can be delivered, whereas plant disruption level refers to available fraction of full capacity, c_j. Level $l = 0$ represents complete shutdown of a supplier or a plant, while level $l = \overline{L_i}$ and $l = \overline{L^j}$ represents normal conditions with no disruption, i.e., full order delivery and full capacity available, respectively.

There are in total, $\overline{S} = \prod_{i\in I}(\overline{L_i}+1)\prod_{j\in J}(\overline{L^j}+1)$, potential scenarios, where each scenario $s \in S$ is represented by an $(\overline{I}+\overline{J})$- dimensional vector $\lambda_s = \{\lambda_{1s},\ldots,\lambda_{\overline{I},s},$

Table 11.1 Notation: two-period approach

	Indices
i	= supplier, $i \in I$
j	= assembly plant, $j \in J$
l	= disruption level, $l \in L_i, i \in I$, $l \in L^j, j \in J$
r	= region, $r \in R$
s	= disruption scenario, $s \in S$
	Input parameters
c_j	= per period capacity of non-disrupted plant j
D	= total demand for parts/products
e_i	= fixed ordering cost for supplier i
g	= per unit penalty cost of unfulfilled demand for products
o_i	= per unit price of parts purchased from supplier i
ε_j	= additional per unit production cost at recovery plant $j > 1$
φ_j	= fixed production setup cost at recovery plant $j > 1$
ψ_j	= per unit transshipment cost from primary plant $j = 1$ to plant $j > 1$
p_{il}	= probability of disruption level l for supplier i
π_{jl}	= probability of disruption level l for plant j
p^r	= regional disruption probability for region r
t_s	= start time period of disruption event s
γ_{il}, γ_i^s	= fraction of an order delivered by supplier i under disruption level l, under scenario s, respectively
δ_{jl}, δ_j^s	= fraction of capacity of plant j available under disruption level l, under scenario s, respectively
τ_{ij}	= delivery lead time from supplier i to plant j
σ_j	= transshipment time from primary plant $j = 1$ to plant $j > 1$
$TTR(i,l)$, $PRT(j,l)$	= time-to-recover, respectively for supplier i, plant j hit by disruption at level l
θ_{is}, ϑ_{js}	= time-to-recover, respectively of supplier i, plant j from disruption under scenario s
$CTR(i,l)$, $PRC(j,l)$	= cost-to-recover, respectively for supplier i, plant j hit by disruption at level l
ρ_{is}, ρ_{js}	= cost-to-recover, respectively of supplier i, plant j from disruption under scenario s

$\lambda_{\bar{I}+1,s}, \ldots, \lambda_{\bar{I}+\bar{J},s}\}$. $\lambda_{is} \in L_i$ is the disruption level of supplier $i \in I$ and $\lambda_{\bar{I}+j,s} \in L^j$, the disruption level of plant $j \in J$, under scenario $s \in S$.

The probability P_s of disruption scenario $s \in S$ is $P_s = \prod_{r \in R} P_s^r$, where P_s^r is the probability of realizing disruption scenario s in region r

$$P_s^r = \begin{cases} (1-p^r)(\prod_{i \in I^r} \prod_{l \in L_i : \lambda_{is}=l} p_{il})(\prod_{j \in J^r} \prod_{l \in L^j : \lambda_{\bar{I}+j,s}=l} \pi_{jl}), \\ \text{if } \sum_{i \in I^r} \lambda_{is} + \sum_{j \in J^r} \lambda_{\bar{I}+j,s} > 0 \\ p^r + (1-p^r)(\prod_{i \in I^r} p_{i0})(\prod_{j \in J^r} \pi_{j0}), \\ \text{if } \sum_{i \in I^r} \lambda_{is} + \sum_{j \in J^r} \lambda_{\bar{I}+j,s} = 0, \end{cases} \tag{11.1}$$

where p_{il} and π_{jl} denote probability of disruption level l for supplier i and plant j, respectively.

The basic assumptions used to formulate the decision-making problem are provided in Sect. 10.2. In particular:

- Orders for parts are placed at the beginning of the planning horizon and, under normal conditions, parts ordered from supplier i are delivered to primary assembly plant $j = 1$ in period τ_{i1}, where τ_{i1} is the total of manufacturing lead time and transportation time to plant $j = 1$.
- Demand for products is period-independent, i.e., is not related to planning periods and should be satisfied by the end of the planning horizon with a penalty cost charged for total unfulfilled demand.
- At most one disruptive event is assumed to occur over the entire planning horizon.
- Disruption to primary supplier (primary assembly plant) under scenario $s \in S$ occurs in period t_s and recovery process starts in period $t_s + 1$, so that the disrupted supplier $i \in I$ (disrupted plant $j \in J$) returns to its full capacity in period $t = t_s + \theta_{is}$ ($t = t_s + \vartheta_{js}$).

The simultaneously determined decisions to be implemented, before and after a disruptive event include:

- selection of primary suppliers and allocation of orders for parts before disruption, to hedge against all potential disruption scenarios;
- selection of recovery suppliers and assembly plants after disruption, for each disruption scenario;
- allocation of unfulfilled demand for parts among recovery suppliers after disruption, for each disruption scenario;
- allocation of unfulfilled customer demand for products among recovery plants after disruption, for each disruption scenario;
- transshipment of parts from primary assembly plant to recovery plants after disruption, for each disruption scenario;
- production planning at assembly plants before disruption and after disruption, for each disruption scenario.

11.3 A Two-Period Model for Risk-Neutral Decision-Making

In this section a two-period SMIP model **DSupport2bc_E** is presented for the integrated risk-neutral selection of primary and recovery supply and demand portfolios and production scheduling under disruption risks.

The two-period model **DSupport2bc_E** is a simplified version of the multi-period model **DSupport_E**, developed in Chap. 10 for a single disruptive event and period-independent demand. The multi-period planning horizon made up of $\overline{T}$ unit-length planning periods is replaced by the two-period horizon of total length, $\overline{T}$. The planning horizon is divided into two aggregate periods: pre-disruption Period 1 and post-disruption Period 2. Accordingly, the time-indexed scheduling variables, x^s_{jt}, that denote production in assembly plant j in period t under disruption scenario s are replaced by the two variables, ξ^s and x^s_j, cumulative production in primary assembly plant $j = 1$, before disruption under scenario s and cumulative production in each recovery assembly plant $j \in J$, after disruption under scenario s, respectively.

The corresponding risk mitigation and recovery decisions are:

1. Selection of primary supply and demand portfolios and production scheduling before disruption,
2. Selection of recovery supply, transshipment and demand portfolios, and production scheduling after disruption.

Using the proposed multi-portfolio approach, both primary and recovery portfolio decisions can be integrated and made simultaneously. The different portfolio types are defined below (see, Sect. 1.4 and Table 11.2).

The Primary Supply Portfolio

$$(v_1, \ldots, v_{\bar{I}}),$$

specifies supplies of parts, Dv_i, from selected primary suppliers $i \in I$ (such that $u_i = 1$) to primary assembly plant $j = 1$, where

$$v_i \in [0,1];\ i \in I,$$
$$\sum_{i \in I} v_i = 1.$$

The Recovery Supply Portfolio for Scenario s

$$(V_{1j}^s, \ldots, V_{\bar{I},j}^s);\ j \in J,$$

specifies supplies of parts, DV_{ij}^s, from selected recovery suppliers, $i \in I$ (such that, $U_i^s = 1$), to selected recovery assembly plants, $j \in J$ (such that $q_j^s = 1$), where

$$V_{ij}^s \in [0,1];\ i \in I, j \in J$$
$$\sum_{i \in I}(\gamma_i^s v_i + \sum_{j \in J} V_{ij}^s) = 1,$$

and $\sum_{i \in I} \gamma_i^s v_i$ denotes delivery of parts from primary suppliers to primary assembly plant. $\gamma_i^s = \gamma_{i,\lambda_{is}}$, and λ_{is} is disruption level of supplier i under scenario s.

The Recovery Transshipment Portfolio for Scenario s

$$(w_1^s, \ldots, w_{\bar{J}}^s),$$

specifies transshipment of inventory of parts, Dw_j^s, from primary assembly plant $j = 1$ to selected recovery plants, $j \in J$ (such that $q_j^s = 1$), where

$$w_j^s \in [0,1];\ j \in J,$$
$$\sum_{j \in J} w_j^s = \sum_{i \in I} \gamma_i^s v_i - \xi^s / D,$$

and ξ^s denotes cumulative production (as well as the usage of parts) at primary assembly plant $j = 1$, before disruption under scenario s.

The Recovery Demand Portfolio for Scenario s

$$(Q_1^s, \ldots, Q_J^s),$$

specifies an allocation of unfulfilled demand for products, DQ_j^s, among selected recovery plants $j \in J$ (such that $q_j^s = 1$), where

$$Q_j^s \in [0,1];\ j \in J,$$
$$\xi^s/D + \sum_{j \in J} Q_j^s = 1.$$

In addition to cumulative production at primary assembly plant $j = 1$, before disruption, ξ^s, cumulative production after disruption, x_j^s, is determined for each recovery plant j (such that $q_j^s = 1$) and each scenario s.

Table 11.2 Second stage variables: two-period approach

q_j^s	=	1, if assembly plant j is selected as a recovery plant under disruption scenario s; otherwise $q_j^s = 0$ (recovery plant selection)
Q_j^s	$\in$	$[0,1]$ the fraction of total demand for products to be completed by recovery plant j under disruption scenario s (recovery demand portfolio)
U_i^s	=	1, if supplier i is selected as a recovery supplier under disruption scenario s; otherwise $U_i^s = 0$ (recovery supplier selection)
V_{ij}^s	$\in$	$[0,1]$, the fraction of total demand for parts ordered from recovery supplier i to recovery plant j, under disruption scenario s (recovery supply portfolio)
w_j^s	$\in$	$[0,1]$, the fraction of total demand for parts, transshipped from the primary plant $j = 1$ to recovery plant j, under disruption scenario s, where w_1^s represents the parts that remain in the primary plant $j = 1$ (recovery transshipment portfolio)
ξ^s	$\geq$	0, cumulative production in primary assembly plant $j = 1$, before disruption under scenario s (pre-disruption production scheduling variables)
x_j^s	$\geq$	0, cumulative production in recovery assembly plant j, after disruption under scenario s (post-disruption production scheduling)
		Auxiliary variables
μ_i^s	=	1, if $u_i = U_i^s = 1$; otherwise $\mu_i^s = 0$ (elimination of double fixed ordering costs)

Let E^c be the expected cost per product to be minimized, and E^{sl}, the expected service level, i.e., the expected fraction of the total fulfilled demand for products, to be maximized.

$$\begin{aligned} E^c = \sum_{s \in S} \sum_{i \in I} P_s e_i (u_i + U_i^s - \mu_i^s)/D + \sum_{s \in S} \sum_{i \in I} P_s \rho_{is} U_i^s / D \\ + \sum_{s \in S} \sum_{i \in I} P_s o_i (\gamma_i^s v_i + \sum_{j \in J} V_{ij}^s) + \sum_{s \in S} \sum_{j \in J} P_s \psi_j w_j^s \\ + \sum_{s \in S} \sum_{j \in J} P_s ((\varphi_j + \rho_{js}) q_j^s + \varepsilon_j x_j^s)/D + g(1 - E^{sl}), \end{aligned} \tag{11.2}$$

$$E^{sl} = \sum_{s \in S} P_s (\xi^s + \sum_{j \in J} x_j^s)/D. \tag{11.3}$$

The expected cost per product, E^c, (11.2), consists of various fixed and variable costs per product. The fixed cost per product includes expected cost: of ordering parts from primary and recovery suppliers; of recovery process for impacted suppliers selected as recovery suppliers; of moving production from primary plant to recovery plants; and of recovery process for impacted plants selected as recovery plants, (cf. Table 11.1). The variable cost per product includes expected cost: of purchasing parts from primary and recovery suppliers; of transshipment parts from primary plant to recovery plants; of additional production cost in recovery plants different from primary plant; and of penalty for unfulfilled demand, $g(1-\sum_{s\in S}P_s(\xi^s+\sum_{j\in J}x_j^s)/D)=g(1-E^{sl})$, (cf. Table 11.1).

DSupport2bc_E: *Risk-neutral selection of primary and recovery* **D***emand and* **Supp***ly* **port***folios:* **2***-period model with* **b***est-***c***ase capacity constraints*

Minimize (11.2)

subject to

Primary supply portfolio selection constraints:

the total demand for parts must be fully allocated among primary suppliers, (11.4), and parts cannot be ordered from non-selected primary suppliers, (11.5),

$$\sum_{i\in I} v_i = 1 \tag{11.4}$$

$$v_i \le u_i;\ i\in I. \tag{11.5}$$

Recovery supply and demand portfolio selection constraints:

- the unfulfilled demand for parts cannot be assigned to non-selected recovery suppliers, (11.6),
- for non-selected recovery plants, parts can neither be ordered from recovery suppliers, (11.7), nor transshipped to, from primary plant, (11.8), and the unfulfilled demand for products cannot be assigned to, (11.9),
- the unfulfilled demand for parts must be fully allocated among recovery suppliers, (11.10), and the unfulfilled demand for products must be fully allocated among recovery plants, (11.11),
- the supply and demand portfolio flow conservation equations ensure that for each recovery plant, the recovery supplies and transshipment of parts are in balance with the demand for products to be fulfilled by that plant, (11.12),
- the parts balance constraints ensure that the total transshipment of parts from primary plant is equal to the partially fulfilled supplies from primary suppliers less the usage of parts for production at primary plant before disruption, (11.13),

(continued)

- each supplier selected to both primary and recovery portfolio is charged exactly once with fixed ordering cost in the objective function, (11.14),

$$V_{ij}^s \leq U_i^s;\ i \in I, j \in J, s \in S \tag{11.6}$$

$$V_{ij}^s \leq q_j^s;\ i \in I, j \in J, s \in S \tag{11.7}$$

$$w_j^s \leq q_j^s;\ j \in J, s \in S \tag{11.8}$$

$$Q_j^s \leq q_j^s;\ j \in J, s \in S \tag{11.9}$$

$$\sum_{i \in I} (\gamma_i^s v_i + \sum_{j \in J} V_{ij}^s) = 1;\ s \in S \tag{11.10}$$

$$\xi^s / D + \sum_{j \in J} Q_j^s = 1;\ s \in S \tag{11.11}$$

$$\sum_{i \in I} V_{ij}^s + w_j^s = Q_j^s;\ j \in J, s \in S \tag{11.12}$$

$$\sum_{j \in J} w_j^s = \sum_{i \in I} \gamma_i^s v_i - \xi^s / D;\ s \in S \tag{11.13}$$

$$\mu_i^s \leq (u_i + U_i^s)/2;\ i \in I, s \in S. \tag{11.14}$$

Production capacity constraints:

- the cumulative capacity of primary plant before disruption, (11.15), and the cumulative capacity of each selected recovery plant after disruption, (11.16), cannot be exceeded,

$$\xi^s \leq c_1(t_s - 1);\ s \in S \tag{11.15}$$

$$x_j^s \leq c_j(\delta_j^s \vartheta_{js} + (\overline{T} - t_s - \vartheta_{js}))q_j^s;\ j \in J, s \in S. \tag{11.16}$$

Supply-transshipment-production coordinating constraints:

- for each disruption scenario s, cumulative production before disruption in primary assembly plant $j = 1$ cannot exceed cumulative deliveries by period $t_s - 2$ from primary suppliers, (11.17),
- for each disruption scenario s, cumulative production after disruption in recovery assembly plant j cannot exceed cumulative deliveries from all recovery suppliers, plus transhipment of parts from primary assembly plant $j = 1$, (11.18),

$$\xi^s / D \leq \sum_{i \in I: \tau_{i1} \leq t_s - 2} \gamma_i^s v_i;\ s \in S \tag{11.17}$$

$$x_j^s / D \leq \sum_{i \in I} V_{ij}^s + w_j^s;\ j \in J, s \in S. \tag{11.18}$$

Non-negativity and integrality conditions:

$$q_j^s \in \{0, 1\};\ j \in J, s \in S \tag{11.19}$$

(continued)

$$Q_j^s \in [0,1];\ j \in J, s \in S \tag{11.20}$$
$$u_i \in \{0,1\};\ i \in I \tag{11.21}$$
$$v_i \in [0,1];\ i \in I \tag{11.22}$$
$$U_i^s \in \{0,1\};\ i \in I, s \in S \tag{11.23}$$
$$V_{ij}^s \in [0,1];\ i \in I, j \in J, s \in S \tag{11.24}$$
$$w_j^s \in [0,1];\ j \in J, s \in S \tag{11.25}$$
$$\xi^s \geq 0;\ s \in S \tag{11.26}$$
$$x_j^s \geq 0;\ j \in J, s \in S \tag{11.27}$$
$$\mu_i^s \in \{0,1\};\ i \in I, s \in S, \tag{11.28}$$

where $\delta_j^s = \delta_{j,\lambda_{\overline{I}+j,s}}$, and $\lambda_{\overline{I}+j,s}$ is disruption level of plant j under scenario s.

Notice that the supply-transshipment-production coordinating constraints (11.17)–(11.18) are incapable of accounting for supply and transshipment delivery timing. In the two-period setting, parts delivered from the recovery suppliers and transshipped from the primary plant are assumed to be available for processing during the entire period after the disruption. As a result the cumulative production capacity, Eq. (11.16), can be fully utilized and hence the available capacity of recovery plants may be overestimated. The two-period model **DSupport2bc_E** yields the best-case solutions.

In order to better account for time-varying availability of parts, Eq. (11.16) can be replaced by the worst-case capacity constraints defined in Proposition 11.1

Proposition 11.1. *The worst-case capacity constraints under disruption scenario s, are*

$$x_j^s \leq c_j(\delta_j^s \max\{0, \vartheta_{js} - \max_{i \in I}(\theta_{is} + \tau_{ij})\}$$
$$+(\overline{T} - t_s - \max\{\max_{i \in I}(\theta_{is} + \tau_{ij}), \vartheta_{js}\}))q_j^s;\ j \in J, s \in S. \tag{11.16wc}$$

Proof. Equation (11.16wc) accounts for the latest delivery of recovery supplies to recovery plants, $t_s + \max_{i \in I}(\theta_{is} + \tau_{ij})$, and by that the production capacity available after disruption is maximally reduced.

Denote by **DSupport2wc_E** (*Risk-neutral selection of primary and recovery* **D***emand and* **Sup***ply* **port***folios:* **2***-period model with* **w***orst-***c***ase capacity constraints*), model **DSupport2bc_E** with best-case capacity constraints (11.16) replaced by worst-case capacity constraints (11.16wc). In contrast to **DSupport2bc_E**, the two-period model **DSupport2wc_E** may underestimate the available production capacity and yields worst-case solutions.

Notice that for each disruption scenario, models **DSupport2bc_E** and **DSupport2wc_E** eliminate the end of horizon inventory of products. The last assertion can be additionally validated by the following simple Proposition 11.2.

Proposition 11.2. *No end of horizon inventory of products.*

$$\xi^s/D + \sum_{j\in J} x_j^s/D \le 1;\ s \in S. \tag{11.29}$$

Proof. Equations (11.18), (11.12) and (11.11) imply

$$\xi^s/D + \sum_{j\in J} x_j^s/D \le \xi^s/D + \sum_{j\in J}(\sum_{i\in I} V_{ij}^s + w_j^s)$$
$$= \xi^s/D + \sum_{j\in J} Q_j^s = \xi^s/D + 1 - \xi^s/D = 1;\ s \in S.$$

11.3.1 Multi-Period Model

In the multi-period model **DSupport_E** presented in Sect. 10.3.1, the two-period planning horizon of length, $\overline{T}$, is replaced by the set $T = \{1,\ldots,\overline{T}\}$ of $\overline{T}$ unit-length planning periods. Accordingly, the cumulative production variables ξ^s and x_j^s are replaced by time-indexed scheduling variables, x_{jt}^s, that denote production in plant j in period t under disruption scenario s.

DSupport_E: *Risk-neutral selection of primary and recovery* **D***emand and* **Supp***ly* **port***folios and production scheduling*

Minimize

$$E^c = \sum_{s\in S} P_s \sum_{i\in I}((e_i(u_i + U_i^s - \mu_i^s) + \rho_{is}U_i^s)/D + o_i(\gamma_i^s v_i + \sum_{j\in J} V_{ij}^s))$$
$$+ \sum_{s\in S} P_s \sum_{j\in J}(\psi_j w_j^s + (\varphi_j + \rho_{js})q_j^s + \sum_{t\in T} \varepsilon_j x_{jt}^s)/D + g(1 - E^{sl}), \tag{11.30}$$

where

$$E^{sl} = \sum_{s\in S}\sum_{j\in J}\sum_{t\in T} P_s x_{jt}^s/D \tag{11.31}$$

(continued)

subject to (11.4)–(11.10), (11.12), (11.14), (11.19)–(11.25), (11.28) and

Recovery supply and demand portfolio selection constraints:

$$\sum_{t\in T:t<t_s} x_{1t}^s/D + \sum_{j\in J} Q_j^s = 1;\ s\in S \tag{11.32}$$

$$\sum_{j\in J} w_j^s = \sum_{i\in I} \gamma_i^s v_i - \sum_{t\in T:t<t_s} x_{1t}^s/D;\ s\in S. \tag{11.33}$$

Production capacity constraints:

$$x_{1t}^s \le c_1;\ t\in T, s\in S: t<t_s \tag{11.34}$$

$$x_{jt}^s \le c_{jt}^s q_j^s;\ j\in J, t\in T, s\in S: t\ge t_s \tag{11.35}$$

$$\sum_{t\in T:t\ge t_s} x_{jt}^s/D \le Q_j^s;\ j\in J, s\in S \tag{11.36}$$

$$x_{jt}^s \ge 0;\ j\in J, t\in T, s\in S, \tag{11.37}$$

where

$$c_{jt}^s = \begin{cases} \delta_j^s c_j, & \text{if } t_s \le t \le t_s + \vartheta_{js} - 1 \\ c_j, & \text{if } t \le t_s - 1, t \ge t_s + \vartheta_{js}. \end{cases} \tag{11.38}$$

Supply-transshipment-production coordinating constraints:

$$\sum_{t'\in T:t'\le t} x_{1t'}^s \le \sum_{i\in I:\tau_{i1}\le t-1} \gamma_i^s v_i + \sum_{i\in I:t_s+\theta_{is}+\tau_{i1}\le t-1} V_{i1}^s - (\text{if } t\le \bar{t}_s \text{ then } 0 \text{ else } \sum_{j\in J:j>1} w_j^s);$$
$$t\in T, s\in S \tag{11.39}$$

$$\sum_{t'\in T:t'\le t} x_{jt'}^s \le \sum_{i\in I:t_s+\theta_{is}+\tau_{ij}\le t-1} V_{ij}^s + (\text{if } t\le \bar{t}_s+\sigma_j \text{ then } 0 \text{ else } w_j^s);$$
$$j\in J, t\in T, s\in S: j>1, \tag{11.40}$$

where $\bar{t}_s = \max\{t_s, \max_{i\in I}\tau_{i1}\}$ and, $\bar{t}_s+1$, is the start period of transshipment of parts to recovery plants.

In a similar way to Proposition 11.2, Eqs. (11.32) and (11.36) imply no end of horizon inventory of products for each disruption scenario

$$\sum_{t\in T:t<t_s} x_{1t}^s/D + \sum_{j\in J}\sum_{t\in T:t\ge t_s} x_{jt}^s/D \le 1;\ s\in S. \tag{11.41}$$

While in the two-period models, **DSupport2bc_E** and **DSupport2wc_E**, the two aggregate planning periods, before and after a disruption, are considered only, in the above multi-period formulation, **DSupport_E**, a detailed timing of supplies and production over the entire planning horizon can be modeled.

11.4 Computational Examples

This section presents some computational examples to compare two- and multi-period modeling approaches. Although the input data for the examples are in part hypothetical, their relations to each other are real and the computational examples can be linked with a real-world problem in the electronics supply chains, e.g., Sawik (2011a, 2017a, 2019a). The computational experiments were performed using the AMPL programming language and the Gurobi 7.5.0 solver on a MacBookPro laptop with Intel Core i7 processor running at 2.8GHz and with 16GB RAM. The basic input parameters (see, Table 11.3) are taken from the computational examples presented in Sect. 10.5.

Table 11.3 Input parameters: two-period approach

$\overline{I}=4$ suppliers, $\overline{J}=2$ assembly plants, $\overline{T}=30$ planning periods, $\overline{R}=2$ two geographic regions	
$I^1=\{1,2\}$, $I^2=\{3,4\}$, $J^1=\{1,2\}$	Two suppliers and two plants located in region $r=1$ and two suppliers in region $r=2$
$D=300{,}000$	Total demand for parts/products
Disruption levels:	
$\overline{L}_i=3$	Four disruption levels for each supplier $i\in I$
$\overline{L}^j=1$	Two disruption levels for each plant $j\in J$
Local disruption probabilities of suppliers:	
$p_{i0}=0.1(1-p_{i3})$	Complete shutdown (level $l=0$)
$p_{i1}=0.3(1-p_{i3})$	Major disruption (level $l=1$)
$p_{i2}=0.6(1-p_{i3})$	Minor disruption (level $l=2$)
$p_{i3}\in[0.89,0.99]$; $i\in I^1$, $p_{i3}\in[0.79,0.89]$; $i\in I^2$	Non disruption (level $l=3$)
Local non disruption probabilities of plants:	$\pi_{11}=0.85$, $\pi_{21}=0.95$
Regional disruption probability:	$p^1=0.001$, $p^2=0.01$
Fulfillment rates of suppliers:	
$\gamma_{i0}=0\ \forall i\in I$, $\gamma_{i3}=1\ \forall i\in I$	
$\gamma_{i1}\in[0.01,0.50]\ \forall i\in I^1$ $\gamma_{i1}\in[0.01,0.30]\ \forall i\in I^2$, $\gamma_{i2}\in[0,51,0.99]\ \forall i\in I^1$ $\gamma_{i2}\in[0,31,0.99]\ \forall i\in I^2$	
Plant capacity:	
$c_1=10{,}000$, $c_2=5000$	
Cost parameters:	
$o=(14,12,8,9)$, $e=(8000,6000,12000,13{,}000)$	
$g\in\{1,10,100,1000,10{,}000,100{,}000,\infty\}$	Unit penalties for unfulfilled demand for products, where $g=\infty$ denotes maximization of service level
$\varepsilon_2=1$, $\varphi_2=100$, $\psi_2=0.1$	
$CTR(i,l)$=if $l=0$ then $100e_i$; if $l=1$ then $10e_i$; if $l=2$ then e_i $\forall i\in I$	
$PRC(1,0)=PRC(2,0)=10{,}000$	
Time parameters:	
$\tau_{1j}=\tau_{2j}=2$, $\tau_{3j}=\tau_{4j}=4, \forall j\in J$, $\sigma_2=2$	
$TTR(i,l)$=if $l=0$ then 12; if $l=1$ then 10; if $l=2$ then 8 $\forall i\in I$	
$PRT(1,0)=10$, $PRT(2,0)=5$	

The total number of all potential disruptive events was $(\overline{L}+1)^m(\overline{L^1}+1)(\overline{L^2}+1) = (4^4)(2^2) = 1024$. In the computational examples a disruption scenario was defined as a combination of disruptive event and its start time. The start time t_s of each disruptive event was assumed to be not greater than the maximum delivery lead time to primary assembly plant $j = 1$, $\max_{i \in I}(\tau_{i1}) = 4$. Thus $t_s \in \{1,2,3,4\}$, and the total number of all potential scenarios to be considered was 1024×4=4096, since each disruptive event $s = 1,\ldots,1024$ may occur at four different start times.
Notice that $\gamma_i^s = 1$, if $t_s > \tau_{i1}$, i.e., full order is delivered by supplier i, if disruption occurs after delivery lead time to primary assembly plant.

The probability P_s of realizing each disruption scenario $s \in S$ was calculated as follows:

$P_s = 0.1P_s^1 P_s^2$ for $s \leq 1024$,
$P_s = 0.2P_{s-1024}^1 P_{s-1024}^2$ for $1025 \leq s \leq 2048$,
$P_s = 0.3P_{s-2048}^1 P_{s-2048}^2$ for $2049 \leq s \leq 3072$,
$P_s = 0.4P_{s-3072}^1 P_{s-3072}^2$ for $3073 \leq s \leq 4096$,

where probabilities P_s^r; $r = 1,2$, $s \leq 1024$ were obtained using formula (11.1).
Notice that disruptive events with later start times were modeled to be more likely.

The solution results for the two-period models **DSupport2bc_E** with best-case capacity constraints (11.16) and **DSupport2wc_E** with worst-case capacity constraints (11.16wc) as well as for the multi-period model **DSupport_E**, are summarized and compared in Table 11.4.

For different values of unit penalty cost g, Table 11.4 presents the expected cost E^c along with the associated expected service level E^{sl}. In addition, the size of the SMIP models is presented. The table also shows primary supply portfolio, expected recovery demand portfolio, and expected fraction of demand fulfilled before and after disruption. Notice that expected service level is the sum of expected fraction of demand fulfilled before and after disruption, cf., Eqs. (11.3), (11.31). In a similar way, expected demand portfolio and expected fraction of demand fulfilled before disruption sum up to 100%, cf., Eqs. (11.11), (11.32).

As g increases to reduce the unfulfilled demand, more expensive and diversified supply portfolios are selected to hedge against all potential disruption scenarios and a higher service level is attained. The results indicate that supplier $i = 4$ (the more expensive among most unreliable suppliers $i \in I^2$) was never selected for the cost objective function. The supply portfolio is more diversified for the multi-period model. The table clearly shows that neglecting detailed timing of supply and transshipment of parts and timing of production in the two-period model results in slightly overestimated solution values (slightly lower expected cost and higher expected service) for model **DSupport2bc_E** and underestimated solution values (greater expected cost and lower expected service) for model **DSupport2wc_E**. Moreover, the lack of detailed timing of recovery supplies and the corresponding production volumes leads to distorted recovery demand portfolio with the overloaded recovery plant $j = 2$, in particular for model **DSupport2wc_E**, which is clearly illustrated in Fig. 11.1. For different values of unit penalty g and the two-period models **DSupport2bc_E** and **DSupport2wc_E**, Fig. 11.1a and b, respectively, show the expected

Table 11.4 Solution results: two-period vs. multi-period models

g	1	10	10^2	10^3	10^4	10^5	∞[a]
Best-case solutions							
Two-period model **DSupport2bc_E**: Var.=99,264, Bin.=40,964, Cons.=136,133, Nonz.=425,212[b]							
Exp.Cost E^c, (11.2)	8.79	9.84	14.55	16.04	18.34	41.27	–
Exp.Service E^{sl}, (11.3)×100%	76.90	91.11	99.58	99.90	99.97	99.97	99.97
Primary	1(0)	1(0)	1(0)	1(0)	1(0)	1(0)	1(19)
Supply portfolio	2(0)	2(0)	2(3)	2(10)	2(10)	2(10)	2(38)
(% of total demand)	3(100)	3(100)	3(97)	3(90)	3(90)	3(90)	3(13)
	4(0)	4(0)	4(0)	4(0)	4(0)	4(0)	4(30)
Exp.Recovery demand	1(85.76)	1(77.96)	1(85.38)	1(85.06)	1(84.99)	1(84.99)	1(72.44)
Portfolio (% of total demand)[c]	2(14.24)	2(22.04)	2(13.28)	2(10.94)	2(11.01)	2(11.01)	2(26.48)
Exp.Production in Period 1[d]	0	0	1.34	4.00	4.00	4.00	1.08
Exp.Production in Period 2[e]	76.90	91.11	98.25	95.90	95.97	95.97	98.89
LP[f]	8.77	9.76	14.37	15.05	18.15	41.08	99.97[g]
CPU[h]	10	16	61	49	31	32	3
Worst-case solutions							
Two-period model **DSupport2wc_E**							
Exp.Cost E^c, (11.2)	8.98	10.94	23.31	95.63	812.98	7986	–
Exp.Service E^{sl}, (11.3)×100%	57.57	81.41	91.06	92.03	92.03	92.03	92.03
Primary	1(0)	1(0)	1(0)	1(0)	1(0)	1(0)	1(7)
Supply portfolio	2(0)	2(0)	2(10)	2(10)	2(10)	2(10)	2(11)
(% of total demand)	3(100)	3(100)	3(90)	3(90)	3(90)	3(90)	3(77)
	4(0)	4(0)	4(0)	4(0)	4(0)	4(0)	4(11)
Exp.Recovery demand	1(85.76)	1(65.93)	1(71.43)	1(73.20)	1(73.20)	1(73.20)	1(72.23)
Portfolio (% of total demand)[c]	2(14.24)	2(34.07)	2(24.57)	2(22.80)	2(22.80)	2(22.80)	2(26.04)
Exp.Production in Period 1[d]	0	0	4.00	4.00	4.00	4.00	1.73
Exp.Production in Period 2[e]	57.57	81.41	87.06	88.03	88.03	88.03	90.30
LP[f]	8.96	10.90	23.22	95.54	812.90	7986	92.03[(g)]
CPU[h]	9	11	34	35	13	8	2
Optimal solutions							
Multi-period model **DSupport_E**: Var.=286,214, Bin.=40,964, Cons.=526,499, Nonz.=4,388,276[b]							
Exp.Cost Ec, (11.30)	8.82	9.95	14.65	16.70	26.76	126.98	–
Exp.Service Esl, (11.31)×100%	79.65	90.58	99.46	99.89	99.89	99.89	99.89
Primary	1(0)	1(0)	1(0)	1(0)	1(0)	1(0)	1(33)
Supply portfolio	2(0)	2(0)	2(7)	2(3)	2(3)	2(15)	2(13)
(% of total demand)	3(100)	3(100)	3(93)	3(97)	3(97)	3(85)	3(27)
	4(0)	4(0)	4(0)	4(0)	4(0)	4(0)	4(27)
Exp.Recovery demand	1(85.76)	1(74.84)	1(87.51)	1(85.40)	1(85.40)	1(87.24)	1(64.04)
Portfolio (% of total demand)[c]	2(14.24)	2(25.16)	2(12.28)	2(14.50)	2(14.50)	2(12.65)	2(35.86)
Exp.Production before disruption[i]	0	0	0.21	0.10	0.10	0.10	0.10
Exp.Production after disruption[j]	79.65	90.58	99.25	99.78	99.78	99.78	99.78
LP[k]	8.80	9.78	14.41	15.61	25.67	125.83	99.89[l]
CPU[h]	28	50	99	84	92	96	14

[a]Maximization of service level

[b]Var.=total number of variables, Bin.=number of binary variables, Cons.=number of constraints, Nonz.= number of nonzero coefficients

[c]$1(\sum_{s\in S} P_s Q_1^s \times 100\%), 2(\sum_{s\in S} P_s Q_2^s \times 100\%)$

[d]Expected fraction of demand fulfilled before disruption, $\sum_{s\in S} P_s \xi^s / D \times 100\%$

[e]Expected fraction of demand fulfilled after disruption, $\sum_{s\in S} P_s \sum_{j\in J} x_j^s / D \times 100\%$

[f]LP relaxation solution value of (11.2), [g]LP relaxation solution value of (11.3)

[h]Elapsed time in seconds for proven optimal solution

[i]Expected fraction of demand fulfilled before disruption, $\sum_{s\in S} P_s \sum_{j\in J} \sum_{t\in T: t<t_s} x_{jt}^s / D \times 100\%$

[j]Expected fraction of demand fulfilled after disruption, $\sum_{s\in S} P_s \sum_{j\in J} \sum_{t\in T: t\geq t_s} x_{jt}^s / D \times 100\%$

[k]LP relaxation solution value of (11.30), [l]LP relaxation solution value of (11.31)

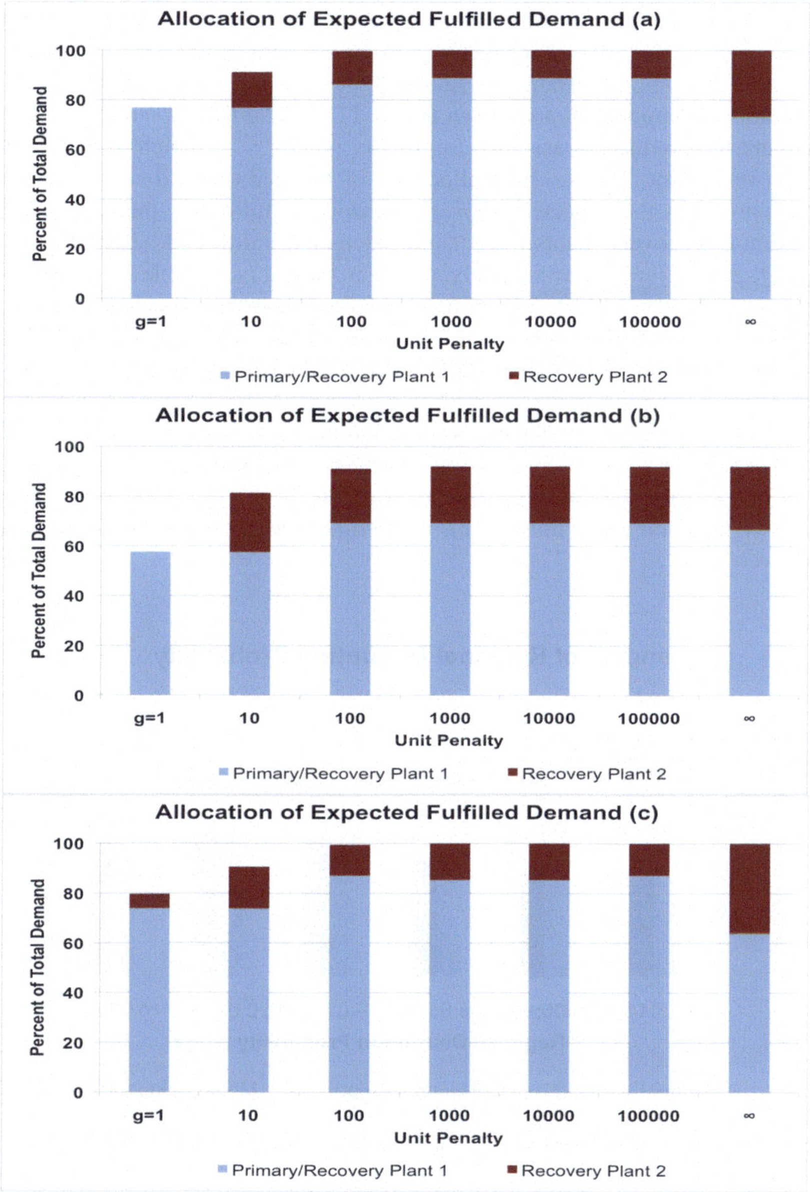

Fig. 11.1 Allocation of demand among assembly plants: (**a**) model **DSupport2bc_E**, (**b**) model **DSupport2wc_E**, (**c**) model **DSupport_E**

fraction of total demand for products fulfilled by primary (and recovery) assembly plant $j = 1$, $\sum_{s\in S} P_s(\xi^s + x_1^s)/D$, and by recovery assembly plant $j = 2$, $\sum_{s\in S} P_s x_2^s/D$. For comparison, the corresponding values, $\sum_{s\in S}\sum_{t\in T} P_s x_{jt}^s$, $j = 1,2$, for the multi-period model **DSupport_E** are shown in Fig. 11.1c. The figures indicate that most of demand for products is assigned to primary plant $j = 1$, which is also selected as a recovery plant. The resulting allocation of fulfilled demand among assembly plants approaches the expected recovery demand portfolio, i.e., the expected allocation among recovery plants of remaining demand, unfulfilled before disruption, $\sum_{s\in S} P_s Q_1^s$ and $\sum_{s\in S} P_s Q_2^s$, respectively, for plant 1 and 2 (see, Table 11.4). However, the expected demand fulfilled at each plant is less than the expected recovery demand portfolio (that does not include demand fulfilled before disruption) and the total fulfilled demand is less than 100%, even for the service level objective (11.3) or (11.31).

Table 11.4 also indicates that the proposed multi-portfolio approach leads to very strong LP relaxation solution values for both models. The solver was capable of finding proven optimal solutions for all examples with CPU time not exceeding several seconds and a few minutes for the two-period and the multi-period model, respectively.

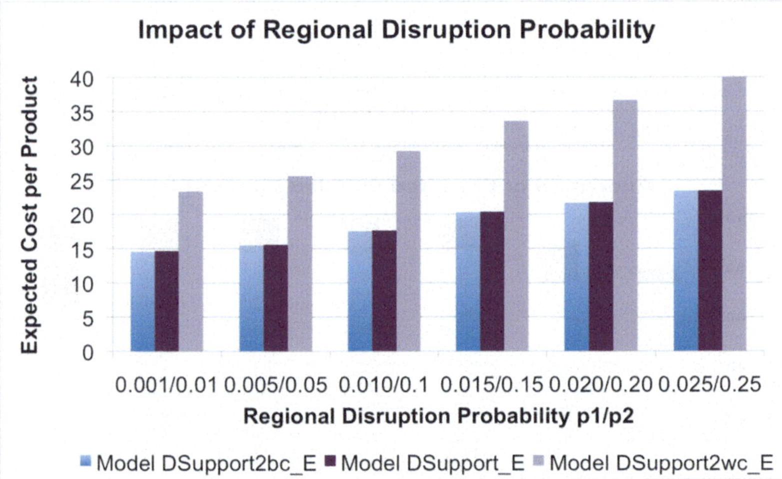

Fig. 11.2 Impact of regional disruption probability on the expected cost: $g = 100$

It is well known that a fair estimation of disruption probability is not an easy task and even small probability errors may severely impact the results. Figure 11.2 illustrates the impact of regional disruption probability $p^1 = 0.001, 0.005, 0.01, 0.015, 0.020, 0.025$ and $p^2 = 0.01, 0.05, 0.1, 0.15, 0.20, 0.25$ on expected cost per product, E^c, for all models with unit penalty $g = 100$. The figure clearly demonstrates that the expected cost increases with probability of regional disruption. For smaller probabilities, the cost increase is due to increasing unfulfilled demand, since over 90%

of parts are ordered from the cheapest, unreliable primary supplier, $i = 3$, in region $r = 2$. For greater probabilities, all parts are ordered from reliable and expensive primary supplier, $i = 2$, in region $r = 1$, and penalty for unfulfilled demand is reduced, however, purchasing cost dominates the other costs. In addition, Fig. 11.2 shows that solution values for models **DSupport2bc_E** and **DSupport_E** are very similar.

Table 11.5 Managerial insights: two-period approach

Decision-making issue	Decision-making impact
DSupport2bc_E vs. **DSupport2wc_E**	While the simplified two-period setting may overestimate (model **DSupport2bc_E**) or underestimate (model **DSupport2wc_E**) the available production capacity, it can be applied for a fast, rough-cut evaluation of disruption mitigation and recovery policy. For example, to obtain lower and upper bounds on the optimal values of cost and service or approximate primary and recovery supply and demand portfolios for different disruption scenarios
DSupport2bc_E vs. **DSupport_E**	If period-independent demand and single disruptive event are considered, then the two-period model, **DSupport2bc_E**, with best-case capacity constraints and the multi-period model, **DSupport_E**, may yield very close solution results

The major managerial insights are summarized in Table 11.5. Overall, the results of this chapter indicate that *owing to an embedded network flow structure, both the two-period and the multi-period SMIP formulations developed using the proposed multi-portfolio approach have a strong LP relaxation. The proposed SMIP models require very short computational time to yield proven optimal solutions for reasonable size problems, using commercially available solvers.*

11.5 Notes

The supply chain disruption management utilizes various quantitative methods to optimize disruption mitigation and recovery strategies. Rajagopal et al. (2017) presented a systematic literature review and a comprehensive analysis of the decision-making models for supply chain risk. 538 research articles published from 2005 to 2016 in the Academic Journal Guide quality rated journals were collected and 126 articles were shortlisted for the final review. They identified mitigation strategies and the decision-making models most addressed by scholars working on supply chain risk mitigation. Stochastic programming and mixed integer linear programming were found to be the commonly studied modeling methods. For example, Ruiz-Torres et al. (2013) studied the optimal supply allocation and contingency planning in supply networks made up of multiple suppliers with different cost and reliability characteristics, and a set of separate demand points. Suppliers have production flexibility that allows them to deliver a contingency quantity in case other suppliers fail. The problem objective was to minimize the total network costs. The problem was formulated as a mixed integer program, readily solvable by commercial mathematical programming software. Namdar et al. (2017) developed a scenario-based stochastic mixed integer programming model for single and multiple sourcing in the presence of disruption and operational risks. To achieve supply chain resilience under disruptions they considered single and multiple sourcing, backup supplier contracts, spot purchasing, and collaboration and visibility.

Most approaches, including the above mentioned, used in supply chain disruption management assume that some estimation of disruptive events probability and potential losses is available. At the same time, a fair probability estimation of rare events is a complicated problem and even small errors in those estimations may significantly impact the modeling results (e.g., Chan and Kroese 2011; Bukowski 2019).

The material presented in this chapter is based on results presented in Sawik (2019b), where two-period SMIP models were developed. In contrast to the above briefly reviewed works that focus on selected decisions independently of the others, the two-period modeling approach allows for optimization of integrated decision-making under disruption risks. The developed two-period model for the multi-portfolio approach leads to a strong SMIP formulation with a network flow structure embedded that requires a very short computational time to yield proven optimal solutions for reasonable size problems, using commercially available solvers. The future research should concentrate on relaxations of the various simplified assumptions used to formulate the two-period model. A straightforward enhancement of the developed models is a multiple part types and product types setting with subsets of suppliers available for each part type and assembly plants for each product type. While the problem size may considerably increase, the enhanced models for multiple part and product types will retain their network flow structure and the associated strength of the LP relaxation. Another assumption that can be relaxed is a partial recovery with the fraction of capacity recovered and available in each period. The proposed models can also be easily modified for the risk-averse decision-making, e.g., using conditional value-at-risk as a risk measure (see, Chap. 10). In some practical cases, the inventory of parts at the impacted primary assembly plant may be nonavailable for transshipment or the transshipment itself can be nonavailable or prohibitively expensive. Then, the recovery supply portfolio should be selected, even after full delivery of all required parts to the impacted primary assembly plant.

Problems

11.1. In the two-period models presented in this chapter introduce multi-level penalty cost for unfulfilled demand, with unit penalty dependent on the level of unfulfilled demand.

11.2. Enhance the two-period models presented in this chapter for the case in which the impacted primary suppliers and primary assembly plants cannot be recovered by the end of the planning horizon.

11.3. Formulate the two-period risk-averse models **DSupport2bc_CV(c)** and **DSupport2bc_CV(sl)** to minimize CVaR of cost and maximize CVaR of service level, respectively.

11.4. Identify best-case and worst-case scenarios for each of the two-period models **DSupport2bc_E** and **DSupport2wc_E**.

11.5. Enhance the two-period models presented in this chapter for a three-period setting: Period 1 before a disruption, Period 2 during recovery, and Period 3 after full recovery.

Chapter 12
Selection of Supply and Demand Portfolios and Production and Inventory Scheduling

12.1 Introduction

This chapter enhances SMIP models **DSupport_CV(c)** and **DSupport_CV(sl)**, developed in Chap. 10 for the risk-averse integrated selection of supply and demand portfolios and production scheduling with the period-dependent demand. The enhancements allow for a delayed satisfaction of demand using product inventory. To this end, production and inventory are scheduled simultaneously and additional cost of the inventory holding and penalty for delayed demand is introduced in the objective function. The primary portfolios to be implemented before a disruptive event are optimized simultaneously with recovery portfolios and production and inventory scheduling in the aftermath period. Moreover, the portfolios for both part suppliers and product manufacturers in different geographic regions are determined simultaneously. Using conditional cost-at-risk and conditional service-at-risk as risk measures, the risk-averse solutions are obtained for different customer demand patterns. The following two time-indexed SMIP models are presented in this chapter:

SupportPIS_CV(c) for risk-averse selection of primary and recovery supply and demand portfolios and production and inventory scheduling to minimize CVaR of cost;
SupportPIS_CV(sl) for risk-averse selection of primary and recovery supply and demand portfolios and production and inventory scheduling to maximize CVaR of service level.

Numerical examples, computational results, and some comparison of solution results for the two different risk-averse objective functions and different demand

T. Sawik, *Supply Chain Disruption Management*,
International Series in Operations Research & Management Science 291,
https://doi.org/10.1007/978-3-030-44814-1_12

patterns are provided in Sect. 12.4, where the major managerial insights drawn from the theoretical and computational studies are summarized.

12.2 Problem Description

This section provides description of the integrated selection of primary and recovery supply and demand portfolios and production and inventory scheduling under disruption risks.

Table 12.1 Notation: supply portfolio selection and production and inventory scheduling

	Indices
i	= supplier, $i \in I$
j	= assembly plant, $j \in J$
l	= disruption level, $l \in L_i, i \in I, l \in L^j, j \in J$
r	= region, $r \in R$
s	= disruption scenario, $s \in S$
t	= planning period, $t \in T$
	Input Parameters
ξ	= per unit and per period product inventory holding cost
ζ	= per unit and per period penalty cost of delayed demand for products
c_j	= per period capacity of non-disrupted plant j
c^s_{jt}	= capacity in period t of plant j under disruption scenario s
d_t	= demand for products in period t
e_i	= fixed ordering cost for supplier i
g	= per unit penalty cost of unfulfilled demand for products
ε_j	= additional per unit production cost at recovery plant $j > 1$
φ_j	= fixed production setup cost at recovery plant $j > 1$
ψ_j	= per unit transshipment cost from primary plant $j = 1$ to plant $j > 1$
o_i	= per unit price of parts purchased from supplier i
p_{il}	= probability of disruption level l for supplier i
π_{jl}	= probability of disruption level l for plant j
p^r	= regional disruption probability for region r
t_s	= start time period of disruption event s
γ_{il}	= fraction of an order delivered by supplier i under disruption level l
γ^s_i	= fraction of an order delivered by supplier i under scenario s
δ_{jl}	= fraction of capacity of plant j available under disruption level l
δ^s_j	= fraction of capacity of plant j available under scenario s
τ_{ij}	= delivery lead time from supplier i to plant j
σ_j	= transshipment time from primary plant $j = 1$ to plant $j > 1$
θ_{is}	= time-to-recover of supplier i from disruption under scenario s
ϑ_{js}	= time-to-recover of plant j from disruption under scenario s
ρ_{is}	= cost-to-recover of supplier i from disruption under scenario s
ρ_{js}	= cost-to-recover of plant j from disruption under scenario s

Consider a supply chain in which a single producer assembles one product type in several assembly plants to meet customer demand, using a critical part type that can be manufactured and provided by several suppliers. (for notation used, see Table 12.1). Let $I = \{1,\ldots,\overline{I}\}$ be the set of $\overline{I}$ suppliers, $J = \{1,\ldots,\overline{J}\}$, the set of $\overline{J}$ assembly plants, and $T = \{1,\ldots,\overline{T}\}$ the set of $\overline{T}$ planning periods.

If we denote by d_t, the demand for product in period t, then $D_t = \sum_{t' \in T: t' \leq t} d_{t'}$ is the cumulative demand for products by period t and $D = \sum_{t \in T} d_t$ is the total demand for the entire planning horizon. Without loss of generality, we assume that one unit of critical part is required to manufacture one unit of finished product. Then, D, is the total demand for finished products as well as total demand for critical parts. Let $j = 1$ be the primary plant, where the total demand for products, D, is initially assigned.

The suppliers of parts and assembly plants are located in different geographic regions, subject to potential regional disasters that may result in complete shutdown of all suppliers and plants in the same region simultaneously. Denote by I^r and J^r, respectively, the subsets of suppliers and plants in region $r \in R$, and by p^r, the regional disruption probability for region r.

In addition, each supplier $i \in I$ is subject to random local disruptions of different levels, $l \in L_i = \{0, \ldots, \overline{L}_i\}$, where disruption level refers to the fraction of an order that can be delivered. Level $l = 0$ represents complete shutdown of a supplier, i.e., no order delivery, while level $l = \overline{L}_i$ represents normal conditions with no disruption, i.e., full order delivery. Denote by p_{il}, the probability of disruption level l for supplier i, and by γ_{il}, the fraction of an order that can be delivered by supplier i under disruption level l (fulfillment rate),

$$\gamma_{il} = \begin{cases} 0 & \text{if } l = 0 \\ \in (0,1) & \text{if } l = 1, \ldots, \overline{L}_i - 1 \\ 1 & \text{if } l = \overline{L}_i. \end{cases} \tag{12.1}$$

Similarly to suppliers, each plant $j \in J$ is subject to random local disruptions of different levels, $l \in L^j = \{0, \ldots, \overline{L}^j\}$, where disruption level refers to available fraction of full capacity, c_j, available per period under normal conditions. Level $l = 0$ represents complete shutdown of an assembly plant, while level $l = \overline{L}^j$ represents normal conditions, i.e., full capacity, c_j, available. Denote by π_{jl}, the probability of disruption level l for plant j, and by δ_{jl}, the fraction of available capacity of plant j under disruption level l.

$$\delta_{jl} = \begin{cases} 0 & \text{if } l = 0 \\ \in (0,1) & \text{if } l = 1, \ldots, \overline{L}^j - 1 \\ 1 & \text{if } l = \overline{L}^j. \end{cases} \tag{12.2}$$

The total number of all potential scenarios is $\prod_{i \in I}(\overline{L}_i + 1) \prod_{j \in J}(\overline{L}^j + 1)$. Each scenario $s \in S$ is represented by an $(\overline{I} + \overline{J})$-dimensional vector $\lambda_s = \{\lambda_{1s}, \ldots, \lambda_{\overline{I}}, \lambda_{\overline{I}+1,s}, \ldots, \lambda_{\overline{I}+\overline{J},s}\}$, where $\lambda_{is} \in L_i$ is the disruption level of supplier $i \in I$ and $\lambda_{\overline{I}+j,s} \in L^j$ is the disruption level of plant $j \in J$, under scenario $s \in S$. Disruption s is assumed to occur in period t_s, and the corresponding suppliers $i \in I$ (such that $\lambda_{is} < \overline{L}_i$) and plants $j \in J$ (such that $\lambda_{\overline{I}+j,s} < \overline{L}^j$) are assumed to be simultaneously hit by the disruption.

The probability P_s of disruption scenario $s \in S$ is $P_s = \prod_{r \in R} P_s^r$, where P_s^r is the probability of realizing disruption scenario s in region r (cf. (10.3))

$$P_s^r = \begin{cases} (1-p^r)(\prod_{i \in I^r} \prod_{l \in L_i : \lambda_{is}=l} p_{il})(\prod_{j \in J^r} \prod_{l \in L^j : \lambda_{\bar{I}+j,s}=l} \pi_{jl}), \\ \text{if } \sum_{i \in I^r} \lambda_{is} + \sum_{j \in J^r} \lambda_{\bar{I}+j,s} > 0 \\ p^r + (1-p^r)(\prod_{i \in I^r} p_{i0})(\prod_{j \in J^r} \pi_{j0}), \\ \text{if } \sum_{i \in I^r} \lambda_{is} + \sum_{j \in J^r} \lambda_{\bar{I}+j,s} = 0. \end{cases} \tag{12.3}$$

When supplier i is hit by disruption at level l, its recovery process to normal conditions takes $TTR(i,l)$ time periods (time-to-recover) and let $CTR(i,l)$ be the cost-to-recover. For each supplier i, denote by θ_{is} and ρ_{is}, respectively, time-to-recover and cost-to recover from disruption under scenario s

$$\theta_{is} = TTR(i,l);\ i \in I, s \in S : l = \lambda_{is}, \tag{12.4}$$

$$\rho_{is} = CTR(i,l);\ i \in I, s \in S : l = \lambda_{is}. \tag{12.5}$$

Similarly, when plant j is hit by disruption at level l, its recovery process to normal conditions takes $PRT(j,l)$ time periods (Plant Recovery Time) and cost $PRC(j,l)$ (Plant Recovery Cost). For each plant j, denote by ϑ_{js} and ρ_{js}, respectively, time-to-recover and cost-to-recover from disruption under scenario s

$$\vartheta_{js} = PRT(j,l);\ j \in J, s \in S : l = \lambda_{\bar{I}+j,s}, \tag{12.6}$$

$$\rho_{js} = PRC(j,l);\ j \in J, s \in S : l = \lambda_{\bar{I}+j,s}. \tag{12.7}$$

The manufacturer decisions on supply and demand portfolio selection are integrated with production and inventory scheduling in the assembly plants to optimize the supply chain performance under disruption risks. In particular, if production and demand for products do not match, the manufacturer is charged with additional cost of surplus or shortage of product inventory, i.e., the inventory holding cost (ξ, per unit and per period) or penalty cost for demand fulfilled with delay (ζ, per unit and per period) or not fulfilled at all (g, per unit). Moreover, the manufacturer who moves production to an alternate assembly plant j incurs a fixed cost, φ_j, and encounters additional per unit cost of production, ε_j, and per unit cost, ψ_j, of transshipment of parts from the primary plant, where $\varphi_1 = 0$, $\varepsilon_1 = 0$, and $\psi_1 = 0$. A recovery plant can be a disrupted primary plant $j = 1$ with reduced capacity during recovery process and then with its full capacity or a new plant, non-disrupted or disrupted less severely than the primary plant.

The following assumptions are made to formulate the problem:

- Each supplier has sufficient capacity to meet total demand for parts.
- One unit of a critical part is needed to assemble one unit of product.
- Orders for parts are placed at the beginning of the planning horizon and, under normal conditions, parts ordered from supplier i are delivered to assembly plant j in period τ_{ij}, where τ_{ij} is the total of manufacturing lead time and transportation time.

- If supplier $i \in I$ (assembly plant $j \in J$) is hit by disruption s in period t_s, its recovery process starts in period $t_s + 1$, so that the disrupted supplier (disrupted plant) returns to its full capacity in period $t = t_s + \theta_{is}$ ($t = t_s + \vartheta_{js}$).
- A single disruption scenario is assumed to realize over the entire planning horizon. Multiple disruptions, one after the other in a series, during the recovery process are not considered.
- Period-dependent demand for products is considered that should be satisfied during the planning horizon using build up inventory of products and a penalty cost is charged for demand fulfilled with delay and not fulfilled at all.
- A recovery supplier can be a disrupted primary supplier after its recovery to full capacity or a new supplier.
- A recovery assembly plant can be a disrupted primary assembly plant during and after its recovery to full capacity or a new assembly plant.

12.3 Models for Risk-Averse Decision-Making

In this section two SMIP models **SupportPIS_CV(c)** and **SupportPIS_CV(sl)** are proposed for integrated, risk-averse selection of primary and recovery supply portfolios, recovery demand and transshipment portfolios and production and inventory scheduling. The objective is to reduce the risk of worst-case cost by minimizing conditional cost-at-risk, CVaR^c, and reduce the risk of worst-case service level by maximizing conditional service-at-risk, CVaR^{sl}, respectively. Problems **SupportPIS_CV(c)** and **SupportPIS_CV(sl)** are stochastic multi-portfolio selection problems in which primary supply portfolio is selected to hedge against all potential disruption scenarios. At the same time, recovery supply, transshipment, and demand portfolios are simultaneously selected for each scenario. The portfolios are defined below (for definitions of first and second stage variables, see Table 12.2).

The Primary Supply Portfolio

$$(v_1, \ldots, v_m),$$

specifies supplies of parts from primary suppliers to primary assembly plant $j = 1$, where

$$v_i \in [0,1]; i \in I, \ \sum_{i \in I} v_i = 1.$$

The Recovery Supply Portfolio for Scenario s

$$(V_{1j}^s, \ldots, V_{mj}^s); \ j \in J,$$

specifies supplies of parts from recovery suppliers to recovery assembly plants, where

$$V_{ij}^s \in [0,1]; i \in I, j \in J, \ \sum_{i \in I}(\gamma_i^s v_i + \sum_{j \in J} V_{ij}^s) = 1,$$

Table 12.2 Problem variables: supply portfolio selection and production and inventory scheduling

First stage variables	
u_i	$\in \{0,1\}$, such that the value 1 means that supplier i is selected as a primary supplier; otherwise $u_i = 0$ (primary supplier selection)
v_i	$\in [0,1]$, fraction of total demand for parts, ordered from primary supplier i, to be delivered to primary plant $j = 1$ (primary supply portfolio)
Second stage variables	
q_j^s	$\in \{0,1\}$, such that the value 1 means that assembly plant j is selected as a recovery plant under disruption scenario s; otherwise $q_j^s = 0$ (recovery plant selection)
Q_j^s	$\in [0,1]$ the fraction of total demand for products to be completed by recovery plant j under disruption scenario s (recovery demand portfolio).
U_i^s	$\in \{0,1\}$, such that the value 1 means that supplier i is selected as a recovery supplier under disruption scenario s; otherwise $U_i^s = 0$ (recovery supplier selection)
V_{ij}^s	$\in [0,1]$, fraction of total demand for parts, ordered from recovery supplier i to recovery plant j, under disruption scenario s (recovery supply portfolio)
w_j^s	$\in [0,1]$, fraction of total demand for parts, transshipped from the primary plant $j = 1$ to recovery plant, j, under scenario s, where w_1^s represents parts that remain in the primary plant $j = 1$ (recovery transshipment portfolio)
x_{jt}^s	≥ 0, production in plant j in period t under disruption scenario s (production scheduling)
y_t^s	≥ 0, inventory of products at the beginning of period t under disruption scenario s (inventory scheduling)
z_t^s	≥ 0, shortage of products at the beginning of period t under disruption scenario s (inventory scheduling)
Auxiliary variable	
a_t^s	$\in \{0,1\}$, such that value 1 means that there exists inventory of products at the beginning of period t under disruption scenario s; otherwise $a_t^s = 0$ (inventory scheduling)
b_t^s	$\in \{0,1\}$, such that value 1 means that there exists shortage of products at the beginning of period t under disruption scenario s; otherwise $b_t^s = 0$ (inventory scheduling)
μ_i^s	$\in \{0,1\}$, such that the value 1 means that $u_i = U_i^s = 1$; otherwise $\mu_i^s = 0$ (elimination of double fixed ordering costs)

$\sum_{i \in I} \gamma_i^s v_i$ denotes actual delivery of parts from primary suppliers to primary plant under disruption scenarios s.

The Recovery Transshipment Portfolio for Scenario s

$$(w_1^s, \ldots, w_n^s),$$

specifies transshipment of parts from primary assembly plant to recovery plants, where

$$w_j^s \in [0,1]; j \in J, \ \sum_{j \in J} w_j^s = \sum_{i \in I} \gamma_i^s v_i - \sum_{t \in T: t < t_s} x_{1t}^s / D,$$

and $\sum_{t \in T: t < t_s} x_{1t}^s$ denotes production at primary assembly plant $j = 1$, before disruption.

The Recovery Demand Portfolio for Scenario s

$$(Q_1^s, \ldots, Q_n^s),$$

specifies an allocation of unfulfilled demand for products among recovery plants, where

$$Q_j^s \in [0,1]; j \in J, \quad \sum_{t \in T: t < t_s} x_{1t}^s / D + \sum_{j \in J} Q_j^s = 1.$$

Models **SupportPIS_CV(c)** and **SupportPIS_CV(sl)** presented below are deterministic equivalent mixed integer programs of the two-stage stochastic mixed integer programs with recourse (e.g., Birge and Louveaux 2011). The primary supply portfolio selection variables, u_i, v_i, are referred to as first stage decisions, and the scenario-dependent recovery supply, transshipment, and demand portfolio selection variables, U_i^s, V_{ij}^s, w_j^s, q_j^s, Q_j^s, and production and inventory scheduling variables, a_t^s, b_t^s, x_{jt}^s, y_t^s, z_t^s, are referred to as recourse or second stage decisions (cf. Table 12.2).

SupportPIS_CV(c): *Selection of primary and recovery* **Sup***ply and demand* **port***folios and* **P***roduction and* **I***nventory* **S***cheduling to minimize* **CV***aR of* **c***ost*

Minimize

$$CVaR^c = VaR^c + (1-\alpha)^{-1} \sum_{s \in S} P_s \mathscr{C}_s, \tag{12.8}$$

subject to

Primary Supply Portfolio Selection Constraints:

$$\sum_{i \in I} v_i = 1, \tag{12.9}$$

$$v_i \le u_i;\ i \in I. \tag{12.10}$$

Recovery Supply and Demand Portfolio Selection Constraints:

$$V_{ij}^s \le U_i^s;\ i \in I, j \in J, s \in S, \tag{12.11}$$

$$V_{ij}^s \le q_j^s;\ i \in I, j \in J, s \in S, \tag{12.12}$$

$$Q_j^s \le q_j^s;\ j \in J, s \in S, \tag{12.13}$$

$$w_j^s \le q_j^s;\ j \in J, s \in S, \tag{12.14}$$

$$\sum_{i \in I} (\gamma_i^s v_i + \sum_{j \in J} V_{ij}^s) = 1;\ s \in S, \tag{12.15}$$

$$\sum_{t \in T: t < t_s} x_{1t}^s / D + \sum_{j \in J} Q_j^s = 1;\ s \in S, \tag{12.16}$$

$$\sum_{i \in I} V_{ij}^s + w_j^s = Q_j^s;\ j \in J, s \in S, \tag{12.17}$$

$$\sum_{j \in J} w_j^s = \sum_{i \in I} \gamma_i^s v_i - \sum_{t \in T: t < t_s} x_{1t}^s / D;\ s \in S, \tag{12.18}$$

$$\mu_i^s \le (u_i + U_i^s)/2;\ i \in I, s \in S. \tag{12.19}$$

(continued)

Production Capacity Constraints:

$$x^s_{1t} \le c_1;\ t \in T, s \in S : t < t_s, \quad (12.20)$$

$$x^s_{jt} \le c^s_{jt} q^s_j;\ j \in J, t \in T, s \in S : t \ge t_s, \quad (12.21)$$

$$\sum_{t \in T: t \ge t_s} x^s_{jt}/D \le Q^s_j;\ j \in J, s \in S. \quad (12.22)$$

Supply-Transshipment-Production Coordinating Constraints:

$$\sum_{t' \in T: t' \le t} x^s_{1t'} \le \sum_{i \in I: \tau_{i1} \le t-1} \gamma^s_i v_i + \sum_{i \in I: t_s + \theta_{is} + \tau_{i1} \le t-1} V^s_{i1}$$
$$-(\text{if } t \le \bar{t}_s \text{ then } 0 \text{ else } \sum_{j \in J: j > 1} w^s_j);$$
$$t \in T, s \in S \quad (12.23)$$

$$\sum_{t' \in T: t' \le t} x^s_{jt'} \le \sum_{i \in I: t_s + \theta_{is} + \tau_{ij} \le t-1} V^s_{ij} + (\text{if } t \le \bar{t}_s + \sigma_j \text{ then } 0 \text{ else } w^s_j);$$
$$j \in J, t \in T, s \in S : j > 1, \quad (12.24)$$

where $\bar{t}_s = \max\{t_s, \max_{i \in I} \tau_{i1}\}$ and, $\bar{t}_s + 1$, is the start period of transshipment of parts to recovery plants.

Product Inventory Constraints:

$$y^s_{t+1} - z^s_{t+1} = \sum_{j \in J} \sum_{t' \in T: t' \le t} x^s_{jt'} - D_t;\ t \in T, s \in S, \quad (12.25)$$

$$y^s_{t+1} \ge \sum_{j \in J} \sum_{t' \in T: t' \le t} x^s_{jt'} - D_t;\ t \in T, s \in S, \quad (12.26)$$

$$z^s_{t+1} \ge D_t - \sum_{j \in J} \sum_{t' \in T: t' \le t} x^s_{jt'};\ t \in T, s \in S, \quad (12.27)$$

$$a^s_t + b^s_t \le 1;\ t \in T, s \in S, \quad (12.28)$$

$$y^s_t/D \le a^s_t;\ t \in T, s \in S, \quad (12.29)$$

$$z^s_t/D \le b^s_t;\ t \in T, s \in S, \quad (12.30)$$

$$y^s_{h+1} = 0;\ s \in S. \quad (12.31)$$

Risk Constraints:

– the tail cost for scenario s, $\mathscr{C}_s$, is defined as the nonnegative amount by which cost in scenario s exceeds VaR^c,

$$\mathscr{C}_s \ge \sum_{i \in I} (e_i(u_i + U^s_i - \mu^s_i) + \rho_{is} U^s_i + o_i(\gamma^s_i v_i + \sum_{j \in J} V^s_{ij}))/D$$
$$+ \sum_{j \in J} (\psi_j w^s_j + (\varphi_j + \rho_{js}) q^s_j + \sum_{t \in T} \varepsilon_j x^s_{jt})/D$$
$$+ \sum_{t \in T} (\xi y^s_t + \zeta z^s_t)/D + g z^s_{h+1}/D - VaR^c;\ s \in S, \quad (12.32)$$

(continued)

Non-negativity and Integrality Conditions:

$$\mathscr{C}_s \geq 0;\ s \in S, \tag{12.33}$$

$$a_t^s \in \{0,1\};\ t \in T, s \in S, \tag{12.34}$$

$$b_t^s \in \{0,1\};\ t \in T, s \in S, \tag{12.35}$$

$$q_j^s \in \{0,1\};\ j \in J, s \in S, \tag{12.36}$$

$$Q_j^s \in [0,1];\ j \in J, s \in S, \tag{12.37}$$

$$u_i \in \{0,1\};\ i \in I, \tag{12.38}$$

$$v_i \in [0,1];\ i \in I, \tag{12.39}$$

$$U_i^s \in \{0,1\};\ i \in I, s \in S, \tag{12.40}$$

$$V_{ij}^s \in [0,1];\ i \in I, j \in J, s \in S, \tag{12.41}$$

$$w_j^s \in [0,1];\ j \in J, s \in S, \tag{12.42}$$

$$x_{jt}^s \geq 0;\ j \in J, t \in T, s \in S, \tag{12.43}$$

$$y_t^s \geq 0;\ t \in T, s \in S, \tag{12.44}$$

$$z_t^s \geq 0;\ t \in T, s \in S, \tag{12.45}$$

$$\mu_i^s \in \{0,1\};\ i \in I, s \in S, \tag{12.46}$$

where

$$c_{jt}^s = \begin{cases} \delta_j^s c_j, & \text{if } t_s \leq t \leq t_s + \vartheta_{js} - 1 \\ c_j, & \text{if } t \leq t_s - 1, t \geq t_s + \vartheta_{js}. \end{cases} \tag{12.47}$$

Equation (12.9) is the primary supply portfolio selection constraint and Eq. (12.10) ensures that demand for parts cannot be assigned to non-selected primary suppliers. Equations (12.11) and (12.12) ensure that the unfulfilled demand for parts cannot be assigned to non-selected recovery suppliers and cannot be ordered for non-selected recovery plants, respectively. Equation (12.13) ensures that the unfulfilled demand for products cannot be assigned to non-selected recovery plants, and Eq. (12.14) ensures that parts from primary plant cannot be transshipped to non-selected recovery plants. Equations (12.15) and (12.16) are the recovery supply portfolio and the recovery demand portfolio selection constraints, respectively. The supply and demand portfolio flow conservation equations (12.17) ensure that for each recovery plant, the recovery supplies, and transshipment of parts are in balance with the demand for products to be fulfilled by that plant. The balance constraints for parts, (12.18), ensure that the total transshipments of parts from primary supplier is equal to the partially fulfilled supplies from primary suppliers less the usage of parts for production before disruption. Equation (12.19) ensures that each supplier selected to both primary and recovery portfolio is charged exactly once with fixed ordering cost in the objective function. Equations (12.20) and (12.21) ensure that before a disruptive event, the production at primary plant in every period cannot exceed its capacity, and after a disruptive event, the production at each selected recovery plant in every period cannot exceed the plant available capacity, respectively. Equation (12.22) guarantees that the total production at each recovery plant cannot exceed the assigned portion of total demand for products.

The supply-transshipment-production coordinating constraints (12.23)–(12.24) ensure that (12.23), in primary assembly plant, the cumulative demand for parts cannot exceed the cumulative deliveries from primary and recovery suppliers, less transshipment of parts to recovery plants, if $t > \bar{t}_s$, (12.24), in recovery assembly plants, the cumulative demand for parts cannot exceed the cumulative deliveries from recovery suppliers, plus transshipment of parts from primary plant, if $t > \bar{t}_s + \sigma_j$.

The product inventory constraints (12.25)–(12.27) ensure that inventory or shortage of products in period $t+1$ are, respectively, positive or negative difference between cumulative production and demand by period t. Equations (12.28)–(12.30) indicate that in each period, there is either inventory or shortage or production and demand are in balance. Finally, Eq. (12.31) ensures that no inventory of products remains at the end of the planning horizon.

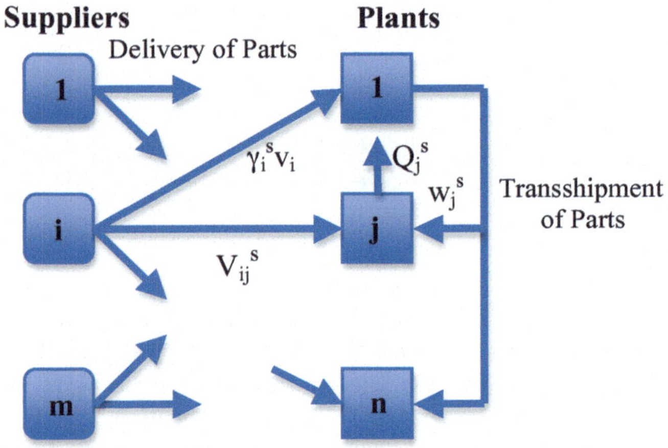

Fig. 12.1 Primary ($\gamma_i^s v_i$) and recovery (V_{ij}^s, w_j^s) flow of parts, and products (Q_j^s)

The cost per product (see, (12.32)) consists of different fixed and variable cost per product. The fixed cost includes cost of ordering parts from primary and recovery suppliers, cost of recovery processes of suppliers and plants, and cost of moving production from primary to recovery plants. The variable cost includes cost of purchasing parts from primary and recovery suppliers, cost of transshipment of parts, additional production cost in recovery plants, cost of holding inventory of products, cost of penalty for demand fulfilled with delay, and cost of penalty for unfulfilled demand.

Notice that the supply and demand portfolio selection constraints form an embedded network flow problem, (see, Fig. 12.1). In particular, Eqs. (12.17) and (12.18) are flow conservation equations under scenario s for each node j (assembly plant) and node $j = 1$ (primary assembly plant), respectively. $\gamma_i^s v_i$ represents flow of parts from supplier i (source node) to primary plant $j = 1$ (sink/transshipment node), V_{ij}^s represents flow of parts from recovery supplier i (source node) to recovery plant j (sink node), and w_j^s represents flow of parts from primary plant $j = 1$ (transshipment node) to recovery plant j (sink node). Finally, Q_j^s represents outflow of products from recovery plant j.

SupportPIS_CV(sl): *Selection of primary and recovery* **Supp***ly and demand* **port***folios and* **P***roduction and* **I***nventory* **S***cheduling to to maximize* **CV***aR of* **s***ervice* **l***evel*

Maximize

$$CVaR^{sl} = VaR^{sl} - (1-\alpha)^{-1} \sum_{s \in S} P_s \mathscr{S}_s, \qquad (12.48)$$

subject to (12.9)–(12.31), (12.34)–(12.47) and

Risk Constraints

– the tail service level for scenario s, $\mathscr{S}_s$, is defined as the nonnegative amount by which VaR^{sl} exceeds service level in scenario s,

$$\mathscr{S}_s \geq VaR^{sl} + z_{max}^s / D - 1; \; s \in S, \qquad (12.49)$$

$$\mathscr{S}_s \geq 0; \; s \in S, \qquad (12.50)$$

where, $z_{max}^s = \max_{t \in T} z_t^s$ is the maximum shortage of products under disruption scenario s.

In the proposed models CVaR is represented by auxiliary functions (12.8) and (12.48), where $\alpha \in (0,1)$ is the confidence level, VaR$^{(c)}$ (VaR^{sl}) is the targeted cost (targeted service level) such that for a given confidence level α, for $100\alpha\%$ of the scenarios, the outcome is below VaR$^{(c)}$ (above VaR^{sl}).

In the risk-averse decision-making, the confidence level α is fixed by the decision maker to control the risk of losses due to supply disruptions. The higher the confidence level α, the more risk-aversive is the decision maker and the smaller percent of the highest cost (lowest service) outcomes is focused on. The decision maker is willing to accept only solutions for which the the total probability of scenarios with cost greater than VaR^c (service smaller than VaR^{sl}) is not greater than $1 - \alpha$.

Worst-Case Service Level

Proposition 12.1. *The lowest service level,* $\underline{SL}$*, can be calculated as below.*

$$\underline{SL} = \min\{\underline{SL1}/D, \underline{SL2}/D\}. \quad (12.51)$$

$\underline{SL1} = D - \sum_{t \in T: t \le t^{SL}} d_t$, *is the greatest unfulfilled demand due to non available production capacity in disabled plants, where,*

$$t^{SL} = \max_{s \in S}\{t_s + \max_{j \in J}(\max(\max_{i \in I}\{\theta_{is} + \tau_{ij}\}, \vartheta_{js}, \sigma_j))\},$$

is the longest time required to resume production after disruption.

$\underline{SL2} = \min_{j \in J}\{(\overline{T} - \max_{s \in S}\{t_s + \max(\max_{i \in I}\{\theta_{is} + \tau_{ij}\}, \vartheta_{js}, \sigma_j)\})c_j\}$ *is minimum production capacity available after recovery.*

Proof. The lowest service level is associated with worst-case disruption scenario $s \in S$ for which primary suppliers and primary assembly plant are all shut down by disruption at time t_s before any production was started. Then, production resumes at plant j with the latest recovery time (after $t_s + \vartheta_{js}$ periods), after the latest delivery of parts by a recovery supplier (i.e., after $t_s + \max_{i \in I}\{\theta_{is} + \tau_{ij}\}$ periods) and after transshipment of parts from the primary plant (i.e., after $t_s + \sigma_j$ periods), whichever occurs later. Thus, $\sum_{t \in T: t \le t^{SL}} d_t$, is the maximum shortage due to disruption, i.e., shortage of products under worst-case scenario. On the other hand, the lowest service level is associated with minimum recovery capacity. The smallest recovery capacity of each plant j is determined by the number of periods, $|\{t \in T : t > \max_{s \in S}\{t_s + \max(\max_{i \in I}\{\theta_{is} + \tau_{ij}\}, \vartheta_{js}, \sigma_j)\}\}|$, remaining for production after its full recovery.

12.4 Computational Examples

The basic input data for the computational examples presented in this section are in part hypothetical. However, their relations to each other are real and in part they have been taken from a real case study. In particular, the case studies of Toyota supply chain disruption and recovery after the Great East Japan earthquake and tsunami of March 11, 2011 and Thailand's floods in October 2011 (e.g., Fujimoto and Park 2013, Park et al. 2013, MacKenzie et al. 2014, Matsuo 2015, Haraguchi and Lall 2015) have been analyzed. Moreover, demand patterns are based on a real-world data from a high-tech manufacturer (Sawik 2011a). The computational experiments were performed using the AMPL programming language and the Gurobi 7.5.0 solver on a MacBookPro laptop with Intel Core i7 processor running at 2.8GHz and with 16GB RAM. The basic input parameters are shown in Table 12.3 (see also Table 10.4).

In order to emphasize the impact of disruptions on the primary assembly plant $j = 1$, the plant has greater capacity and is modeled to be less reliable than plant $j = 2$ ($c_1 = 10{,}000 > c_2 = 5000$, $\pi_{11} = 0.75 < \pi_{21} = 0.85$).

The total number of all potential disruptive events was $(\overline{L}+1)^m(\overline{L^1}+1)(\overline{L^2}+1) = (4^4)(2^2) = 1024$. In the computational examples a disruption scenario was defined as a combination of disruptive event and its start time. In order to consider

Table 12.3 Input parameters: supply portfolio selection and production and inventory scheduling

$\overline{I} = 4$ suppliers, $\overline{J} = 2$ assembly plants, $\overline{T} = 30$ planning periods	
$R = \{1,2\}$	two geographic regions
$I^1 = \{1,2\}$, $J^1 = \{1,2\}$	two suppliers and two plants in region $r = 1$
$I^2 = \{3,4\}$	two suppliers in region $r = 2$
Total demand for parts/products	$D = 300{,}000$
Demand per period for parts/products	
Constant demand pattern:	$d_t = 10{,}000, \forall t \in T$
Increasing demand pattern:	
$d = (0,2,7,0,10,8,6,0,9,8,9,11,2,5,5,10,11,13,17,7,17,13,18,10,7,19,28,18,15,15) \times 1000$	
Decreasing demand pattern:	
$d = (0,23,21,14,15,20,13,9,18,18,16,7,10,12,11,10,5,5,2,13,9,10,9,0,6,8,0,3,7,6) \times 1000$	
Disruption levels	
$\overline{L}_i = 3$	four disruption levels for each supplier $i \in I$
$\overline{L}^j = 1$	two disruption levels for each plant $j \in J$
Local disruption probabilities of suppliers	
$p_{i0} = 0.1(1 - p_{i3})$	complete shutdown (level $l = 0$)
$p_{i1} = 0.3(1 - p_{i3})$	major disruption (level $l = 1$)
$p_{i2} = 0.6(1 - p_{i3})$	minor disruption (level $l = 2$)
$p_{i3} \in [0.75, 0.95]$; $i \in I^1$, $p_{i3} \in [0.65, 0.75]$; $i \in I^2$	non disruption (level $l = 3$)
Local non disruption probabilities of plants:	$\pi_{11} = 0.75$, $\pi_{21} = 0.85$
Regional disruption probability:	$p^1 = 0.001$, $p^2 = 0.01$
Fulfillment rates of suppliers	
$\gamma_{i0} = 0\ \forall i \in I$, $\gamma_{i3} = 1\ \forall i \in I$	
$\gamma_{i1} \in [0.01, 0.50]\ \forall i \in I^1$ $\gamma_{i1} \in [0.01, 0.30]\ \forall i \in I^2$, $\gamma_{i2} \in [0,51,0.99]\ \forall i \in I^1$ $\gamma_{i2} \in [0,31,0.99]\ \forall i \in I^2$	
Plant capacity	
$c_1 = 10{,}000$, $c_2 = 5000$	
Cost parameters	
$o = (14,12,8,9)$, $e = (8000, 6000, 12{,}000, 13{,}000)$	
$\varepsilon_2 = 1$, $\varphi_2 = 100$, $\psi_2 = 0.1$, $\xi = \max_{i \in I} o_i/10 = 1.4$, $\zeta = 10$, $g = 100$	
$CTR(i,l)$=if $l = 0$ then $100e_i$; if $l = 1$ then $10e_i$; if $l = 2$ then e_i $\forall i \in I$	
$PRC(1,0) = PRC(2,0) = 10{,}000$	
Time parameters	
$\tau_{1j} = \tau_{2j} = 2$, $\tau_{3j} = \tau_{4j} = 4, \forall j \in J$, $\sigma_2 = 2$	
$TTR(i,l)$=if $l = 0$ then 12; if $l = 1$ then 10; if $l = 2$ then 8 $\forall i \in I$	
$PRT(1,0) = 10$, $PRT(2,0) = 5$	

non-trivial supply disruptions, the start time t_s of each disruptive event was assumed to be not greater than the maximum delivery lead time to primary assembly plant $j = 1$, $\max_{i \in I}(\tau_{i1}) = 4$. Thus, $t_s \in \{1,2,3,4\}$, and the total number of all potential scenarios to be considered was $1024 \times 4 = 4096$, since each disruptive event, $s = 1, \ldots, 1024$, may occur at four different start times.

Notice that $\gamma_i^s = 1$, if $t_s > \tau_{i1}$, i.e., full order is delivered by supplier i, if disruption occurs after delivery lead time to primary assembly plant.

The probability P_s of realizing each disruption scenario, $s \in S$, was calculated as follows:

$P_s = 0.1P_s^1 P_s^2$ for $s \leq 1024$,
$P_s = 0.2P_{s-1024}^1 P_{s-1024}^2$ for $1025 \leq s \leq 2048$,
$P_s = 0.3P_{s-2048}^1 P_{s-2048}^2$ for $2049 \leq s \leq 3072$, and
$P_s = 0.4P_{s-3072}^1 P_{s-3072}^2$ for $3073 \leq s \leq 4096$,

where probabilities P_s^k; $k = 1,2$, $s \leq 1024$ were obtained using formula (12.3). Notice that disruptive events with later start times were modeled to be more likely.

The solution results are summarized in Tables 12.4 and 12.5, respectively, for model **SupportPIS_CV(c)** and **SupportPIS_CV(sl)**. The tables show primary supply portfolio and expected demand portfolios for different values of confidence level $\alpha = 0.5, 0.75, 0.9, 0.95, 0.99$. The solution values, CVaR and VaR, are presented along with the associated expected values of cost per product,

$$
\begin{aligned}
E^c = \sum_{s \in S} P_s \sum_{i \in I} (e_i(u_i + U_i^s - \mu_i^s) + \rho_{is} U_i^s + o_i(\gamma_i^s v_i + \sum_{j \in J} V_{ij}^s))/D \\
+ \sum_{s \in S} P_s \sum_{j \in J} (\psi_j w_j^s + (\varphi_j + \rho_{js}) q_j^s + \sum_{t \in T} \varepsilon_j x_{jt}^s)/D \\
+ \sum_{s \in S} \sum_{t \in T} P_s (\xi y_t^s + \zeta z_t^s)/D \\
+ g \sum_{s \in S} P_s z_{h+1}^s / D,
\end{aligned}
\qquad (12.52)
$$

and expected service level,

$$
E^{sl} = 1 - \sum_{s \in S} P_s z_{max}^s / D, \qquad (12.53)
$$

as well as with solution value of LP relaxation of stochastic MIP models **SupportPIS_CV(c)** and **SupportPIS_CV(sl)**.

Table 12.4 shows that while CVaR^c, VaR^c and E^c increase with the confidence level α, the expected service level, E^{sl}, decreases as the decision maker becomes more risk-aversive with respect to cost. Similar results for model **SupportPIS_CV(sl)** are shown in Table 12.5. While CVaR^{sl}, VaR^{sl}, and E^{sl} decrease with the confidence level α, E^c increases as the decision maker becomes more risk-aversive with respect to service level.

Table 12.4 indicates that the primary supply portfolio for model **SupportPIS_CV(c)** becomes less diversified as the confidence level increases and for the highest $\alpha = 0.99$ a single supplier $i = 2$ is selected only, the cheapest among most reliable suppliers. In contrast, for model **SupportPIS_CV(sl)** a diversified supply portfolio with all suppliers is selected for all confidence levels, see Table 12.5. The expected recovery demand portfolios are similar for both models and all confidence levels, with the primary plant $j = 1$ selected as a major recovery supplier.

For model **SupportPIS_CV(c)** and three different demand patterns, Fig. 12.2 presents expected cumulative production at each plant $j \in J$, by each period,

$$\sum_{s \in S} P_s \sum_{t' \in T : t' \leq t} x^s_{jt'};\ t \in T,$$

and expected shortage of products, at the end of each period,

Table 12.4 Solution results for model **SupportPIS_CV(c)**

Confidence level α	0.50	0.75	0.90	0.95	0.99
Constant demand pattern					
Var.=365,531, Bin.=56,064, Cons.=693,807, Nonz.=7,804,322[a]					
CVaR^c	64.20	92.25	104.80	120.01	136.71
VaR^c	34.53	69.26	87.49	90.87	134.92
Exp.Cost E^c, (12.52)	49.32	59.67	71.63	72.67	90.40
Exp.Service E^{sl}, (12.53)$\times$100%	87.90	85.02	81.69	81.84	80.22
Primary supply portfolio:	1(5)				
Supplier(% of total demand)[b]	2(8)	2(5)	2(4)	2(13)	2(100)
	3(78)	3(82)	3(83)	3(74)	
	4(9)	4(13)	4(13)	4(13)	
Exp.Recovery demand portfolio:	1(90)				
Plant (% of total demand)[c]	2(10)				
LP[d]	63.95	92.06	104.46	119.34	134.96
Increasing demand pattern					
Var.=596,787, Bin.=207,504, Cons.=1,003,583, Nonz.=9,671,394[a]					
CVaR^c	39.00	57.45	67.71	83.54	96.41
VaR^c	19.39	46.61	51.57	53.43	94.58
Exp.Cost E^c, (12.52)	29.19	44.27	47.78	50.77	56.00
Exp.Service E^{sl}, (12.53)$\times$100%	93.42	82.47	80.90	79.45	79.83
Primary supply portfolio:	2(9)	2(3)	2(4)	2(16)	2(100)
Supplier(% of total demand)[b]	3(83)	3(88)	3(87)	3(75)	
	4(8)	4(9)	4(9)	4(9)	
Exp.Recovery demand portfolio:	1(89)				
Plant (% of total demand)[c]	2(10)				
LP[d]	38.85	56.85	67.37	82.08	94.80
Decreasing demand pattern					
Var.=375,267, Bin.=59,824, Cons.=699,143, Nonz.=7,778,870[a]					
CVaR^c	96.97	125.04	137.69	152.87	169.64
VaR^c	67.46	100.42	120.41	123.80	167.86
Exp.Cost E^c, (12.52)	82.21	98.35	104.28	106.33	128.97
Exp.Service E^{sl}, (12.53)$\times$100%	69.12	64.52	62.99	62.59	60.58
Primary supply portfolio:	1(5)				
Supplier(% of total demand)[b]	2(8)	2(5)	2(3)	2(14)	2(100)
	3(78)	3(82)	3(84)	c(76)	
	4(9)	4(13)	4(13)	4(10)	
Exp.Recovery demand portfolio:	1(89)	1(90)	1(89)	1(90)	1(87)
Plant (% of total demand)[c]	2(10)	2(10)	2(10)	2(10)	2(13)
LP[d]	96.80	124.85	137.36	152.20	167.90

[a]Var.=total number of variables, Bin.=number of binary variables
Cons.=number of constraints, Nonz.= number of nonzero coefficients

[b]$1(v_1 \times 100), 2(v_2 \times 100), 3(v_3 \times 100), 4(v_4 \times 100)$

[c]$1(\sum_{s \in S} P_s Q^s_1 \times 100), 2(\sum_{s \in S} P_s Q^s_2 \times 100)$

[d]LP relaxation solution value of (12.8)

Table 12.5 Solution results for model **SupportPIS_CV(sl)**

Confidence level α	0.50	0.75	0.90	0.95	0.99
Constant demand pattern					
Var.=365,531, Bin.=56,064, Cons.=693,807, Nonz.=7,544,834[a]					
CVaR^{sl}	82.55	71.83	70.17	69.10	68.32
VaR^{sl}	93.33	78.33	71.67	70.00	68.33
Exp.Service E^{sl}, (12.53)×100%	87.94	82.70	82.67	82.34	82.16
Exp.Cost E^{c}, (12.52)	534.65	554.09	553.47	556.06	556.38
Primary supply portfolio:	1(33)	1(28)	1(30)	1(25)	1(28)
Supplier(% of total demand)[b]	2(27)	2(25)	2(21)	2(27)	2(27)
	3(13)	3(21)	3(24)	3(21)	3(21)
	4(27)	4(26)	4(25)	4(27)	4(24)
Exp.Recovery demand portfolio:	1(89)	1(92)	1(90)		
Plant (% of total demand)[c]	2(10)	2(8)	2(10)		
LP[d]	82.55	71.83	70.17	69.10	68.32
Increasing demand pattern					
Var.=596,787, Bin.=207,504, Cons.=1,003,583, Nonz.=9,332,090[a]					
CVaR^{sl}	91.24	89.14	87.12	86.27	84.96
VaR^{sl}	93.33	93.33	88.67	86.67	85.00
Exp.Service E^{sl}, (12.53)×100%	93.55	93.55	90.66	88.78	87.10
Exp.Cost E^{c}, (12.52)	513.63	513.62	516.70	519.23	521.70
Primary supply portfolio:	1(35)	1(35)	1(31)	1(32)	1(33)
Supplier(% of total demand)[b]			2(32)		
	3(11)	3(11)	3(17)	3(16)	3(12)
	4(22)	4(22)	4(20)	4(20)	4(23)
Exp.Recovery demand portfolio:	1(88)		1(89)		
Plant (% of total demand)[c]	2(10)		2(11)		
LP [d]	91.24	89.14	87.12	86.27	84.96
Decreasing demand pattern					
Var.=375,267, Bin.=59,824, Cons.=699,143, Nonz.=7,513,406[a]					
CVaR^{sl}	63.99	53.69	51.92	51.10	50.31
VaR^{sl}	74.33	61.00	53.67	52.00	50.33
Exp.Service E^{sl}, (12.53)×100%	69.16	63.92	63.87	63.62	63.32
Exp.Cost E^{c}, (12.52)	569.41	588.41	588.60	590.24	593.00
Primary supply portfolio:	1(33)	1(26)	1(30)	1(27)	1(26)
Supplier(% of total demand)[b]	2(28)	2(26)	2(21)	2(22)	2(28)
	3(12)	3(23)	3(24)	3(23)	3(21)
	4(27)	4(25)	4(25)	4(28)	4(25)
Exp.Recovery demand portfolio:	1(88)	1(90)			
Plant (% of total demand)[c]	2(10)	2(10)			
LP[d]	63.99	53.69	51.92	51.10	50.31

[a] Var.=total number of variables, Bin.=number of binary variables, Cons.=number of constraints, Nonz.= number of nonzero coefficients

[b] $1(v_1 \times 100), 2(v_2 \times 100), 3(v_3 \times 100), 4(v_4 \times 100)$

[c] $1(\sum_{s \in S} P_s Q_1^s \times 100), 2(\sum_{s \in S} P_s Q_2^s \times 100)$

[d] LP relaxation solution value of (12.48)

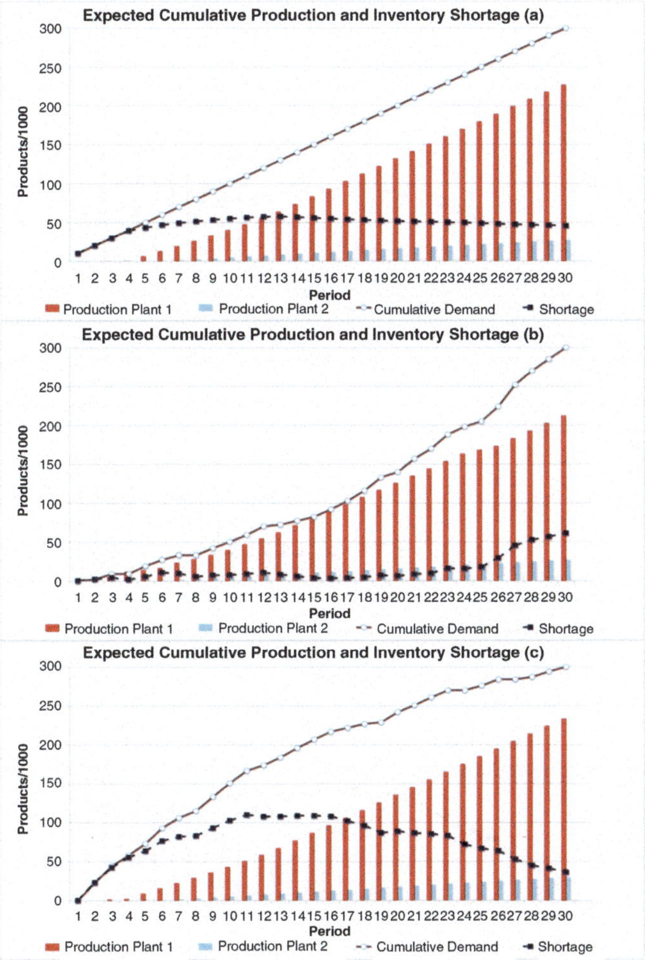

Fig. 12.2 Expected cumulative production and shortage of products for model **Support-PIS_CV(c)**, $\alpha = 0.99$: (**a**) constant demand pattern, (**b**) increasing demand pattern, (**c**) decreasing demand pattern

$$\sum_{s \in S} P_s z_{t+1}^s;\ t \in T,$$

associated with the risk-averse solution for $\alpha = 0.99$. In addition, Fig. 12.2 shows cumulative demand for products by each period $t \in T$,

$$D_t = \sum_{t' \in T: t' \leq t} d_{t'}.$$

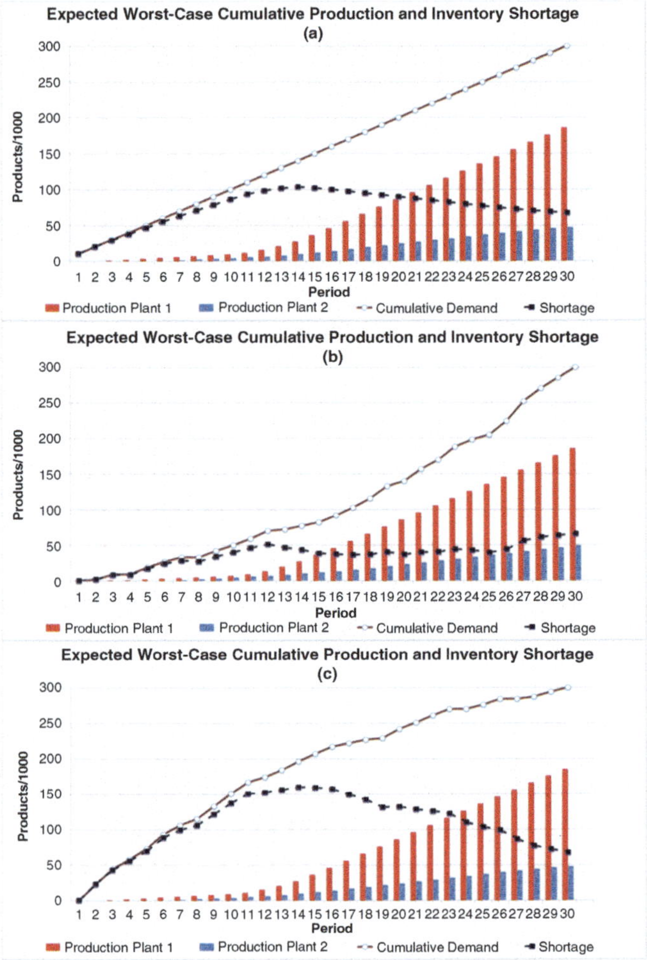

Fig. 12.3 Expected worst-case cumulative production and shortage of products for model **SupportPIS_CV(c)**, $\alpha = 0.99$: (**a**) constant demand pattern, (**b**) increasing demand pattern, (**c**) decreasing demand pattern

For the top $100(1-\alpha)\% = 1\%$ of the worst-case scenarios, $s \in S$, with cost,

$$\sum_{i\in I}(e_i(u_i + U_i^s - \mu_i^s) + \rho_{is}U_i^s + o_i(\gamma_i^s v_i + \sum_{j\in J} V_{ij}^s))/D$$
$$+ \sum_{j\in J}(\psi_j w_j^s + (\varphi_j + \rho_{js})q_j^s + \sum_{t\in T}\varepsilon_j x_{jt}^s)/D$$
$$+ \sum_{t\in T}(\xi y_t^s + \zeta z_t^s)/D + g z_{h+1}^s/D,$$

exceeding VaR^c, Fig. 12.3 presents expected worst-case cumulative production at each plant $j \in J$, by each period,

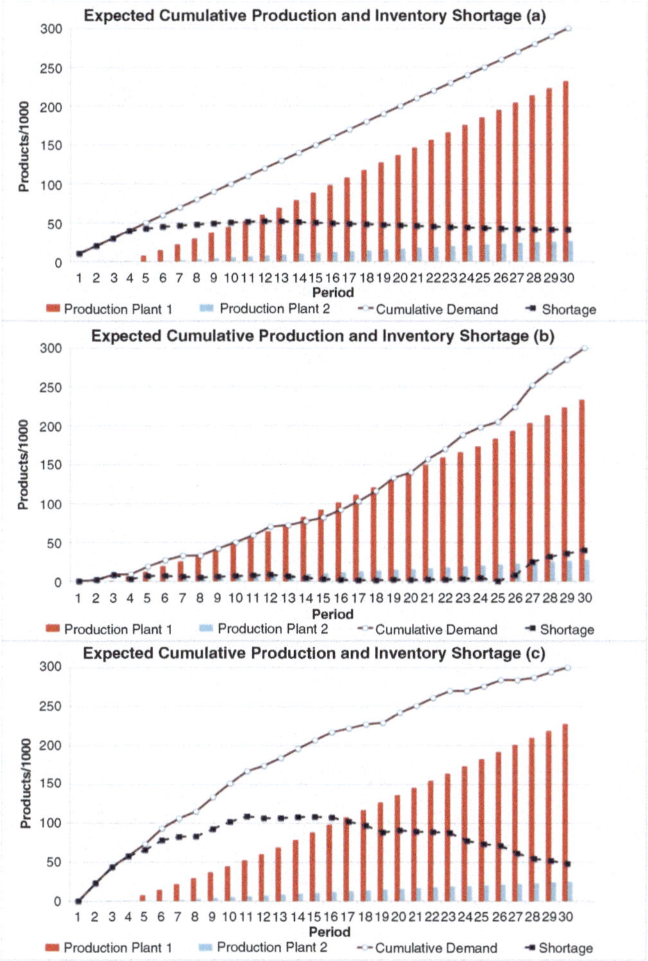

Fig. 12.4 Expected cumulative production and shortage of products for model **SupportPIS_CV(sl)**, $\alpha = 0.99$: (**a**) constant demand pattern, (**b**) increasing demand pattern, (**c**) decreasing demand pattern

$$\frac{\sum_{s \in S: \mathscr{C}_s > 0} P_s \sum_{t' \in T: t' \leq t} x^s_{jt'}}{\sum_{s \in S: \mathscr{C}_s > 0} P_s}; \; t \in T,$$

and expected worst-case shortage of products, at the end of each period,

$$\frac{\sum_{s \in S: \mathscr{C}_s > 0} P_s z^s_{t+1}}{\sum_{s \in S: \mathscr{C}_s > 0} P_s}; \; t \in T.$$

Similar diagrams for model **SupportPIS_CV(sl)**, three different demand patterns and $\alpha = 0.99$ are shown in Figs. 12.4 and 12.5. Figure 12.4 presents expected cumulative production at each plant and expected shortage of products and Fig. 12.5 shows expected worst-case cumulative production,

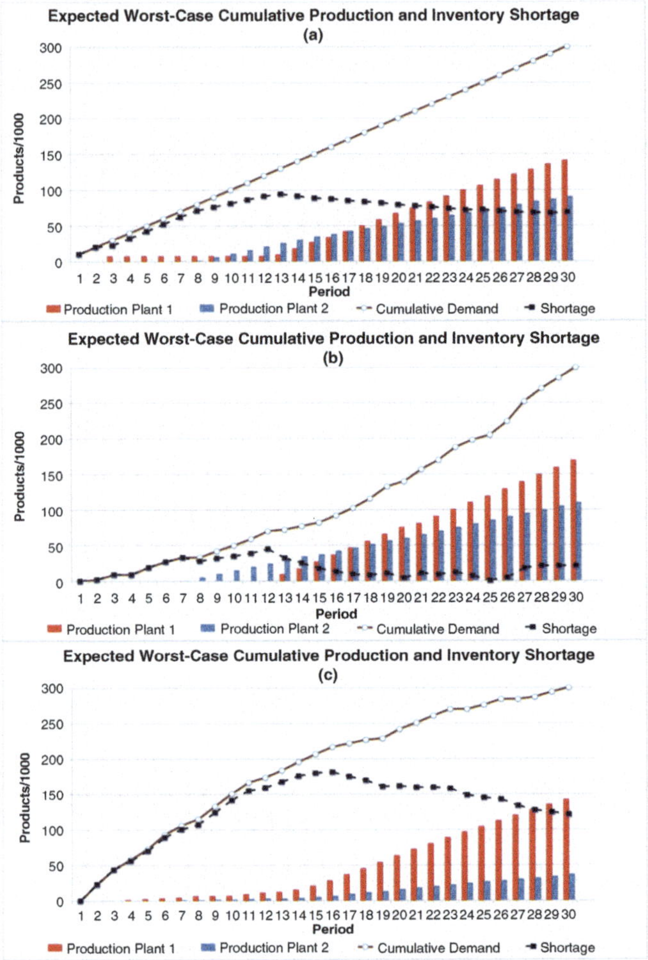

Fig. 12.5 Expected worst-case cumulative production and shortage of products for model **SupportPIS_CV(sl)**, $\alpha = 0.99$: (**a**) constant demand pattern, (**b**) increasing demand pattern, (**c**) decreasing demand pattern

$$\frac{\sum_{s\in S:\mathscr{S}_s>0} P_s \sum_{t'\in T:t'\leq t} x^s_{jt'}}{\sum_{s\in S:\mathscr{S}_s>0} P_s}; \ t \in T,$$

and expected worst-case shortage of products,

$$\frac{\sum_{s\in S:\mathscr{S}_s>0} P_s z^s_{t+1}}{\sum_{s\in S:\mathscr{S}_s>0} P_s}; \ t \in T,$$

for the top $100(1-\alpha)\% = 1\%$ of the worst-case scenarios, $s \in S$, with service level, $1 - z^s_{max}/D$, below VaR^{sl}.

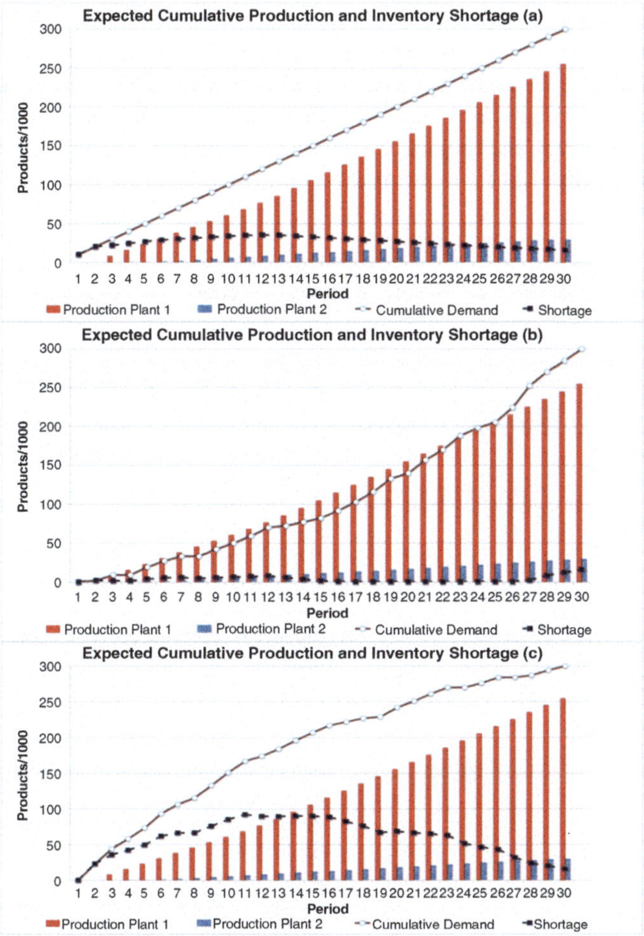

Fig. 12.6 Expected cumulative production and shortage of products for model **SupportPIS_CV(c)**, $\alpha = 0.5$: (**a**) constant demand pattern, (**b**) increasing demand pattern, (**c**) decreasing demand pattern

Comparison of solution results for different demand patterns demonstrate that decreasing demand pattern with most demand for products concentrated at the beginning of the planning horizon leads to highest cost and lowest service level for both expected and expected worst-case scenarios, see Tables 12.4, 12.5 and Figs. 12.2, 12.3, 12.4 and 12.5. In contrast to increasing demand pattern, for which the impact of disruption risks can be mitigated much better. Comparison of solution results for the two different models indicates that when the primary objective is to optimize service level with no regard to costs, both supply and demand portfolios are more diversified. Both demand for parts and unfulfilled demand for products are more evenly allocated among all primary suppliers and recovery plants, respectively.

For comparison, Fig. 12.6 presents expected results for model **SupportPIS_CV(c)** and the lowest confidence level $\alpha = 0.5$. The figure demonstrates that for the risk-aversive primary supply portfolio focusing on the highest 50% of cost outcomes, i.e., oriented more on business-as-usual, the expected results are better than for a more risk-aversive portfolio, i.e., for a higher confidence level. Clearly, the best expected results can be achieved for the risk-neutral decision-making, i.e., for $\alpha = 0$. Then, however, the impact of worst-case scenarios cannot be mitigated satisfactorily. Thus, the right choice of the confidence level by decision maker is a crucial decision.

Table 12.6 Managerial insights: supply portfolio selection and production and inventory scheduling

Decision-making issue	Decision-making impact
Cost- vs. service-oriented decision-making	For the service-oriented decision-making, both supply and demand portfolios are more diversified
Expected vs. expected worst-case performance vs. confidence level	In contrast to worst-case performance, the impact of disruption risks on business-as-usual supply chain performance is better mitigated by risk-averse decision-making with lower confidence levels. The right choice of the confidence level by the decision maker is a crucial decision
Demand pattern vs. disruption time	When a disruptive event occurs at the beginning of the planning horizon, then decreasing demand pattern with most demand for products concentrated also at the beginning of the planning horizon leads to highest cost and lowest service level for both expected and expected worst-case scenarios. In contrast, the impact of disruption risks is much better mitigated for the increasing demand pattern with most demand of products concentrated at the end of the planning horizon

The major managerial insights are summarized in Table 12.6. Overall, the results of computational experiments for different demand patterns demonstrate that *the proposed multi-portfolio approach and developed SMIP models with an embedded network flow problem can be a flexible and efficient tool for supply chain disruption management. The multi-portfolio approach leads to computationally very efficient SMIP models with a very strong LP relaxation.* Proven optimal solutions were obtained for all examples with CPU time ranging from around 100 s for model **SupportPIS_CV(sl)** to over 600 s for model **SupportPIS_CV(c)**.

12.5 Notes

The supply chain disruption management utilizes various quantitative methods to optimize disruption mitigation and recovery strategies. Typical methods include scenario-based stochastic programming, mixed integer programming, fuzzy set theory, game theory, control theory, simulation, genetic algorithms, etc. A systematic literature review and a comprehensive analysis of the decision-making models for supply chain risk were presented by Rajagopal et al. (2017). Stochastic programming and mixed integer linear programming were found to be the commonly studied modeling methods. The literature on disruption recovery in supply chains was also reviewed by Ivanov et al. (2017). The paper structured and classified existing research streams and application areas of different quantitative methods subject to different disruption risks and recovery measures. An interesting model for severe disruptions was proposed by MacKenzie et al. (2014) and applied to a simulation based on the Great East Japan earthquake and tsunami in 2011. In the model a disruption simultaneously impacts several suppliers. The model incorporated decisions made by both suppliers and manufacturers during the disruption of random duration. The decisions may include whether or not suppliers move production to an alternate facility, hold parts inventory; a manufacturer purchases parts from alternate suppliers that are not impacted, helps a primary supplier recover more quickly, or holds finished products inventory. Meena et al. (2014) studied the problem of determining the number of suppliers under risks of supplier failure due to catastrophic events. A simple heuristic algorithm was proposed to determine the number of suppliers to minimize total costs subject to a target service level and to maximize the service level subject to a total costs constraint. Schmitt et al. (2015) investigated optimal design of a multi-location supply chain under supply disruptions. They examined the expected costs and cost variances of the system in both a centralized and a decentralized inventory system. They showed that when demand is deterministic and supply is disrupted, a decentralized inventory reduces cost variance through the risk diversification effect, and that a decentralized inventory may also be selected when both supply is disrupted and demand is stochastic. A bi-objective stochastic mixed integer programming model that integrates supplier selection and risk mitigation strategy selection was proposed by Yoon et al. (2018). The authors suggested that a combination of upstream and downstream risk mitigation strategies should be jointly considered with supplier selection rather than considering these decisions separately. A similar integrated approach was also proposed by Sawik (2013d,e, 2014a,b, 2015b, 2016a).

The material presented in this chapter has been updated using results presented in Sawik (2019c), where the multi-portfolio approach and SMIP models were developed for the spatial and temporal integrated risk-averse planning in a supply chain under disruption risks. The future research should concentrate on relaxations of the various simplified assumptions used to formulate the problem (see, Sect. 12.2). Most approaches used in supply chain disruption management including the proposed scenario-based modeling approach, assume that some estimation of disruptive events probability and potential losses is available. However, a fair probability estimation of rare events is a complicated problem and even small errors in those estimations may significantly impact the modeling results. Simchi-Levi et al. (2015) proposed a novel risk-exposure model for analyzing operational-disruption risk with no need to estimate the probability of any specific disruptive events. The model is capable of assessing the impact of a disruption originating anywhere in a supply chain and has been applied by Ford Motor Company to identify risk exposures, evaluate risk mitigation actions, and develop optimal contingency plans. The future research should concentrate on development the probability independent approaches for supply chain disruption management, for example, by focusing on worst-case scenarios.

Problems

12.1. Formulate the risk-neutral models **SupportPIS_E(c)** and **SupportPIS_E(sl)** to minimize expected cost and to maximize expected service level, respectively.

12.2. Formulate the mean-risk models **SupportPIS_ECV(c)** and **SupportPIS_ECV (sl)** to minimize expected cost and CVaR of cost and to maximize expected service level and CVaR of service level, respectively.

12.3. In the SMIP models presented in this chapter introduce delay penalty when demand d_t for period t is not satisfied in a due window $[t-\Delta, t+\Delta]$, given the window length, Δ.

12.4. Modify the problem formulation to account for selection of primary demand portfolio, i.e., with the demand for products initially allocated among multiple primary assembly plants.

12.5. Assuming that a disruptive event hits the primary assembly plant after the maximum delivery lead time from suppliers, what type of demand pattern might have a negative impact on supply chain worst-case performance?

Part V
Selection of Supply Portfolio in Multi-Tier Supply Chain Networks

Chapter 13
Selection of Resilient Multi-Tier Supply Portfolio

13.1 Introduction

This chapter develops a multi-portfolio approach to disruption mitigation and recovery in a geographically dispersed multi-tier supply chain network. In order to mitigate the impact of material flow disruptions and improve resilience of their supply chains, the manufacturers apply different disruption management strategies, such as multi-sourcing, backup supplier contracts, recovery and transshipment supplies, production, and logistics flexibility. The resilience can also be improved by (e.g., Chap. 4) suppliers fortification, pre-positioning of risk mitigation inventory, spot purchasing, collaboration, and visibility. However, the manufacturers often neglect a multi-tier structure of their supply chains. They primarily focus on their immediate suppliers and do not even know their higher tier suppliers. For example, in a multi-tier supply chain network, where multiple suppliers may share a common supplier (which Toyota refers to as a 'diamond-shaped' supply chain), the resilient multi-sourcing strategy may not prove its practical usefulness. Moreover disruptions of the inter- and intra-tier flows may have an impact on the entire network and may lead to the ripple effect, i.e., supply chain disruption propagation. Therefore, to minimize potential losses in a multi-tier supply chain network geographically dispersed over multiple regions, the integrated disruption management strategies need to be developed for coordinated recovery of disrupted flows over the entire network.

The multi-portfolio approach proposed in this chapter is an enhancement of a two-tier supply portfolio approach introduced in Part I to mitigate the impact of disruption risks, extended later in Part IV for a combined risk mitigation and recovery in the supply chains. In addition to multi-sourcing and recovery and transshipment supplies, to improve the resilience of a multi-tier supply chain network and to hedge against all disruption scenarios, a risk mitigation inventory of critical parts is pre-positioned at primary suppliers of parts of different tiers and at primary assembly plant. The risk mitigation inventory is used to compensate for the loss of capacity

T. Sawik, *Supply Chain Disruption Management*,
International Series in Operations Research & Management Science 291,
https://doi.org/10.1007/978-3-030-44814-1_13

of the suppliers at immediately higher tier hit by disruptions, and to fulfill non-delivered orders placed on the disrupted suppliers.

The following SMIP models are presented in this chapter:

ResMTSupport_E(c) for risk-neutral selection of resilient multi-tier supply portfolio to minimize expected cost;
ResMTSupport_E(sl) for risk-neutral selection of resilient multi-tier supply portfolio to maximize expected service level;
ResMTSupport_CV(c) for risk-averse selection of resilient multi-tier supply portfolio to minimize CVaR of cost;
ResMTSupport_CV(sl) for risk-averse selection of resilient multi-tier supply portfolio to maximize CVaR of service level.

Numerical examples and computational results are reported in Sects. 13.6.1 and 13.6.2, respectively, for risk-neutral and risk-averse decision-making. The major managerial insights drawn from the theoretical and computational studies are summarized in Sect. 13.6.

13.2 Problem Description

Consider a multi-tier supply chain network, geographically dispersed over multiple regions, with multiple facilities available at each tier, see Fig. 13.1. Let $I_k = \{1,\dots m_k\}$ be the index set of m_k facilities of tier k and $K = \{0,1,\dots n\}$—the index set of all tiers, where tiers $k \geq 1$ represent suppliers of critical parts and m_k denotes the number of available suppliers of tier k. Tier, $k = 0$, represents final manufacturer of finished products and m_0 denotes the number of available assembly plants. The manufacturer sources critical parts from first-tier suppliers. The kth-tier suppliers provides with critical parts the $(k-1)^{th}$-tier suppliers (or assembly plants, if $k = 1$).

Without loss of generality, we assume that each tier supplier and the final manufacturer require one unit of input parts from the immediately upper tier supplier to produce one unit of its part and finished product, respectively. Denote by D, total demand for finished products (as well as total demand for critical parts of each tier) and let g be the per unit penalty cost of unfulfilled demand for finished products.

Denote by c_{ik}, the limited capacity of supplier (or assembly plant, if $k = 0$), $i \in I_k$ of tier $k \in K$, e.g., the total number of parts (or finished products) that can be manufactured and supplied over the planning horizon. Let o_{ik} ($\overline{o}_{ik}$) be the price per part for primary (recovery, expedited) supplies (if $k \geq 1$), respectively, where $\overline{o}_{ik} > o_{ik}$. The fixed cost of ordering parts from supplier $i \in I_k$, $k \geq 1$ is denoted by e_{ik} and by e_{i0}, the fixed setup cost for assembly plant $i \in I_0$ (Table 13.1).

The suppliers of parts and assembly plants are geographically dispersed over $\overline{R}$ regions, subject to potential regional disruptions which may arise from regional natural disasters (such as earthquakes, fires, floods, hurricanes, etc.) and may result in

Table 13.1 Notation: resilient multi-tier supply portfolio

Indices	
k	$=$ supply chain tier, $k \in K = \{0, 1, \ldots, n\}$
i	$=$ supplier ($k \geq 1$) or assembly plant ($k = 0$), $i \in I_k = \{1, \ldots, m_k\}$
r	$=$ geographic region, $r \in R = \{1, \ldots, \overline{R}\}$
s	$=$ disruption scenario, $s \in S = \{1, \ldots, \overline{S}\}$
Input parameters	
D	$=$ total demand for parts/products
g	$=$ unit penalty cost of unfulfilled demand for products
c_{ik}	$=$ capacity of supplier i of tier k (plant i, if $k = 0$)
e_{ik}	$=$ cost of ordering parts from supplier i of tier k (setup cost for plant i, if $k = 0$)
f_{ik}	$=$ unit price of parts pre-positioned at supplier i of tier k (plant i, if $k = 0$)
h_{ik}	$=$ unit cost of pre-positioning risk mitigation inventory of parts from tier $k+1$ at primary supplier $i \in I_k$ of tier k (plant i, if $k = 0$)
o_{ik} ($\overline{o}_{ik}$)	$=$ unit price of parts purchased from primary (recovery) supplier i of tier k
p_{ik}	$=$ local disruption probability for supplier i of tier k
p^r	$=$ regional disruption probability for geographic region r
ψ_{ijk}	$=$ unit cost of transshipment parts from supplier $i \in I_k$ to supplier $j \in I_k$ of tier k

complete shutdown of all facilities in the same region simultaneously. In order to address geographic proximity, denote by $I^{(r)}$ the subsets of suppliers and/or assembly plants (of the same and/or of different tiers, see Fig. 13.1) in region $r \in R$, and by p^r, the regional disruption probability for region r.

In addition, the supplies of parts (delivery of finished products) are subject to random local disruptions that are uniquely associated with a particular supplier (assembly plant), which may arise from equipment breakdowns, local labor strike, bankruptcy, terrorist attack, etc. Denote by p_{ik} the local disruption probability for facility $i \in I_k$ of tier k, i.e., the ordered parts (finished products) are delivered without disruptions with probability $(1 - p_{ik})$, or not at all with probability p_{ik}.

Let P_s be the probability that disruption scenario s is realized, where each scenario $s \in S$ is comprised of a unique subset $I^s \subset I = \bigcup_{k \in K} I_k$ of facilities that are in non-disrupted state, and $S = \{1, \ldots, \overline{S}\}$ is the index set of all disruption scenarios, where $\overline{S} = 2^M$ is total number of all potential scenarios and, $M = \sum_{k \in K} m_k$ is total number of all facilities. Each scenario $s \in S$ can be represented by an M-dimensional, 0-1 vector, $\lambda_s = \{\lambda_{1,n,s}, \ldots, \lambda_{m_0,0,s}\}$, where $\lambda_{i,k,s} = 0$ and $\lambda_{i,k,s} = 1$ denote, respectively, disruption and non-disruption state of facility $i \in I_k$ under scenario $s \in S$. For each scenario $s \in S$, the facilities $i \in I_k \setminus I^s$, $k \in K$, are disrupted either by a local disruptive event, regional disruptive event, or by both events.

The probability P_s of each disruption scenario $s \in S$ is derived as follows. First, the probability P_s^r of realizing disruption scenario s for facilities in $I^{(r)}$ is determined. If there are non-disrupted facilities in region r, i.e., $I^{(r)} \bigcap I^s \neq \emptyset$, then P_s^r is the product of regional non-disruption probability, $(1 - p^r)$, local probabilities of non-disrupted facilities, $(1 - p_{ik})$, $i \in I_k \bigcap I^{(r)} \bigcap I^s, k \in K$, and local probabilities of disrupted facilities, p_{ik}, $i \in I_k \bigcap I^{(r)} \setminus I^s, k \in K$. Otherwise, i.e., if all facilities in region r are disrupted, $I^{(r)} \bigcap I^s = \emptyset$, then either the entire region is disrupted with probability, p^r, or the region is non-disrupted with probability, $(1 - p^r)$, and every

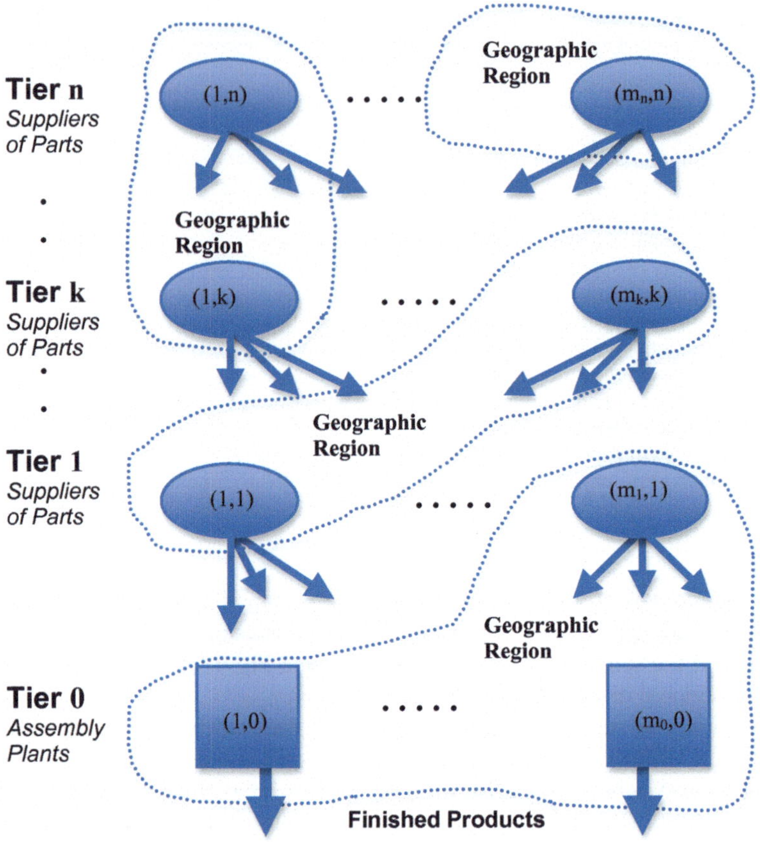

Fig. 13.1 A multi-tier, geographically dispersed supply chain network

facility $i \in I_k \bigcap I^{(r)}, k \in K$ is locally disrupted with probability, p_{ik}. Thus, the probability P_s^r is

$$P_s^r = \begin{cases} (1-p^r)\prod_{k\in K}\prod_{i\in I_k \bigcap I^{(r)} \bigcap I^s}(1-p_{ik})\prod_{i\in I_k \bigcap I^{(r)} \setminus I^s} p_{ik} & \text{if } I^{(r)} \bigcap I_s \neq \emptyset \\ p^r + (1-p^r)\prod_{k\in K}\prod_{i\in I_k \bigcap I^{(r)}} p_{ik} & \text{if } I^{(r)} \bigcap I^s = \emptyset. \end{cases} \tag{13.1}$$

The probability P_s of each disruption scenario $s \in S$ is

$$P_s = \prod_{r\in R} P_s^r. \tag{13.2}$$

In order to improve the resilience of a multi-tier supply chain network and to hedge against all disruption scenarios, a risk mitigation inventory of critical parts can be pre-positioned at primary suppliers of parts of different tiers and at primary plant of the final manufacturer. The risk mitigation inventory is used to compensate for the loss of capacity of the suppliers at immediately higher tier hit by disruptions, and to fulfill non-delivered orders placed on the disrupted suppliers. The final manufacturer can carry risk mitigation inventory of parts from first-tier suppliers, while k^{th}-tier suppliers ($k < n$) can carry risk mitigation inventory of parts from $(k+1)^{th}$-tier suppliers. Denote by, h_{ik}, unit cost of pre-positioning risk mitigation inventory of parts from tier $k+1$ at supplier $i \in I_k$ of tier k, and by, f_{ik}, unit price of those parts.

The following assumptions are made to formulate the problem:

- Suppliers have different limited capacity and differ in price of offered parts.
- Suppliers of the same tier may share their immediate upper tier suppliers, which may result in overlap in the higher tiers.
- The highest tier, $k = n$, suppliers use their own raw materials to manufacture critical parts required for suppliers of tier $k = n - 1$.
- Suppliers of each tier, $k < n$, and the final manufacturer are assumed to require one unit of input critical parts from the immediately upper tier supplier to produce one unit of its part and finished product, respectively.
- The regional disruptive events in each region and the local disruptive events are independent events.
- At most a single disruption scenario is realized only and once disabled, suppliers and plants are not available for assignment in the same planning horizon.
- Primary supplies from non-disrupted to disrupted suppliers (or plants) are considered fulfilled and parts delivered to disrupted suppliers (or plants) can be later transshipped to recovery suppliers (or plants).
- Recovery suppliers include non-disrupted primary suppliers.
- For every tier of the supply chain network and each disruption scenario, all parts supplied to non-disrupted facilities from immediately upper tier are processed and delivered to immediately lower tier. In particular, parts delivered from tier $k = 1$ to non-disrupted assembly plants of tier $k = 0$ are processed into finished products and delivered to customers.
- The risk mitigation inventory pre-positioned at a supplier (or a plant) is linked to its capacity and cannot be greater than the supplier (or plant) normal capacity.
- Depending on disruption scenario realized, the pre-positioned risk mitigation inventory can be used fully, partially, or not at all.
- The risk mitigation inventory is replenished once it is used and required parts can be purchased at higher prices on the spot market.

The above assumptions are in line with other works in the field as well as with industrial practice, e.g., Park et al. (2013), MacKenzie et al. (2014), Matsuo (2015), Ang et al. (2017).

We propose a stylized, static decision-making model (for a dynamic, two-period model, see Chap. 14). The decision maker selects ahead of time primary suppliers and assembly plants, allocates among them orders quantity, and pre-positions risk mitigation inventory to mitigate the impact of disruption risks. Simultaneously, for each potential disruption scenario the decision maker selects recovery suppliers and assembly plants, allocates among them unfulfilled orders quantity, and decides on transshipment of parts and usage of pre-positioned inventory. The objective of the risk-neutral decision-making is to minimize total expected cost (sum of all costs over the entire supply chain) or maximize expected service level of final manufacturer, whereas the risk-averse decision maker aims at minimizing CVaR of total cost or maximizing CVaR of service level. While total expected cost and expected service level are typical average performance measures of supply chains, and are usually in conflict, the risk-averse objectives that refer to worst-case performance of a multi-tier supply chain network, are also in conflict.

13.3 Models for Risk-Neutral Decision-Making

Table 13.2 Problem variables: resilient multi-tier supply portfolio

	First stage variables
u_{ik}	= 1, if supplier $i \in I_k$ of tier k (plant, if $k=0$) is selected as primary supplier (plant); otherwise $u_{ik}=0$ (primary supplier or plant selection)
v_{ijk}	= the fraction of total demand for parts ordered from primary supplier $i \in I_{k+1}$ of tier $k+1$ to primary supplier $j \in I_k$ of tier k (plant, if $k=0$) (primary supply portfolio)
x_{ik}	= risk mitigation inventory of parts manufactured at tier $k+1$ pre-positioned at primary supplier $i \in I_k$ of tier k (plant, if $k=0$), in fraction of its capacity c_{ik} (risk mitigation inventory allocation)
	Second stage variables
U_{ik}^s	= 1, if supplier $i \in I_k$ of tier k (plant, if $k=0$) is selected as recovery supplier (plant) under disruption scenario s; otherwise $U_{ik}^s=0$ (recovery supplier or plant selection)
V_{ijk}^s	= the fraction of total demand for parts ordered under disruption scenario s from recovery supplier $i \in I_{k+1}$ of tier $k+1$ to non-disrupted supplier $j \in I_k$ of tier k (plant, if $k=0$) (recovery supply portfolio)
w_{ijk}^s	= transshipment of parts of tier $k+1$, from disabled primary supplier $i \in I_k$ of tier k (plant, if $k=0$) to non-disrupted supplier (plant, if $k=0$) $j \in I_k$ of the same tier under disruption scenario s (recovery transshipment portfolio)
y_{ik}^s	= risk mitigation inventory of parts pre-positioned at primary supplier $i \in I_k$ of tier k (plant, if $k=0$), used under disruption scenario s (risk mitigation inventory usage)
z^s	= unfulfilled demand for products under disruption scenario s

In this section two stochastic mixed integer programs **ResMTSupport_E(c)** and **ResMTSupport_E(sl)** are presented for selection of resilient supply portfolio in a multi-tier supply chain under disruption risks to minimize total expected cost over the entire supply chain and maximize expected service level of final manufacturer, respectively.

The resilient portfolio aims at reducing potential losses from non-delivered parts through (for definition of first and second stage variables of the two-stage stochastic program, see Table 13.2):

(i) optimal selection of primary suppliers and assembly plants and allocation of primary orders and pre-position of risk mitigation inventory (first stage decisions) and

(ii) optimal selection of recovery suppliers and assembly plants, allocation of recovery orders and transshipments and usage of pre-positioned risk mitigation inventory, for each disruption scenario (second stage decisions).

If a primary supplier of tier $k+1$ (or primary assembly plant of tier $k=0$) is hit by disruption and fails to deliver parts to tier k (or products to customers), then non-delivered amount of parts (or products) can be fully or partially replaced by:

- recovery supplies of parts from recovery suppliers of tier $k+1$ to recovery suppliers of tier k;
- recovery supplies of products to customers from recovery plants of tier $k=0$;
- transshipment of parts from disabled primary suppliers of tier k (primary plants of tier $k=0$), to recovery suppliers (plants) of the same tier;
- usage of risk mitigation inventory of parts pre-positioned at non-disrupted primary suppliers (or plants) of tier k.

In model **ResMTSupport_E(c)** the overall quality of the resilient supply portfolio will be measured by the total expected cost per product over the entire supply chain, E^c,

$$
\begin{aligned}
E^c = & \sum_{k\in K}\sum_{i\in I_k} e_{ik}\big(u_{ik} + \sum_{s\in S: i\in I^s} P_s(U^s_{ik} - u_{ik})\big)/D \\
& + \sum_{k\in K: k<n}\sum_{j\in I_k}\Big(\sum_{i\in I_{k+1}} o_{i,k+1}v_{ijk} - \sum_{s\in S}\sum_{i\in I_{k+1}\setminus I^s} P_s o_{i,k+1}v_{ijk}\Big) \\
& + \sum_{s\in S}\sum_{k\in K: k<n}\sum_{i\in I_{k+1}\bigcap I^s}\sum_{j\in I_k\bigcap I^s} P_s \bar{o}_{i,k+1}V^s_{ijk} \\
& + \sum_{k\in K: k<n}\sum_{i\in I_k} c_{ik}\big(h_{ik}x_{ik} + \sum_{s\in S}\sum_{i\in I^s} P_s f_{ik} y^s_{ik}\big)/D \\
& + \sum_{s\in S}\sum_{k\in K: k<n}\sum_{i\in I_k\setminus I^s}\sum_{j\in I_k\bigcap I^s} P_s \psi_{ijk} w^s_{ijk} \\
& + g(1-E^{sl}),
\end{aligned}
\tag{13.3}
$$

where

$$
E^{sl} = 1 - \sum_{s\in S} P_s z^s, \tag{13.4}
$$

denotes expected service level, i.e., the expected fraction of fulfilled demand for products.

The expected cost consists of different fixed and variable cost.

The fixed cost per product includes the expected cost of ordering parts from primary and recovery suppliers as well as setup cost of primary and recovery assembly plants (double charge of ordering/setup cost is eliminated, if primary supplier/plant is selected as a recovery supplier/plant),

$\sum_{k\in K}\sum_{i\in I_k} e_{ik}(u_{ik} + \sum_{s\in S: i\in I^s} P_s(U^s_{ik} - u_{ik}))/D$.

The variable cost per product includes the expected:

(a) cost of purchasing parts from primary suppliers less cost of ordered and non-delivered parts,
$\sum_{k\in K:k<n}\sum_{j\in I_k}(\sum_{i\in I_{k+1}} o_{i,k+1}v_{ijk} - \sum_{s\in S}\sum_{i\in I_{k+1}\setminus I^s} P_s o_{i,k+1}v_{ijk})$;
(b) cost of purchasing parts from recovery suppliers,
$\sum_{k\in K:k<n}\sum_{s\in S}\sum_{i\in I_{k+1}\cap I^s}\sum_{j\in I_k\cap I^s} P_s\overline{o}_{i,k+1}V^s_{ijk}$;
(c) cost of pre-positioning risk mitigation inventory of parts,
$\sum_{k\in K:k<n}\sum_{i\in I_k} h_{ik}c_{ik}x_{ik}/D$;
(d) cost of usage pre-positioned risk mitigation inventory of parts,
$\sum_{s\in S}\sum_{k\in K:k<n}\sum_{i\in I_k\cap I^s} P_s f_{ik}c_{ik}y^s_{ik}/D$;
(e) cost of transshipment of parts,
$\sum_{s\in S}\sum_{k\in K:k<n}\sum_{i\in I_k\setminus I^s}\sum_{j\in I_k\cap I^s} P_s\psi_{ijk}w^s_{ijk}$;
(f) cost of unfulfilled demand for products due to supply disruptions,
$g\sum_{s\in S}P_s z^s = g(1-E^{sl})$.

The SMIP model **ResMTSupport_E(c)** is formulated below.

ResMTSupport_E(c): *Selection of* **Res***ilient* **M***ulti-***T***ier* **Sup***ply* **port***folio to minimize* **E***xpected* **c***ost*

Minimize (13.3)
subject to (13.4) and
Primary supply portfolio selection constraints:

- parts can be ordered from, (13.5), and delivered to, (13.6), selected primary suppliers only,
- the total demand for parts at each tier must be fully met by primary supplies from the immediately higher tier suppliers, i.e., the primary supply portfolio must be selected, (13.7).

$$v_{ijk} \le u_{i,k+1};\ k\in K, i\in I_{k+1}, j\in I_k : k<n, \tag{13.5}$$

$$v_{ijk} \le u_{jk};\ k\in K, i\in I_{k+1}, j\in I_k : k<n, \tag{13.6}$$

$$\sum_{i\in I_{k+1}}\sum_{j\in I_k} v_{ijk} = 1; k\in K : k<n. \tag{13.7}$$

Recovery supply portfolio selection constraints:

- a non-disrupted primary supplier is selected as a recovery supplier, (13.8),
- recovery flows of parts can be realized between non-disrupted facilities only and cannot be sourced from non-selected recovery suppliers, (13.9), and sunk at non-selected recovery suppliers (plants, if $k=0$), (13.10),
- recovery supplies from tier $k+1$ to tier k are not required, when primary suppliers at both tiers k and $k+1$ were not disrupted, (13.11),
- the parts from disabled primary suppliers can be transshipped to recovery suppliers only, (13.12), (13.13),

(continued)

– total transshipment of parts from disrupted primary supplier of tier k cannot exceed supplies of parts to this supplier from primary suppliers of tier $k+1$, (13.14), (transshipment constraints).

$$U_{ik}^s \geq u_{ik};\ s \in S, k \in K, i \in I_k \bigcap I^s, \quad (13.8)$$

$$V_{ijk}^s \leq U_{i,k+1}^s;\ s \in S, k \in K, i \in I_{k+1} \bigcap I^s, j \in I_k \bigcap I^s : k < n, \quad (13.9)$$

$$V_{ijk}^s \leq U_{jk}^s;\ s \in S, k \in K, i \in I_{k+1} \bigcap I^s, j \in I_k \bigcap I^s : k < n, \quad (13.10)$$

$$V_{ijk}^s \leq \sum_{l \in I_{k+1} \setminus I^s} u_{l,k+1} + \sum_{l \in I_k \setminus I^s} u_{lk};$$

$$s \in S, k \in K, i \in I_{k+1} \bigcap I^s, j \in I_k \bigcap I^s : k < n, \quad (13.11)$$

$$w_{ijk}^s \leq u_{ik};\ s \in S, k \in K, i \in I_k \setminus I^s, j \in I_k \bigcap I^s : k < n, \quad (13.12)$$

$$w_{ijk}^s \leq U_{jk}^s;\ s \in S, k \in K, i \in I_k \setminus I^s, j \in I_k \bigcap I^s : k < n, \quad (13.13)$$

$$\sum_{j \in I_k \bigcap I^s} w_{ijk}^s \leq \sum_{j \in I_{k+1} \bigcap I^s} v_{jik};\ s \in S, k \in K, i \in I_k \setminus I^s : k < n. \quad (13.14)$$

Primary and recovery capacity constraints:

– the capacity of each primary supplier, $(k > 0)$, (13.15), and final manufacturer, $(k = 0)$, (13.16), cannot be exceeded,
– the capacity of each recovery supplier, (13.17), and each recovery plant, (13.18), cannot be exceeded.

Total production over the entire planning horizon by a non-disrupted supplier, $i \in I_{k+1} \bigcap I^s$ $(k < n)$, and plant, $j \in I_0 \bigcap I^s$, is limited by its total capacity, $c_{i,k+1}$ and c_{j0}, respectively.

Nonzero primary supplies and usage of risk mitigation inventory may appear in the left-hand side of capacity constraints (13.17) and (13.18) only if primary supplier (or plant) is selected as a recovery supplier (or plant), i.e., if $U_{ik}^s = u_{ik}$, cf. (13.8).

$$\sum_{j \in J_k} v_{ijk} \leq c_{i,k+1} u_{i,k+1}/D;\ k \in K, i \in I_{k+1} : k < n, \quad (13.15)$$

$$\sum_{i \in I_1} v_{ij0} \leq c_{j0} u_{j0}/D;\ j \in I_0, \quad (13.16)$$

$$\sum_{j \in I_k \bigcap I^s} (v_{ijk} + V_{ijk}^s) \leq c_{i,k+1} U_{i,k+1}^s/D;$$

$$s \in S, k \in K, i \in I_{k+1} \bigcap I^s : k < n, \quad (13.17)$$

$$\sum_{i \in I_1 \bigcap I^s} (v_{ij0} + V_{ij0}^s) + \sum_{i \in I_0 \setminus I^s} w_{ij0}^s + c_{j0} y_{j0}^s/D \leq c_{j0} U_{j0}^s/D;$$

(continued)

$$j \in I_0 \bigcap I^s, s \in S. \quad (13.18)$$

Pre-positioning and usage of risk mitigation inventory constraints:

- risk mitigation inventory can be pre-positioned at selected primary suppliers only, (13.19),
- the risk mitigation inventory used to replace non-delivered parts cannot exceed the pre-positioned inventory, (13.20),
- for each tier k and disruption scenario s, the unfulfilled demand for parts from tier $k+1$ can be fully or partially replaced by the pre-positioned inventory and recovery supplies, (13.21).

$$x_{ik} \leq u_{ik};\ k \in K, i \in I_k : k < n, \quad (13.19)$$

$$y^s_{ik} \leq x_{ik};\ s \in S, k \in K, i \in I_k : k < n, \quad (13.20)$$

$$\sum_{i \in I_k \bigcap I^s} c_{ik} y^s_{ik}/D + \sum_{i \in I_{k+1} \bigcap I^s} \sum_{j \in I_k \bigcap I^s} V^s_{ijk} \leq \sum_{i \in I_{k+1} \setminus I^s} \sum_{j \in I_k} v_{ijk};$$
$$s \in S, k \in K : k < n. \quad (13.21)$$

Flow balance constraints:

- total primary supplies of parts manufactured by primary supplier of tier, $0 < k < n$, to primary suppliers of tier $k-1$ is in balance with primary supplies of parts to this supplier from primary suppliers of tier $k+1$, (13.22), (primary flow conservation equations),
- total recovery supplies of parts manufactured by recovery supplier of tier, $0 < k < n$, to non-disrupted suppliers of tier $k-1$ is in balance with recovery supplies of parts to this supplier from suppliers of tier $k+1$, transshipment of parts to this supplier from disrupted primary suppliers of tier k and usage of risk mitigation inventory pre-positioned at this supplier, (13.23), (recovery flow conservation equations),
- for each disruption scenario s, demand for products fulfilled by non-disrupted assembly plants, $(1 - z^s)$, is equal to primary and recovery supplies of parts from tier $k = 1$ and transshipment of parts and usage of pre-positioned inventory at tier $k = 0$, (13.24), (demand fulfillment constraints).

$$\sum_{j \in I_{k-1}} v_{ij,k-1} = \sum_{j \in I_{k+1}} v_{jik};\ k \in K, i \in I_k : 0 < k < n, \quad (13.22)$$

$$\sum_{j \in I_{k-1} \bigcap I^s} V^s_{ij,k-1} = \sum_{j \in I_{k+1} \bigcap I^s} V^s_{jik} + \sum_{j \in I_k \setminus I^s} w^s_{jik} + c_{ik} y^s_{ik}/D;$$
$$s \in S, k \in K, i \in I_k \bigcap I^s : 0 < k < n, \quad (13.23)$$

$$\sum_{i \in I_1 \bigcap I^s} \sum_{j \in I_0 \bigcap I^s} (v_{ij0} + V^s_{ij0}) + \sum_{i \in I_0 \setminus I^s} \sum_{j \in I_0 \bigcap I^s} w^s_{ij0}$$

(continued)

$$+ \sum_{i \in I_0 \bigcap I^s} c_{i0} y^s_{i0}/D + z^s = 1; \; s \in S. \quad (13.24)$$

5. Non-negativity and integrality conditions:

$$u_{ik} \in \{0,1\}; \; k \in K, i \in I_k, \quad (13.25)$$

$$v_{ijk} \in [0,1]; \; k \in K, i \in I_{k+1}, j \in I_k : k < n, \quad (13.26)$$

$$U^s_{ik} \in [0,1]; \; s \in S, k \in K, i \in I_k \bigcap I^s, \quad (13.27)$$

$$V^s_{ijk} \in [0,1]; \; s \in S, k \in K, i \in I_{k+1} \bigcap I^s, j \in I_k \bigcap I^s : k < n, \quad (13.28)$$

$$w^s_{ijk} \in [0,1]; \; s \in S, k \in K, i \in I_k \setminus I^s, j \in I_k \bigcap I^s : k < n, \quad (13.29)$$

$$x_{ik} \in [0,1]; \; k \in K, \in I_k : k < n, \quad (13.30)$$

$$y^s_{ik} \in [0,1]; \; s \in S, k \in K, i \in I_k : k < n, \quad (13.31)$$

$$z^s \geq 0; \; s \in S. \quad (13.32)$$

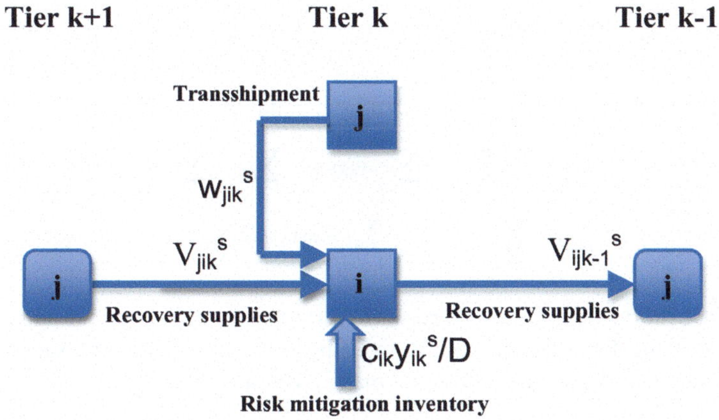

Fig. 13.2 Recovery flows of parts

Proposition 13.1. *Model* **ResMTSupport_E(c)** *contains an embedded network flow problem.*

Proof. In model **ResMTSupport_E(c)**, the primary, recovery, and transshipment portfolio selection constraints form an embedded network flow problem (see, Fig. 13.2) where:

(continued)

- v_{ijk} represents inter-tier primary flow of parts from primary supplier $i \in I_{k+1}$ of tier $k+1$ to primary supplier (or assembly plant, if $k=0$), $j \in I_k$ of tier k;
- V^s_{ijk} represents inter-tier recovery flow of parts from recovery supplier $i \in I_{k+1}$ of tier $k+1$ to recovery supplier (or assembly plant, if $k=0$), $j \in I_k$ of tier k;
- w^s_{ijk} represents intra-tier transshipment flow of parts from disrupted primary supplier $i \in I_k \setminus I^s$ of tier k to recovery supplier (or assembly plant, if $k=0$), $j \in I_k$ of the same tier.

Equations (13.22) and (13.23) are primary and recovery flow conservation equations, respectively, and (13.24) are demand for products fulfillment constraints.

Notice that the nonnegative product shortage variables z^s can be eliminated from the model using demand fulfillment constraints (13.24). Then, variables z^s in the objective functions (13.3) and (13.4) should be replaced by

$$1 - \sum_{i \in I_1 \bigcap I^s} \sum_{j \in I_0 \bigcap I^s} v_{ij0} - \sum_{i \in I_1 \bigcap I^s} \sum_{j \in I_0 \bigcap I^s} V^s_{ij0} - \sum_{i \in I_0 \setminus I^s} \sum_{j \in I_0 \bigcap I^s} w^s_{ij0} - \sum_{i \in I_0 \bigcap I^s} c_{i0} y^s_{i0} / D,$$

and equality constraints (13.24), replaced by inequality constraints

$$\sum_{i \in I_1 \bigcap I^s} \sum_{j \in I_0 \bigcap I^s} v_{ij0} + \sum_{i \in I_1 \bigcap I^s} \sum_{j \in I_0 \bigcap I^s} V^s_{ij0}$$
$$+ \sum_{i \in I_0 \setminus I^s} \sum_{j \in I_0 \bigcap I^s} w^s_{ij0} + \sum_{i \in I_0 \bigcap I^s} c_{i0} y^s_{i0} / D \le 1; \; s \in S.$$

The product shortage variables z^s are slack variables for the last constraints.

The following **Proposition 13.2** derived from the model **ResMTSupport_E(c)** and presented below provide additional managerial insights.

Proposition 13.2. *The recovery supplies of parts to the assembly plants (the lowest tier, $k=0$, of supply chain) are strictly dependent on recovery supplies of parts from the highest tier, $k=n$.*

$$\sum_{i \in I_1 \bigcap I^s} \sum_{j \in I_0 \bigcap I^s} V^s_{ij0} = \sum_{i \in I_n \bigcap I^s} \sum_{j \in I_{n-1} \bigcap I^s} V^s_{ij,n-1}$$
$$+ \sum_{k \in K: 0<k<n} \sum_{i \in I_k \setminus I^s} \sum_{j \in I_k \bigcap I^s} w^s_{ijk} + \sum_{k \in K: 0<k<n} \sum_{i \in I_k \bigcap I^s} c_{ik} y^s_{ik} / D; \; s \in S. \quad (13.33)$$

Equation (13.33) indicates that for each disruption scenario s, recovery supplies of parts from tier $k=1$ to the lowest tier $k=0$ are in balance

(continued)

with recovery supplies of parts from the highest tier $k = n$ to tier $k = n-1$ and transshipment of parts and usage of pre-positioned inventory at all intermediate tiers $k = 1,\ldots,n-1$.

Proof. By summing up recovery flow conservation equations (13.23) over $i \in I_k \bigcap I^s$ and then substituting, $\sum_{i\in I_k\bigcap I^s}\sum_{j\in I_{k-1}\bigcap I^s} V^s_{ij,k-1}$, iteratively for, $k = 1,\ldots,n$, in Eq. (13.23), the demand fulfillment constraints can be rewritten as

$$\sum_{i\in I_1\bigcap I^s}\sum_{j\in I_0\bigcap I^s} v_{ij0} + \sum_{i\in I_n\bigcap I^s}\sum_{j\in I_{n-1}\bigcap I^s} V^s_{ij,n-1}$$
$$+\sum_{k\in K:k<n}\sum_{i\in I_k\setminus I^s}\sum_{j\in I_k\bigcap I^s} w^s_{ijk} + \sum_{k\in K:k<n}\sum_{i\in I_k\bigcap I^s} c_{ik}y^s_{ik}/D$$
$$+z^s = 1;\ s \in S. \qquad (13.34)$$

Equation (13.34) ensures that for each disruption scenario s, fulfilled demand for products, $(1-z^s)$, is in balance with primary supplies of parts from tier $k = 1$ to $k = 0$, recovery supplies of parts from tier n to $n-1$ and transshipment of parts and usage of pre-positioned inventory at all tiers $k = 0,1,\ldots,n-1$. Comparison of Eqs. (13.24) and (13.34) implies equation (13.33).

The implied constraint (13.33) can be added to strengthen model **ResMTSupport_E(c)**.

In model **ResMTSupport_E(sl)** presented below, the overall quality of the resilient supply portfolio will be measured by the expected service level, E^{sl} (13.4).

ResMTSupport_E(sl): *Selection of* **Res***ilient* **M***ulti-***T***ier* **Supp***ly* **port***folio to maximize* **E***xpected* **s***ervice* **l***evel*

Minimize (13.4)
subject to (13.5)–(13.32).

The following **Proposition 13.3** indicates that a resilient supply chain is capable of reaching the highest expected service level.

Proposition 13.3. *The expected service level for resilient supply portfolio is bounded by its highest possible value, one.*

$$E^{sl} \le 1. \qquad (13.35)$$

(continued)

Proof. Equations (13.24), (13.21), (13.14) and (13.7) imply

$$E^{sl} = 1 - \sum_{s\in S} P_s z^s =$$

$$\sum_{s\in S} P_s\Big(\sum_{i\in I_1\bigcap I^s}\ \sum_{j\in I_0\bigcap I^s} v_{ij0} + \sum_{i\in I_0\setminus I^s}\ \sum_{j\in I_0\bigcap I^s} w^s_{ij0}$$

$$+ \sum_{i\in I_0\bigcap I^s} c_{i0} y^s_{i0}/D + \sum_{i\in I_1\bigcap I^s}\ \sum_{j\in I_0\bigcap I^s} V^s_{ij0}\Big) \le$$

$$\sum_{s\in S} P_s\Big(\sum_{i\in I_1\bigcap I^s}\ \sum_{j\in I_0\bigcap I^s} v_{ij0} + \sum_{j\in I_1\bigcap I^s}\ \sum_{i\in I_0\setminus I^s} v_{ji0} + \sum_{i\in I_1\setminus I^s}\ \sum_{j\in I_0} v_{ij0}\Big) =$$

$$\sum_{s\in S} P_s\Big(\sum_{i\in I_1\bigcap I^s}\ \sum_{j\in I_0\bigcap I^s} v_{ij0} + \sum_{i\in I_1\bigcap I^s}\ \sum_{j\in I_0\setminus I^s} v_{ij0} + \sum_{i\in I_1\setminus I^s}\ \sum_{j\in I_0} v_{ij0}\Big) =$$

$$\sum_{s\in S} P_s\Big(\sum_{i\in I_1\bigcap I^s}\ \sum_{j\in I_0} v_{ij0} + \sum_{i\in I_1\setminus I^s}\ \sum_{j\in I_0} v_{ij0}\Big) =$$

$$+ \sum_{s\in S} P_s\Big(\sum_{i\in I_1}\ \sum_{j\in I_0} v_{ij0}\Big) = 1.$$

Notice that for a non-resilient supply portfolio, i.e., with no recovery supplies and risk mitigation inventory, the upper bound on the expected service level is strictly less than one.

Models, **ResMTSupport_E(c)** and **ResMTSupport_E(sl)**, have been formulated under the assumption (see, Sect. 13.2) that: "primary supplies from non-disrupted to disrupted suppliers (or plants) are considered fulfilled and parts delivered to disrupted suppliers (or plants) can be later transshipped to recovery suppliers (or plants)". In some practical situations, however, the inventory of parts at the impacted primary suppliers or assembly plants may be non available for transshipment or the transshipment itself can be non available or prohibitively expensive. For example, transshipment between suppliers or plants in different geographic regions may be non available,

$$w^s_{ijk} = 0;\ s \in S, r1, r2 \in R, k \in K, i \in (I_k \bigcap I^{(r1)}) \setminus I^s, j \in (I_k \bigcap I^{(r2)}) \bigcap I^s :$$
$$r1 \ne r2, k < n,$$

or the unit cost of transshipment, ψ_{ijk}, can be much greater than the unit penalty, g, and the unit cost of recovery supplies, $\max_{i\in I_{k+1}}(\overline{o}_{i,k+1}); k \in K, i \in I_k, j \in I_k : k < n$. Then, the recovery supply portfolio should be considered, even after full delivery of all required parts to the impacted primary suppliers and plants.

13.4 Models for Risk-Averse Decision-Making

In this section, two stochastic MIP models, **ResMTSupport_CV(c)** and **ResMTSupport_CV(sl)**, are proposed for a risk-averse decision-making. The objective of the risk-averse decisions is to reduce the risk of worst-case cost by minimizing CVaR^c (conditional cost-at-risk) for model **ResMTSupport_CV(c)** or to reduce the risk of worst-case service by maximizing CVaR^{sl} (conditional service-at-risk) for model **ResMTSupport_CV(sl)**.

The resilient multi-tier supply portfolio will be optimized by calculating VaR^c (cost-at-risk), VaR^{sl} (service-at-risk) and simultaneously minimizing CVaR^c, maximizing CVaR^{sl}, respectively, for model **ResMTSupport_CV(c)**, **ResMTSupport_CV(sl)**. Given confidence level α, $CVaR^c$ and CVaR^{sl} are calculated as shown below.

$$CVaR^c = VaR^c + (1-\alpha)^{-1}\sum_{s\in S} P_s\mathscr{C}_s \tag{13.36}$$

$$CVaR^{sl} = VaR^{sl} - (1-\alpha)^{-1}\sum_{s\in S} P_s\mathscr{S}_s, \tag{13.37}$$

where $\mathscr{C}_s$ is the tail cost for scenario s, defined as the nonnegative amount by which cost in scenario s exceeds VaR^c, $\mathscr{S}_s$ is the tail service level for scenario s, defined as the nonnegative amount by which VaR^{sl} exceeds service level in scenario s.

The confidence level, $\alpha \in (0,1)$, is fixed by the decision maker and VaR^c (VaR^{sl}) is the targeted cost (service) such that for a given confidence level α, for $100\alpha\%$ of the scenarios, the outcome is below VaR^c (above VaR^{sl}). The decision maker is willing to accept only solutions for which the the total probability of scenarios with costs greater than VaR^c (service lower than VaR^{sl}) is not greater than $1-\alpha$. The higher the confidence level α, the more risk aversive is the decision maker and the smaller percent of the highest cost (the lowest service) outcomes is focused on.

In model **ResMTSupport_CV(c)**, different fixed and variable cost are considered for each disruption scenario, $s \in S$.

The fixed cost per product includes cost of ordering parts from primary and recovery suppliers (double charge of ordering cost is eliminated, if primary supplier is selected as a recovery supplier)

$\sum_{k\in K}\sum_{i\in I_k} e_{ik}(u_{ik} + (\text{if } i \in I_k \bigcap I^s \text{ then } (U^s_{ik} - u_{ik}) \text{ else } 0)/D$.

The variable cost per product includes:

(a) cost of purchasing parts from primary suppliers less cost of ordered and non-delivered parts,
$\sum_{k\in K:k<n}\sum_{j\in I_k}(\sum_{i\in I_{k+1}} o_{i,k+1}v_{ijk} - \sum_{i\in I_{k+1}\setminus I^s} o_{i,k+1}v_{ijk})$;
(b) cost of purchasing parts from recovery suppliers,
$\sum_{k\in K:k<n}\sum_{i\in I_{k+1}\bigcap I^s}\sum_{j\in I_k\bigcap I^s} \overline{o}_{i,k+1}V^s_{ijk}$;
(c) cost of pre-positioning risk mitigation inventory of parts,
$\sum_{k\in K:k<n}\sum_{i\in I_k} h_{ik}c_{ik}x_{ik}/D$;
(d) cost of usage pre-positioned risk mitigation inventory of parts,
$\sum_{k\in K:k<n}\sum_{i\in I_k\bigcap I^s} f_{ik}c_{ik}y^s_{ik}/D$;

(e) cost of transshipment of parts, $\sum_{k\in K:k<n}\sum_{i\in I_k\setminus I^s}\sum_{j\in I_k\bigcap I^s}\psi_{ijk}w^s_{ijk}$;
(f) cost of unfulfilled demand for products due to supply disruptions, gz^s.

The SMIP model **ResMTSupport_CV(c)** is formulated below.

ResMTSupport_CV(c): *Selection of* **Res***ilient* **M***ulti-***T***ier* **Supp***ly* **port***folio to minimize* **CV***aR of* **c***ost*
Minimize (13.36)
subject to (13.5)–(13.32) and
Risk constraints:

- the tail cost for scenario s, $\mathscr{C}_s$, is defined as the nonnegative amount by which cost in scenario s exceeds VaR^c.

$$\mathscr{C}_s \geq \sum_{k\in K}\sum_{i\in I_k} e_{ik}(u_{ik} + (\text{if } i\in I_k\bigcap I^s \text{ then } (U^s_{ik} - u_{ik}) \text{ else } 0))/D$$
$$+ \sum_{k\in K:k<n}\sum_{j\in I_k}(\sum_{i\in I_{k+1}} o_{i,k+1}v_{ijk} - \sum_{i\in I_{k+1}\setminus I^s} o_{i,k+1}v_{ijk})$$
$$+ \sum_{k\in K:k<n}\sum_{i\in I_{k+1}\bigcap I^s}\sum_{j\in I_k\bigcap I^s} \overline{o}_{i,k+1}V^s_{ijk}$$
$$+ \sum_{k\in K:k<n}\sum_{i\in I_k} c_{ik}(h_{ik}x_{ik} + \sum_{i\in I^s} f_{ik}y^s_{ik})/D$$
$$+ \sum_{k\in K:k<n}\sum_{i\in I_k\setminus I^s}\sum_{j\in I_k\bigcap I^s}\psi_{ijk}w^s_{ijk} + bz^s - VaR^c;\ s\in S, \quad (13.38)$$
$$\mathscr{C}_s \geq 0;\ s\in S. \quad (13.39)$$

In the SMIP model **ResMTSupport_CV(sl)** presented below, the overall quality of the resilient supply portfolio will be measured by conditional service-at-risk, CVaR^{sl}, (13.37).

ResMTSupport_CV(sl): *Selection of* **Res***ilient* **M***ulti-***T***ier* **Supp***ly* **port***folio to minimize* **CV***aR of* **s***ervice* **l***evel*
Maximize (13.37)
subject to (13.5)–(13.32) and
Risk constraints:

- the tail service level for scenario s, $\mathscr{S}_s$, is defined as the nonnegative amount by which VaR^{sl} exceeds service level in scenario s,

(continued)

$$\mathcal{S}_s \geq VaR^{sl} - (1 - z_s);\ s \in S, \tag{13.40}$$

$$\mathcal{S}_s \geq 0;\ s \in S. \tag{13.41}$$

The following **Proposition 13.4** shows relation between optimal conditional service-at-risk, CVaR^{sl}, (13.37), and the associated expected service level, E^{sl}, (13.4), i.e., the expected fraction of fulfilled demand for products

Proposition 13.4. *In the risk-averse solution of* **ResMTSupport_CV(sl)** *for a given confidence level α, conditional service-at-risk, $CVaR^{sl}$, and the associated expected service level, E^{sl}, are related as below.*

$$CVaR^{sl} \geq \frac{E^{sl} - \alpha}{1 - \alpha}. \tag{13.42}$$

Proof. Equations (13.4), (13.37), (13.40)–(13.41) imply

$$\begin{aligned}
CVaR^{sl} &= VaR^{sl} - \sum_{s \in S} P_s \mathcal{S}_s/(1-\alpha) = VaR^{sl} - \sum_{s \in S: \mathcal{S}_s > 0} P_s \mathcal{S}_s/(1-\alpha) \\
&= VaR^{sl} - \sum_{s \in S: 1 - z_s < VaR^{sl}} P_s (VaR^{sl} - 1 + z_s)/(1-\alpha) \\
&= VaR^{sl} - (VaR^{sl} - 1) \sum_{s \in S: 1 - z_s < VaR^{sl}} P_s/(1-\alpha) \\
&\quad - \sum_{s \in S: 1 - z_s < VaR^{sl}} P_s z_s/(1-\alpha) \\
&\geq VaR^{sl} - (VaR^{sl} - 1) - (1 - E^{sl})/(1-\alpha) \\
&= 1 - (1 - E^{sl})/(1-\alpha) = (E^{sl} - \alpha)/(1-\alpha),
\end{aligned} \tag{13.43}$$

where definition of VaR^{sl} for a given confidence level α, and definition of expected service level, (13.4), imply the following inequalities:

$$\sum_{s \in S: 1 - z_s < VaR^{sl}} P_s \leq 1 - \alpha, \tag{13.44}$$

$$\sum_{s \in S: 1 - z_s < VaR^{sl}} P_s z_s \leq \sum_{s \in S} P_s z_s = 1 - E^{sl}. \tag{13.45}$$

The inequality (13.42) indicates that $CVaR^{sl}$ approaches E^{sl} when α goes to zero.

Using results of **Proposition 13.4** for a chosen confidence level, the decision maker may simply estimate worst-case service level of the final manufacturer in a multi-tier supply chain network, based on the available value of average service level.

13.5 The Impact of Unit Penalty for Unfulfilled Demand

This section defines risk-neutral and risk-averse thresholds for unit penalty of unfulfilled demand for products above which pre-positioning of the risk mitigation inventory is profitable, respectively, under risk-neutral and risk-averse decision-making.

First, **Proposition 13.5** defines a threshold for unit penalty of unfulfilled demand for products above which the risk mitigation inventory is pre-positioned under risk-neutral decision-making. The proposition indicates that the cost of improving supply chain resilience by pre-positioning of risk mitigation inventory cannot be greater than potential losses due to unfulfilled demand for products with no risk mitigation inventory. If expected reduction of penalty cost for unfulfilled demand for products owing to usage of risk mitigation inventory is greater than the expected cost of pre-positioning and using the inventory, then pre-positioning of the inventory is profitable.

Proposition 13.5. *If unit penalty for unfulfilled demand for products, g, is greater than* $\underline{g}$*, then pre-positioning of risk mitigation inventory at assembly plants (the lowest tier, k = 0, of supply chain) is selected, where*

$$\underline{g} = \frac{\sum_{i\in I_0} c_{i0}h_{i0}x_{i0} + \sum_{s\in S}\sum_{i\in I_0\bigcap I^s} P_s c_{i0} f_{i0} y^s_{i0}}{\sum_{s\in S}\sum_{i\in I_0\bigcap I^s} P_s c_{i0} y^s_{i0}}. \quad (13.46)$$

Proof. Denote by, $z^s_{x=0}$ and $z^s_{x>0}$, unfulfilled demand under scenario s, when the risk mitigation inventory is not pre-positioned and is pre-positioned, respectively. According to demand for products fulfillment constraints (13.24)

$$z^s_{x=0} = 1 - \sum_{i\in I_1\bigcap I^s}\sum_{j\in I_0\bigcap I^s} v_{ij0} - \sum_{i\in I_1\bigcap I^s}\sum_{j\in I_0\bigcap I^s} V^s_{ij0} - \sum_{i\in I_0\setminus I^s}\sum_{j\in I_0\bigcap I^s} w^s_{ij0};\ s\in S,$$

$$z^s_{x>0} = 1 - \sum_{i\in I_1\bigcap I^s}\sum_{j\in I_0\bigcap I^s} v_{ij0} - \sum_{i\in I_1\bigcap I^s}\sum_{j\in I_0\bigcap I^s} V^s_{ij0} - \sum_{i\in I_0\setminus I^s}\sum_{j\in I_0\bigcap I^s} w^s_{ij0} - \sum_{i\in I_0\bigcap I^s} c_{i0}y^s_{i0}/D;\ s\in S.$$

The expected reduction of penalty cost for unfulfilled demand for products owing to usage of risk mitigation inventory is

$$g\sum_{s\in S} P_s(z^s_{x=0} - z^s_{x>0}) = g\sum_{s\in S}\sum_{i\in I_0\bigcap I^s} P_s c_{i0} y^s_{i0}/D.$$

(continued)

The pre-positioning of risk mitigation inventory is profitable only if the above reduction of penalty cost is greater than the expected cost of pre-positioning and using the inventory, that is, if

$$g\sum_{s\in S}\sum_{i\in I_0\bigcap I^s} P_s c_{i0} y_{i0}^s > \sum_{i\in I_0} c_{i0} h_{i0} x_{i0} + \sum_{s\in S}\sum_{i\in I_0\bigcap I^s} P_s c_{i0} f_{i0} y_{i0}^s.$$

Dividing both sides of the above inequality by, $\sum_{s\in S}\sum_{i\in I_0\bigcap I^s} P_s c_{i0} y_{i0}^s > 0$, implies the threshold (13.46).

The threshold (13,46) is tight and can be computed by using a feasible solution to model **ResMTSupport_E(c)** with a nonzero usage of risk mitigation inventory at the lowest tier $k = 0$.

Note that if $f_{i0} = f_0, h_{i0} = h_0; i \in I_0$, the threshold (13.46) reduces to

$$f_0 + h_0 \frac{\sum_{i\in I_0} c_{i0} x_{i0}}{\sum_{s\in S}\sum_{i\in I_0\bigcap I^s} P_s c_{i0} y_{i0}^s}.$$

In the risk-averse decision-making, pre-positioning of the risk mitigation inventory depends on both the value of unit penalty for unfulfilled demand, g, and the confidence level, α. Given confidence level, α, the minimum unit penalty, $\underline{g}(\alpha)$, required to pre-position risk mitigation inventory of parts from tier $k = 1$ at assembly plants of tier $k = 0$, satisfies the following general condition:

$$\underline{g}(\alpha) = \min\{g(\alpha) : CVaR^c(g(\alpha), x > 0) < CVaR^c(g(\alpha), x = 0)\}, \quad (13.47)$$

where x denotes pre-positioned risk mitigation inventory.

Given confidence level α, $CVaR^c(g(\alpha), x > 0)$ and $CVaR^c(g(\alpha), x = 0)$ denote optimal solution value CVaR^c, (13.36) of model **ResMTSupport_CV(c)** with unit penalty, $g(\alpha)$, respectively, with pre-positioned risk mitigation inventory and with none inventory, i.e., with constraints (13.19) replaced by $x_{ik} = 0;\ k \in K, i \in I_k : k < n$.

Notice that $\underline{g}(\alpha)$ does not need to uniformly vary with α. However, when $\alpha \to 0$, then $\underline{g}(\alpha)$, goes to its greatest value, $\underline{g}(0)$. The latter denotes a risk-neutral threshold for unit penalty, above which risk mitigation inventory is pre-positioned under risk-neutral decision-making, defined in (13.46) and simply denoted by $\underline{g}$.

Both the risk-averse, $\underline{g}(\alpha)$, and the risk-neutral, $\underline{g}$, thresholds allow the decision maker to profitably pre-position risk mitigation inventory, based on contractual unit penalty for unfulfilled demand.

The following **Proposition 13.6** defines another threshold for unit penalty of unfulfilled demand for products above which a recovery assembly plant is selected in a supply chain with dual plants.

Proposition 13.6. *In a supply chain with two assembly plants at tier $k = 0$, if primary assembly plant ($i = 1$ of tier $k = 0$) is the only disrupted facility, then plant $i = 2$ is selected as a recovery plant only if unit penalty for unfulfilled demand for products, g, is greater than*

$$g_{min} = \psi_{1,2,0} + \frac{e_{2,0}}{\min\{c_{2,0}, D\}}. \tag{13.48}$$

Proof. Plant, $i = 2$ is selected as a recovery assembly plant if setup, transshipment and recovery supply costs are smaller than savings on penalty cost owing to demand fulfilled by the recovery plant, that is, if

$$g(1 - z_s) > e_{2,0}/D + \psi_{1,2,0} w^s_{1,2,0} + \sum_{i \in I_1 \bigcap I^s} \overline{o}_{i1} V^s_{i,2,0}, \tag{13.49}$$

where $D(1 - z_s)$ is fulfilled demand.

For the scenario considered, where the primary assembly plant ($i = 1$ of tier $k = 0$) is the only disrupted facility, Eq. (13.24) implies $1 - z_s = w^s_{1,2,0} + \sum_{i \in I_1 \bigcap I^s} V^s_{i,2,0} > 0$. Thus,

$$g > \frac{e_{2,0}/D + \psi_{1,2,0} w^s_{1,2,0} + \sum_{i \in I_1 \bigcap I^s} \overline{o}_{i1} V^s_{i,2,0}}{w^s_{1,2,0} + \sum_{i \in I_1 \bigcap I^s} V^s_{i,2,0}} \geq$$
$$\frac{e_{2,0}/D + \psi_{1,2,0}(w^s_{1,2,0} + \sum_{i \in I_1 \bigcap I^s} V^s_{i,2,0})}{w^s_{1,2,0} + \sum_{i \in I_1 \bigcap I^s} V^s_{i,2,0}} \geq$$
$$\psi_{1,2,0} + \frac{e_{2,0}/D}{w^s_{1,2,0} + \sum_{i \in I_1 \bigcap I^s} V^s_{i,2,0}}, \tag{13.50}$$

where $\psi_{1,2,0} \leq \overline{o}_{i1}$, $i \in I_1$.

According to Eqs. (13.14) and (13.18), transshipment of parts from disrupted primary plant, $i = 1$, to recovery plant $i = 2$, $w^s_{1,2,0}$, satisfies inequalities: $w^s_{1,2,0} \leq v_{1,1,0} = 1$ and $w^s_{1,2,0} + \sum_{i \in I_1 \bigcap I^s} V^s_{i,2,0} \leq c_{2,0}/D$. If $c_{2,0} > D$, then the right-hand side of the last inequality should be replaced by $\min\{c_{2,0}/D, 1\}$. Thus $w^s_{1,2,0} + \sum_{i \in I_1 \bigcap I^s} V^s_{i,2,0} \leq \min\{c_{2,0}/D, 1\}$. Substituting in (13.50), $w^s_{1,2,0} + \sum_{i \in I_1 \bigcap I^s} V^s_{i,2,0}$ by its upper bound $\min\{c_{2,0}/D, 1\}$, the threshold (13.48) is obtained.

Notice that the decision maker may select recovery plant, based on contractual unit penalty for unfulfilled demand, with threshold (13.48) depending on plant setup per capacity and transshipment costs only.

13.6 Computational Examples

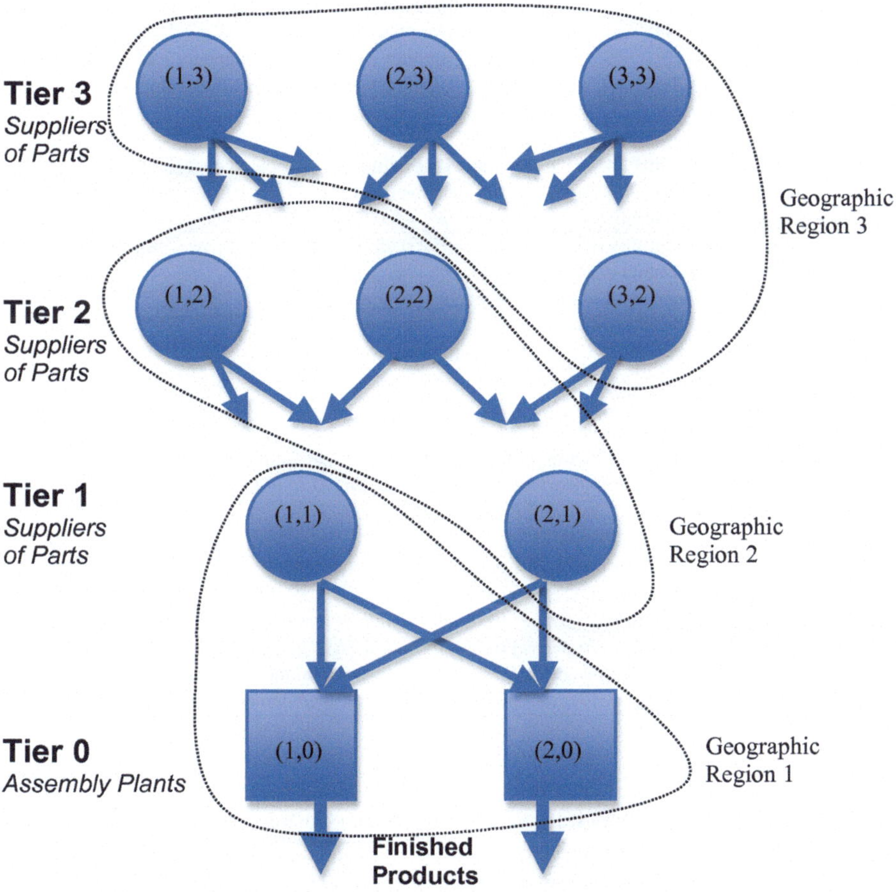

Fig. 13.3 A four-tier supply chain dispersed over three geographic regions

This section provides simple computational examples to demonstrate that the proposed multi-tier portfolio approach and the two-period stochastic MIP formulation with an embedded network flow structure lead to computationally highly efficient decision-making tool. The input data for the examples are in part hypothetical, however, their relations to each other are real and the computational example can be linked with real-world problems in automotive supply chains, e.g., MacKenzie et al. (2014), Matsuo (2015). For example, final manufacturers are located in geographic regions less exposed to natural disasters. Regional disruption probabilities are greater for more distant regions, where higher tier, less reliable, and mainly low-cost suppliers are located. Suppliers of the same tier may share their immedi-

Table 13.3 Input parameters: resilient multi-tier supply portfolio

$R = \{1,2,3\}$	three geographic regions
$K = \{0,1,2,3\}$	four-tier supply chain
$m_0 = m_1 = 2,\ m_2 = m_3 = 3$	two assembly plants at tier 0, two suppliers at tier 1, and three suppliers at each tier 2 and 3
$I^{(1)} = \{(1,0),(2,0),(1,1)\}$	in region 1: two plants and supplier 1 of tier 1
$I^{(2)} = \{(2,1),(1,2),(2,2)\}$	in region 2: supplier 2 of tier 1 and suppliers 1 and 2 of tier 2
$I^{(3)} = \{3,2),(1,3),(2,3),(3,3)\}$	in region 3: supplier 3 of tier 2 and all three suppliers of tier 3
$D = 10000$	Total demand for parts/products
	Disruption probability:
Local disruption probability	$p_{ik},\ k \in K, i \in I_k$ uniformly distributed over [0.05,0.10] for $(i,k) \in I^{(1)}$, over [0.10,0.15] for $(i,k) \in I^{(2)}$, and over [0.15,0.25] for $(i,k) \in I^{(3)}$
Regional disruption probability (i):	$p^1 = 0.005,\ p^2 = 0.025,\ p^3 = 0.05$
Regional disruption probability (ii):	$p^1 = 0.01,\ p^2 = 0.05,\ p^3 = 0.1$
Regional disruption probability (iii):	$p^1 = 0.05,\ p^2 = 0.25,\ p^3 = 0.5$
	Capacity of assembly plants ($k = 0$) and suppliers ($k > 0$):
$c_{1k} = D,\ k \in K,\ c_{ik} = D/2,\ k \in K, i \in I_k : i > 1$	
	Cost parameters:
$o_{11} = 25, o_{21} = 26, o_{12} = 13, o_{22} = 14, o_{32} = 15, o_{13} = 4, o_{23} = 6, o_{33} = 5,\ \overline{o}_{ik} = 1.5 o_{ik}$	
$f_{ik} = f_k = \max_{i \in I_{k+1}}(1.2 o_{i,k+1}),\ k \in K, i \in I_k : k < 3$	
$h_{ik} = h_k = \max_{i \in I_{k+1}}(0.2 o_{i,k+1}),\ k \in K, i \in I_k : k < 3$	
$e_{1k} = 0,\ k \in K$, i.e., no ordering (setup) cost for preselected primary suppliers (plant)	
$e_{ik},\ k \in K, i \in I_k : i > 1$ uniformly distributed over [5D,10D] for $(i,k) \in I^{(1)}$, over [10D,15D] for $(i,k) \in I^{(2)}$, and over [15D,20D] for $(i,k) \in I^{(3)}$	
$\psi_{ijk} = 1,\ k \in K, i \in I_k, j \in I_k : k < 3$	
	Unit penalty for risk-neutral decision-making:
Regional disruption probability (i):	$g = 110.00$ or $g = 128.87$ or $g = \underline{g} = 128.88$ or $g = 150.00$
Regional disruption probability (ii):	$g = 110.00$ or $g = 129.36$ or $g = \underline{g} = 129.37$ or $g = 150.00$
Regional disruption probability (iii):	$g = 110.00$ or $g = 136.46$ or $g = \underline{g} = 136.47$ or $g = 150.00$
	Unit penalty for risk-averse decision-making:
$g = 10, 16.99, 17.01, 50\ (g_{min} = 17)$	unit penalty cost of unfulfilled demand for products

ate upper tier suppliers, which may result in overlap in the higher tiers, and which Toyota refers to as a "diamond-shaped" supply chain (e.g., Ang et al. 2017).

The computational experiments were performed using the AMPL programming language and the Gurobi 7.5.0 solver on a MacBookPro laptop with Intel Core i7 processor running at 2.8 GHz and with 16 GB RAM. Table 13.3 shows basic input parameters for a four-tier supply chain dispersed over three geographic regions (see, Fig. 13.3), where multiple suppliers/assembly plants of the same and/or different tiers are located in different geographic region. The total number of disruption scenarios is $2^{\sum_{k \in K} m_k} = 1024$.

Three different variants (i), (ii), and (iii) of regional disruption probabilities are considered. For each variant, region $g = 1$, where final manufacturers are located, is the most reliable. The more distant is the geographic region r from region 1, the greater its regional disruption probability: $p^3 > p^2 > p^1$. In addition, local disruption probabilities are also higher for suppliers in more vulnerable regions. Finally, less reliable suppliers of the same tier offer lower prices of parts, and Fig. 13.4 illustrates the correlation between suppliers price and reliability. For each supplier,

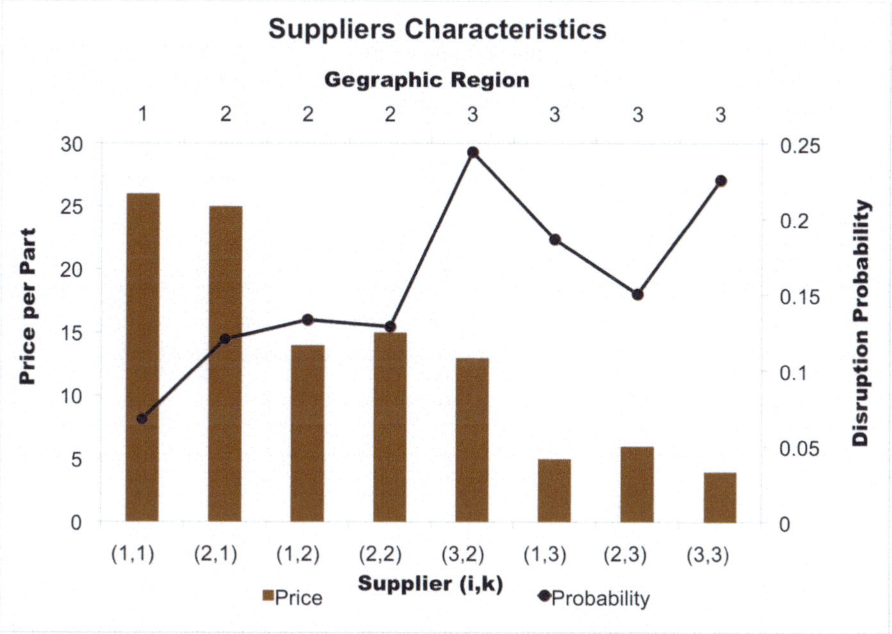

Fig. 13.4 Suppliers characteristics for regional disruption probability (ii)

$(i,k) : i \in I_k \bigcap I^{(r)}, k \in K, r \in R$, Fig. 13.4 shows its price per part, o_{ik}, and disruption probability, $p^r + (1 - p^r)p_{ik}$.

13.6.1 Risk-Neutral Decision-Making

Solution results for regional disruption probabilities (ii) are shown in Table 13.4, where primary, expected recovery and expected transshipment portfolios and pre-positioned and expected used risk mitigation inventory are presented along with solution values and their LP relaxations. The expected recovery portfolios refer to the expected supplies to meet demand unfulfilled by primary supplies. Since the primary supplies meet demand for most of the scenarios, the expected recovery supplies are a small percentage of total demand. For model **ResMTSupport_E(c)** and different values of unit penalty, g, Table 13.4 shows the expected cost per product, E^c, (13.3), along with the associated expected service level, E^{sl}, (13.4).

Notice that (see, (13.4) and (13.24) and compare with results in Table 13.4)

$$E^{sl} = \sum_{s \in S} P_s \Big(\sum_{i \in I_1 \bigcap I^s} \sum_{j \in I_0 \bigcap I^s} v_{ij0} + \sum_{i \in I_1 \bigcap I^s} \sum_{j \in I_0 \bigcap I^s} V^s_{ij0} + \sum_{i \in I_0 \setminus I^s} \sum_{j \in I_0 \bigcap I^s} w^s_{ij0} + \sum_{i \in I_0 \bigcap I^s} c_{i0} y^s_{i0} / D \Big).$$

Table 13.4 Solution results for regional disruption probability (ii)

Model	**ResMTSupport_E(c)**				**ResMTSupport_E(sl)**
Model size [a]	Var. = 22,232, Bin. = 10,250, Con. = 34,552, Nonz. = 98,258				
Unit penalty, $g =$	110.00	129.36	$\underline{g} = 129.37$	150.00	∞
Exp.Cost, (13.3)	48.74	50.45	50.46	51.64	–
Exp.Service ×100%, (13.4)	91.14	91.18	94.26	97.66	98.27
Primary Portfolio					
outflow from (i,k) [b]					
(1,0)	1				0.5
(2,0)	0				0.5
(1,1)	1				0.5
(2,1)	0				0.5
(1,2)	1				0
(2,2)	0				0.5
(3,2)	0				0.5
(1,3)	1				0
(2,3)	0				0.5
(3,3)	0				0.5
inflow to (j,k) [c]					
(1,0)	1				0.5
(2,0)	0				0.5
(1,1)	1				0.5
(2,1)	0				0.5
(1,2)	1				0
(2,2)	0				0.5
(2,3)	0				0.5
Pre-positioned Inventory	–	–	$x_{1,0} = 0.5$	$x_{1,0} = 0.5$	$x_{ik} = 1, k \in K, i \in I_k : k < 3$
Exp.Usage of Inventory [h]	–	–	0.031	0.031	0.11
Exp.Recovery Portfolio					
outflow from (i,k) [d]					
(1,0)	0.025	0.025	0.056	0.056	0.133
(2,0)	0.035				0.036
(1,1)	0				0.005
(2,1)	0.027				0.001
(2,2)	0.002	0.003	0.003	0.003	0.001
(3,2)	0				0.001
inflow to (j,k) [e]					
(1,0)	0.025				0.002
(2,0)	0.002				0.003
(1,1)	0				0.002
(2,1)	0.002	0.003	0.003	0.003	0
Exp.Transshipment Portfolio					
outflow from (i,k) [f]					
(1,0)	0.034				0.028
(2,0)	0				0.027
(1,1)	0.024				0
(2,1)	0				0.001
(1,2)	0.003				0
(2,2)	0				0.001

(continued)

Table 13.4 (continued)

Model	**ResMTSupport_E(c)**	**ResMTSupport_E(sl)**
Model size [a]	Var. = 22,232, Bin. = 10,250, Con. = 34,552, Nonz. = 98,258	
Unit penalty, $g =$	110.00 129.36 $\underline{g} = 129.37$ 150.00	∞
inflow to (j,k) [g]		
(1,0)	0	0.027
(2,0)	0.034	0.028
(1,1)	0	0.001
(2,1)	0.024	0
(2,2)	0.003	0
(3,2)	0	0.001
LP [j]	48.16 49.40 49.85 49.88	98.27

[a] Number of variables, binary variables, constraints, nonzero coefficients
[b] $\sum_{j\in I_{k-1}} v_{ij,k-1}$, [c] $\sum_{i\in I_{k+1}} v_{ijk}$, [d] $\sum_{s\in S}\sum_{j\in I_{k-1}} P_s V^s_{ij,k-1}$, [e] $\sum_{s\in S}\sum_{i\in I_k} P_s V^s_{ijk}$
[f] $\sum_{s\in S}\sum_{j\in I_k} P_s w^s_{ijk}$, [g] $\sum_{s\in S}\sum_{i\in I_k} P_s w^s_{ijk}$
[h] $\sum_{s\in S}\sum_{k\in K:k<3}\sum_{i\in I_k} P_s c_{ik} y^s_{ik}/D$
[j] LP relaxation solution value of (13.3) and (13.4)

Table 13.5 Impact of regional disruption probability

Model	**ResMTSupport_E(c)**				**ResMTSupport_E(sl)**
	Regional disruption probability (i)				
Unit penalty, $g =$	110.00	128.87	$\underline{g} = 128.88$	150.00	∞
Exp.Cost, (13.3)	48.78	50.34	50.35	51.44	-
Exp.Service $\times 100\%$, (13.4)	91.68	91.73	94.82	94.87	98.77
Pre-positioned Inventory	–	–	$x_{1,0} = 0.5$	$x_{1,0} = 0.5$	$x_{ik} = 1, k \in K, i \in I_k : k < 3$
Exp.Usage of Inventory	–	–	0.031	0.031	0.10
LP	48.18	49.73	49.73	50.81	98.77
	Regional disruption probability (iii)				
Unit penalty, $g =$	110.00	136.46	$\underline{g} = 136.47$	150.00	∞
Exp.Cost, (13.3)	48.45	51.53	51.54	53.22	-
Exp.Service $\times 100\%$, (13.4)	86.85	86.87	89.83	89.85	94.30
Pre-positioned Inventory	–	–	$x_{1,0} = 0.5$	$x_{1,0} = 0.5$	$x_{ik} = 1, k \in K, i \in I_k : k < 3$
Exp.Usage of Inventory	–	–	0.030	0.030	0.19
LP	47.97	51.05	51.05	52.73	94.30

As g increases a higher service level is attained. However, even for $g \to \infty$, i.e., when the objective is to maximize service level, E^{sl} is less than one, because the impact of disruptions cannot be fully mitigated. Notice a significant jump of the expected service level, E^{sl}, from 91.18% for $g = 129.36$ to 94.26% for the threshold, $\underline{g} = 129.37$, while the corresponding expected cost per product, E^c, increases only by 0.01. These are the results of pre-positioning and usage of risk mitigation inventory to reduce unfulfilled demand for products.

Similar basic solution results are presented in Table 13.5 for regional disruption probabilities (i) and (iii). The table indicates that higher regional disruption probability leads to higher threshold $\underline{g}$ and lower expected service level.

13.6.2 Risk-Averse Decision-Making

The solution results for the risk-averse decision-making are summarized in Tables 13.6–13.10, for model **ResMTSupport_CV(c)** with unit penalty, $g \in \{10, 16.99, 17.01, 50\}$, and model **ResMTSupport_CV(sl)**. The tables show solution values of conditional cost-at-risk, CVaR^c, and cost-at-risk, VaR^c, for model **ResMTSupport_CV(c)** and conditional service-at-risk, CVaR^{sl} and service-at-risk, VaR^{sl}, for model **ResMTSupport_CV(sl)**. The tables also provide associated solution values of expected service level, E^{sl}, (13.4), and expected cost per product, E^c, (13.3).

Table 13.6 Risk-averse solutions for model **ResMTSupport_CV(c)**: $g = 10$

Confidence level α	0.50	0.75	0.90	0.95	0.99
Var. = 22,087, Bin.=10,250, Cons.=38,149, Nonz.=145,725 [a]					
Regional disruption probability (i)					
CVaR^c, (13.36)	43.11	44.21	47.53	51.76	**52**[c]
VaR^c	42	42	42	48	**52**[c]
Exp.Cost, (13.3)	39.21	39.43	38.74	39.42	38.72
Exp.Service ×100%, (13.4)	85.65	85.63	85.59	85.71	85.59
LP [b]	43.05	44.10	47.25	51.30	51.65
Regional disruption probability (ii)					
CVaR^c, (13.36)	43.05	44.11	47.27	51.45	**52**[c]
VaR^c	42	42	42	48	**52**[c]
Exp.Cost, (13.3)	39.96	38.71	38.20	38.20	38.20
Exp.Service ×100%, (13.4)	85.22	85.26	85.16	85.16	85.16
LP [b]	43	44	46.99	50.99	51.64
Regional disruption probability (iii)					
CVaR^c, (13.36)	40.79	43.36	45.41	48.81	**52**[c]
VaR^c	38	42	42	42	**52**[c]
Exp.Cost, (13.3)	34.05	35.52	34.51	36.71	34.05
Exp.Service ×100%, (13.4)	81.72	81.75	81.72	81.77	81.72
LP [b]	40.74	43.28	45.20	48.40	51.56

[a] number of variables, binary variables, constraints, nonzero coefficients
[b] LP relaxation solution value of (13.36)
[c] VaR^c = CVaR^c = 52 is the highest cost that may occur

Table 13.6 indicates that for $g = 10$ and all regional disruption probabilities, (i), (ii), and (iii), the highest cost that may occur is 52 and it is attained for the highest confidence level, $\alpha = 0.99$. The probability mass functions for the risk-averse solutions and the highest confidence level, $\alpha = 0.99$, are shown in Fig. 13.5. The figure demonstrates that the probability is concentrated in a few points, which is typical for the scenario-based optimization under uncertainty, where the probability measure is concentrated in finitely many points. The probability of the highest cost, 52, is greater than $1 - \alpha$, and hence, $CVaR^c = VaR^c = 52$. For comparison, Fig. 13.5 also shows probability mass functions for unit penalty, $g = 50$, for which, $VaR^c < CVaR^c$ and for regional disruption probabilities: (i), $VaR^c = 75.50$, $CVaR^c = 81.68$; (ii), $VaR^c = 75.50$, $CVaR^c = 81.45$; (iii), $VaR^c = 75.50$, $CVaR^c = 79.70$.

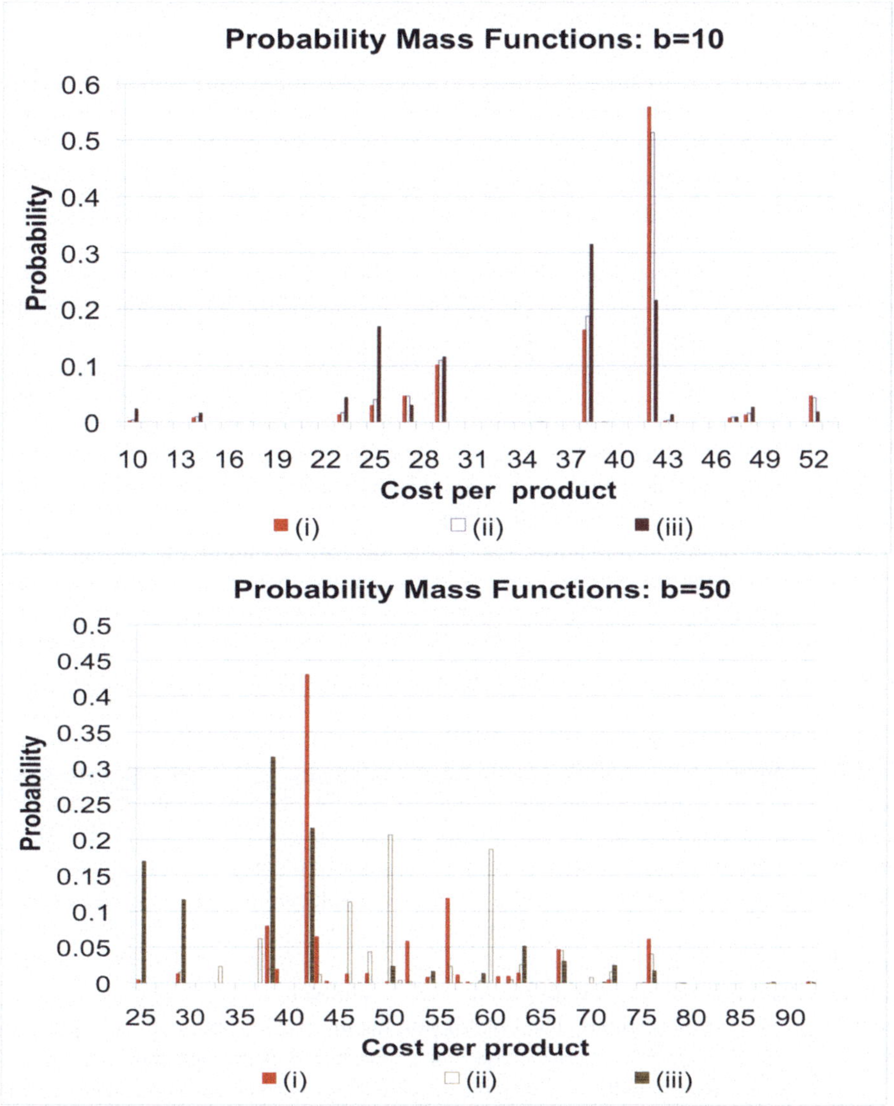

Fig. 13.5 Probability mass functions for model **ResMTSupport_CV(c)**: $\alpha = 0.99$

The Impact of Unit Penalty for Unfulfilled Demand

For unit penalty $g = 10$ and all confidence levels, risk mitigation inventory of parts from tier $k = 1$ is not pre-positioned at primary assembly plant of tier $k = 0$. The result indicates that for $g = 10$, cost of pre-positioning and usage the inventory would be greater than potential savings in penalty for unfulfilled demand for products. However, the risk mitigation inventory can be pre-positioned and used for higher values of g.

For the presented example, the values of $\underline{g}(\alpha)$, (13.47), computationally determined for regional disruption probabilities (i) and different confidence level, $\alpha \in \{0.5, 0.75, 0.90, 0.95, 0.99\}$, are:

$$\underline{g}(0.5) = 80.44;\ \underline{g}(0.75) = 56.22;\ \underline{g}(0.9) = 50.87;\ \underline{g}(0.95) = 79.63;\ \underline{g}(0.99) = 76.01.$$

Notice that $\underline{g}(0.5)$ is greater than $\underline{g}(\alpha)$ for the other confidence levels. For the presented example, the risk-neutral threshold, (13.46), is $\underline{g}(0) = 128.88$, which is much greater than the above risk-averse thresholds.

The above results demonstrate that a risk-neutral threshold for unit penalty, $\underline{g}(0)$, required to pre-position risk mitigation inventory under risk-neutral decision-making is much larger than the corresponding risk-averse thresholds, $\underline{g}(\alpha)$, for different confidence levels, $\alpha \in (0,1)$. Thus, for a risk-averse decision-making, smaller values of unit penalty, g, are sufficient to profitably pre-position risk mitigation inventory.

The impact of unit penalty for unfulfilled demand is additionally illustrated below, by comparison of example results for $g = 10$ and $g = 50$, regional disruption scenario (ii) and the highest confidence level, $\alpha = 0.99$. For unit penalty, $g = 10$, the highest cost 52, is achieved, for example, for scenario $\lambda_{512} = \{1,1,1,1,1,1,1,1,0,1\}$, where the only disrupted facility is primary assembly plant, $i = 1$. However, non-disrupted plant, $i = 2$ with capacity $D/2$, was not selected as recovery plant since ordering cost, $e_{2,0}/D = 8$, and transshipment cost, $\psi_{1,2,0}w^s_{1,2,0} = 0.5$, in total 8.5, would be greater than savings on penalty cost, $g/2 = 5$, if 50% of demand was fulfilled by that plant. As a result no product was assembled, $z_{512} = 1$, and the highest penalty $gz_{512} = 10$ is attained. The cost of purchasing parts from primary suppliers is $\sum_{k \in K: k<n} o_{1,k+1}v_{11k} = o_{11} + o_{12} + o_{13} = 42$, which makes the optimal cost, 52.

However, for unit penalty, $g = 50$, the second plant is selected as recovery plant since setup and transshipment costs, 8.5, are smaller than penalty savings owing to demand fulfilled by that plant. Because the capacity of recovery plant is $D/2$, the service level is only 50%, i.e., $1 - z_{512} = 0.5$, and hence the penalty for unfulfilled demand is $gz_{512} = 25$. Thus, penalty savings are, 50-25=25, which are greater than setup and transshipment costs, 8.5. The cost of purchasing parts from primary suppliers is $\sum_{k \in K: k<n} o_{1,k+1}v_{11k} = 42$, which makes the total cost for $g = 50$, equal to $VaR = 42 + 8.5 + 25 = 75.50$.

For the presented examples, the threshold, (13.48) (cf. **Proposition 13.6**), is $g_{min} = 17$.

Table 13.7 Solution results for model **ResMTSupport_CV(c)**: $g = 16.99$ and $g = 17.01$

Confidence level α	0.50	0.75	0.90	0.95	0.99
Regional disruption probability (ii): $g = 16.99$					
CVaR^c, (13.36)	43.95	45.90	51.76	58.45	58.99
VaR^c	42	42	42	54.99	58.99
Exp.Cost, (13.3)	39.53	39.55	39.30	47.54	39.23
Exp.Service ×100%, (13.4)	85.20	85.16	85.20	85.38	85.16
Regional disruption probability (ii): $g = 17.01$					
CVaR^c, (13.36)	43.95	45.91	51.77	58.46	59.006
VaR^c	42	42	42	55.005	59.005
Exp.Cost, (44)	39.60	40.45	40.45	45.88	43.73
Exp.Service ×100%, (38)	88.51	88.51	88.51	88.26	87.92

The impact of unit penalty, g, on service level is illustrated in Table 13.7. Table 13.7 compares solution results for model **ResMTSupport_CV(c)** with regional disruption probability (ii) and unit penalty $g = 16.99 < b_{min}$ and $g = 17.01 > b_{min}$. The table shows that a slight increase of unit penalty from $g = 16.99$ to $g = 17.01$, results in slightly increasing CVaR^c. At the same time, however, a significant improvement of the associated expected service level, E^{sl}, owing to opening a recovery plant $i = 2$ for $g = 17.01 > b_{min}$, is observed.

Table 13.8 Impact of non available transshipment: **ResMTSupport_CV(c)**, $g = 50$

Confidence level α	0.50	0.75	0.90	0.95	0.99
Regional disruption probability (ii), $\psi_{ijk} = 1$					
CVaR^c, (13.36)	49.79	57.58	72.08	76.29	81.45
VaR^c	42	42	67	71.50	75.50
Exp.Cost, (13.3)	43.38	43.03	43.17	43.15	58.00
Exp.Service ×100%, (13.4)	88.51	88.51	88.50	88.49	88.48
Regional disruption probability (ii), $\psi_{ijk} = 100$					
CVaR^c, (13.36)	52.00	62.01	75.98	76.96	84.78
VaR^c	42	42	75	75	75
Exp.Cost, (13.3)	44.22	44.42	51.81	51.84	51.85
Exp.Service ×100%, (13.4)	85.16	85.16	85.82	85.82	85.82

The impact of non available transshipment is illustrated in Table 13.8, using solution results for model **ResMTSupport_CV(c)** with unit penalty, $g = 50$, and low unit transshipment cost, $\psi_{ijk} = 1$, vs prohibitively high, $\psi_{ijk} = 100$. A general observation is that non available transshipment of inventory leads to higher conditional cost-at-risk, CVaR^c, and lower expected service level, E^{sl}, which is due to additional cost of expedited recovery supplies and/or unfulfilled demand for products.

Solution results for model **ResMTSupport_CV(sl)** are presented in Table 13.9, and the corresponding probability mass functions for different regional disruption probabilities are shown in Table 13.10. Table 13.9 indicates that for regional disruption probability (ii) and $\alpha = 0.99$ and for regional disruption probability (iii) and

Table 13.9 Risk-averse solutions for model **ResMTSupport_CV(sl)**

Confidence level α	0.50	0.75	0.90	0.95	0.99
Var. = 22,087, Bin. = 10250, Cons. = 38,149, Nonz. = 10,6523 [a]					
Regional disruption probability (i)					
$CVaR^{sl} \times 100\%$, (13.37)	90.79	81.59	53.97	40.12	0.62
$VaR^{sl} \times 100\%$	100			50	
Exp.Service $\times 100\%$, (13.4)	95.39			95.37	
LP [b]	90.79	81.59	53.97	40.12	0.62
Regional disruption probability (ii)					
$CVaR^{sl} \times 100\%$, (13.37)	89.83	79.67	49.17	35.15	**0**[c]
$VaR^{sl} \times 100\%$	100			50	**0**[c]
Exp.Service $\times 100\%$, (13.4)	94.91			94.71	85.16
LP [b]	89.83	79.67	49.17	35.15	0
Regional disruption probability (iii)					
$CVaR^{sl} \times 100\%$, (13.37)	82.16	64.33	22.67	**0**[c]	
$VaR^{sl} \times 100\%$	100		50	**0**[c]	
Exp.Service $\times 100\%$, (13.4)	91.08		90.55	81.72	
LP [b]	82.16	64.33	22.67	0	

[a] Number of variables, binary variables, constraints, nonzero coefficients
[b] LP relaxation solution value of (13.37)
[c] VaR^{sl}=$CVaR^{sl}$=0 is the lowest service level that may occur

Table 13.10 Probability of service level for model **ResMTSupport_CV(sl)**

Service level %	Confidence level α				
	0.50	0.75	0.90	0.95	0.99
Regional disruption probability (i)					
0	0.0099				
50	0.0723			0.0727	
100	0.9178			0.9174	
Regional disruption probability (ii)					
0	0.0149			0.0149	0.1484
50	0.0719			0.0761	0
100	0.9132			0.9090	0.8516
Regional disruption probability (iii)					
0	0.0547			0.1828	
50	0.0690		0.0796	0	
100	0.8763		0.8657	0.8172	

$\alpha = 0.95$ and 0.99, the lowest service level, 0%, is attained. Table 13.10 demonstrates that the probability is concentrated in at most three points: 0, 50, and 100% of service level (there are at most three capacity levels, 0, $D/2$ or D). For regional disruption probability (ii) and (iii) and the highest confidence level $\alpha = 0.99$, the probability of the lowest service level, 0%, respectively, 0.1484 and 0.1828, is greater than $1 - \alpha$, and hence $CVaR^{sl} = VaR^{sl} = 0\%$. Similarly, probability, 0.1828, for $\alpha = 0.95$ and regional disruption probability (iii).

The major managerial insights are summarized in Table 13.11. Overall, the results of computational experiments indicate that *the developed multi-portfolio ap-*

proach simplifies coordinated decision-making in the multi-tier supply chain networks. The embedded network flow problem leads to computationally efficient stochastic mixed integer program with a very strong LP relaxation.
The approach has proven to be computationally very efficient.

Table 13.11 Managerial insights: resilient multi-tier portfolio selection

Decision-making issue	Decision-making impact
Unit penalty cost for unfulfilled demand for products.	In the risk-neutral decision-making, a higher unit penalty leads to a higher expected service level.
Regional disruption probability.	In the risk-neutral decision-making, a higher regional disruption probability leads to a lower expected service level and a higher threshold for unit penalty cost, required to pre-position risk mitigation inventory.
Risk-averse vs risk-neutral thresholds for unit penalty.	When conditional cost-at-risk (worst-case performance) is optimized, the risk-averse threshold sufficient to profitably pre-position risk mitigation inventory is much lower than the risk-neutral one required to optimize expected cost (average performance).
Unit penalty threshold for selection recovery assembly plant.	Threshold for unit penalty for unfulfilled demand for products required to select recovery assembly plant depends on plant setup per capacity and transshipment costs only. Thus, in a multi-tier supply chain with dual plants and disrupted primary plant, the decision maker may select recovery plant, based on contractual unit penalty for unfulfilled demand.
Balancing of inter- and intra-tier recovery flows in a disrupted multi-tier supply chain network.	In order to minimize potential losses in a multi-tier supply chain network, e.g., due to the ripple effect, integrated flow balancing actions are required to simultaneously recover disrupted flows over the entire network.
A tight lower bound on optimal conditional service-at-risk is a linear function of associated expected service level and selected confidence level.	In a multi-tier supply chain network, the decision maker may simply estimate worst-case service level of the final manufacturer for selected confidence levels, using the available value of average service level.

13.7 Notes

Different disruption management strategies that are applied in practice to minimize losses due to the shortage of material supplies include: multi-sourcing, backup supplier contracts, recovery and transshipment supplies, production and logistics flexibility, etc. The resilience of a supply chain, known as its ability to maintain, execute and recover the planned execution under disruption risks (e.g., Cardoso et al. 2015, Ribeiro and Barbosa-Povoa 2018) can be improved, for example, by suppliers fortification, pre-positioning of risk mitigation inventory, spot purchasing, collaboration, and visibility (e.g., Sheffi 2005, Namdar et al., 2017, Ivanov 2018). A backup resilient supply portfolio might also be considered with in-house manufacturing of parts in a recovery assembly plant, e.g., Freeman et al. (2018).

The recent literature reviews on disruption risk mitigation strategies, supply chain resilience, and quantitative methods were provided, for example, in Narasimhan and Talluri (2009), Van Mieghem (2012), Tomlin and Wang (2012), Rajagopal et al. (2017). Multi-tier case studies were presented in Choi and Hong (2002), Holweg and Pil (2008), Thome et al. (2014). However, supplier selection problem in a multi-tier supply chain network under disruption risks is nearly unreported in the literature, e.g., DeCroix (2013), Mena et al. (2013). The literature has been concentrated mainly on two-tier (manufacturer–supplier dyad) or three-tier (manufacturer–supplier–supplier's supplier triad) supply chains. The complex multi-tier supply chain structures have been also neglected by manufacturers. For example, after the Great East Japan earthquake and tsunami in March 2011, Toyota faced immediate shortages on over 400 parts, and production capacity was reduced for 6 months following the disaster (e.g., Ang et al. 2017). Toyota and other manufacturers, who prior to the disaster had primarily focused on their immediate suppliers and did not even know their higher tier suppliers, realized that multi-tier supply chains require more comprehensive strategies.

In Ang et al. (2017) sourcing was studied in a three-tier supply chain, which consists of a manufacturer, tier 1 suppliers and tier 2 suppliers prone to disruption from natural disasters. However, tier 1 suppliers disruption risk was ignored. The paper showed that the manufacturer's optimal strategy depends on the degree of overlap in the supply chain. Manufacturer's procurement decision in a three-tier supply chain was investigated by Yoon et al. (2019), where the manufacturer cannot directly obtain information on disruption risk for tier 2 supplier except through tier 1 supplier. The paper demonstrated that information sharing makes the manufacturer's decisions more conservative, i.e., carrying more inventories, while tier 1 supplier decision is contingent on tier 2 supplier reliability.

Among various approaches proposed on the literature to improve resilience of supply chains, the most common are mathematical programming, simulation, and recently machine learning, e.g., Hosseini et al. (2019a,b). For example, in Hosseini et al. (2019b), a stochastic bi-objective mixed integer programming model is proposed that considers several proactive and reactive resilience strategies including supplier segregation, backup supplier contracts, supplier reliability, surplus inventory, and supplier restoration capability. The problem objective is to maximize the distances between all pairs of supplier locations and to minimize the total cost of supplier selection. In Cavalcante et al. (2019, a combined machine learning and simulation approach was developed to create digital supply chain twins and select resilient suppliers in digital manufacturing with the on-time delivery as a resilience measure.

The material presented in this chapter extends literature in three significant ways. First, a comprehensive model was proposed for coordinated decision-making to select both primary suppliers and assembly plants, allocate among them orders quantity, and pre-position risk mitigation inventory to hedge against all potential disruption scenarios in a multi-tier supply chain network. Second, simultaneously for each disruption scenario, both recovery suppliers and assembly plants are selected and unfulfilled orders quantity are allocated among them and decisions on transshipment of parts and usage the pre-positioned inventory are made. Finally, this study investigates two equally important, conflicting objective functions: minimization of total expected cost over the entire supply chain and maximization of expected service level of final manufacturer for risk-neutral decision-making or minimization of CVaR of cost or maximization of CVaR of service level for risk-averse decision-making, see Sawik (2020).

The serial-parallel structure of supply chain considered in this chapter may represent a critical path of a multi-tier supply chain network, for example, identified using the risk-exposure index introduced in Simchi-Levi et al. (2015).

In the models proposed, the supply chain resilience is achieved by multi-sourcing, recovery and transshipment supplies and pre-positioning and usage of risk mitigation inventory. The resilience can be further improved by fortification of selected primary suppliers and assembly plants. Depending on the level of fortification, the capacity of each protected facility remains fully or partially available under a disruptive event. Accordingly, the protected suppliers (plants) are capable of fully or partially supplying parts (products) in the face of disruptive events.

In the future research the various simplified assumptions used to formulate the problem can be relaxed. A straightforward enhancement of the developed model is a multiple part types and product types setting with subsets of suppliers available for each part type and assembly plants for each product type. In addition to penalty cost of final manufacturer for unfulfilled demand for products also suppliers can be directly charged with a penalty for unfulfilled orders for parts due to supply disruptions. The proposed static model can be further enhanced for development of a multi-period decision-making model (for a two-period setting, see Chap. 14) to determine a dynamic supply portfolio to account for time-varying demand, lead times from suppliers, transportation and transshipment times, disruption times and duration of recovery processes, etc. A possible application of a dynamic multi-tier decision-making model for investigation of the ripple effect (e.g., Ivanov 2018) in supply chains could be a further area of the future research.

Problems

13.1. Modify the SMIP models presented in this chapter to account for recovery process of impacted primary suppliers and assembly plants.

13.2. Enhance the SMIP models developed in this chapter to improve the resilience of a multi-tier supply chain network by fortification of primary suppliers and primary assembly plants with multiple fortification levels available for selection.

13.3. Try to identify a threshold for the maximum fortification investment cost that cannot be exceeded to profitably fortify a primary assembly plant.

13.4. Try to identify a threshold for the minimum unit penalty cost for unfulfilled demand for products required to profitably fortify a primary assembly plant.

13.5. Explain whether or not the threshold g_{min} (13.48) is valid for both risk-neutral and risk-averse decision-making.

Chapter 14
Selection of Resilient Multi-Tier Supply Portfolio: A Two-Period Approach

14.1 Introduction

In this chapter the static (single-period) decision-making model proposed in Chap. 13 for selection of resilient multi-tier supply portfolio is enhanced for a dynamic (two-period) setting. The planning horizon is divided into two aggregate periods: pre-disruption Period 1 and post-disruption Period 2. Using the two-period approach developed in Chap. 11 for selection of supply and demand portfolios in a two-tier supply chain, a non-trivial enhancement for a multi-tier supply chain network is proposed in this chapter. The two-period model supports the integrated decision-making in a multi-tier supply chain network to be implemented before a disruptive event to hedge against all potential disruption scenarios and for each scenario in the aftermath period. The decisions to be made ahead of time include selection of primary suppliers at different tiers and assembly plants, allocation among them orders quantity, and production volumes and pre-positioning of risk mitigation inventory. Simultaneously, for each disruption scenario the aftermath decisions are made: recovery suppliers and assembly plants are selected, unfulfilled demands allocated among them, and transshipment of parts and the level of usage pre-positioned risk mitigation inventory are determined. The following two-period SMIP models are presented in this chapter:

ResMTSupport2_E(c) two-period model for risk-neutral selection of resilient multi-tier supply portfolio to minimize expected cost;

ResMTSupport2_E(sl) two-period model for risk-neutral selection of resilient multi-tier supply portfolio to maximize expected service level.

T. Sawik, *Supply Chain Disruption Management*, International Series in Operations Research & Management Science 291,
https://doi.org/10.1007/978-3-030-44814-1_14

Numerical examples, computational results, and major managerial insights drawn from the theoretical and computational study presented in this chapter are summarized in Section 14.6.

14.2 Problem Description

Table 14.1 Notation: resilient multi-tier supply portfolio, two-period approach

	Indices
k	$=$ supply chain tier, $k \in K = \{0,1,\ldots,n\}$
i	$=$ supplier ($k \geq 1$) or assembly plant ($k=0$), $i \in I_k = \{1,\ldots,m_k\}$
r	$=$ geographic region, $r \in R = \{1,\ldots,\overline{R}\}$
s	$=$ disruption scenario, $s \in S = \{1,\ldots,\overline{S}\}$
	Input parameters
D	$=$ total demand for parts/products
T	$=$ planning horizon
c_{ik}	$=$ per period capacity of supplier i of tier k (plant i, if $k=0$)
e_{ik}	$=$ cost of ordering parts from supplier i of tier k (setup cost for plant i, if $k=0$)
f_{ik}	$=$ unit price of parts pre-positioned at supplier i of tier k (plant i, if $k=0$)
g	$=$ unit penalty cost of unfulfilled demand for products
h_{ik}	$=$ unit cost of pre-positioning risk mitigation inventory of parts from tier $k+1$ at primary supplier $i \in I_k$ of tier k (plant i, if $k=0$)
o_{ik} ($\overline{o}_{ik}$)	$=$ unit price of parts purchased from primary (recovery) supplier i of tier k
p_{ik}	$=$ local disruption probability for supplier i of tier k
p^r	$=$ regional disruption probability for geographic region g
t_s	$=$ start time period of disruption under scenario s
ψ_{ijk}	$=$ unit cost of transshipment parts from supplier $i \in I_k$ to supplier $j \in I_k$ of tier k (plant, if $k=0$)
τ_{ijk}	$=$ delivery lead time from supplier $i \in I_{k+1}$ of tier $k+1$ to supplier $j \in I_k$ of tier k (plant, if $k=0$)

This section provides a brief description of the decision-making problem. A multi-tier supply chain is considered with multiple facilities available at each tier and dispersed over multiple geographic regions. Let $R = \{1,\ldots\overline{R}\}$ be the index set of all regions and denote by $I^{(r)}$, the subset of suppliers and/or assembly plants (of the same and/or of different tiers, see Fig. 13.1) in geographic region $r \in R$.

Let $I_k = \{1,\ldots m_k\}$ be the index set of m_k facilities of tier k, and $K = \{0,1,\ldots n\}$, the index set of all tiers. The tiers $k \geq 1$ represent suppliers of critical parts and m_k denotes the number of available suppliers of tier k. Tier $k=0$ represents final manufacturer of finished products and m_0 denotes the number of available assembly plants. The manufacturer sources critical parts from first-tier suppliers. The kth-tier suppliers provides with critical parts the $(k-1)^{th}$-tier suppliers (or assembly plants, if $k=1$).

Without loss of generality, we assume that each tier supplier and the final manufacturer require one unit of input parts from the immediately upper tier supplier to produce one unit of its part and finished product, respectively. Let D be total demand for finished products as well as total demand for critical parts of each tier. Demand for products should be satisfied by the end the planning horizon with a penalty cost charged for total unfulfilled demand and denote by, g, the per unit penalty cost of unfulfilled demand for products (Table 14.1).

The suppliers have different limited capacity and differ in price of offered parts. Let c_{ik} be the per period capacity of supplier (or assembly plant, if $k = 0$), $i \in I_k$ of tier $k \in K$, i.e., the total number of parts (or finished products) that can be manufactured in one period. The purchasing price per part for primary and recovery (expedited) supplies is denoted, respectively, by o_{ik} and $\overline{o}_{ik}$, $i \in I_k$, $k \geq 1$. The fixed cost of ordering parts from supplier $i \in I_k$, $k \geq 1$ is denoted by e_{ik} and by e_{i0} the fixed setup cost for assembly plant $i \in I_0$.

Denote by P_s the probability that disruption scenario s is realized, where each scenario $s \in S$ is comprised of a unique subset $I^s \subset I = \bigcup_{k\in K} I_k$ of facilities that are in state "on" (non-disrupted), and $S = \{1,\ldots,\overline{S}\}$ is the index set of all disruption scenarios, where, $\overline{S} = 2^M$, is the total number of all potential scenarios and, $M = \sum_{k\in K} m_k$, is the total number of all facilities. For each scenario $s \in S$, the facilities $i \in I_k \setminus I^s$, $k \in K$, are disrupted either by a local disruptive event, regional disruptive event or by both events.

The probability P_s of each disruption scenario $s \in S$ is provided below by formulae (14.1) and (14.2) (cf. (13.1) and (13.2) in Sect. 13.2)

$$P_s = \prod_{g\in G} P_s^r, \tag{14.1}$$

where P_s^r is the probability of realizing disruption scenario s for facilities in $I^{(r)}$

$$P_s^r = \begin{cases} (1-p^r)\prod_{k\in K}\prod_{i\in I_k\bigcap I^{(r)}\bigcap I^s}(1-p_{ik})\prod_{i\in I_k\bigcap I^{(r)}\setminus I^s} p_{ik} & \text{if } I^{(r)}\bigcap I^s \neq \emptyset \\ p^r + (1-p^r)\prod_{k\in K}\prod_{i\in I_k\bigcap I^{(r)}} p_{ik} & \text{if } I^{(r)}\bigcap I^s = \emptyset. \end{cases} \tag{14.2}$$

In the above formulae, p_{ik} is the local disruption probability for facility $i \in I_k$ of tier k, and p^r is the regional disruption probability for geographic region r.

We assume that only a single disruption scenario is realized and once disabled, suppliers and plants are not available for assignment during the planning horizon (time-to-recover is assumed to be longer than the length of planning horizon, T). Disruption to primary suppliers (plants) under scenario $s \in S$ occurs in period t_s. A primary supplier (plant) hit by disruption s is capable of producing parts (products) by period $t_s - 1$ at the latest.

The resilience of a multi-tier supply chain can be improved by pre-positioning of risk mitigation inventory at primary suppliers of parts at different tiers that will hedge against all disruption scenarios. We assume that the manufacturer of finished products can carry risk mitigation inventory of parts from first-tier suppliers, while k^{th}-tier suppliers ($k < n$) can carry risk mitigation inventory of parts from $(k+1)^{th}$-

tier suppliers. The risk mitigation inventory is used to compensate for the loss of capacity of the suppliers at immediately higher tier, hit by disruptions, and to fulfill non-delivered orders placed on the disrupted suppliers. Depending on the realized disruption scenario, the pre-positioned inventory can be used fully, partially or not at all. The risk mitigation inventory pre-positioned at a supplier (or a plant) is limited, e.g., by available budget, by available storage space, etc. The inventory is replenished once it is used and required parts can be purchased at higher prices on the spot market. We assume that the inventory quantity is linked to capacity and cannot be greater than the supplier (or plant) full capacity, Tc_{ik}. Denote by, h_{ik}, unit cost of pre-positioning risk mitigation inventory of parts from tier $k+1$ at supplier $i \in I_k$ of tier k, and by, f_{ik}, unit purchasing price of those parts.

The basic assumptions used to formulate the decision-making problem are summarized below.

- Suppliers have different limited capacity per period and differ in price of offered parts.
- Suppliers of the same tier may share their immediate upper tier suppliers, which may result in overlap in the higher tiers.
- The highest tier, $k = n$, suppliers use their own raw materials to manufacture critical parts required for suppliers of tier $k = n-1$.
- Suppliers of each tier, $k < n$, and the final manufacturer are assumed to require one unit of input critical parts from the immediately upper tier supplier to produce one unit of its part and finished product, respectively.
- Primary supplies from non-disrupted to disrupted suppliers (or plants) are considered fulfilled and parts delivered to disrupted suppliers (or plants) can be later transshipped to recovery suppliers (or plants).
- Primary suppliers of each tier, $k < n$, hold regular inventory of parts from the immediately upper tier, sufficient to continue production during maximum lead time from the upper tier primary suppliers. The regular inventory is replenished by primary supplies.
- The regional disruptive events in each region and the local disruptive events are independent events.
- At most a single disruptive event may occur only.
- Time-to-recover is longer than the length of the planning horizon. Hence, the recovery of disabled suppliers and plants is not considered.
- Recovery suppliers include non-disrupted primary suppliers.
- For every tier of the supply chain network and each disruption scenario, all parts supplied to non-disrupted facilities from immediately upper tier are processed and delivered to immediately lower tier. In particular, parts delivered from tier $k = 1$ to non-disrupted assembly plants of tier $k = 0$ are processed into finished products and delivered to customers.
- The risk mitigation inventory pre-positioned at a supplier (or a plant) is linked to its capacity and cannot be greater than the supplier (or plant) full capacity.
- Depending on realized disruption scenario, the pre-positioned risk mitigation inventory can be used fully, partially or not at all.

- The risk mitigation inventory is replenished once it is used and required parts can be purchased at higher prices on the spot market.
- Demand for products is period-independent and should be satisfied by the end the planning horizon with a penalty cost charged for total unfulfilled demand.

We propose a stylized, two-period decision-making model, where the planning horizon is divided into two aggregate periods: pre-disruption Period 1 and post-disruption Period 2. The decision maker selects ahead of time primary suppliers and assembly plants, allocates among them orders quantity and pre-positions risk mitigation inventory to hedge against all potential disruption scenarios. Simultaneously, the decision maker selects recovery suppliers and assembly plants for each disruption scenario, allocates among them unfulfilled orders quantity, and decides on transshipment of parts and usage of pre-positioned inventory to minimize expected cost or maximize expected service.

14.3 Models for Risk-Neutral Decision-Making

Table 14.2 Problem variables: resilient multi-tier supply portfolio, two-period approach

	First stage variables
u_{ik}	= 1, if supplier $i \in I_k$ of tier k (plant, if $k = 0$) is selected as primary supplier (plant); otherwise $u_{ik} = 0$ (primary supplier or plant selection)
v_{ijk}	= the fraction of total demand for parts ordered from primary supplier $i \in I_{k+1}$ of tier $k+1$ to primary supplier $j \in I_k$ of tier k (plant, if $k = 0$) (primary supply portfolio)
x_{ik}	= risk mitigation inventory of parts manufactured at tier $k+1$ pre-positioned at primary supplier $i \in I_k$ of tier k (plant, if $k = 0$), in fraction of its total capacity, Tc_{ik} (risk mitigation inventory allocation)
	Second stage variables
U^s_{ik}	= 1, if supplier $i \in I_k$ of tier k (plant, if $k = 0$) is selected as recovery supplier (plant) under disruption scenario s; otherwise $U^s_{ik} = 0$ (recovery supplier or plant selection)
V^s_{ijk}	= the fraction of total demand for parts ordered under disruption scenario s from recovery supplier $i \in I_{k+1}$ of tier $k+1$ to non-disrupted supplier $j \in I_k$ of tier k (plant, if $k = 0$) (recovery supply portfolio)
w^s_{ijk}	= transshipment of parts of tier $k+1$, from disabled primary supplier $i \in I_k$ of tier k (plant, if $k = 0$) to non-disrupted supplier (plant, if $k = 0$) $j \in I_k$ of the same tier under disruption scenario s (recovery transshipment portfolio)
y^s_{ik}	= risk mitigation inventory of parts pre-positioned at primary supplier $i \in I_k$ of tier k (plant, if $k = 0$), used under disruption scenario s (risk mitigation inventory usage)
z^s	= unfulfilled demand for products under disruption scenario s
ρ^s_{ijk}	= the fraction of total demand for parts produced by primary supplier $i \in I_{k+1}$ of tier $k+1$ before disruption under scenario s and delivered to supplier $j \in I_k$ of tier k (plant, if $k = 0$) (supplies of parts by primary suppliers in pre-disruption Period 1)
ρ^s_i	= the fraction of total demand for products manufactured by primary plant $i \in I_0$ before disruption under scenario s (production of products by primary plants in pre-disruption Period 1)

In this section two two-period SMIP models **ResMTSupport2_E(c)** and **ResMT-Support2_E(sl)** are presented for selection of resilient supply portfolio in a multi-tier supply chain under disruption risks to minimize total expected cost over the entire supply chain and maximize expected service level of final manufacturer, respectively.

For each disruption scenario $s \in S$, the planning horizon T is divided into two aggregate periods:

1. Pre-disruption Period 1, from time 0 through time t_s when the disruption occurs;
2. Post-disruption Period 2, from time $t_s + 1$ until the end of the planning horizon, T.

The integrated decisions for the entire planning horizon and all potential disruption scenarios are made at the beginning of Period 1, i.e., at time 0.

The resilient portfolio aims at reducing potential losses from non-delivered parts through:

1. Optimal selection of primary suppliers and assembly plants and primary flows of part orders as well as pre-position of risk mitigation inventory to be implemented in Period 1 (first stage decisions).
2. Optimal selection of recovery suppliers and assembly plants and recovery flow of parts, transshipment of parts, and usage of risk mitigation inventory for each disruption scenario $s \in S$, to be implemented in Period 2 (second stage decisions).

If a primary supplier of tier $k+1$ is hit by disruption and fails to deliver all ordered parts to tier k, then non-delivered amount of parts can be fully or partially replaced by:

- recovery supplies from recovery suppliers of tier $k+1$ to recovery suppliers of tier k;
- transshipment of parts from disabled primary suppliers of tier k (or plants, if $k = 0$), to recovery suppliers (or plants, if $k = 0$) of the same tier;
- usage of risk mitigation inventory pre-positioned at non-disrupted primary suppliers of tier k.

The primary supplies of parts between primary facilities, $i, j \in I^s$, that are not disabled by disruption under scenario s, are continued over the two aggregate Periods 1 and 2, until completion, while the other supplies can be discontinued in planning period t_s without completion. Then, the uncompleted primary supplies should be replaced in Period 2 by recovery supplies and risk mitigation inventory.

For each primary supplier (primary plant), we assume that the regular inventory of parts allows production in Period 1 to be started at time 0. Depending on supplies, production in Period 1 can be continued until period $t_s - 1$ or t_s, if the supplier is or is not hit by disruption s, respectively. The parts delivered in Period 1 to suppliers or plants disrupted in period t_s can be later (in Period 2) transshipped to recovery facilities.

In model **ResMTSupport2_E(c)** the overall quality of the resilient supply portfolio will be measured by the total expected cost per product over the entire supply chain, E^c (for definition of first and second stage variables of the two-stage stochastic program, see Table 14.2)

$$E^c = \sum_{k\in K}\sum_{i\in I_k} e_{ik}(u_{ik} + \sum_{s\in S: i\in I^s} P_s(U_{ik}^s - u_{ik}))/D$$
$$+ \sum_{k\in K:k<n}\sum_{j\in I_k}(\sum_{i\in I_{k+1}} o_{i,k+1}v_{ijk} - \sum_{s\in S}\sum_{i\in I_{k+1}\setminus I^s} P_s o_{i,k+1}(v_{ijk} - \rho_{ijk}^s))$$
$$+\sum_{s\in S}\sum_{k\in K:k<n}\sum_{i\in I_{k+1}\bigcap I^s}\sum_{j\in I_k\bigcap I^s} P_s\overline{e}_{i,k+1}V_{ijk}^s$$
$$+ \sum_{k\in K:k<n}\sum_{i\in I_k} Tc_{ik}(h_{ik}x_{ik} + \sum_{s\in S}\sum_{i\in I^s} P_s f_{ik}y_{ik}^s)/D$$
$$+\sum_{s\in S}\sum_{k\in K:k<n}\sum_{i\in I_k\setminus I^s}\sum_{j\in I_k\bigcap I^s} P_s\psi_{ijk}w_{ijk}^s$$
$$+g(1-E^{sl}), \tag{14.3}$$

where

$$E^{sl} = 1 - \sum_{s\in S} P_s z^s, \tag{14.4}$$

denotes expected service level, i.e., the expected fraction of fulfilled demand for products.

The expected cost consists of different fixed and variable cost.

The fixed cost per product includes the expected cost of ordering parts from primary and recovery suppliers as well as setup cost of primary and recovery assembly plants (double charge of ordering/setup cost is eliminated, if primary supplier/plant is selected as a recovery supplier/plant),

$$\sum_{k\in K}\sum_{i\in I_k} e_{ik}(u_{ik} + \sum_{s\in S: i\in I^s} P_s(U_{ik}^s - u_{ik}))/D.$$

The variable cost per product includes the expected cost of:

(a) purchasing parts from primary suppliers less cost of ordered and non-delivered parts,

$$\sum_{k\in K:k<n}\sum_{j\in I_k}(\sum_{i\in I_{k+1}} o_{i,k+1}v_{ijk} - \sum_{s\in S}\sum_{i\in I_{k+1}\setminus I^s} P_s o_{i,k+1}(v_{ijk} - \rho_{ijk}^s));$$

(b) purchasing parts from recovery suppliers,

$$\sum_{k\in K:k<n}\sum_{s\in S}\sum_{i\in I_{k+1}\bigcap I^s}\sum_{j\in I_k\bigcap I^s} P_s\overline{o}_{i,k+1}V_{ijk}^s;$$

(c) pre-positioning risk mitigation inventory of parts,

$$\sum_{k\in K:k<n}\sum_{i\in I_k} h_{ik}Tc_{ik}x_{ik}/D;$$

(d) usage pre-positioned risk mitigation inventory of parts,

$$\sum_{s\in S}\sum_{k\in K:k<n}\sum_{i\in I_k\bigcap I^s} P_s f_{ik} T c_{ik} y^s_{ik}/D;$$

(e) transshipment of parts,

$$\sum_{s\in S}\sum_{k\in K:k<n}\sum_{i\in I_k\setminus I^s}\sum_{j\in I_k\bigcap I^s} P_s \psi_{ijk} w^s_{ijk};$$

(f) unfulfilled demand for products due to supply disruptions,

$$g\sum_{s\in S} P_s z^s = g(1-E^{sl}).$$

The stochastic mixed integer program **ResMTSupport2_E(c)** is formulated below.

ResMTSupport2_E(c): *Selection of* **Res***ilient* **M***ulti*-**T***ier* **Supp***ly* **port***folio:* **2***-period approach to minimize* **E***xpected* **c***ost*

Minimize (14.3)

subject to

$$v_{ijk} \le u_{i,k+1};\ k\in K, i\in I_{k+1}, j\in I_k : k<n \quad (14.5)$$

$$v_{ijk} \le u_{jk};\ k\in K, i\in I_{k+1}, j\in I_k : k<n \quad (14.6)$$

$$\sum_{i\in I_{k+1}}\sum_{j\in I_k} v_{ijk} = 1; k\in K : k<n \quad (14.7)$$

$$\sum_{j\in I_{k-1}} v_{ij,k-1} = \sum_{j\in I_{k+1}} v_{jik};\ k\in K, i\in I_k : 0<k<n \quad (14.8)$$

$$\sum_{j\in J_k} v_{ijk} \le Tc_{i,k+1}u_{i,k+1}/D;\ k\in K, i\in I_{k+1} : k<n \quad (14.9)$$

$$\sum_{i\in I_1} v_{ij0} \le Tc_{j0}u_{j0}/D;\ j\in I_0 \quad (14.10)$$

$$U^s_{ik} \ge u_{ik};\ s\in S, k\in K, i\in I_k\bigcap I^s \quad (14.11)$$

$$V^s_{ijk} \le U^s_{i,k+1};\ s\in S, k\in K, i\in I_{k+1}\bigcap I^s, j\in I_k\bigcap I^s : k<n \quad (14.12)$$

$$V^s_{ijk} \le U^s_{jk};\ s\in S, k\in K, i\in I_{k+1}\bigcap I^s, j\in I_k\bigcap I^s : k<n \quad (14.13)$$

$$V^s_{ijk} \le \sum_{l\in I_{k+1}\setminus I^s} u_{l,k+1} + \sum_{l\in I_k\setminus I^s} u_{lk};$$

$$s\in S, k\in K, i\in I_{k+1}\bigcap I^s, j\in I_k\bigcap I^s : k<n \quad (14.14)$$

(continued)

$$w^s_{ijk} \le u_{ik};\ s \in S, k \in K, i \in I_k \setminus I^s, j \in I_k \bigcap I^s : k < n \quad (14.15)$$

$$w^s_{ijk} \le U^s_{jk};\ s \in S, k \in K, i \in I_k \setminus I^s, j \in I_k \bigcap I^s : k < n \quad (14.16)$$

$$x_{ik} \le u_{ik};\ k \in K, i \in I_k : k < n \quad (14.17)$$

$$y^s_{ik} \le x_{ik};\ s \in S, k \in K, i \in I_k : k < n \quad (14.18)$$

$$\sum_{i \in I_k \bigcap I^s} T c_{ik} y^s_{ik}/D + \sum_{i \in I_{k+1} \bigcap I^s} \sum_{j \in I_k \bigcap I^s} V^s_{ijk} \le \sum_{i \in I_{k+1} \setminus I^s} \sum_{j \in I_k} (v_{ijk} - \rho^s_{ijk});\ s \in S, k \in K : k < n \quad (14.19)$$

$$\sum_{j \in I_k \bigcap I^s} w^s_{ijk} \le \sum_{j \in I_{k+1} : \tau_{jik} \le t_s} \rho^s_{jik} - (\text{if } k > 0 \text{ then } \sum_{j \in I_{k-1}} \rho^s_{i,j,k-1} \text{ else } \rho^s_i);\ s \in S, k \in K, i \in I_k \setminus I^s : k < n \quad (14.20)$$

$$\rho^s_{ijk} \le v_{ijk};\ s \in S, k \in K, i \in I_{k+1}, j \in I_k : k < n \quad (14.21)$$

$$\sum_{j \in I_k} \rho^s_{ijk} \le (\text{if } i \in I_{k+1} \bigcap I^s \text{ then } t_s \text{ else if } i \in I_{k+1} \setminus I^s \text{ then} (t_s - 1)) \times c_{i,k+1} u_{i,k+1}/D;\ s \in S, k \in K, i \in I_{k+1} : k < n \quad (14.22)$$

$$\rho^s_i \le \sum_{j \in I_1} v_{ji0};\ s \in S, i \in I_0 \quad (14.23)$$

$$\rho^s_i \le (\text{if } i \in I_0 \bigcap I^s \text{ then } t_s \text{ else if } i \in I_0 \setminus I^s \text{ then } (t_s - 1)) \times c_{i0} u_{i0}/D;\ s \in S, i \in I_0 \quad (14.24)$$

$$\sum_{j \in I_{k+1} \bigcap I^s} v_{jik} + \sum_{j \in I_{k+1} \setminus I^s} \rho^s_{jik} + \sum_{j \in I_{k+1} \bigcap I^s} V^s_{jik} + \sum_{j \in I_k \setminus I^s} w^s_{jik} + T c_{ik} y^s_{ik}/D \le ((T - t_s) U^s_{ik} + t_s u_{ik}) c_{ik}/D;\ s \in S, k \in K, i \in I_k \bigcap I^s : k < n \quad (14.25)$$

$$\sum_{j \in I_{k-1} \bigcap I^s} V^s_{ij,k-1} = \sum_{j \in I_{k+1} : \tau_{jik} \ge t_s} \rho^s_{jik} + \sum_{j \in I_{k+1} \bigcap I^s} V^s_{jik} + \sum_{j \in I_k \setminus I^s} w^s_{jik} + T c_{ik} y^s_{ik}/D;\ s \in S, k \in K, i \in I_k \bigcap I^s : 0 < k < n \quad (14.26)$$

$$\sum_{i \in I_0 \setminus I^s} \rho^s_i + \sum_{i \in I_1 \setminus I^s} \sum_{j \in I_0 \bigcap I^s} \rho^s_{ij0} + \sum_{i \in I_1 \bigcap I^s} \sum_{j \in I_0 \bigcap I^s} v_{ij0} + \sum_{i \in I_1 \bigcap I^s} \sum_{j \in I_0 \bigcap I^s} V^s_{ij0} + \sum_{i \in I_0 \setminus I^s} \sum_{j \in I_0 \bigcap I^s} w^s_{ij0} + \sum_{i \in I_0 \bigcap I^s} T c_{i0} y^s_{i0}/D + z^s = 1;\ s \in S \quad (14.27)$$

$$u_{ik} \in \{0,1\};\ k \in K, i \in I_k \quad (14.28)$$

(continued)

$$v_{ijk} \in [0,1];\ k \in K, i \in I_{k+1}, j \in I_k : k < n \quad (14.29)$$

$$U^s_{ik} \in [0,1];\ s \in S, k \in K, i \in I_k \bigcap I^s \quad (14.30)$$

$$V^s_{ijk} \in [0,1];\ s \in S, k \in K, i \in I_{k+1} \bigcap I^s, j \in I_k \bigcap I^s : k < n \quad (14.31)$$

$$w^s_{ijk} \in [0,1];\ s \in S, k \in K, i \in I_k \setminus I^s, j \in I_k \bigcap I^s : k < n \quad (14.32)$$

$$x_{ik} \in [0,1];\ k \in K, \in I_k : k < n \quad (14.33)$$

$$y^s_{ik} \in [0,1];\ s \in S, k \in K, i \in I_k : k < n \quad (14.34)$$

$$z^s \geq 0;\ s \in S \quad (14.35)$$

$$\rho^s_{ijk} \geq 0;\ s \in S, k \in K, i \in I_{k+1}, j \in I_k : k < n \quad (14.36)$$

$$\rho^s_i \geq 0;\ s \in S, i \in I_0. \quad (14.37)$$

The following five groups of constraints are introduced in model **ResMTSupport2_E(c)**.

1. Primary supply portfolio selection constraints (14.5)–(14.10):
   - parts can be ordered from (14.5) and delivered to (14.6) selected primary suppliers only,
   - the total demand for parts at each tier must be fully met by primary supplies from the immediately higher tier suppliers, i.e., the primary supply portfolio must be selected, (14.7),
   - total primary supplies of parts manufactured by primary supplier of tier $0 < k < n$ to primary suppliers of tier $k-1$ is in balance with primary supplies of parts to this supplier from primary suppliers of tier $k+1$, (14.8), (primary flow conservation equations),
   - the capacity of each primary supplier, ($k > 0$), (14.9), and final manufacturer, ($k = 0$), (14.10), cannot be exceeded.
2. Recovery supply portfolio selection constraints (14.11)–(14.16):
   - a non-disrupted primary supplier is selected as a recovery supplier, (14.11),
   - recovery flows of parts can be realized between non-disrupted facilities only and cannot be sourced from non-selected recovery suppliers, (14.12) and sunk at non-selected recovery suppliers (plants, if $k = 0$), (14.13),
   - recovery supplies from tier $k+1$ to tier k are not required, when primary suppliers at both tiers k and $k+1$ were not disrupted, (14.14),
   - the parts from disabled primary suppliers can be transshipped to recovery suppliers only, (14.15), (14.16).
3. Pre-positioning and usage of risk mitigation inventory constraints (14.17)–(14.20):
   - risk mitigation inventory can be pre-positioned at selected primary suppliers only, (14.17),

- the risk mitigation inventory used to replace non-delivered parts cannot exceed the pre-positioned inventory, (14.18),
- for each tier k and disruption scenario s, the unfulfilled demand for parts from tier $k+1$ can be fully or partially replaced by the pre-positioned inventory and recovery supplies, (14.19),
- total transshipment of parts from primary supplier of tier k disrupted at t_s cannot exceed remaining inventory of parts delivered by primary suppliers of tier $k+1$ by t_s, and unprocessed in Period 1, (14.20), (transshipment constraints).

4. Production planning constraints (14.21)–(14.25):

- total production in Period 1 of primary supplier of tier $k+1$ to be shipped to suppliers of tier k, is limited by: (14.21), planned primary supplies, v_{ijk}; (14.22), supplier capacity available by period, t_s (for non-disrupted supplier) or $t_s - 1$ (for disrupted supplier),
- total production in Period 1 of primary plant is limited by: (14.23),planned primary supplies from tier 1; (14.24), plant capacity available by period, t_s (for non-disrupted plant) or $t_s - 1$ (for disrupted plant),
- total production over the entire planning horizon by a non-disrupted supplier or plant, $i \in I_k \bigcap I^s$, (recovery only, if $U^s_{ik} = 1$, $u_{ik} = 0$, or both primary and recovery, if $U^s_{ik} = u_{ik} = 1$, cf. (14.11)) is limited by its total capacity, $((T - t_s)U^s_{ik} + t_s u_{ik})c_{ik}$, and is determined by primary supplies from non-disabled suppliers, primary supplies of parts manufactured in Period 1 by disabled suppliers, recovery and transshipment supplies plus used up risk mitigation inventory, if $U^s_{ik} = u_{ik} = 1$, or by recovery and transshipment supplies only, if $U^s_{ik} = 1$, $u_{ik} = 0$, (14.25).

5. Recovery flow balance constraints (14.26)–(14.27):

- total recovery supplies of parts from recovery supplier of tier k to recovery suppliers of tier $k-1$ are in balance with supplies of parts manufactured in Period 1 by primary suppliers of tier $k+1$ and delivered in Period 2, recovery supplies by recovery suppliers of tier $k+1$, transshipment of parts from disrupted primary suppliers of tier k and usage of risk mitigation inventory pre-positioned at this supplier, (14.26), (recovery flow conservation equations),
- for each disruption scenario s, the fulfilled demand for products, $(1 - z^s)$, is equal to production and supplies in Period 1, respectively, by disrupted primary plants and from disrupted primary suppliers of tier 1, primary and recovery supplies of parts from tier 1 and transshipment of parts and usage of pre-positioned inventory at tier 0, (14.27), (demand fulfillment constraints).

Proposition 14.1. *Model* **ResMTSupport2_E(c)** *contains an embedded network flow problem.*

Proof. In model **ResMTSupport2_E(c)**, the primary, recovery, and transshipment portfolio selection constraints form an embedded network flow problem (see, Fig. 14.1) where:

- v_{ijk} represents inter-tier primary flow of parts from primary supplier $i \in I_{k+1}$ of tier $k+1$ (primary source node) to primary supplier (or assembly plant, if $k = 0$), $j \in I_k$ of tier k (primary sink node);
- ρ^s_{ijk} and ρ^s_i represent inter-tier primary flow of parts and primary flow of products manufactured in Period 1 (before disruption) by primary suppliers and plants, respectively,
- V^s_{ijk} represents inter-tier recovery flow of parts from recovery supplier $i \in I_{k+1}$ of tier $k+1$ (recovery source node) to recovery supplier (or assembly plant, if $k = 0$), $j \in I_k$ of tier k (recovery sink node);
- w^s_{ijk} represents intra-tier transshipment flow of parts from disrupted primary supplier $i \in I_k \setminus I^s$ of tier k (transshipment source node) to recovery supplier (or assembly plant, if $k = 0$), $j \in I_k$ of the same tier (transshipment sink node).

(continued)

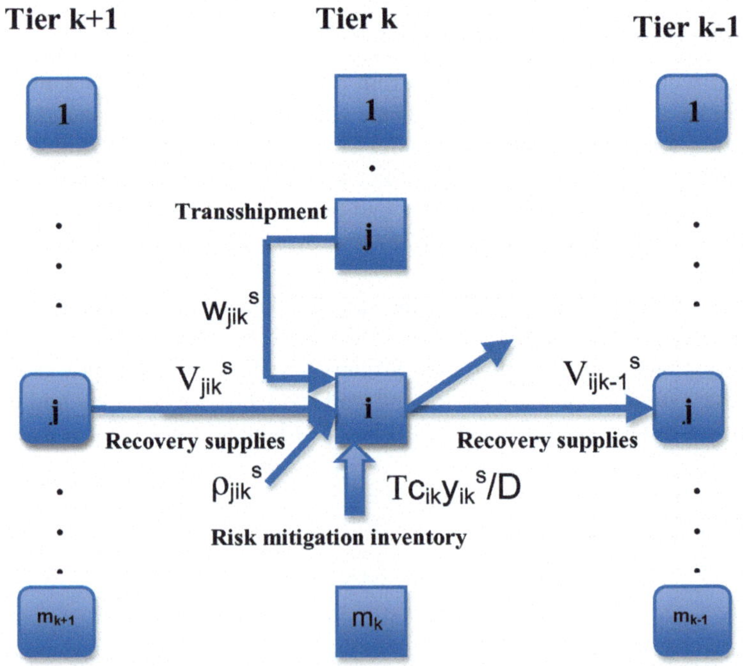

Fig. 14.1 Recovery flows of parts

Equations (14.8) and (14.26) are primary and recovery flow conservation equations, respectively, and (14.27) are demand for products fulfillment constraints.

We assume that for every tier of supply chain and each disruption scenario, all parts supplied to non-disrupted facilities from immediately upper tier are processed and delivered to immediately lower tier. In particular parts delivered from tier $k = 1$ to non-disrupted assembly plants of tier $k = 0$ are processed into finished products and delivered to customers. Note that the nonnegative product shortage variables z^s can be eliminated from the model using demand fulfillment constraints (14.27). Then, variables z^s in the objective functions (14.3) and (14.4) should be replaced by

$$1 - \sum_{i\in I_0\setminus I^s} \rho_i^s - \sum_{i\in I_1\setminus I^s} \sum_{j\in I_0\bigcap I^s} \rho_{ij0}^s - \sum_{i\in I_1\bigcap I^s} \sum_{j\in I_0\bigcap I^s} v_{ij0}$$
$$\sum_{i\in I_1\bigcap I^s} \sum_{j\in I_0\bigcap I^s} V_{ij0}^s - \sum_{i\in I_0\setminus I^s} \sum_{j\in I_0\bigcap I^s} w_{ij0}^s - \sum_{i\in I_0\bigcap I^s} Tc_{i0}y_{i0}^s/D,$$

and equality constraints (14.27), replaced by inequality constraints

$$\sum_{i\in I_0\setminus I^s} \rho_i^s + \sum_{i\in I_1\setminus I^s} \sum_{j\in I_0\bigcap I^s} \rho_{ij0}^s + \sum_{i\in I_1\bigcap I^s} \sum_{j\in I_0\bigcap I^s} v_{ij0}$$
$$+ \sum_{i\in I_1\bigcap I^s} \sum_{j\in I_0\bigcap I^s} V_{ij0}^s + \sum_{i\in I_0\setminus I^s} \sum_{j\in I_0\bigcap I^s} w_{ij0}^s + \sum_{i\in I_0\bigcap I^s} Tc_{i0}y_{i0}^s/D \le 1;\ s \in S.$$

The product shortage variables z^s are slack variables for the last constraints.

The two propositions derived from the model **ResMTSupport2_E(c)** and presented below provide additional managerial insights.

Proposition 14.2. *The recovery supplies of parts to the assembly plants (the lowest tier, $k = 0$, of supply chain) are strictly dependent on recovery supplies of parts from the highest tier, $k = n$.*

$$\sum_{i\in I_1\bigcap I^s} \sum_{j\in I_0\bigcap I^s} V_{ij0}^s = \sum_{i\in I_n\bigcap I^s} \sum_{j\in I_{n-1}\bigcap I^s} V_{ij,n-1}^s$$
$$+ \sum_{k\in K:0<k<n} \sum_{i\in I_k\setminus I^s} \sum_{j\in I_k\bigcap I^s} w_{ijk}^s + \sum_{k\in K:0<k<n} \sum_{i\in I_k\bigcap I^s} Tc_{ik}y_{ik}^s/D$$
$$+ \sum_{k\in K:0<k<n} \sum_{i\in I_{k+1}:\tau_{ijk}\ge t_s} \sum_{j\in I_k\bigcap I^s} \rho_{ijk}^s;\ s \in S. \quad (14.38)$$

Equation (14.38) indicates that for each disruption scenario s, recovery supplies of parts from tier $k = 1$ to the lowest tier $k = 0$ are in balance

(continued)

with recovery supplies of parts from the highest tier $k = n$ to tier $k = n-1$ and transshipment of parts and usage of pre-positioned inventory at all intermediate tiers $k = 1,\ldots,n-1$, plus supplies of parts manufactured in Period 1 and delivered in Period 2.

Proof. Equation (14.38) can be achieved by summing up recovery flow conservation equations (14.26) over $i \in I_k \bigcap I^s$ and then substituting, $\sum_{i\in I_{k+1}\bigcap I^s}\sum_{j\in I_k\bigcap I^s} V_{ijk}^s$, in the right-hand side of Eq. (14.26), iteratively for $k = 1,\ldots,n-1$.

The implied constraint (14.38) can be added to strengthen model **ResMTSupport2_E(c)**.

The following **Proposition 14.3** indicates that the cost of improving supply chain resilience by pre-positioning of risk mitigation inventory cannot be greater than potential losses due to unfulfilled demand for products with no risk mitigation inventory. If, owing to usage of risk mitigation inventory, expected reduction of penalty cost for unfulfilled demand for products is greater than the expected cost of pre-positioning and using the inventory, then pre-positioning of the inventory is profitable. **Proposition 14.3** defines a threshold for unit penalty of unfulfilled demand for products above which the risk mitigation inventory is pre-positioned.

Proposition 14.3. *If unit penalty for unfulfilled demand for products, g, is greater than $\underline{g}$, then pre-position of risk mitigation inventory at assembly plants (the lowest tier, $k = 0$, of supply chain) is selected, where*

$$\underline{g} = \frac{\sum_{i\in I_0} c_{i0}h_{i0}x_{i0} + \sum_{s\in S}\sum_{i\in I_0\bigcap I^s} P_s c_{i0} f_{i0} y_{i0}^s}{\sum_{s\in S}\sum_{i\in I_0\bigcap I^s} P_s c_{i0} y_{i0}^s}. \tag{14.39}$$

If $f_{i0} = f_0, h_{i0} = h_0; i \in I_0$, then

$$\underline{g} = f_0 + h_0 \frac{\sum_{i\in I_0} c_{i0}x_{i0}}{\sum_{s\in S}\sum_{i\in I_0\bigcap I^s} P_s c_{i0} y_{i0}^s}.$$

Proof. Denote by, $z_{x=0}^s$ and $z_{x>0}^s$, unfulfilled demand under scenario s, when the risk mitigation inventory is not pre-positioned and is pre-positioned, respectively. According to demand for products fulfillment constraints (14.27)

$$z_{x=0}^s = 1 - \sum_{i\in I_0\setminus I^s}\rho_i^s - \sum_{i\in I_1\setminus I^s}\sum_{j\in I_0\bigcap I^s}\rho_{ij0}^s - \sum_{i\in I_1\bigcap I^s}\sum_{j\in I_0\bigcap I^s} v_{ij0}$$
$$\sum_{i\in I_1\bigcap I^s}\sum_{j\in I_0\bigcap I^s} V_{ij0}^s - \sum_{i\in I_0\setminus I^s}\sum_{j\in I_0\bigcap I^s} w_{ij0}^s;\ s \in S,$$

(continued)

$$z^s_{x>0} = 1 - \sum_{i \in I_0 \setminus I^s} \rho^s_i - \sum_{i \in I_1 \setminus I^s} \sum_{j \in I_0 \bigcap I^s} \rho^s_{ij0} - \sum_{i \in I_1 \bigcap I^s} \sum_{j \in I_0 \bigcap I^s} v_{ij0}$$

$$\sum_{i \in I_1 \bigcap I^s} \sum_{j \in I_0 \bigcap I^s} V^s_{ij0} - \sum_{i \in I_0 \setminus I^s} \sum_{j \in I_0 \bigcap I^s} w^s_{ij0} - \sum_{i \in I_0 \bigcap I^s} Tc_{i0} y^s_{i0}/D;\ s \in S.$$

The expected reduction of penalty cost for unfulfilled demand for products owing to usage of risk mitigation inventory is

$$g \sum_{s \in S} P_s (z^s_{x=0} - z^s_{x>0}) = g \sum_{s \in S} \sum_{i \in I_0 \bigcap I^s} P_s T c_{i0} y^s_{i0}/D.$$

The pre-positioning of risk mitigation inventory is profitable only if the above reduction of penalty cost is greater than the expected cost of pre-positioning and using the inventory, that is, if

$$g \sum_{s \in S} \sum_{i \in I_0 \bigcap I^s} P_s T c_{i0} y^s_{i0} > \sum_{i \in I_0} T c_{i0} h_{i0} x_{i0} + \sum_{s \in S} \sum_{i \in I_0 \bigcap I^s} P_s T c_{i0} f_{i0} y^s_{i0}.$$

Dividing both sides of the above inequality by, $\sum_{s \in S} \sum_{i \in I_0 \bigcap I^s} P_s T c_{i0} y^s_{i0} > 0$, implies the threshold (14.39).

The threshold (14.39) is tight and can be computed by using a feasible solution to model **ResMTSupport2_E(c)** with a nonzero usage of risk mitigation inventory at the lowest tier $k = 0$.

In model **ResMTSupport2_E(sl)** presented below, the overall quality of the resilient supply portfolio will be measured by the expected service level, E^{sl} (14.4).

ResMTSupport2_E(sl): *Selection of* **Res***ilient* **M***ulti-***T***ier* **Supp***ly* **port***folio:* **2***-period approach to minimize* **E***xpected* **s***ervice* **l***evel*
Minimize (14.4)
subject to (14.5)–(14.37).

The following **Proposition 14.4** indicates that a resilient supply chain is capable of reaching the highest expected service level.

Proposition 14.4. *The expected service level for resilient supply portfolio is bounded by its highest possible value, one.*

$$E^{sl} \leq 1. \qquad (14.40)$$

(continued)

Proof. Equations (14.27), (14.19)–(14.21), and (14.7) imply

$$E^{sl} = 1 - \sum_{s\in S} P_s z^s =$$

$$\sum_{s\in S} P_s\Big(\sum_{i\in I_0\setminus I^s} \rho_i^s + \sum_{i\in I_1\setminus I^s} \sum_{j\in I_0\bigcap I^s} \rho_{ij0}^s + \sum_{i\in I_1\bigcap I^s} \sum_{j\in I_0\bigcap I^s} v_{ij0}$$

$$+ \sum_{i\in I_1\bigcap I^s} \sum_{j\in I_0\bigcap I^s} V_{ij0}^s + \sum_{i\in I_0\bigcap I^s} Tc_{i0}y_{i0}^s/D + \sum_{i\in I_0\setminus I^s} \sum_{j\in I_0\bigcap I^s} w_{ij0}^s\Big) \le$$

$$\sum_{s\in S} P_s\Big(\sum_{i\in I_0\setminus I^s} \rho_i^s + \sum_{i\in I_1\setminus I^s} \sum_{j\in I_0\bigcap I^s} \rho_{ij0}^s + \sum_{i\in I_1\bigcap I^s} \sum_{j\in I_0\bigcap I^s} v_{ij0}$$

$$+ \sum_{i\in I_1\setminus I^s} \sum_{j\in I_0} (v_{ij0} - \rho_{ij0}^s) + \sum_{i\in I_0\setminus I^s} \sum_{j\in I_1:\tau_{ji0}\le t_s} \rho_{ji0}^s - \sum_{i\in I_0\setminus I^s} \rho_i^s\Big) =$$

$$\sum_{s\in S} P_s\Big(\sum_{i\in I_1\bigcap I^s} \sum_{j\in I_0\bigcap I^s} v_{ij0} + \sum_{i\in I_1\setminus I^s} \sum_{j\in I_0} v_{ij0}$$

$$+ \sum_{i\in I_1\setminus I^s} \sum_{j\in I_0\bigcap I^s} \rho_{ij0}^s - \sum_{i\in I_1\setminus I^s} \sum_{j\in I_0} \rho_{ij0}^s + \sum_{i\in I_1:\tau_{ij0}\le t_s} \sum_{j\in I_0\setminus I^s} \rho_{ij0}^s\Big) \le$$

$$\sum_{s\in S} P_s\Big(\sum_{i\in I_1\bigcap I^s} \sum_{j\in I_0\bigcap I^s} v_{ij0} + \sum_{i\in I_1\setminus I^s} \sum_{j\in I_0} v_{ij0} + \sum_{i\in I_1\bigcap I^s} \sum_{j\in I_0\setminus I^s} v_{ij0}\Big) =$$

$$\sum_{s\in S} P_s\Big(\sum_{i\in I_1\bigcap I^s} \sum_{j\in I_0} v_{ij0} + \sum_{i\in I_1\setminus I^s} \sum_{j\in I_0} v_{ij0}\Big) =$$

$$+ \sum_{s\in S} P_s\Big(\sum_{i\in I_1} \sum_{j\in I_0} v_{ij0}\Big) = 1.$$

The above upper bound on the expected service level can be strictly less than one, if recovery supplies are nonavailable and risk mitigation inventory cannot be pre-positioned, i.e., for a non-resilient supply chain.

Notice that Propositions 14.2, 14.3, and 14.4 for the two-period modeling approach are similar or identical with Propositions, respectively, 13.2, 13.5, and 13.3 for a single-period setting. The corresponding formulae (14.38), (14.39), and (14.40) are similar or identical with formulae, respectively (13.33), (13.46), and (13.35).

14.4 Computational Examples

This section provides simple computational examples to demonstrate that the proposed multi-tier portfolio approach and the two-period stochastic MIP formulation with an embedded network flow structure lead to computationally highly efficient decision-making tool. The input data for the examples are similar to that used for computational study in Chap. 13.

Table 14.3 shows basic input parameters for a four-tier supply chain dispersed over three geographic regions, where multiple suppliers/assembly plants of the same and/or different tiers are located in the same geographic region (see, Fig. 13.3). Three different variants (i), (ii), and (iii) of regional disruption probabilities are considered. For each variant, region $r = 1$, where final manufacturers are located, is the most reliable. The more distant is the geographic region r from region 1, the greater its regional disruption probability: $p^3 > p^2 > p^1$. In addition, local disruption probabilities are also higher for suppliers in more vulnerable regions. Finally, less reliable suppliers of the same tier offer lower prices of parts. The correlation between suppliers price and reliability is illustrated in Fig. 13.4, that shows price per part, o_{ik}, and disruption probability, $p^r + (1 - p^r)p_{ik}$, for each supplier $(i,k) : i \in I_k \bigcap I^{(r)}, k \in K, r \in R$.

In order to consider non-trivial supply disruptions, start times of disruptive events are assumed to be not later than the latest delivery time, that is, period 4 (the maximum delivery lead time, τ_{ijk}, is four periods). Hence, $t_s \in \{1,2,3,4\}$, and the total number of all potential scenarios to be considered is $2^M \times 4 = 4096$, since each of the $2^M = 1024$ disruptive events may occur at four different start times, where $M = \sum_{k \in K} m_k = 10$.

Table 14.3 Input parameters: resilient multi-tier supply portfolio, two-period approach

$R = \{1,2,3\}$	three geographic regions
$K = \{0,1,2,3\}$	four-tier supply chain
$m_0 = m_1 = 2,\ m_2 = m_3 = 3$	two assembly plants at tier 0, two suppliers at tier 1, and three suppliers at each tier 2 and 3
$I^{(1)} = \{(1,0),(2,0),(1,1)\}$	in region 1: two plants and supplier 1 of tier 1
$I^{(2)} = \{(2,1),(1,2),(2,2)\}$	in region 2: supplier 2 of tier 1 and suppliers 1 and 2 of tier 2
$I^{(3)} = \{3,2),(1,3),(2,3),(3,3)\}$	in region 3: supplier 3 of tier 2 and all three suppliers of tier 3
$\tau_{ijk} = 2$, if i and j in the same region; else $\tau_{ijk} = 4$	delivery lead times from suppliers
$D = 10000$	total demand for parts/products
$T = 20$	planning horizon
	Disruption probability:
Local disruption probability	$p_{ik},\ k \in K, i \in I_k$ uniformly distributed over [0.05,0.10]
for $(i,k) \in I^{(1)}$, over [0.10,0.15] for $(i,k) \in I^{(2)}$, and over [0.15,0.25] for $(i,k) \in I^{(3)}$	
Regional disruption probability	(i) $p^1 = 0.005,\ p^2 = 0.025,\ p^3 = 0.05$
	(ii) $p^1 = 0.01,\ p^2 = 0.05,\ p^3 = 0.1$
	(iii) $p^1 = 0.05,\ p^2 = 0.25,\ p^3 = 0.5$
$t_s \in \{1,2,3,4\}$	Start time period for disruptive event s
Capacity per period of assembly plants ($k = 0$) and suppliers ($k > 0$):	
$c_{1k} = D/T,\ k \in K,\ c_{ik} = c_{1k}/2,\ k \in K, i \in I_k : i > 1$	
Cost parameters:	
$o_{11} = 26, o_{21} = 25, o_{12} = 14, o_{22} = 15, o_{32} = 13, o_{13} = 5, o_{23} = 6, o_{33} = 4, \overline{o}_{ik} = 1.5 o_{ik}$	
$f_{ik} = f_k = \max_{i \in I_{k+1}}(1.2 o_{i,k+1}),\ k \in K, i \in I_k : k < 3$	
$h_{ik} = h_k = \max_{i \in I_{k+1}}(0.2 o_{i,k+1}),\ k \in K, i \in I_k : k < 3$	
$e_{1k} = 0,\ k \in K$, i.e., no ordering (setup) cost for preselected primary suppliers (plant)	
$e_{ik},\ k \in K, i \in I_k : i > 1$ uniformly distributed over [5D,10D] for $(i,k) \in I^{(1)}$, over [10D,15D] for $(i,k) \in I^{(2)}$, and over [15D,20D] for $(i,k) \in I^{(3)}$	
$\psi_{ijk} = 1,\ k \in K, i \in I_k, j \in I_k : k < 3$	
Unit penalty cost of unfulfilled demand for products for different regional disruption probability	(i) $b = 100.00$ or $b = 128.87$ or $b = \underline{g} = 128.88$ or $b = 150.00$
	(ii) $b = 100.00$ or $b = 129.36$ or $b = \underline{g} = 129.37$ or $b = 150.00$
	(iii) $b = 100.00$ or $b = 133.46$ or $b = \underline{g} = 133.47$ or $b = 150.00$

In order to simplify disruption modeling and analyses, disruption scenarios are numbered in nondecreasing order of their start times: the first 1024 scenarios are associated with $t_s = 1$; the second 1024 scenarios with $t_s = 2$; the third 1024 scenarios with $t_s = 3$; and the fourth 1024 scenarios with $t_s = 4$. Given probabilities P^r_s; $g = 1,2,\ s \le 1024$, (14.2), the probability P_s of realizing each disruption scenario $s \in \{1,\dots,4096\}$ is calculated as follows:

$P_s = \beta_1 P^1_s P^2_s$ for $s \le 1024$,
$P_s = \beta_2 P^1_{s-1024} P^2_{s-1024}$ for $1025 \le s \le 2048$,
$P_s = \beta_3 P^1_{s-2048} P^2_{s-2048}$ for $2049 \le s \le 3072$, and
$P_s = \beta_4 P^1_{s-3072} P^2_{s-3072}$ for $3073 \le s \le 4096$,

Table 14.4 Solution results for regional disruption probability (ii)

Model	**ResMTSupport2_E(c)**				**ResMTSupport2_E(sl)**
Model size[a]	Var. = 12,7530, Bin. = 40,970, Con. = 204,135, Nonz. = 573,891				
Unit penalty, $g =$	100.00	129.36	129.37($=\underline{g}$)	150.00	∞
Exp.Cost, (14.3)	52.99	57.00	57.00	58.74	–
Exp.Service ×100%, (14.4)	86.33	86.34	91.57	91.57	94.59
Primary Portfolio outflow from (i,k)[b], exp. production in Period 1[c]					
(1,0)	1, 0.008				0.650, 0.0670
(2,0)	0				0.350, 0.0010
(1,1)	1, 0.013				0.700, 0.0180
(2,1)	0				0.300, 0.0050
(1,2)	1, 0.006				0.275, 0.0072
(2,2)	0				0.225, 0.0022
(3,2)	0				0.500, 0.0022
(1,3)	1, 0.004				0.275, 0.0001
(2,3)	0				0.500, 0.0001
(3,3)	0				
Exp.Recovery Portfolio outflow from (i,k)[d]					
(1,0)	0	0	0.052	0.052	0.078
(2,0)	0				0.0330
(1,1)	0				0.0110
(2,1)	0				0.0036
(1,2)	0				0
(2,2)	0				0
(3,2)	0				0.0001
Pre-positioned Inventory	–	–	$x_{1,0} = 0.85$	$x_{1,0} = 0.85$	$x_{ik} = 1, k \in K, i \in I_k : k < 3$
Exp.Usage of Inventory[e]	–	–	0.052	0.052	0.084
LP[f]	52.72	56.60	56.60	58.32	94.59

[a] number of variables, binary variables, constraints, nonzero coefficients
[b] $\sum_{j\in I_{k-1}} v_{ij,k-1}$, [c] $\sum_{s\in S}\sum_{j\in I_{k-1}} P_s \rho^s_{ij,k-1}$,
[d] $\sum_{s\in S}\sum_{j\in I_{k-1}} P_s V^s_{ij,k-1}$,
[e] $\sum_{s\in S}\sum_{k\in K:k<3}\sum_{i\in I_k} P_s T c_{ik} y^s_{ik}/D$
[f] LP relaxation solution value of (14.3) and (14.4)

where coefficients, $\beta_1, \beta_2, \beta_3$, and β_4 such that $\beta_1+\beta_2+\beta_3+\beta_4 = 1$ (i.e., $\sum_{s=1}^{4096} P_s = 1$) assigned to probabilities of disruptive events with different start times, respectively, for $t_s = 1$, $t_s = 2$, $t_s = 3$, and $t_s = 4$, reflect likelihood of their occurrence.

In the computational examples $\beta_1 = 0.1$, $\beta_2 = 0.2$, $\beta_3 = 0.3$, and $\beta_4 = 0.4$, i.e. disruptive events with a later start time are modeled to be more likely.

Solution results for regional disruption probability (ii) are shown in Table 14.4, where primary and expected recovery portfolios and pre-positioned and expected used risk mitigation inventory are presented as well as expected production in Period 1 along with solution values and their LP relaxations. The expected recovery portfolios refer to the expected supplies to meet demand unfulfilled by primary supplies, cf. Eq. (14.19). Since the primary supplies meet demand for most of the scenarios, the expected recovery supplies are a small percentage of total demand. For model **ResMTSupport2_E(c)** and different values of unit penalty, g, Table 14.4 shows the expected cost per product, E^c, (14.3), along with the associated expected service level, E^{sl}, (14.4). Notice that

$$E^{sl} = \sum_{s\in S} P_s\Big(\sum_{i\in I_0\setminus I^s} \rho_i^s + \sum_{i\in I_1\setminus I^s} \sum_{j\in I_0\bigcap I^s} \rho_{ij0}^s + \sum_{i\in I_1\bigcap I^s} \sum_{j\in I_0\bigcap I^s} v_{ij0}$$

$$+ \sum_{i\in I_1\bigcap I^s} \sum_{j\in I_0\bigcap I^s} V_{ij0}^s + \sum_{i\in I_0\setminus I^s} \sum_{j\in I_0\bigcap I^s} w_{ij0}^s + \sum_{i\in I_0\bigcap I^s} Tc_{i0} y_{i0}^s/D\Big).$$

Table 14.5 Impact of regional disruption probability

Model	ResMTSupport2_E(c)				ResMTSupport2_E(sl)
	Regional disruption probability (i)				
Unit penalty, $g =$	100.00	128.87	128.88($=\underline{g}$)	150.00	∞
Exp.Cost, (14.3)	53.20	57.03	57.03	58.72	-
Exp.Service $\times 100\%$, (14.4)	86.74	86.74	92.00	92.01	95.00
Pre-positioned Inventory	–	–	$x_{1,0} = 0.85$	$x_{1,0} = 0.85$	$x_{ik} = 1, k \in K, i \in I_k : k < 3$
Exp.Usage of Inventory $^{(h)}$	–	–	0.053	0.053	0.079
LP	52.93	56.62	56.62	58.29	95.00
	Regional disruption probability (iii)				
Unit penalty, $g =$	100.00	133.46	133.47($=\underline{g}$)	150.00	∞
Exp.Cost, (14.3)	51.30	56.96	56.96	58.92	-
Exp.Service $\times 100\%$ (14.4),	83.09	83.10	88.12	88.12	91.31
Pre-positioned Inventory	–	–	$x_{1,0} = 0.85$	$x_{1,0} = 0.85$	$x_{ik} = 1, k \in K, i \in I_k : k < 3$
Exp.Usage of Inventory $^{(h)}$	–	–	0.050	0.050	0.152
LP	51.05	56.58	56.58	58.53	91.31

As g increases a higher service level is attained. However, even for $g \to \infty$, i.e., when the objective is to maximize service level, E^{sl} is less than one, because the impact of disruptions cannot be fully mitigated. Notice a significant jump of the expected service level, E^{sl}, from 86.34% for $g = 129.36$ to 91.57% for the threshold, $\underline{g} = 129.37$, (14.39), while the corresponding expected cost per product, E^c, remains at the same level, 57.00. These are the results of improved supply chain resilience by pre-positioning and usage of risk mitigation inventory to reduce unfulfilled demand for products. Table 14.4 demonstrates that the primary supply portfolio and expected production by primary suppliers in Period 1 are independent of unit penalty cost, g. Notice that the only recovery flow is outflow of finished products from the assembly plant with a nonzero risk mitigation inventory of parts from tier 1 suppliers, for which the total unit cost of risk mitigation inventory, $f_0 + h_0$ ($f_0 = \max_{i \in I_1}(1.2o_{i1})$ $h_0 = \max_{i \in I_1}(0.2o_{i1})$), is smaller than unit cost of expedited supplies of parts, $\overline{o}_{i1} = 1.5o_{i1}$.

Table 14.5 presents similar solution results for regional disruption probabilities (i) and (iii). Tables 14.4 and 14.5 indicate that higher regional disruption probability leads to higher threshold $\underline{g}$, (14.39) and lower expected service level, E^{sl}. While the decrease of expected service level with regional disruption probability is intuitive, the simultaneous increase of unit penalty threshold for pre-positioning of risk mitigation inventory is rather counterintuitive.

In the above examples disruptive events with a later start time were modeled to be more likely ($\beta_1 = 0.1$, $\beta_2 = 0.2$, $\beta_3 = 0.3$, and $\beta_4 = 0.4$). For regional disruption probability (ii), the above scenarios are compared in Table 14.6 with scenarios for which earlier disruptive events are more likely ($\beta_1 = 0.4$, $\beta_2 = 0.3$, $\beta_3 = 0.2$, and $\beta_4 = 0.1$) and with scenarios for equal likelihood of all disruption start times ($\beta_1 = \beta_2 = \beta_3 = \beta_4 = 0.25$). Table 14.6 indicates that the more likely are earlier disruptive events, the lower is expected service level and the higher is expected cost due to higher expected penalty for unfulfilled demand and usage of risk mitigation inventory. The computational study shows that disruptions at the beginning of the planning horizon, i.e., before ordered parts delivery, may deteriorate performance of a multi-tier supply network more severely. Thus, reliable suppliers with a shorter delivery lead time are more preferable.

Table 14.6 Impact of disruption start times: regional disruption probability (ii)

Model	**ResMTSupport2_E(c)**				**ResMTSupport2_E(sl)**
Unit penalty, $g =$	100.00	129.36	129.37($=\underline{b}$)	150.00	∞
β	$\beta_1 = 0.1$, $\beta_2 = 0.2$, $\beta_3 = 0.3$ and $\beta_4 = 0.4$				
Exp.Cost, (14.3)	52.99	57.00	57.00	58.74	–
Exp.Service ×100%, (14.4)	86.33	86.34	91.57	91.57	94.59
β	$\beta_1 = \beta_2 = \beta_3 = \beta_4 = 0.25$				
Exp.Cost, (14.3)	53.29	57.41	57.41	59.22	–
Exp.Service ×100%, (14.4)	85.98	85.98	91.22	91.22	93.98
β	$\beta_1 = 0.4$, $\beta_2 = 0.3$, $\beta_3 = 0.2$ and $\beta_4 = 0.1$				
Exp.Cost, (3)	53.59	57.81	57.81	59.61	–
Exp.Service ×100%, (4)	85.63	85.63	90.87	90.87	93.40

The computational experiments were performed using the AMPL programming language and the Gurobi 8.0 solver on a MacBookPro laptop with Intel Core i7 processor running at 2.8 GHz and with 16 GB RAM. The approach has proven to be computationally very efficient. Proven optimal solutions were found in a few seconds of CPU time.

Table 14.7 Managerial insights: resilient multi-tier supply portfolio, two-period approach

Decision-making issue	Decision-making impact
A tight threshold for unit penalty of unfulfilled demand for products.	The level of contractual penalty for unfulfilled customer demand may have a direct impact on resilience of a multi-tier supply chain network. Below the threshold, pre-positioning of risk mitigation inventory of parts from tier 1 suppliers at assembly plants is not profitable.
Unit penalty vs. service level.	A higher unit penalty leads to a higher expected service level.
Regional disruption probability.	A higher regional disruption probability leads to a lower expected service level and a higher threshold for unit penalty of unfulfilled demand for products, required to pre-position risk mitigation inventory.
Disruptions at the beginning of the planning horizon.	The reliable suppliers with a shorter delivery lead time are more preferable. Disruptions before delivery of ordered parts may deteriorate performance of a multi-tier supply network more severely.
Variation of unit penalty at the threshold.	A slight increase of unit penalty at the threshold may result in a significant increase (discontinuous vertical jump) of the expected service level, while the expected cost remains nearly unchanged.
Upper bound on the expected service level of resilient multi-tier supply chain network.	A resilient multi-tier supply chain network is capable of fulfilling customer demand under disruption risks.
Balancing of inter- and intra-tier recovery flows in a disrupted multi-tier supply chain network.	In order to minimize potential losses in a multi-tier supply chain network, e.g., due to the ripple effect, integrated flow balancing actions are required to simultaneously recover disrupted flows over the entire network.
Two-period vs. single-period modeling approach.	Unlike the single-period modeling approach (see, Chap. 13), the two-period setting can be applied to determine a dynamic supply portfolio, to account for time-varying demand, lead times from suppliers, transportation and transshipment times, disruption times and duration of recovery processes, etc. (cf. Chap. 11)

The major managerial insights are summarized in Table 14.7 (cf. Table 13.11). Overall, the results of computational experiments emphasize *the importance of spatially and temporally coordinated decision-making in a multi-tier supply chain network under disruption risks and show how the decision maker may improve its resilience. The new features in decision-making added to the existing methods by the above results include unit penalty threshold for pre-positioning of risk mitigation inventory, discontinuity (vertical jump) of expected service level at the threshold, counterintuitive higher unit penalty threshold for higher regional disruption probability, preference for selection of reliable suppliers with shorter delivery lead times, and possibility of fulfilling customer demand under disruption risks by resilient supply networks.*

14.5 Notes

The literature on supply chain disruption management mainly focuses on manufacturer-supplier dyad and at best on manufacturer-supplier-supplier's supplier triad, while the more challenging issues involving disruption management of longer supply chains, dealing with many tiers and their interactions are rarely investigated. Multi-tier case studies were presented in Choi and Hong (2002), Holweg and Pil (2008), Mena et al. (2013), Thome et al. (2014). Recent literature reviews on disruption risk mitigation strategies in supply chains were provided, for example, in Narasimhan and Talluri (2009), Tomlin and Wang (2011), Rajagopal et al. (2017), Yoon et al. (2018). However, supplier selection in a multi-tier supply chain under disruption risks is very rarely reported in the literature, e.g., Kull and Closs (2008), DeCroix (2013), Ang et al. (2017), Yoon et al. (2019). For example, the importance of considering disruption risk of tier 2 suppliers was shown by Kull and Closs (2008). DeCroix (2013) analyzed an assembly system with a single end product and a general assembly structure, where one or more of the component suppliers is subject to random supply disruptions. The author explored a variety of strategic questions. For example, he found that choosing a supplier with a longer lead time can sometimes yield lower system costs and because of component complementarities, choosing suppliers whose disruptions are perfectly correlated yields lower system costs than choosing suppliers whose disruptions are independent. Ang et al. (2017) studied sourcing in a three-tier supply chain, which consists of a manufacturer, tier 1 suppliers and tier 2 suppliers prone to disruption from natural disasters. However, they ignored tier 1 suppliers disruption risk. They showed that the manufacturer's optimal strategy depends on the degree of overlap in the supply chain: if tier 1 suppliers share tier 2 suppliers, the manufacturer relies less on direct mitigation (procuring excess inventory and multi sourcing in tier 1) and more on indirect mitigation (penalizing tier 1 suppliers for a failure to deliver ordered units). Yoon et al. (2019) investigated manufacturer's procurement decision in a three-tier supply chain, where the manufacturer cannot directly obtain information on disruption risk for tier 2 supplier except through tier 1 supplier. The authors showed that information sharing makes the manufacturer's decisions more conservative, i.e. carrying more inventories, while tier 1 supplier decision is contingent on tier 2 supplier reliability.

The material presented in this chapter is based on results presented in Sawik (2019d), where the two-period SMIP model was developed for selection of resilient multi-tier supply portfolio. In the developed model, the supply chain resilience is achieved by multi sourcing, recovery and transshipment supplies and pre-positioning, and usage of risk mitigation inventory. The resilience can be further improved by fortification of selected primary suppliers and assembly plants. Depending on the level of fortification, the capacity of each protected facility remains fully or partially available under a disruptive event. Accordingly, the protected suppliers (plants) are capable of fully or partially supplying parts (products) in the face of disruptive events. However, the two-period model has some limitations. For example, it cannot be directly applied in case of multiple disruptions, one after the other in a series, during the recovery process. Then, the two-stage stochastic setting should be replaced by a multi-stage one. The future research should concentrate on relaxations of the various simplified assumptions used to formulate the problem. The risk-neutral objectives, expected cost or service, can be easily replaced by the risk-averse conditional cost-at-risk or conditional service-at-risk. In the future research the proposed two-period model can be used as a basis for development of a multi-period decision-making model to determine a dynamic supply portfolio to account for time-varying demand, duration of recovery processes, etc. A possible application of a dynamic multi-tier decision-making model for investigation of the ripple effect (e.g., Ivanov et al. 2014, Ivanov et al. 2019) in supply chains could be a further area of the future research.

Problems

14.1. Modify the two-period model developed in this chapter to account for recovery time of impacted primary suppliers and assembly plants shorter than the length of the planning horizon. Consider the two types of modified models: with best-case and with worst-case capacity constraints (cf. Chap. 11).

14.2. Enhance the two-period model developed in this chapter to support decision on fortification of primary suppliers and primary assembly plants with multiple fortification levels and partially available capacity of impacted facilities in Period 2.

14.3. Formulate the two-period models for risk-averse decision-making, **ResMTSupport2_CV(c)** and **ResMTSupport2_CV(sl)**, to minimize CVaR of cost and to maximize CVaR of service level, respectively.

14.4. Try to explain the background of the counterintuitive higher unit penalty threshold for higher regional disruption probability.

14.5. When the unit transshipment cost can be considered prohibitively expensive with respect to unit recovery supply cost?

Part VI
Information Flow Disruption Management

Chapter 15
Selection of Cybersecurity Safequards Portfolio

15.1 Introduction

Nowadays information technology (IT) becomes one of the core elements of global supply chains that are increasingly at risk of a disruption of their information systems. Contemporary supply chains are cyber-physical systems, in which IT may improve their robustness and resilience, e.g., Pereira (2009). For example, an important determinant of supply chain competitiveness is supply chain visibility defined as "the extent to which actors within a supply chain have access to or share information which they consider as key or useful to their operations and which they consider will be of mutual benefit" (Lee et al. 2014). The supply chain visibility is accomplished through interorganizational information systems, which should be very well protected against cyber-attacks. In supply chain risk management, cybersecurity focuses on gaining visibility and control over the entire multi-tier supply chain network, e.g., Boyson (2014).

This chapter deals with the selection of cybersecurity safeguards portfolio to prevent and mitigate the impact of information flow disruptions in a supply chain. The objective of IT security planning in supply chains is to protect information flows against disruptions and the supply chain assets against a compromise in the area of confidentiality, possession or control, integrity of information or information systems, authenticity, availability and utility, (the Parkerian hexad, e.g., Falco et al. 2019). The asset types may include systems and applications, networks, end-user systems, and offline media and devices. The various actions developed to prevent intrusions or to mitigate the impact of successful breaches and information flow disruptions are called security safeguards, controls or countermeasures. In view of the variety of methods used by attackers to infiltrate a supply chain IT infrastructure and disrupt operations, a wide range of different countermeasures are developed. Some countermeasures are used to limit physical access to an IT infrastructure, (e.g., key entry systems, retinal or finger print scans), other block access or pro-

T. Sawik, *Supply Chain Disruption Management*,
International Series in Operations Research & Management Science 291,
https://doi.org/10.1007/978-3-030-44814-1_15

tect privacy over networks (e.g., firewalls, data encryption, or virus and spyware scanners), while additional countermeasures are designed to permit recovery after a successful intrusion. A common practice is information redundancy. Firms back up supplier data, customer data, bill-of-material data, etc. However, backup of hard drives may not be sufficient for a supply chain to continue its operations, more often backup of the system itself is required. For example, if a widespread power outage severely impacts business continuity by compromising information systems.

In practice, even the most sophisticated countermeasures cannot be expected to completely block attacks as new attack profiles proliferate and the time and cost required to adapt to them the implemented countermeasures are not negligible. The potential threats, however, cannot be ignored and without countermeasures the losses resulting from many more breaches and disruptions of information flows could easily excess the costs of countermeasure implementations. In this chapter a scenario-based SMIP approach with CVaR as a risk measure is proposed for the decision-making. Given a set of potential threats and a set of available countermeasures, the decision maker needs to decide which countermeasure to implement under limited budget to minimize potential losses from successful cyber-attacks. The selection of countermeasures is based on their effectiveness of blocking different threats, implementation costs, and probability of potential attack scenarios. The risk-neutral, risk-averse models, and, in particular, mean-risk trade-off model provide the decision maker with a simple tool for balancing expected and worst-case losses and for shaping of the resulting cost distribution through the selection of optimal subset of countermeasures for implementation, i.e., the selection of optimal cybersecurity safeguards portfolio.

The following SMIP models are presented in this chapter:

NCyberport_E for risk-neutral selection of cybersecurity safeguards portfolio to minimize expected loss (nonlinear model);
Cyberport_E for risk-neutral selection of cybersecurity safeguards portfolio to minimize expected loss (linear model);
Cyberport_CV for risk-averse selection of cybersecurity safeguards portfolio to minimize CVaR of loss;
Cyberport_EB for risk-neutral selection of cybersecurity safeguards portfolio to minimize expected loss and required budget;
Cyberport_CVB for risk-averse selection of cybersecurity safeguards portfolio to minimize CVaR of loss and required budget;
Cyberport_EBCV for mean-risk selection of cybersecurity safeguards portfolio to optimize trade-off between expected loss plus required budget and CVaR of loss.

In the computational experiments described in Sect. 15.6, an application of the developed models is illustrated with numerical examples and major managerial insights are summarized at the end of this chapter.

15.2 Problem Description

Table 15.1 Notation: cybersecurity safeguards portfolio

	Indices
i	= threat, $i \in I$
j	= countermeasure, $j \in J$
s	= attack scenario, $s \in S$
l	= countermeasure implementation level, $l \in L = \{0,1\}$ ($l = 0$ - off, $l = 1$ - on)
	Input parameters
a_i	= loss caused by a successful attack episode of threat i
B	= available budget for countermeasures
c_j	= cost of countermeasure j
I_s	= subset of threats in scenario s
p_i	= probability of threat i
q_{ijl}	= $\max\{1-l, r_{ij}\}$ - proportion of threats i that survive if countermeasure j is implemented at level l ($q_{ij0} = 1$ and $q_{ij1} = r_{ij}$)
r_{ij}	= proportion of threats i that survive if countermeasure j is implemented
α	= confidence level

Let $I = \{1,\dots m\}$ be the set of m threats and $J = \{1,\dots n\}$ the set of n countermeasures (for notation used, see Table 15.1). Denote by p_i the probability of threat i, i.e., attack episode of threat i occurs with probability p_i, or not at all with probability $(1-p_i)$. Let P_s be the probability that attack scenario s is realized, where each scenario $s \in S$ is comprised of a unique subset $I_s \subset I$ of threats that appear in the cyber-attack, and $S = \{1,\dots,\overline{S}\}$ is the index set of all scenarios (note that there are a total of $\overline{S} = 2^m$ potential attack scenarios). The probability of attack scenario s in the presence of independent threat events is

$$P_s = \prod_{i \in I_s} p_i \cdot \prod_{i \notin I_s} (1-p_i). \tag{15.1}$$

Let $r_{ij} \in [0,1]$ be the proportion of threats i that survive if countermeasure j is implemented, where $r_{ij} = 0$ indicates that countermeasure j totally prevents successful attacks of threat i, whereas $r_{ij} = 1$ denotes that countermeasure j is totally incapable of mitigating threat i.

Note, that even the most sophisticated countermeasures, in practice are rarely expected to completely block attacks, and hence the survival proportions r_{ij} are rarely equal to 0.

The blocking effectiveness of each countermeasure is assumed to be independent whether or not it is used alone or together with other countermeasures, and the proportion of successful attacks of threats type i that survive all countermeasures in the subset $JS \subseteq J$ of selected countermeasures is a multiplication of proportions $r_{ij}, j \in JS$. Let $\prod_{j \in JS} r_{ij}$ be the proportion of threats i that survive if subset JS of countermeasures is implemented. Note that the expected proportion of successful attacks of threat type i for the subset JS of selected countermeasures is $p_i \prod_{j \in JS} r_{ij}$.

Denote by c_j the cost of implementing countermeasure j, and by a_i the cost of (loss from) a successful attack episode of threat i. The available budget B for selected countermeasures implementation is limited. The subset of selected countermeasures $JS \subseteq J$ must satisfy the available budget constraint, $\sum_{j \in JS} c_j \leq B$, i.e., the total expenditures on the selected countermeasures cannot exceed the available budget B. Furthermore, the selected countermeasures are assumed to be feasible with respect to various additional constraints, e.g., they cannot contain mutually exclusive countermeasures or they must contain all countermeasures contingent on each other, etc.

The decision maker needs to decide which countermeasures to select to minimize losses from surviving occurrences of threats under limited budget for countermeasures implementation.

15.3 Models for Risk-Neutral Decision-Making

In this section SMIP models are proposed for a risk-neutral selection of optimal cybersecurity safeguards portfolio, i.e., the risk-neutral selection of optimal subset of countermeasures for implementation. The problem variables are defined in Table 15.2.

Let $L = \{0, 1\}$ be the set of implementation levels of each countermeasure, where $l = 0$ denotes that a particular countermeasure is not selected for implementation, otherwise $l = 1$. The decision whether or not to select a particular countermeasure will be represented by a binary variable $u_{jl}, j \in J, l \in L$, where $u_{jl} = 1$, if countermeasure j is implemented at level l, otherwise $u_{jl} = 0$, that is, countermeasure j is selected for implementation if $u_{j1} = 1$ and $u_{j0} = 0$, otherwise $u_{j1} = 0$ and $u_{j0} = 1$. The above definition implies that each countermeasure j is selected at exactly one level, i.e., $\sum_{l \in L} u_{jl} = 1$.

Denote by $q_{ijl} = \max\{1 - l, r_{ij}\}$ the proportion of threats i that survive if countermeasure j is implemented at level l, that is $q_{ij0} = 1$ and $q_{ij1} = r_{ij}$.

15.3.1 Nonlinear Model for Risk-Neutral Decision-Making

For a risk-neutral decision-making the overall quality of the selected countermeasure portfolio can be measured by the expected cost of losses from successful attacks.

The proportion of successful attacks of threats type i that survive all selected countermeasures is a multiplication of individual proportions of the selected countermeasures and can be expressed as

$$\prod_{j \in J} \Big(\sum_{l \in L} q_{ijl} u_{jl}\Big).$$

Table 15.2 Problem variables: cybersecurity safeguards portfolio

	First stage variables
u_{jl}	= 1, if countermeasure j is implemented at level l, otherwise $u_{jl} = 0$ (countermeasure selection)
v_{ijl}	= proportion of surviving occurrences of threat i to be addressed by countermeasure j at level l
w_{ij}	= proportion of surviving occurrences of threat i that passed countermeasures 1 through j
W_i	= proportion of successful attacks of type i
	Auxiliary variables
VaR	loss-at-risk, the targeted loss such that for a given confidence level α, for $100\alpha\%$ of the scenarios, the outcome is below VaR
$\mathscr{C}_s$	≥ 0, the tail loss for attack scenario s, i.e., the amount by which losses in scenario s exceed VaR

As a result, the expected loss from successful attacks is given by a nonlinear formula

$$\sum_{s\in S}\sum_{i\in I_s} P_s a_i \Big(\prod_{j\in J}\big(\sum_{l\in L} q_{ijl}u_{jl}\big)\Big). \tag{15.2}$$

The 0-1 nonlinear stochastic programming program **NCyberport_E** for risk-neutral selection of optimal subset of security safeguards is formulated below.

NCyberport_E: **N***onlinear model for risk-neutral selection of* **Cyber***security safeguards* **port***folio to minimize* **E***xpected loss*

Minimize (15.2)

subject to

1. Countermeasure selection constraints:

– each countermeasure is selected at exactly one level (i.e., implemented or not implemented) (15.3),

– the expenditures on selected countermeasures cannot exceed available budget (15.4),

$$\sum_{l\in L} u_{jl} = 1; j \in J \tag{15.3}$$

$$\sum_{j\in J} c_j u_{j1} \leq B. \tag{15.4}$$

3. Integrality conditions:

$$u_{jl} \in \{0,1\};\ j \in J, l \in L. \tag{15.5}$$

The 0-1 nonlinear stochastic program **NCyberport_E** is computationally hard for solving, even for small size instances of the problem. The next subsection describes a recursive procedure that is capable of computing the nonlinear objective function (15.2) using a set of linear equations.

15.3.2 Linear Model for Risk-Neutral Decision-Making

This subsection shows how the nonlinear stochastic binary programming model **NCyberport_E** can be linearized using a recursive procedure applied to prove the following proposition.

Proposition 15.1. *The nonlinear objective function (15.2) can be replaced with a formula*

$$E^c = \sum_{s \in S} \sum_{i \in I_s} P_s a_i W_i, \tag{15.6}$$

where

$$W_i = \prod_{j \in J} (\sum_{l \in L} q_{ijl} u_{jl}); \; i \in I, \tag{15.7}$$

is the proportion of successful attacks of type i. The proportion of successful attacks W_i for each threat i can be calculated recursively, using linear equations.

Proof. In order to compute W_i for each threat i, a recursive procedure is proposed below.

Denote by

$$w_{ij} = \prod_{k \in J: k \leq j} (\sum_{l \in L} q_{ikl} u_{kl}); \; i \in I, j \in J, \tag{15.8}$$

the proportion of surviving occurrences of threat i that passed countermeasures 1 through j. Note that (cf. (15.7), (15.8))

$$W_i = w_{in}; \; i \in I. \tag{15.9}$$

For each threat $i \in I$ and countermeasure $j \in J$, w_{ij} can be calculated recursively as follows.

The initial condition that explicitly prescribes the first term is

$$w_{i1} = \sum_{l \in L} q_{i1l} u_{1l}; \; i \in I. \tag{15.10}$$

(continued)

The recurrence formula by means of which the remaining terms are determined inductively is

$$w_{ij} = (\sum_{l \in L} q_{ijl} u_{jl}) w_{i,j-1},\ i \in I, j \in J. \tag{15.11}$$

In order to eliminate nonlinear terms in the right-hand side of Eq. (15.11), define an auxiliary variable

$$v_{ijl} = u_{jl} w_{ij-1};\ i \in I, j \in J, l \in L, \tag{15.12}$$

where $w_{i0} = 1$ for all i, i.e., all threat events of each type i are capable of attacking supply chain IT infrastructure. Note that v_{ijl} represents the proportion of surviving occurrences of threat i to be addressed by countermeasure j at level l

Substituting $v_{ijl} = u_{jl} w_{ij-1}$ into Eq. (15.11) yields

$$w_{ij} = \sum_{l \in L} q_{ijl} v_{ijl};\ i \in I, j \in J, \tag{15.13}$$

and, in particular, for $j = n$,

$$w_{in} = \sum_{l \in L} q_{inl} v_{inl};\ i \in I, \tag{15.14}$$

where $w_{in} = W_i$ for all $i \in I$, (15.9).

On the other hand, replacing j with $j+1$ in Eq. (15.12) gives

$$v_{i,j+1,l} = u_{j+1,l} w_{ij}; i \in I, j \in J : j < n, \tag{15.15}$$

Summing both sides of Eq. (15.15) on l for each i and j, and using the fact that $\sum_{l \in L} u_{j+1,l} = 1; j \in J$, (15.3), yields

$$\sum_{l \in L} v_{i,j+1,l} = w_{ij};\ i \in I, j \in J : j < n. \tag{15.16}$$

Comparison of Eqs. (15.13) and (15.16) produces to the following relation:

$$\sum_{l \in L} q_{ijl} v_{ijl} = \sum_{l \in L} v_{i,j+1,l};\ i \in I, j \in J : j < n, \tag{15.17}$$

that is, for each threat i, the proportion of occurrences which survive countermeasure j is addressed by countermeasure $j+1$.

(continued)

Finally, setting $j = 1$ and summing both sides of Eq. (15.12) on l, using the fact that $w_{i0} = 1$ for all i and $\sum_{l \in L} u_{1l} = 1$, (15.3), yields

$$\sum_{l \in L} v_{i1l} = 1; \; i \in I, \tag{15.18}$$

that is, all threat events of each type i are addressed by countermeasure $j = 1$.

The above recursive procedure eliminates all variables w_{ij} for each i, except for w_{in}. Summarizing, the proportion of successful attacks $W_i = w_{in}$ for each threat i can be calculated recursively, using Eqs. (15.18), (15.17), and (15.14) with w_{in} replaced by W_i.

Note that the linearizing procedure would not be possible if the implementation level $l \in L = \{0,1\}$ of each countermeasure $j \in J$ was not introduced and the variable u_{jl} was replaced with a simple binary selection variable $u_j \in \{0,1\}$, denoting whether or not countermeasure j is selected. In the latter case the proportion of threats i that survive whether or not countermeasure j is selected, is $\max\{1-u_j, r_{ij}\}$, i.e., the survival proportions would depend nonlinearly on the selection variables u_j. In contrast to constant survival proportions $q_{ijl} = \max\{1-l, r_{ij}\}$ for variables u_{jl}, the survival proportions for variables u_j could not be expressed by constant coefficients in the constraints.

The linearized version **Cyberport_E** of model **NCyberport_E** is shown below (see also, network flow model in Deane et al. 2009 and Rakes et al. 2012).

Cyberport_E: *Risk-neutral selection of* **Cyber***security safeguards* **port***folio to minimize* **E***xpected loss*

Minimize (15.6)

subject to

1. Countermeasure selection constraints: (15.3), (15.4)

2. Surviving threats balance constraints:

$$\sum_{l \in L} v_{i1l} = 1; \; i \in I \tag{15.19}$$

$$\sum_{l \in L} q_{ijl} v_{ijl} = \sum_{l \in L} v_{ij+1l}; \; i \in I, j \in J : j < n \tag{15.20}$$

$$\sum_{l \in L} q_{inl} v_{inl} = W_i; \; i \in I \tag{15.21}$$

$$v_{ijl} \le u_{jl}; \; i \in I, j \in J, l \in L. \tag{15.22}$$

3. Non-negativity and integrality conditions:

$$u_{jl} \in \{0,1\}; \; j \in J, l \in L \tag{15.23}$$

$$v_{ijl} \ge 0; \; i \in I, j \in J, l \in L \tag{15.24}$$

$$W_i \ge 0; \; i \in I. \tag{15.25}$$

Constraint (15.22) (cf. (15.12)) ensures that threats can only by addressed by the implemented countermeasures.

Note that the expected proportion of successful attacks of threat type i is $p_i W_i$.

Model **Cyberport_E** will be used to compare the risk-neutral results with those obtained by applying a risk aversive decision-making model described in the next section.

15.4 Model for Risk-Averse Decision-Making

In this section a risk-averse selection of optimal cybersecurity safeguards portfolio is considered with the two popular in financial engineering percentile measures of risk, VaR and CVaR, applied to control the risk of high losses.

When selecting the optimal risk-averse countermeasure portfolio, the decision maker controls the risk of high losses caused by operational disruptions by choosing the confidence level α. The greater the confidence level α, the more risk aversive is the decision maker and the smaller percent of the highest loss outcomes is focused on. The risk aversive decision maker wants to minimize the expected worst-case losses exceeding VaR, by minimizing CVaR, given available budget B for selected countermeasures.

Define $\mathscr{C}_s$ as the tail loss for scenario s, where tail loss is defined as the amount by which losses in scenario s exceed VaR. The cybersecurity safeguards portfolio will be optimized by calculating VaR and minimizing CVaR simultaneously. By measuring CVaR, the magnitude of the tail loss is considered to achieve a more accurate estimate of the risks of minimizing loss. In the proposed model, CVaR is represented by an auxiliary function (15.26) introduced by Rockafellar and Uryasev (2000). The SMIP model **Cyberport_CV** for selection of risk-averse cybersecurity safeguards portfolio to reduce the risk of high losses is formulated below.

Cyberport_CV: *Risk-averse selection of* **Cyber***security safeguards* **port***folio to minimize* **CV***aR of loss*

Minimize

$$CVaR = VaR + (1-\alpha)^{-1} \sum_{s \in S} P_s \mathscr{C}_s \tag{15.26}$$

subject to

1. Countermeasure selection constraints: (15.3)–(15.4)

2. *Surviving threats balance constraints*: (15.19)–(15.22)

2. *Risk constraints*:

– the tail loss for scenario s is defined as the nonnegative amount by which losses in scenario s exceed VaR,

$$\mathscr{C}_s \geq \sum_{i \in I_s} a_i W_i - VaR; \; s \in S \tag{15.27}$$

(continued)

3. Non-negativity and integrality conditions: (15.23)–(15.25) and

$$\mathscr{C}_s \geq 0;\ s \in S. \tag{15.28}$$

Models **Cyberport_E** and **Cyberport_CV** can be enhanced for simultaneous optimization of the expenditures on countermeasures and the loss from successful attacks. Then, the budget constraints (15.4) should be removed from the models and the fixed cost of selected countermeasures, $\sum_{j\in J} c_j u_{j1}$, added to the objective function.

Define the fixed cost of selected countermeasures as the required budget, $B(u)$,

$$B(u) = \sum_{j\in J} c_j u_{j1}. \tag{15.29}$$

The enhanced models **Cyberport_EB** and **Cyberport_CVB** for simultaneous optimization of fixed cost of countermeasures and variable loss from successful attacks are presented below.

Cyberport_EB: *Risk-neutral selection of* **Cyber***security safeguards* **port***folio to minimize* **E***xpected loss and required* **B***udget*
Minimize

$$E^c + B(u) \tag{15.30}$$

subject to (15.3), (15.6), (15.19)–(15.25), (15.29).

Cyberport_CVB: *Risk-averse selection of* **Cyber***security safeguards* **port***folio to minimize* **CV***aR of loss and required* **B***udget*
Minimize

$$CVaR + B(u) \tag{15.31}$$

subject to (15.3), (15.19)–(15.29).

15.5 Model for Mean-Risk Decision-Making

In the single objective approach the countermeasure portfolio is selected by minimizing either the expected loss (plus the required budget) or the expected worst-case loss (plus the required budget). Since the probability of worst-case loss outcomes is usually very low, the expected cost function that aims at optimizing an average performance of IT security system, virtually neglects the worst-case losses. In contrast,

CVaR that aims at optimizing worst-case performance, focuses on the low probability, high loss outcomes, and as the confidence level α increases, a more risk-averse decision-making focuses on a smaller set of the highest loss outcomes.

In this section the two conflicting objectives are considered simultaneously, and a bi-objective selection of countermeasures is presented aimed at minimizing both objective functions to balance the required budget and expected cost with the risk tolerance. This trade-off model, known as the mean-risk model, is formulated as the optimization of a convex combination of the expected loss and the CVaR of loss as a risk measure.

The nondominated solution set of the bi-objective countermeasure portfolio can be found by the parameterization on λ the weighted-sum program **Cyberport_EBCV**. The scalarizing mixed integer program is based on **Cyberport_CVB** model with the addition of objective (15.6) of model **Cyberport_EB**.

Cyberport_EBCV: *Mean-risk selection of* **Cyber***security safeguards* **port***folio to optimize trade-off between* **E***xpected loss plus required* **B***udget and* **CV***aR of loss*

Minimize

$$\lambda(E^c + B(u)) + (1-\lambda)CVaR, \tag{15.32}$$

where $0 \le \lambda \le 1$,

subject to (15.3), (15.6), (15.19)–(15.29).

Now, the decision maker controls both the risk of high losses from successful cyber-attacks by choosing the confidence level α as well as the trade-off between expected and worst-case losses by choosing the trade-off parameter λ. Using the latter parameter, the level of risk allowed into the solution can be controlled. In particular, a decision maker may decide to minimize expected loss while requiring the percentile worst-case loss to be not greater than some parameter. In other words, model **Cyberport_EBCV** with $\lambda = 1$ may include an upper bound on CVaR. However, when α increases so that the worst possible outcomes are considered only, imposing a strict bound on CVaR may result in an infeasible solution, e.g., Chahara and Taaffe (2009).

15.6 Computational Examples

In this section some computational examples are presented to illustrate possible applications of the proposed portfolio approach for the selection of countermeasures to mitigate the impact of information flow disruptions caused by IT security incidents. The following parameters have been used for the example problems:

- $m = n$, the number of threats and the number of countermeasures were equal to 10, and the corresponding number $\overline{S} = 2^m$ of potential attack scenarios was equal to 1024;
- a_i, loss from a successful attack of each threat i (in \$1000): $a_1 = 24$, $a_2 = 122$, $a_3 = 350$, $a_4 = 5$, $a_5 = 250$, $a_6 = 20$, $a_7 = 20$, $a_8 = 25$, $a_9 = 30$, $a_{10} = 10000$;
- c_j, cost of each countermeasure j (in \$1000): $c_1 = 40$, $c_2 = 28$, $c_3 = 80$, $c_4 = 24$, $c_5 = 70$, $c_6 = 50$, $c_7 = 40$, $c_8 = 45$, $c_9 = 50$, $c_{10} = 80$;
- p_i, probability of each threat i: $p_1 = 0.35$, $p_2 = 0.25$, $p_3 = 0.15$, $p_4 = 0.25$, $p_5 = 0.20$, $p_6 = 0.25$, $p_7 = 0.50$, $p_8 = 0.35$, $p_9 = 0.40$, $p_{10} = 0.003$;
- r_{ij}, surviving proportions for each threat i and countermeasure j:

i/j	1	2	3	4	5	6	7	8	9	10
1	0.01	0.5	1	1	1	1	1	1	1	1
2	1	0.04	1	1	0.6	1	1	1	1	1
3	1	1	0.01	0.8	1	0.8	0.9	0.95	0.9	1
4	1	1	1	0.25	1	0.8	0.9	0.95	0.9	0.8
5	1	0.5	1	0.8	0.02	0.8	0.9	0.95	0.9	1
6	1	0.6	1	1	1	0.1	0.6	0.65	0.6	1
7	1	0.5	1	1	1	0.5	0.15	0.2	0.25	1
8	1	0.5	1	1	1	0.55	0.2	0.25	0.3	1
9	1	0.5	1	1	1	0.5	0.15	0.2	0.35	1
10	1	1	1	1	1	1	1	1	1	0.2;

- α, the confidence level, was equal to 0.50, 0.75, 0.90, 0.95 or 0.99;
- B, available budget for countermeasures (in \$1000) was equal to 150, 300 or $\sum_{j \in J} c_j = 507$.

The above data set is similar to the one presented in Rakes et al. (2012), which was based on the threat set reported on IT security forum EndpointSecurity.org. Note that threat $i = 10$ is a high-impact and low-probability threat with the loss from successful attack equal to $a_{10} = \$10M$ and the probability of occurrence equal to $p_{10} = 0.003$. Such a serious and rarely occurring event might be a compromise of some important business data. The remaining threats are more typical such as viruses, data damages, or information disclosures, occurring more frequently and resulting in lower loss from successful attacks. The off-diagonal surviving proportions $r_{ij}, i \neq j$, less than one, indicate that some countermeasures work against their primary threat as well as against other threats. For example, countermeasures $j = 2,6,7,8,9$ may represent strong security policies focused on particular threats and simultaneously showing benefits across the other threats. In contrast, countermeasures $j = 1,3$ reduce the survival proportions of a single threat only.

For the risk-neutral models **Cyberport_E** and **Cyberport_EB**, solution results are shown in Table 15.3, and for the risk-averse models **Cyberport_CV** and **Cyberport_CVB** with different confidence levels, in Tables 15.4 and 15.5. In addition to VaR and CVaR, for comparison with risk-neutral solutions, Tables 15.4 and 15.5 show the corresponding expected loss for the optimal risk-averse countermeasure portfolios. In the tables all costs are expressed in thousands of dollars.

For the risk-neutral models, Table 15.3 indicates that when **Cyberport_E** model is applied, the greater the available budget, the more countermeasures are selected to minimize expected loss. If, however, model **Cyberport_EB** is used, a single countermeasure is selected only to minimize expenditures on countermeasures and expected loss from successful attacks. For **Cyberport_E** model, Fig. 15.1 shows the probability mass function of optimal cost of losses for the available budget $B = 150$, with the tail of cost distribution presented in a separate chart. Similarly, for **Cyberport_EB** model with a variable budget, the probability mass function of optimal cost is shown in Fig. 15.2, where the tail of cost distribution is also presented in a separate chart.

For the risk-averse models, Tables 15.4 and 15.5 demonstrate that the number of selected countermeasures increases with the available budget, which indicates that the impact of disruption risks is mitigated by providing a more secure environment. When α increases, a more risk-averse decision-making focuses on a smaller set of outcomes; however, the number of selected countermeasures is not increasing. For a limited budget, when α increases, the more expensive countermeasures against high-impact, low-probability threats are more frequently selected. As a result the total number of all selected countermeasures decreases with α. The latter result is partly due to the parameter settings for which the countermeasure implementation costs do not always dominate the losses from successful cyber-threats. For example, costs of countermeasures 1,4,6,7,8, and 9 are higher than the losses from a single successful attack of the corresponding threats.

Note that for **Cyberport_CV** model, VaR becomes smaller than expected loss for all budget levels B, when $\alpha = 0.50$ and $\alpha = 0.75$, and for all confidence levels α, when the highest budget level $B = \$507,000$ allows for the selection of all countermeasures, (see, Table 15.4).

Table 15.3 Solution results for risk-neutral models **Cyberport_E** and **Cyberport_EB**

Model	**Cyberport_E**			**Cyberport_EB**
Budget B	150	300	507	28
Expected cost	63,842	17,079	7,589	132,545
Selected countermeasures	2,3,7	2,3,5,7,10	1–10	2

Table 15.4 Solutions results for risk-averse model **Cyberport_CV**

Confidence level α	0.50			0.75		
Budget B	150	300	507	150	300	507
CVaR	121,130	29,154	14,839	224,294	44,849	27,798
VaR	13,500	10,128	1,078	23,780	16,428	2,965
Expected cost	63,842	17,079	7,589	63,842	17,079	7,589
Selected countermeasures	2,3,7	2,3,5,7,10	1–10	2,3,7	2,3,5,7,10	1–10

Confidence level α	0.90			0.95			0.99		
Budget B	150	300	507	150	300	507	150	300	507
CVaR	393,775	84,185	64,597	478,204	145,744	124,985	921,449	624,627	604,858
VaR	302,500	21,450	3,769	318,880	24,178	4,663	414,500	29,028	5,882
Expected cost	92,045	17,079	7,589	92,045	17,079	7,589	92,045	17,079	7,589
Selected countermeasures	2,4,10	2,3,5,7,10	1–10	2,4,10	2,3,5,7,10	1–10	2,4,10	2,3,5,7,10	1–10

Table 15.5 Solutions results for risk-averse model **Cyberport_CVB**

Confidence level α	0.50	0.75	0.90	0.95	0.99
CVaR	144,008	67,089	109,323	172,808	652,214
VaR	29,380	34,500	42,928	49,500	59,000
Expected cost	80,570	31,332	31,332	31,332	31,332
Required budget	108	258	258	258	258
Selected countermeasures	2,3	2,3,5,10	2,3,5,10	2,3,5,10	2,3,5,10

In the computational experiments with risk-averse models, the confidence level α is set at five levels of 0.5, 0.75, 0.90, 0.95, and 0.99, which means that focus is on minimizing the highest 50%, 25%, 10%, 5%, and 1% of all scenario outcomes, i.e., losses from successful attacks. For **Cyberport_CV** model, Fig. 15.3 shows the probability mass function of optimal cost of losses for the available budget $B = 150$ and confidence level $\alpha = 0.99$, with the tail of loss distribution presented in a separate chart. For **Cyberport_CVB** model with a variable budget and confidence level $\alpha = 0.99$ the probability mass function of optimal cost is shown in Fig. 15.4, where the tail of cost distribution is also presented in a separate chart. Figures 15.1, 15.2, 15.3, and 15.4 indicate that the probability measure is concentrated in finitely many points, which is typical for the scenario-based optimization under uncertainty, (e.g., in Fig. 15.1, the probability measure is concentrated in four points). Moreover, the tail of cost distribution shows that the high loss probability is very low. As a result the expected loss for the optimal risk-neutral portfolio and CVaR for the optimal risk-averse portfolio are much lower than the corresponding worst-case loss outcomes.

Table 15.6 Nondominated solutions for mean-risk model **Cyberport_ECV**: $\alpha = 0.9$

λ	0.01	0.10	0.25	0.50	0.75	0.90	0.99
CVaR	65,049	68,145	84,185	109,323	218,193	633,842	710,691
Expected cost	7,908	9,718	17,079	31,332	56,320	116,108	132,545
Required budget	457	388	298	258	188	52	28
Selected countermeasures	1–8,10	1–3,5–7,10	2,3,5,7,10	2,3,5,10	2,3,10	2,4	2

Comparison of tails of the cost distributions for the corresponding risk-neutral and risk-averse solutions (i.e., Fig. 15.1 vs. 15.3 and Fig. 15.2 vs. 15.4) indicates that the optimal risk-averse countermeasure portfolio positively shapes the cost distribution, i.e., significantly reduces the worst-case cost outcomes. For example, the worst-case cost of \$10.6 M for the risk-neutral portfolio determined using **Cyberport_EB** model is reduced to \$2.06 M, when the risk-averse portfolio is applied, determined using **Cyberport_CVB** model.

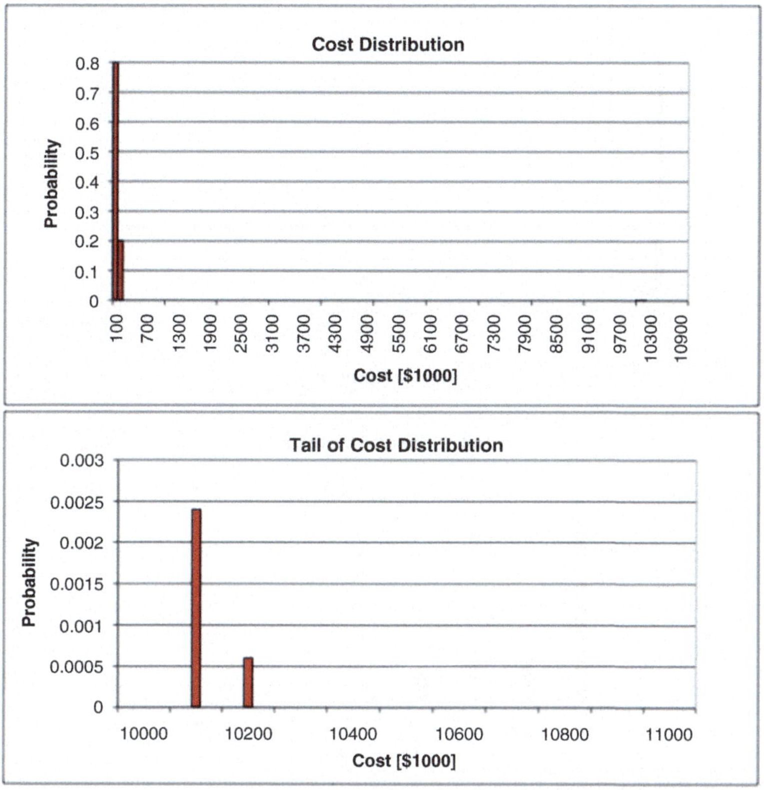

Fig. 15.1 Probability mass function for risk-neutral model **Cyberport_E** with $B = \$150,000$: Expected loss = $63,842

Table 15.7 Examples of mixed integer program size and CPU time

m	2^m [a]	Var.[b]	Bin.	Cons.	CPU[c]
		Risk-neutral model **Cyberport_EB**			
10	1024	231	20	321	<1
15	32768	496	30	766	<1
20	1 048 576	861	40	1241	<1
		Risk-averse model **Cyberport_CVB**			
10	1024	1256	20	1345	<1
15	32768	33265	30	33474	5
20	1 048 576	1 049 438	40	1 049 817	416

[a] m: number of threats (number of countermeasures);
2^m: number of attack scenarios

[b] Var.: number of variables; Bin.: number of binary variables;
Cons.: number of constraints

[c] CPU seconds for proving optimality on a MacBookPro 6.2, Intel Core i7, 2.66 GHz, RAM 8 GB/Gurobi 5.0.1

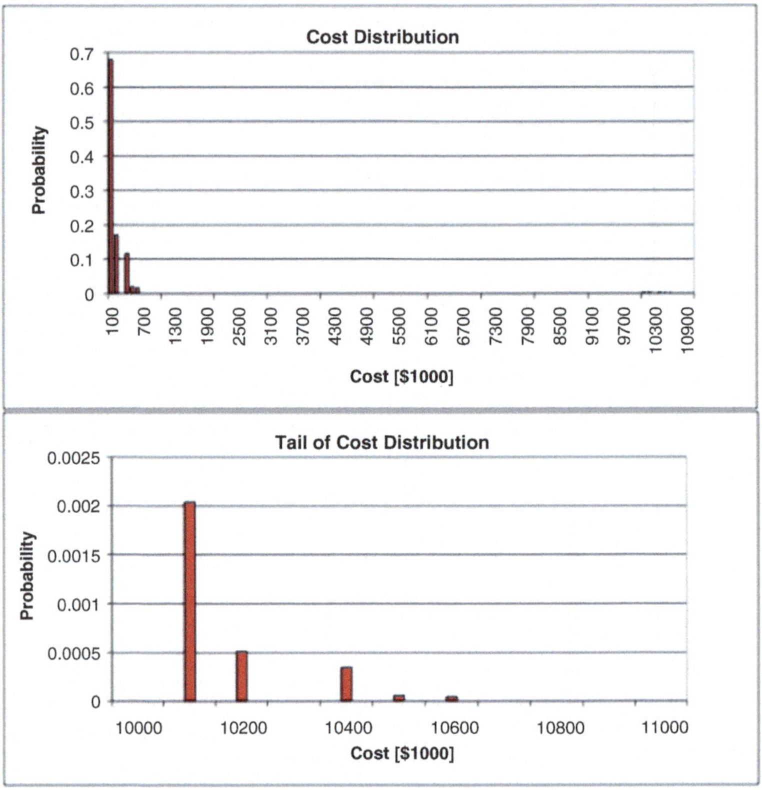

Fig. 15.2 Probability mass function for risk-neutral model **Cyberport_EB**: Required budget = \$28,000, Expected cost = \$132,545

For the bi-objective approach, the subsets of nondominated solutions were computed by parameterization on $\lambda \in \{0.01, 0.10, 0.25, 0.50, 0.75, 0.90, 0.99\}$, the weighted-sum program **Cyberport_EBCV**. The results obtained for the confidence level $\alpha = 0.9$ are presented in Table 15.6. The trade-off between the required budget and expected cost, and CVaR is clearly shown in Fig. 15.5, where the convex efficient frontier of the mean-risk model with $\alpha = 0.9$ is presented. The results emphasize the effect of varying cost/risk preference of the decision maker. The higher the weight λ for the required budget and expected cost, the smaller the number of selected countermeasures. When λ increases from 0.01 to 0.99, the size of the optimal countermeasure portfolio decreases from nine to one selected countermeasure. At the same time the required budget and expected cost decreases from \$464,908 to \$160,545, while CVaR increases from \$65,049 to \$710,691.

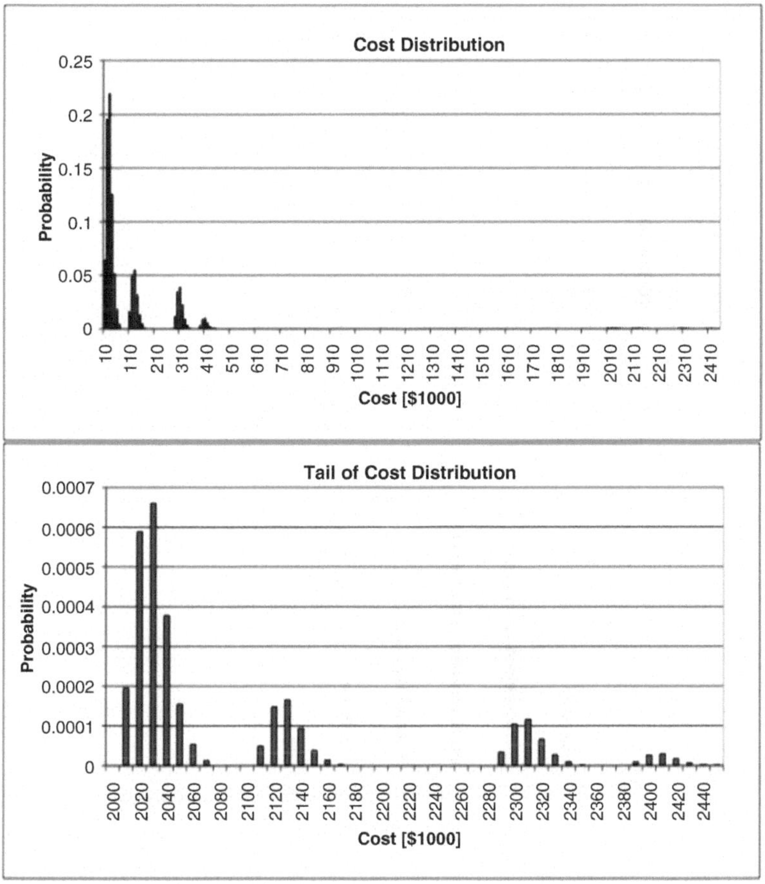

Fig. 15.3 Probability mass function for risk-averse model **Cyberport_CV** with $B = \$150,000$: $\alpha = 0.99$, CVaR = \$921,449, VaR = \$414,500, Expected cost = \$92,045

Note that the nondominated solutions of the weighted-sum program **Cyberport_EBCV** with $\lambda = 1$ and $\lambda = 0$ are identical with the optimal solutions to single objective models **Cyberport_EB** (Table 15.3) and **Cyberport_CV** with unlimited (the highest) budget $B = \$507,000$ (Table 15.4), respectively.

The computational experiments were performed using the AMPL programming language and Gurobi 5.0.1 solver on a laptop MacBookPro 6.2 with Intel Core i7 processor running at 2.66 GHz and with 8 GB RAM. The Gurobi solver was capable of finding proven optimal solutions within CPU seconds for all examples.

Examples of the proposed mixed integer programs size for different numbers $m = n$ of threats and countermeasures are shown in Table 15.7. The size of the risk-neutral model **Cyberport_EB** and the risk-averse model **Cyberport_CVB** is represented by the total number of variables, Var., number of binary variables, Bin.,

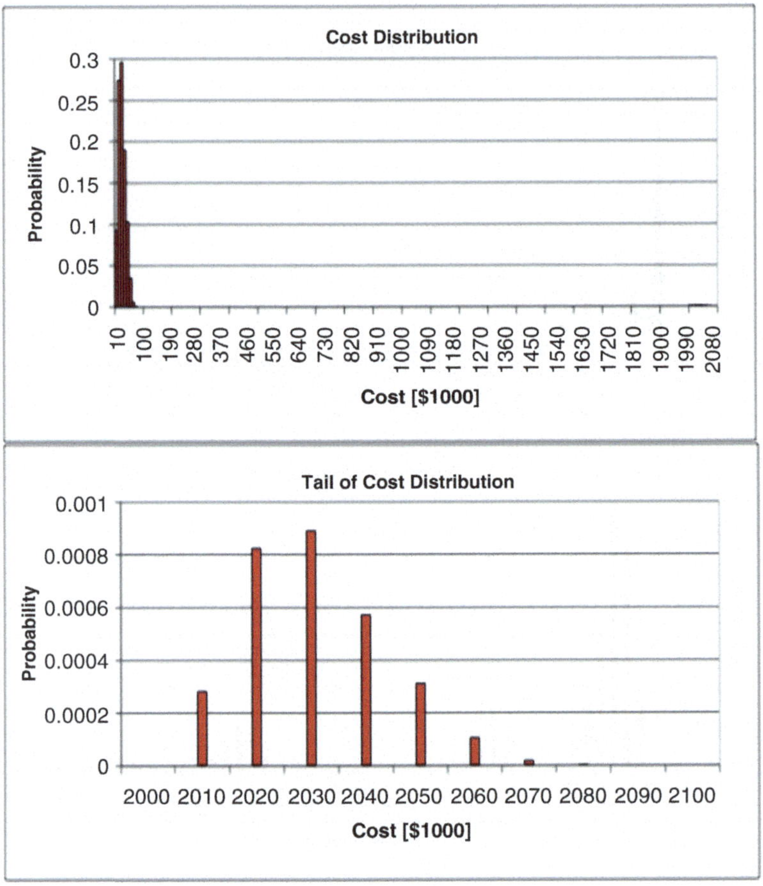

Fig. 15.4 Probability mass function for risk-averse model **Cyberport_CVB**: $\alpha = 0.99$, CVaR = \$652,214, VaR = \$59,000, Required budget = \$258,000, Expected cost = \$31,332

and number of constraints, Cons. The CPU time in seconds required to prove optimality of the solution was ranging from fraction of a second to several hundreds seconds. Note that the number of variables and constraints in the risk-averse model **Cyberport_CVB** grows exponentially in the number m of threats, when all potential attack scenarios are considered. As a result, CPU time increases rapidly when the number of threats increases from $m = 10$ to $m = 20$.

The major managerial insights are summarized in Table 15.8. Overall, the results of computational experiments indicate that *the proposed portfolio approach has proven to be computationally very efficient. For a limited number of attack scenarios considered, the optimal countermeasure portfolio can be found within CPU seconds, using commercially available solvers.*

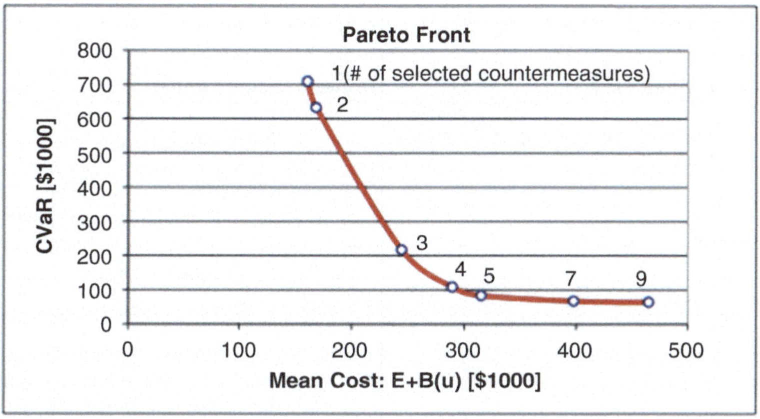

Fig. 15.5 Pareto front for mean-risk model **Cyberport_EBCV**: $\alpha = 0.9$

15.7 Notes

The problem of a right choice of countermeasures for implementation under limited budget is not an easy task. The choice depends not only on reliable data on potential cyber-threats and losses but also requires a good security risk planning tool. The National Institute of Standards and Technology classifies information security controls (NIST SP800-53) into the three categories: technical, operational, and management controls (e.g., Viduto et al. 2012), in which the last two categories can become as critical as the first one. For example, operational countermeasures include monitoring and logging procedures, business continuity/incident response procedures, backup and recovery procedures, while management countermeasures include periodic employee training, periodic testing and review procedures, network auditing, and protection for all sensitive informational assets.

Most of IT security planning models in the literature are qualitative. For example, Bojanc and Jerman-Blazic (2008) introduced methods for identification of the assets, the threats, the vulnerabilities of the ICT systems and proposed a procedure that enables selection of the optimal investment of the necessary security technology based on the quantification of the values of the protected systems. In Egan (2005) a checklist in table form was developed to help decision maker planning a coverage strategy. Chen at al. (2011) discussed current research findings in enterprise risk and security management using web mining techniques.

A study by Gordon and Loeb (2002) uses risk analysis to suggest an optimal budget for a risk-neutral decision maker. In their approach, they compared the loss caused by security incidents to the investment required to reduce the related vulnerability. Based on two general classes of security breach functions, they state that the amount to invest is considerably lower than the expected loss caused by an incident. In fact, they find that the amount to invest in security never exceeds 37% of the expected loss and in most cases will be substantially less. However, these

Table 15.8 Managerial insights: cybersecurity safequards portfolio

Decision-making issue	**Decision-making impact**
Risk-neutral decision-making	
Model **Cyberport_E** vs. model **Cyberport_EB**.	If model **Cyberport_E** model is applied, the greater the available budget, the more countermeasures are selected to minimize expected loss. However, if model **Cyberport_EB** is used, a single countermeasure is selected only to minimize expenditures on countermeasures and expected loss.
Risk-averse decision-making	
Budget vs. confidence level.	The higher the budget available and the higher the confidence level, the more risk-oriented is the countermeasure portfolio selected. The number of selected countermeasures increases with the available budget, and for a limited budget decreases with the confidence level. For a limited budget and lower confidence level, the expensive countermeasures against low-probability, high-impact threats are rarely selected. As a result, the total number of all selected countermeasures is usually greater than that for a higher confidence level, when the more expensive countermeasures against high-impact threats are chosen.
Mean-risk model **Cyberport_EBCV**.	A simple tool for balancing expected and worst-case losses and for shaping of the resulting cost distribution through the selection of optimal countermeasure portfolio. The decision maker is capable of controlling both the risk of high losses from successful cyber-attacks (by choosing the confidence level, α,) as well as the trade-off between expected and worst-case losses (by choosing the trade-off parameter, λ, that represents cost vs. risk preference).

observations only hold true if the security breach functions meet the condition of decreasing marginal returns in case of security investments. Hausken (2006) examined four additional types of security breach functions with different shapes and found that the amount to invest is no longer limited by 37% and different investment strategies should be applied in each case. Wang et al. (2008) proposed a more detailed analysis which makes use of security incident data and value-at-risk to support decision-making.

In contrast to qualitative approaches, the literature on quantitative methods for selection of countermeasures to block or mitigate security attacks is very limited. Based on NIST SP800-30 guidelines, Viduto et al. (2012) developed a risk assessment and optimization model to satisfy organizational security needs in a cost-effective manner. The security countermeasure selection problem was formulated as a multi-objective optimization problem, where variables such as financial cost and risk may affect the final solution. A tailored multi-objective tabu search-based heuristic approach was constructed to solve the proposed multi-objective optimization problem and asses the qualities of its solutions with respect to optimal ones (see also Rees et al. 2011). Deane et al. (2009) developed a linear generalized network flow model that quantifies IT security risk in the supply chain. It was shown how to find solutions for optimal risk reduction under several definitions of optimality: minimizing upstream risk, minimizing downstream risk, and minimizing global supply chain risk. Then, following the mathematical models proposed by Deane et al. (2009), an integer programming model was developed by Rakes et al. (2012), for optimally choosing a subset of countermeasures to block or mitigate security attacks in the presence of a given threat level profile. The model was used to examine the two different types of scenarios: under expected threat levels and under worst-case levels. The authors illustrated the trade-offs in optimal security planning when expected threats are used to parameterize the model versus worst-case values for threat outcomes. To demonstrate the trade-off which occurs if decision makers divert budgets away from planning for ordinary risk in an effort to mitigate the effects of potential high-impact outcomes, budget-dependent risk curves were developed. Schilling and Werners (2016) proposed a combinatorial optimization model to efficiently select security safeguards in order to protect IT infrastructures and systems. The approach is designed to provide decision support for an organization as a whole or separately for specific systems. Fielder et al. (2016) presented an analysis of a hybrid game-theoretic and combinatorial optimization approach to the allocation of cyber security budget. To derive an optimal investment strategy the game solutions were combined with a multi-objective multiple choice knapsack based algorithm.

The material presented in this chapter is based on research reported in Sawik (2013f), where portfolio approach with CVaR as a risk measure was proposed for the selection of countermeasures in IT security planning to prevent and mitigate the impact of cyber-attacks. A critical issue that need to be considered before any practical application of the proposed models is attempted, however, is the estimation of probabilities and the resulting losses associated with each type of threats and countermeasures. In practice, threat likelihood estimates are provided by security experts (e.g., Ryan et al. 2012) and complete distributional information is not available. However, the proposed scenario-based approach does not require such a complete information to be available and only assumes independence of different threat events. In many cases they are independent of each other, but in principle might have a joint probability distribution. The future research should also consider selection of a countermeasure portfolio under the less restrictive assumptions on a joint probability of different cyber-threats. However, relaxation of the assumption of independent threat events may significantly complicate the problem of building potential attack scenarios. For example, modeling of attack scenarios with correlated threat events represented by Bernoulli random variables may require constructing of the corresponding correlated probability distribution, i.e., a correlated binomial distribution instead of the binomial distribution used in this chapter.

Problems

15.1. Design attack scenarios for correlated threat events.

15.2. In practice, different implementation levels can be available for each countermeasure and combined in the countermeasure portfolio, for example, light level with the lowest implementation costs, medium level with the medium costs, and strong level with the highest costs. The higher the implementation level, the greater the degree to which the countermeasure is implemented and the vulnerability reduction. For example, a vulnerability scanning implemented at light, medium, or strong level may correspond to irregular, regular, or real-time scanning, respectively. Enhance the SMIP models presented in this chapter for multiple implementation levels of countermeasures, where different levels are associated with different implementation costs and different effectiveness of blocking different threats.

15.3. Modify the SMIP models presented in this chapter for selection of cybersecurity portfolio such that cannot contain mutually exclusive countermeasures or such that must contain subsets of countermeasures contingent on each other.

15.4. Relax the assumption that the blocking effectiveness of each countermeasure is independent whether or not it is used alone or together with other countermeasures. What would be the implication of that relaxation on the developed models?

15.5. Identify the best-case and the worst-case attack scenarios such that the CVaR of loss is not greater and not less than a fixed threshold loss, respectively.

References

Aissaoui, N., Haouari, M., Hassini, E.: Supplier selection and order lot sizing modeling: a review. Comput. Oper. Res. **34**, 3516–3540 (2007)

Aksoy, A., Ozturk, N.: Supplier selection and performance evaluation in just-in-time production environments. Expert Syst. Appl. **38**, 6351-6359 (2011)

Altendorfer, K., Jodlbauer, H.: An analytical model for service level and tardiness in a single machine MTO production system. Int. J. Prod. Res. **49**(7), 1827–1850 (2011)

Ang, E., Iancu, D.A., Swinney, R.: Disruption risk and optimal sourcing in multitier supply networks. Manag. Sci. **63**(8), 2397–2419 (2017)

Arshinder, Kanda A., Deshmukh S.G.: Supply chain coordination: Perspectives, empirical studies and research directions. Int. J. Prod. Eco **115**, 316–335 (2008)

Bard, J.F., Nananukul, N.: The integrated production-inventory-distribution-routing problem. J. Sched. **12**(3), 257–280 (2009)

Basnet, C., Leung, J.M.Y.: Inventory lot-sizing with supplier selection. Comput. Oper. Res. **32** 1–14 (2005)

Basole, R., Bellamy, M.: Supply network structure, visibility, and risk diffusion: a computational approach. Decis. Sci. **45**(4), 753–789 (2014)

Berger, P.D., Zeng, A.Z.: Single versus multiple sourcing in the presence of risks. J. Operat. Res.Soc. **57**(3), 250–61 (2006)

Berger, P.D., Gerstenfeld, A., Zeng, A.Z.: How many suppliers are best? A decision-analysis approach. OMEGA Int. J. Manag. Sci. **32**(1), 9–15 (2004)

Bijulal, D., Venkateswaran, J., Hemachandra, N.: Service levels, system cost and stability of production–inventory control systems. Int. J. Prod. Res. **49**(23), 7085–7105 (2011)

Birge, J.R., Louveaux, F.: Introduction to Stochastic Programming. New York, Springer (2011)

T. Sawik, *Supply Chain Disruption Management*,
International Series in Operations Research & Management Science 291,
https://doi.org/10.1007/978-3-030-44814-1

Blackhurst, J., Dunn, K.S., Craighead, C.W.: An empirically derived framework of global supply resiliency. J. Bus. Logist. 32(4), 374–391 (2011).

Bojanc, R., Jerman-Blazic, B.: An economic modelling approach to information security risk management. Int. J. Inf. Manag. **28**, 413–422 (2008)

Boyson, S.: Cyber supply chain risk management: revolutionizing the strategic control of critical IT systems. Technovation **34**, 342–353 (2014)

Bukowski, L.: Reliable, Secure and Resilient Logistics Networks. Cham, Springer (2019)

Cakici, E., Mason, S.J., Kurz, M.E.: Multi-objective analysis of an integrated supply chain scheduling problem. Int. J. Prod. Res. **50**(10), 2624–2638 (2012)

Canelas, E., Pinto-Varela, T., Sawik, B.: Electricity portfolio optimization for large consumers: Iberian electricity market case study. Energies **13**, 2249 (2020) DOI:10.3390/en13092249

Cardoso, S.R., Barbosa-Povoa, A.P., Relvas, S., Novais, A.Q.: Resilience metrics in the assessment of complex supply-chains performance operating under demand uncertainty. Omega **56**, 53–73 (2015)

Cavalcante, I.M., Frazzon, E.M., Forcellini, F.A., Ivanov, D.: A supervised machine learning approach to data-driven simulation of resilient supplier selection in digital manufacturing. Int. J. Inf. Manag. **49**, 86–97 (2019)

Celo, S., Nebus, J., Wang, I.K.: The role of internal and external complexity in global factory performance: an NKC application. J. Int. Manag. **24**(1), 65–83 (2018)

Chahara, K., Taaffe, K.: Risk averse demand selection with all-or-nothing orders. OMEGA Int. J. Manag. Sci. **37**(5), 996–1006 (2009)

Chan, J.C.C., Kroese, D.P.: Rare-event probability estimation with conditional Monte Carlo. Ann. Oper. Res. **189**(1), 43–61 (2011)

Che, Z.H., Wang, H.S.: Supplier selection and supply quantity allocation of common and non-common parts with multiple criteria under multiple products. Comput. Ind. Eng. **55**, 110–133 (2008)

Chen, Z.L.: Integrated production and outbound distribution scheduling: Review and extensions. Oper. Res. **58**(1), 130–148 (2010)

Chen, Z.L., Pundoor, G.: Order assignment and scheduling in a supply chain. Oper. Res. **54**(3), 555–572 (2006)

Chen, Z.L., Vairaktarakis, G.L.: Integrated scheduling of production and distribution operations. Manag. Sci. **51**(4), 614–628 (2005)

Chen, D.-S., Batson, R.G., Dang, Y.: Applied Integer Programming: Modeling and Solution. Wiley, Hoboken (2010)

Cheng. B.-Y., Leung, J.Y.-T., Li, K.: Integrated scheduling of production and distribution to minimize total cost using an improved ant colony optimization method. Comput. Ind. Eng. **83**, 217–225 (2015)

Choi, T.Y., Hong, Y.: Unveiling the structure of supply networks: case studies in Honda, Acura and Daimler Chrysler. J. Oper. Manag. **20**(5), 469–493 (2002)

Church, R.L., Scaparra, M.P.: Analysis of facility systems' reliability when subject to attack or a natural disaster. In: Murray, A.T. and Grubesic, T.H. (eds.) Reliability and Vulnerability in Critical Infrastructure: A Quantitative Geographic Perspective. Springer, Berlin (2006)

Cohen, M.A., Kunreuther, H.: Operations risk management: overview of Paul Kleindorfer's contributions. Prod. Oper. Manag. **16**(5): 525–541 (2007)

Cui, L.X.: Joint optimization of production planning and supplier selection incorporating customer flexibility: an improved genetic approach. J. Intell. Manuf. **27**(5), 1017–1035 (2016)

Deane, J.K., Ragsdale, C.T., Rakes, T.R., Rees, L.P.: Managing supply chain risk and disruption from IT security incidents. Oper. Manag. Res. **2**(1), 4–12 (2009)

DeCroix, G.A.: Inventory management for an assembly system subject to supply disruptions. Manag. Sci. **59**(9), 2079–2092 (2013)

Delage, E., Arroyo, S., Ye, Y.: The value of stochastic modeling in two-stage stochastic programs with cost uncertainty. Oper. Res. **62**(6), 1377–1393 (2014)

Demirtas, E.A., Ustun, O.: An integrated multiobjective decision making process for supplier selection and order allocation. OMEGA Int. J. Manag. Sci. **36**, 76–90 (2008)

Dolgui, A., Proth, J.M.: Supply Chain Engineering: Useful Methods and Techniques. Springer, London (2010)

Durbach, I.N.,Stewart, T.J.: Using expected values to simplify decision making under uncertainty. OMEGA Int. J. Manag. Sci. **37**(2), 312–330 (2009)

Eden, C., Williams, T.M., Ackermann, F. and Howick, S.: On the nature of disruption and delay. J. Oper. Res. Soc. **51**, 291–300 (2000)

Egan, M.: The Executive Guide to Information Security. Symantec Press, Indianapolis (2005)

Erenguc, S.S., Simpson, N.C., Vakharia, A.J.: Integrated production/distribution planning in supply chains: an invited review. Eur. J. Oper. Res. **115**(2), 219–236 (1999)

Falasca, M., Zobel, C.W., Cook, D.: A decision support framework to assess supply chain resilience. In: The Proceedings of the 5th International ISCRAM Conference, Washington, pp. 596–605 (2008)

Falco, G., Eling, M., Jablanski, D., Weber, M., Miller, V., Gordon, L. A., Wang, S. S., Schmit, J., Thomas, R., Elvedi, M., Maillart, T., Donavan, E., Dejung, S., Durand, E., Nutter, F., Scheffer, U., Arazi, G., Ohana, G., Lin, H.: Cyber risk research impeded by disciplinary barriers. Sci. **366**(6469), 1066–1069 (2019)

Fattahi, M., Govindan, K., Keyvanshokooh, E.: Responsive and resilient supply chain network design under operational and disruption risks with delivery lead-time sensitive customers. Transp. Res. E **101**, 176–20 (2017)

Fielder, A., Panaousis, E., Malacaria, P., Hankin, C., Smeraldi, F.: Decision support approaches for cyber security investment. Decis. Support. Syst. **86**, 13–23 (2016)

Fiksel, J.: Sustainability and resilience: toward a systems approach. Sustain. Sci. Pract. Policy **2**(2), 14–21 (2006)

Fourer, R., Grayand, D.M., Kernigham, B.W.: AMPL A Modeling Language for Mathematical Programming, Duxbury Thomson, California (2003)

Filippi, C., Guastaroba, G., Speranza, M.G.: Conditional value-at-risk beyond finance: a survey Intl. Trans. Op. Res. **27**, 1277–1319 (2020)

Freeman, N., Mittenthal. J., Keskin, B., Melouk, S.: Sourcing strategies for a capacitated firm subject to supply and demand uncertainty. OMEGA Int. J. Manag. Sci. **77**, 127–142 (2018)

Fujimoto, T., Park, Y.W.: Balancing supply chain competitiveness and robustness through "virtual dual sourcing": lessons from the Great East Japan Earthquake. Int. J. Prod. Econ. **147**, 429–436 (2013)

Fuller, T.: Floodwaters are gone, but supply chain issues linger. The New York Times, New York (2012)

Gabrel, V., Murat, C., Thiele, A.: Recent advances in robust optimization: an overview. Eur. J. Oper. Res. **235**, 471–483 (2014)

Gao, S., Qi, L., Lei, L.: Integrated batch production and distribution scheduling with limited vehicle capacity. Int. J. Prod. Econ. **160** 13–25 (2015)

Ghodsypour, S.H., O'Brien. C.: The total cost of logistics in supplier selection under conditions of multiple sourcing, multiple criteria and capacity constraints. Int. J. Prod. Econ. **73**, 15–27 (2001)

Godinez, V.: How quake tested TI. The Dallas Morning News, Dallas (2011)

Gordon, L. A., Loeb, M. P.: The economics of information security investment. ACM Trans. Inf. Syst. Secur. **5**, 438–457 (2002)

Gurnani, H., Mehrotra, A, Ray, S. (eds.): Supply Chain Disruptions: Theory and Practice of Managing Risk. Springer, London (2012)

Hamdi, F., Ghorbel, A., Masmoudi, F., Dupont L.: Optimization of a supply portfolio in the context of supply chain risk management: literature review. J. Intell. Manuf. **29**(4), 763–788 (2018)

Hammami, R., Temponi, C., Frein, Y.: A scenario-based stochastic model for supplier selection in global context with multiple buyers, currency fluctuation uncertainties, and price discounts. Eur. J. Oper. Res. **233**, 159–170 (2014)

Haraguchi, M., Lall, U.: Flood risks and impacts: a case study of Thailand's floods in 2011 and research questions for supply chain decision making. Int. J. Disaster Risk Reduct. **14**(3), 256–272 (2015)

Hausken, K.: Returns to information security investment: the effect of alternative information security breach functions on optimal investment and sensitivity to vulnerability. Inf. Syst. Front. **8**, 338–349 (2006)

He, H., Hongyan, X.: Dual sourcing and backup production: coexistence versus exclusivity. OMEGA Int. J. Manag. Sci. **57**, 22–33 (2015)

Heckmann, I., Comes, T., Nickel, S.: A critical review on supply chain risk – definition, measure and modeling. OMEGA Int. J. Manag. Sci. **52** 119–132 (2015)

Heese, H.S.: Single versus multiple sourcing and the evolution of bargaining positions. OMEGA Int. J. Manag. Sci. **54** 125–133 (2015)

Hezarkhani, B., Kubiak, W.: Coordinating contracts in SCM: a review of methods and literature. Decis. Making Manuf. Services **4**(1–2), 5–28 (2010)

Ho, W., Xu, X., Dey, P.K.: Multi-criteria decision making approaches for supplier evaluation and selection: a literature review. Eur. J. Oper. Res. **202**, 16–24 (2010)

Holweg, M., Pil, F.: Theoretical perspectives on the coordination of supply chains. J. Oper. Manag. **26**, 389–406 (2008)

Hosseini, S., Ivanov, D., Dolgui A.: Review of quantitative methods for supply chain resilience analysis. Transp. Res. E **125**, 285–307 (2019a)

Hosseini, S., Morshedlou, N., Ivanov, D., Sarder, M.D., Barker, K., Al Khaled, A.: Resilient supplier selection and optimal order allocation under disruption risks. Int. J. Prod. Econ. **213**, 124–137 (2019b)

Ivanov, D., Sokolov, B., Dolgui, A.: The Ripple effect in supply chains: trade-off 'efficiency-flexibility-resilience' in disruption management. Int. J. Prod. Res. **52**(7), 2154–2172 (2014)

Ivanov, D., Sokolov, B., Pavlov, A.: Optimal distribution (re)planning in a centralized multi-stage network under conditions of ripple effect and structure dynamics. Eur. J. Oper. Res. **237**(2), 758–770 (2015)

Ivanov, D., Pavlov, A., Dolgui, A., Pavlov, D., Sokolov, B.: Disruption-driven supply chain (re)-planning and performance impact assessment with consideration of pro-active and recovery policies. Transp. Res. E **90(C)**, 7–24 (2016)

Ivanov, D., Dolgui, A., Sokolov, B., Ivanova, M.: Literature review on disruption recovery in the supply chain. Int. J. Prod. Res. **55**(20), 6158–6174 (2017)

Ivanov, D.: Structural Dynamics and Resilience in Supply Chain Risk Management. Springer, New York (2018)

Ivanov, D., Dolgui, A., Sokolov, B. (eds.): Handbook of Ripple Effects in the Supply Chain. Springer, New York (2019)

Ivanov, D.: Predicting the impacts of epidemic outbreaks on global supply chains: A simulation-based analysis on the coronavirus outbreak (COVID-19/SARS-CoV-2) case. Transp. Res. E. **136**, 101922 (2020)

Izakian, H., Abraham, A.: Fuzzy C-means and fuzzy swarm for fuzzy clustering problem. Expert Syst. Appl. **38**(3), 1835–1838 (2011)

Jenkins, L.: Selecting scenarios for environmental disaster planning. Eur. J. Oper. Res. **121**, 275–286 (2000)

Kall, P., Mayer, J.: Stochastic Linear Programming: Models, Theory, and Computation. Springer, New York (2011)

Kasilingam, R.G., Lee, C.P.: Selection of vendors – a mixed-integer programming approach. Comput. Ind. Eng. **31**, 347–350 (1996)

Khojasteh, Y. (ed.): Supply Chain Risk Management. Springer, Singapore (2018)

Kleindorfer, P.R., Saad, G.H.: Managing disruption risks in supply chains. Prod. Oper. Manag. **14**(1), 53–68 (2005)

Klibi, W., Martel, A.: Modeling approaches for the design of resilient supply networks under disruptions. Int. J. Prod. Econ. **135**, 882–898 (2012)

Klibi, W., Martel, A., Guitouni, A.: The design of robust value-creating supply chain networks: a critical review. Eur. J. Oper. Res. **203**, 283–293 (2010)

Knemeyer, A.M., Zinn, W., Eroglu, C.: Proactive planning for catastrophic events in supply chains. J. Oper. Manag. **27**(2): 141–153 (2009)

Kostreva, M.M., Ogryczak, W., Wierzbicki, A.: Equitable aggregation and multiple criteria analysis. Eur. J. Oper. Res. **158**, 362–377 (2004)

Kouvelis, P., Dong, L., Boyabatli, O., Li, R. (eds.): Handbook of Integrated Risk Management in Global Supply Chains. Wiley, Hoboken (2011)

Kull, T., Closs, D.: The risk of second-tier supplier failures in serial supply chains: implications for order policies and distributor autonomy. Eur. J. Oper. Res. **186**(3), 1158–1174 (2008)

Larsen, C., Thorstenson, A.: A comparison between the order and the volume fill rate for a base stock inventory control system under a compound renewal demand process. J. Oper. Res. Soc. **59**(6), 798–804 (2008)

Larsen, C., Thorstenson, A.: The order and volume fill rates in inventory control systems. Int. J. Prod. Econ. **147**, 13–19 (2014)

Lee, H., Kim, M.S., Kim, K.K.: Interorganizational information systems visibility and supply chain performance. Int. J. Inf. Manag. **34**, 285–295 (2014)

Lei, L., Liu, S., Ruszczynski, A., Park, S.: On the integrated production, inventory and distribution routing problem. IIE Trans. **38**(11), 955–970 (2006)

Li, X., Wang, Q.: Coordination mechanisms of supply chain systems. Eur. J. Oper. Res. **179**, 1–16 (2007)

Li, L., Zabinsky, Z.B.: Incorporating uncertainty into a supplier selection problem. Int. J. Prod. Econ. **134**, 344–356 (2011)

Li, K.P., Ganesan, V.K., Sivakumar, A.I.: Synchronized scheduling of assembly and multi-destination air-transportation in a consumer electronics supply chain. Int. J. Prod. Res. **43**(13), 2671–2685 (2005)

Li, X., Ouyang, Y., Peng, F.: A supporting station model for reliable infrastructure location design under interdependent disruptions. Transp. Res. E **60**, 80–93 (2013)

Liu, S., Papageorgiou, L.G.: Multiobjective optimisation of production, distribution and capacity planning of global supply chains in the process industry. OMEGA Int. J. Manag. Sci. **41**(2), 369–382 (2013)

Liu, S., Sawik, T., Papageorgiou L.: Corrigendum to "Multiobjective optimisation of production, distribution and capacity planning of global supply chains in the process industry" [Omega 41 (2013) 369–382]. OMEGA Int. J. Manag. Sci. **44**(1), 149 (2014)

MacKenzie, C.A., Barker, K., Santos, J.R.: Modeling a severe supply chain disruption and post-disaster decision making with application to the Japanese earthquake and tsunami. IIE Trans. **46**(12), 1243–1260 (2014)

Madadi, A.R., Kurz, M.E., Taaffe, K.M., Sharp, J.L., Mason, S.J.: Supply network design: risk-averse or risk-neutral? Comput. Ind. Eng. **78**, 55–65 (2014)

Maggioni, F., Wallace, S.W.: Analyzing the quality of the expected value solution in stochastic programming. Ann. Oper. Res. **200**(1), 37–54 (2012)

Manavazhi, M.R., Adhikari, D.K.: Material and equipment procurement delays in highway projects in Nepal. Int. J. Proj. Manag. **20**(8), 627–632 (2002)

Marszewska, J.R.: Implications of seismic hazard in Japan on Toyota supply chain disruption risks. In: Proceedings of 13th International Conference on Industrial Logistics, September 28 – October 1, 2016, Zakopane. AGH University of Science & Technology, Kraków, pp. 178–185 (2016)

Matsuo, H.: Implications of the Tohoku earthquake for Toyota's coordination mechanism: supply chain disruption of automotive semiconductors. Int. J. Prod. Econ. **161**, 217–227 (2015)

McGrath, D.: Fujitsu restores its idled fabs, aids in infrastructure rebuild. EE Times **1601**, 18 (2011)
Meena, P.L., Sarmah, S.P., Sarkar, A.: Sourcing decisions under risks of catastrophic event disruptions. Transp. Res. E **47**, 1058–1074 (2014)
Mena, C., A. Humphries, T. Y. Choi.: Toward a theory of multi-tier supply chain management. J. Supply Chain Manag. **49**(2), 58–77 (2013)
Merzifonluoglu Y.: Risk averse supply portfolio selection with supply, demand and spot market volatility. OMEGA Int. J. Manag. Sci. **57**, 40–53 (2015)
Narasimhan, R., Talluri, S.: Perspectives on risk management in supply chains. J. Oper. Manag. **27**(2), 114–118 (2009)
Namdar, J., Li, X., Sawhney, R., Pradhan, N.: Supply chain resilience for single and multiple sourcing in the presence of disruption risks. Int. J. Prod. Res. **56**, 2339–2360 (2017)
Nemhauser, G.L., Wolsey, L.A.: Integer and Combinatorial Optimization. Wiley, New York (1999)
Norrman, A., Jansson, U.: Ericsson's proactive risk management approach after a serious sub-supplier accident. Int. J. Phys. Distrib. Logist. Manag. **34**(5), 434–456 (2004)
Noyan, N.: Risk-averse two-stage stochastic programming with an application to disaster management. Comput. Oper. Res. **39**, 541–559 (2012)
Ogryczak, W., Ruszczynski, A.: Dual stochastic dominance and related mean-risk models. SIAM J. Optim. **13**, 60–78 (2002)
Ogryczak, W., Tamir, A.: Minimizing the sum of the k largest functions in linear time. Inf. Process. Lett. **85**, 117–122 (2003)
Ogryczak, W., Wierzbicki, A., Milewski, M.: A multi-criteria approach to fair and efficient bandwidth allocation. OMEGA Int. J. Manag. Sci. **36**, 451–463 (2008)
Oke, A., Gopalakrishnan, M.: Managing disruptions in supply chains: a case study of a retail supply chain. Int. J. Prod. Econ. **118**, 168–174 (2009)
Panetto, H., Iung, B., Ivanov, D., Weichhart, G., Wang, X.: Challenges for the cyber-physical manufacturing enterprises of the future. Annu. Rev. Control. **47**, 200–213 (2019)
Park, Y., Hong, P., Roh, J.J.: Supply chain lessons from the catastrophic natural disaster in Japan. Bus. Horiz. **56**(1), 75–85 (2013)
Parlar, M., Perry, D.: Inventory models of future supply uncertainty with single and multiple suppliers. Nav. Res. Logist. **43**, 191–210 (1996)
Paul, S.K., Sarker, R., Essam, D.: Managing real–time demand fluctuation under a supplier–retailer coordinated system. Int. J. Prod. Econ. **158**, 231–243 (2014a)
Paul, S.K., Sarker, R., Essam, D.: Real time disruption management for a two-stage batch production–inventory system with reliability considerations. Eur. J. Oper. Res. **237** 113–128 (2014b)
Paul, S.K., Sarker, R., Essam, D.: Managing disruption in an imperfect production–pinventory system. Comput. Ind. Eng. **84**, 101–112 (2015a)
Paul, S.K., Sarker, R., Essam, D.: A disruption recovery plan in a three–stage production–inventory system. Comput. Oper. Res. **57**, 60–72 (2015b)

Paul, S.K., Sarker, R., Essam, D.: Managing risk and disruption in production-inventory and supply chain systems: a review. J. Ind. Manag. Optim. **12**(3), 1009–1029 (2016)

Paul, S.K., Sarker, R., Essam, D.: A quantitative model for disruption mitigation in a supply chain. Eur. J. Oper. Res. **257**(3), 881–895 (2017)

Paul, S.K., Sarker, R.,Essam, D.: A reactive mitigation approach for managing supply disruption in a three–tier supply chain. J. Intell. Manuf. **29**, 1581–1597 (2018)

Pereira, J.V.: The new supply chains frontier: Information management. Int. J. Inf. Manag. **29**, 372–379 (2009)

Ponomarov, S.Y., Holcomb, M.C.: Understanding the concept of supply chain resilience. Int. J. Logist. Manag. **20**(1), 124–143 (2009)

PrasannaVenkatesan, S., Goh, M.: Multi-objective supplier selection and order allocation under disruption risk. Transp. Res. E **95**, 124–142 (2016)

Pundoor, G., Chen, Z.L.: Joint cyclic production and delivery scheduling in a two-stage supply chain. Int. J. Prod. Econ. **119**(1), 55–74 (2009)

Quang, H.T., Hara, Y.: Risks and performance in supply chain: the push effect. Int. J. Prod. Res. **56**(4), 1369–1388 (2018)

Rabbani, M., Khalili, S.H., Janani, H. and Shiripour, M.: Optimization of a dynamic supply portfolio considering risks and discounts constraints. J. Ind. Eng. Manag. **7**(1), 215–253 (2014)

Rajagopal, V., Venkatesan, S.P., Goh, M.: Decision-making models for supply chain risk mitigation: a review. Comput. Ind. Eng. **113**, 646–682 (2017)

Rakes, T.R., Deane, J.K., Rees, L.P.: IT security planning under uncertainty for high-impact events. OMEGA Int. J. Manag. Sci. **40**(1), 79–88 (2012)

Ravindran, A.R., Bilsel, R.U., Wadhwa, V., Yang, T.: Risk adjusted multicriteria supplier selection models with applications. Int. J. Prod. Res. **48**(2), 405–424 (2012)

Rawls, C.G., Turnquist, M.A.: Pre-positioning of emergency supplies for disaster response. Transp. Res. B **44**(4), 521–34 (2010)

Rees, L.P., Deane, J.K., Rakes, T.R., Baker, W.H.: Decision support for cybersecurity risk planning. Decis. Support. Syst. **51**, 493–505 (2011)

Ribeiro, J.P., Barbosa-Povoa, A.P.: Supply chain resilience: definitions and quantitative modelling approaches — a literature review. Comput. Ind. Eng. **115**, 109–122 (2018)

Rockafellar, R.T., Uryasev, S.: Optimization of conditional value-at-risk. J. Risk **2**(3), 21–41 (2000)

Ruiz-Torres, A.J., Mahmoodi, F.: The optimal number of suppliers considering the costs of individual supplier failures. OMEGA Int. J. Manag. Sci. **35**(1), 104–115 (2007)

Ruiz-Torres, A.J., Mahmoodi, F., Zeng, A.Z.: Supplier selection model with contingency planning for supplier failures. Comput. Ind. Eng. **66**, 374–382 (2013)

Ryan, J.J.C.H., Mazzuchi, T.A., Ryan, D.J., de la Cruz, J.L., Cooke, R.: Quantifying information security risks using expert judgment elicitation. Comput. Oper. Res. **39**, 774–784 (2012)

Sarykalin, S., Serraino, G., Uryasev, S.: Value-at-risk vs. conditional value-at-risk in risk management and optimization. Tutorials Oper. Res. INFORMS 270–294 (2008)

Sawik, T.: Stochastic optimal control of a multi-facility, multi-product production scheduling with random times of supplies. Control. Cybern. **6**(3–4), 21–35 (1977)

Sawik, T.: A cyclic versus flexible approach to materials ordering in make-to-order assembly. Math. Comput. Model. **42**(3–4), 279–290 (2005)

Sawik, T.: Hierarchical approach to production scheduling in make-to-order assembly. Int. J. Prod. Res. **44**(4), 801–830 (2006)

Sawik, T.: A lexicographic approach to bi-objective scheduling of single-period orders in make-to-order manufacturing. Eur. J. Oper. Res. **180**(3), 1060–1075 (2007)

Sawik, B.: A three stage lexicographic approach for multi-criteria portfolio optimization by mixed integer programming. Przeglad Elektrotechniczny **84**(9), 108–112 (2008)

Sawik, B.: Lexicographic and weighting approach to multi-criteria portfolio optimization by mixed integer programming. Appl. Manag. Sci. **13**, 3–18 (2009)

Sawik, T.: Coordinated supply chain scheduling. Int. J. Prod. Econ. **120**(2), 437–451 (2009a)

Sawik, T.: Monolithic versus hierarchical approach to integrated scheduling in a supply chain. Int. J. Prod. Res. **47**(21), 5881–5910 (2009b)

Sawik, T.: Multi-objective due-date setting in a make-to-order environment. Int. J. Prod. Res. **47**(22), 6205–6231 (2009c)

Sawik, B.: Selected multi-objective methods for multi-period portfolio optimization by mixed integer programming. Appl. Manag. Sci. **14**, 3–34 (2010)

Sawik, T.: Single vs. multiple objective supplier selection in a make to order environment. OMEGA Int. J. Manag. Sci. **38**(3–4), 203–212 (2010a)

Sawik, T.: An integer programming approach to scheduling in a contaminated area. OMEGA Int. J. Manag. Sci. **38**(3–4), 179–191 (2010b)

Sawik, T.: Scheduling in Supply Chains Using Mixed Integer Programming. Wiley, Hoboken (2011a)

Sawik, T.: Selection of supply portfolio under disruption risks. OMEGA Int. J. Manag. Sci.**39**, 194–208 (2011b)

Sawik, T.: Supplier selection in make-to-order environment with risks. Math. Comput. Model. **53**(9–10), 1670–1679 (2011c)

Sawik, T.: Selection of a dynamic supply portfolio in make-to-order environment with risks. Comput. Oper. Res. **38**(4), 782–796 (2011d)

Sawik, B.: Bi-criteria portfolio optimization models with percentile and symmetric risk measures by mathematical programming. Przeglad Elektrotechniczny **88**(10b): 176–180 (2012a)

Sawik, B.: Downside risk approach for multi-objective portfolio optimization. In: Klatte, D., Luthi, H.-J., Schmedders, K. (eds.) Operations Research Proceedings 2011, pp. 191–196. Springer, Berlin (2012b)

Sawik, B.: Conditional value-at-risk vs. value-at-risk to multi-objective portfolio optimization. Appl. Manag. Sci. **15**, 277–305 (2012c)

Sawik, B.: Survey of multi-objective portfolio optimization by linear and mixed integer programming. Appl. Manag. Sci. **16**, 55–79 (2013a)

Sawik, B.: A review of multi-criteria portfolio optimization by mathematical programming. In: Dash, G.H., Thomaidis, N. (eds.) Recent Advances in Computational Finance, pp. 149–172. Nova Science Publishers, New York (2013b)

Sawik, T.: Selection of resilient supply portfolio under disruption risks. OMEGA Int. J. Manag. Sci.**41**, 259–269 (2013c)

Sawik, T.: Selection and protection of suppliers in a supply chain with disruption risks. Int. J Logis. Syst. Manag. **15**(2–3), 143–159 (2013d)

Sawik, T.: Integrated selection of suppliers and scheduling of customer orders in the presence of supply chain disruption risks. Int. J. Prod. Res. **51**(23–24), 7006–7022 (2013e)

Sawik, T.: Selection of optimal countermeasure portfolio in IT security planning. Decis. Support. Syst. **55**, 156–164 (2013f)

Sawik, T.: Joint supplier selection and scheduling of customer orders under disruption risks: Single vs. dual sourcing. OMEGA Int. J. Manag. Sci. **43**(2), 83–95 (2014a)

Sawik, T.: Optimization of cost and service level in the presence of supply chain disruption risks: single vs. multiple sourcing. Comput. Oper. Res. **51**(1), 11–20 (2014b)

Sawik, T.: On the robust decision-making in a supply chain under disruption risks. Int. J. Prod. Res. **52**(22), 6760–6781 (2014c)

Sawik, T.: Cost vs. customer service level in supply chains under major disruptions. In: Proceedings of LISS 2014, 4th International Conference on Logistics, Informatics and Service Science, July 23–26, 2014, UC Berkeley, pp. 1075–1079. Springer, New York (2014d)

Sawik, T.: On the fair optimization of cost and customer service level in a supply chain under disruption risks. OMEGA Int. J. Manag. Sci.**53**, 58–66 (2015a)

Sawik, T.: Integrated supply chain scheduling under multi-level disruptions. IFAC-Papers On Line **48**(3), 1515–1520 (2015b)

Sawik, B.: Triple-objective models for portfolio optimisation with symmetric and percentile risk measures. Int. J. Logist. Syst. Manag. **25**(1), 96–107 (2016)

Sawik, T.: Integrated supply, production and distribution scheduling under disruption risks. OMEGA Int. J. Manag. Sci.**62**, 131–144 (2016a)

Sawik, T.: On the risk-averse optimization of service level in a supply chain under disruption risks. Int. J. Prod. Res. **54**(1), 97–112 (2016b)

Sawik, T.: Resilient vs. robust supply portfolio under disruption risks. In: Proceedings of 13th International Conference on Industrial Logistics, September 28 – October 1, 2016, Zakopane, pp. 255–263. AGH University of Science & Technology, Kraków (2016c)

Sawik, T.: Mixed integer programming approaches to planning and scheduling in electronics supply chains. Decis. Making Manuf. Services **11**(1–2), 5–17 (2017a)

Sawik, T.: A portfolio approach to supply chain disruption management. Int. J. Prod. Res. **55**(7), 1970–1991 (2017b)

Sawik, T.: Stochastic versus deterministic approach to coordinated supply chain scheduling. Math. Prob. Eng. **2017** 3460721, 15 (2017c) https://doi.org/10.1155/2017/3460721

Sawik, B.: Weighted-sum approach for bi-objective optimization of fleet size with environmental aspects. Appl. Manag. Sci. **19**, 101–116 (2018)

Sawik, T.: Selection of a dynamic supply portfolio under delay and disruption risks. Int. J. Prod. Res. **56**, 760–782 (2018a)

Sawik, T.: Disruption mitigation and recovery in supply chains using portfolio approach. OMEGA Int. J. Manag. Sci.**84**(4), 232–248 (2019a)

Sawik, T.: Two-period vs. multi-period model for supply chain disruption management. Int. J. Prod. Res. **57**(14), 4502–4518 (2019b)

Sawik, T.: A Multi-portfolio approach to integrated risk-averse planning in supply chains under disruption risks. In: Ivanov, D., Dolgui, A., Sokolov, B. (eds.). Handbook of Ripple Effects in the Supply Chain, Chapter 2, pp. 35–63. Springer, New York (2019c)

Sawik, T.: A two-period model for selection of resilient multi-tier supply portfolio. Int. J. Prod. Res. In press (2019d) https://doi.org/10.1080/00207543.2019.1665204

Sawik, T.: On the risk-averse selection of resilient multi-tier supply portfolio. OMEGA Int. J. Manag. Sci. In press (2020) DOI: 10.1016/j.omega.2020.102267

Scaparra, M.P., Church, R.L.: A bilevel mixed-integer program for critical infrastructure protection planning. Comput. Oper. Res. **35**, 1905–1923 (2008)

Scheibe, K.P., Blackhurst, J.: Supply chain disruption propagation: a systemic risk and normal accident theory perspective. Int. J. Prod. Res. **56**(1–2), 43–59 (2018).

Schilling, A., Werners, B.: Optimal selection of IT security safeguards from an existing knowledge base. Eur. J. Oper. Res. **248**(1), 318–327 (2016)

Schmitt, A.J.: Strategies for customer service level protection under multi-echelon supply chain disruption risk. Transp. Res. B **45**(8), 1266–1283 (2011)

Schmitt, A.J., Singh, M.: A quantitative analysis of disruption risk in a multi-echelon supply chain. Int. J. Prod. Econ. **139**, 22–32 (2012)

Schmitt, A.J., Snyder, L.V.: Infinite-horizon models for inventory control under yield uncertainty and disruptions. Comput. Oper. Res. **39**, 850–862 (2012)

Schmitt, A.J., Sun, S.A., Snyder, L.V., Shen, Z.J.M.: Centralization versus decentralization: Risk pooling, risk diversification, and supply chain disruptions. OMEGA Int. J. Manag. Sci.**52**, 201–212 (2015)

Sen, A., Yaman, H., Guller, K. and Korpeoglu, E.: Multi-period supplier selection under price uncertainty. J. Oper. Res. Soc. **65**, 1636–1648 (2014)

Shao, X.F., Dong, M., Comparison of order-fulfilment performance in MTO and MTS systems with an inventory cost budget constraint. Int. J. Prod. Res. **50**(7), 1917–1931 (2012)

Sheffi, Y.: The Resilient Enterprise. MIT Press, Cambridge (2005)

Sheffi, Y., Rice, J.B.: A supply chain view of the resilient enterprise. MIT Sloan Manag. Rev. **47**(1), 41–48 (2005)

Shin, H., Collier, D., Wilson, D.: Supply management orientation and supplier/buyer performance. J. Oper. Manag. **18**, 317–333 (2000)

Simchi-Levi, D., Schmidt, W., Wei, Y., Zhang, P.J., Combs, K., Ge, Y., Gusikhin, O., Sanders, M., Zhang, D.: Identifying risks and mitigating disruptions in the automotive supply chain. Interfaces **45**(5), 375–390 (2015)

Snyder, L.V., Shen, Z.-J.M.: Fundamentals of Supply Chain Theory. Wiley, Hoboken (2011)

Snyder, L.V., Scaparra, P.M, Daskin, M.S., Church, R.L.: Planning for disruptions in supply chain networks. In INFORMS TutORials in Operations Research. Published online: 14 Oct 2014; 234–257 (2014) https://doi.org/10.1287/educ.1063.0025

Snyder, L.V., Atan, Z., Peng, P., Rong, Y., Schmitt, A., Sinsoysal, B.: OR/MS models for supply chain disruptions: a review. IIE Trans. **48**(2), 89–109 (2016)

Sodhi, M.S, Tang, C.S.: Managing Supply Chain Risk. Springer, New York (2012)

Sounderpandian, J., Prasad, S., Madan, M.: Supplies from developing countries: optimal order quantities under loss risks. OMEGA Int. J. Manag. Sci.**36**(1), 122–130 (2008)

Stecke, K.E., Kumar, S.: Sources of supply chain disruptions, factors that breed vulnerability, and mitigating strategies. J. Mark. Channels **16**(3), 193–226 (2009)

Steuer, R.E.: Multiple Criteria Optimization: Theory, Computation and Application. Wiley, New York (1996)

Su, P., Liu, S.G.: Dual sourcing in managing operational and disruption risks in contract manufacturing. Int. J. Prod. Res. **53**(1), 291–306 (2015)

Tang, C.S.: Perspectives in supply chain risk management. Int. J. Prod. Econ. **103**, 451–488 (2006)

Thome, A.M.T, Scavarda, L.F., Pires, S.R.I., Ceryno, P., Klingebiel, K.: A multi-tier study on supply chain flexibility in the automotive industry. Int. J. Prod. Econ. **158**, 91–105 (2014)

Tomlin, B.: On the value of mitigation and contingency strategies for managing supply chain disruption risks. Manag. Sci. **52**(5), 639–657 (2006)

Tomlin, B., Wang, Y.: Operational strategies for managing supply chain disruption risk. In: Kouvelis. P., Boyabatli, O., Dong, L., Li, R. (eds.) Handbook of Integrated Risk Management in Global Supply Chains. Wiley, New York, pp. 79–101 (2012)

Torabi, S.A,, Baghersad, M., Mansouri, S.A.: Resilient supplier selection and order allocation under operational and disruption risks. Transp. Res. E **79**, 22–48 (2015)

Tsai, W.C.: A dynamic sourcing strategy considering supply disruption risks. Int. J. Prod. Res. **54**(7), 2170–2184 (2016)

Uryasev, S.: Conditional value-at-risk: optimization algorithms and applications. Financ. Eng. News **14**(2) (2000)

Ustun, O., Demirtas, E.A.: An integrated multi-objective decision making process for multi-period lot sizing with supplier selection. OMEGA Int. J. Manag. Sci. **36**, 509–521 (2008)

Van Mieghem, J.A.: Risk management and operational hedging: An overview. In: Kouvelis. P., Boyabatli, O., Dong, L., Li, R. (eds.) Handbook of Integrated Risk Management in Global Supply Chains, pp. 13–49. Wiley, New York (2012)

Viduto, V., Maple, C., Huang, W., Lopez-Perez, D.: A novel risk assessment and optimisation model for a multi-objective network security countermeasure selection problem. Decis. Support. Syst. **53**, 599–610 (2012)

Viergutz, C., Knust, S.: Integrated production and distribution scheduling with lifespan constraints. Ann. Oper. Res. **213**, 293–318 (2014)

Walsh, D.: Quake drying up supply of ashy pigment for car paint. Crain's Detroit Business 10 April 2011 (2011)

Wang, G., Lei, L.: Polynomial-time solvable cases of the capacitated multi-echelon shipping network scheduling problem with delivery deadlines. Int. J. Prod. Econ. **137**, 263–271 (2012)

Wang, J., Chaudhury, A., Rao, H. R.: A value-at-risk approach to information security investment. Inform. Syst. Res. **19**, 106–120 (2008)

Ware, N.R., Singh, S.P. and Banwet, D.K.: A mixed-integer non-linear program to model dynamic supplier selection problem. Expert Syst. Appl. **41**, 671–678 (2014)

Weber, C.A., Current, J.R.: A multiobjective approach to vendor selection. Eur. J. Oper. Res. **68**, 173–184 (1993)

Whitney, D.E., Luo, J., Heller, D.A.: The benefits and constraints of temporary sourcing diversification in supply chain disruption and recovery. J. Purch. Supply Manag. **20**, 238–250 (2014)

Wu, D., Olson, D.L.: Supply chain risk, simulation, and vendor selection. Int. J. Prod. Econ. **114**, 646–655 (2008)

Xanthopoulos, A., Vlachos, D., Iakovou, E.: Optimal newsvendor policies for dual-sourcing supply chains: A disruption risk management framework. Comput. Oper. Res. **39**, 350–357 (2012)

Xia, W., Wu, Z.: Supplier selection with multiple criteria in volume discount environments. OMEGA Int. J. Manag. Sci.**35**, 494–504 (2007)

Yager, R.R.: On ordered weighted averaging aggregation operators in multi criteria decision making. IEEE Trans. Syst. Man Cybern. **18**, 183–190 (1988)

Yoon, J., S. Talluri, H. Yildiz, W. Ho.: Models for supplier selection and risk mitigation: a holistic approach. Int. J. Prod. Res. **56**(10), 3636–3661 (2018)

Yoon J, Talluri S, Rosales C.: Procurement decisions and information sharing under multi-tier disruption risk in a supply chain. Int. J. Prod. Res. (2019) https://doi.org/10.1080/00207543.2019.1634296

Yu, H., Zeng, A.Z., Zhao, L.: Single or dual sourcing: decision-making in the presence of supply chain disruption risks. OMEGA Int. J. Manag. Sci.**37**, 788–800 (2009)

Yue, J., Xia, Y., Tran, T.: Selecting sourcing partners for a make-to-order supply chain. OMEGA Int. J. Manag. Sci.**38**, 136–144 (2010)

Zeng, A.Z., Xia, Y., Building a mutually beneficial partnership to ensure backup supply. OMEGA Int. J. Manag. Sci.**52**, 77–91 (2015)

Zobel, C.W., Khansa, L.: Characterizing multi-event disaster resilience. Comput. Oper. Res. **42**, 83–94 (2014)

Index

T. Sawik, *Supply Chain Disruption Management*,
International Series in Operations Research & Management Science 291,
https://doi.org/10.1007/978-3-030-44814-1

D

The manufacturer's authorised representative in the EU is Springer Nature Customer Service Centre GmbH, Europaplatz 3, 69115 Heidelberg, Germany. If you have any concerns regarding our products, please contact ProductSafety@springernature.com

Printed and bound by CPI Group (UK) Ltd, Croydon, CR0 4YY
30/09/2026
02225648-0017